A CITIZEN'S DISCLOSURE ON UFOS AND ETI

VOLUME SIX

THE ROSETTA STONE OF ETI CONTACT AND COMMUNICATIONS

TERENCE M. TIBANDO

A CITIZEN'S DISCLOSURE ON UFOS AND ETI

VOLUME SIX

THE ROSETTA STONE OF ETI
CONTACT AND COMMUNICATIONS

Copyright Page

In writing this book, I sought out the best possible evidence available on this subject whether that was from numerous UFO and ETI related books, networking with other UFO authors, researchers and first-hand witnesses to UFO sightings, from films and TV documentaries, from internet searches, or just from my personal sightings and contact experiences.

When material is quoted in this book full acknowledgement is given to the author or source of that material as indicated by the extensive bibliography, webliography and videography at the back of the book.

When photographic images are used in this book that are obtained from the internet, usually from Google Images, a full search was made to determine copyright information, the author's name, or address or email address or phone number or copyright mark in order to asked permission to use their photographs. In almost all cases where such images are posted to the internet, there was no satisfactory way to identify the owner of the image even through Google because they left no identification of themselves to be found. When an author's name does appear on an image, written permission was sought or it was not used at all; more often than not, there usually was no reply or response back from the owner.

This lack of due diligence to place a copyright mark or the owner's name is all it would take for that person to claim ownership of a picture, yet the lack of it creates major problems for many people, especially for other authors who lawfully seek their permission to use their photo images.

Because this book is one of six volumes in a series created as public educational material and is of a transitional nature, and I have quoted or referenced the websites from where I obtained the photo images and therefore, I am invoking the Fair Use Doctrine also known as Fair Usage Clause to publish these images in my book.

I will of course give full acknowledgement and credit to the author's and owners of such images in all my future book publications in recognition of their work if they come forward to be identified.

A CITIZEN'S DISCLOSURE ON UFOS AND ETI

VOLUME SIX

THE ROSETTA STONE OF ETI CONTACT AND COMMUNICATIONS

TERENCE M. TIBANDO

"Hggna"

A Cosmic Cousin
Publication

Other Publications by the Author Page

Although, this book is the author's first publication it forms a part of six smaller books or volumes that was originally written as one massive tome of UFO and ETI information entitled: **"A Citizen's Disclosure On UFOs And ETI Visiting The Earth"** which began in March 2009 and was completed in August 2016.

Other books/volumes by the author in this series:

1. Book One (Volume One): **"Global Evidence of the UFO and ETI Presence"**

2. Book Two (Volume Two): **"UFO Disclosure and Covert Programs of Deception"**

3. Book Three (Volume Three): **"The Military Industrial Complex, USAPs and Covert Black Projects"**

4. Book Four (Volume four): **"In Search of Extraterrestrial Intelligence"**

5. Book Five (Volume Five): **"Evidence of a Type Two ET Civilization in Our Solar System"**

6. Book Six (Volume Six): **"The Rosetta Stone of ETI Contact and Communications"**

Introduction

This is the final volume in the series "A Citizen's Disclosure on UFOs and ETI". In these volumes, the best available evidence on Unidentified Flying Objects and Extraterrestrial Intelligence was presented while staying clear of unproveable information that is highly controversial in nature with no conclusive evidence or requires more investigation and research.

This book, **The Rosetta stone of ETI Contact and Communications** is an interactive book, with web links and video links used to enhance the research value for the student and the seasoned Ufologist and thus, it extends the six volumes in this series, originally, a 3500-page textbook into a 10,000-page tome!

Please note the colour format in quoting, it is used to distinguish the web links from video links and bibliography which is indicated in the Bibliography section of this book.

The reader will be able to follow the web link references to see where I have collected the information for this book. Unfortunately, at the time of publication of this book, some of these web links may no longer be active, it is the nature of the internet and it ephemeral web links! Try "googling" a word or phrase to help you in your search.

In the first five volumes, "hard evidence" was examined ie. photos, videos, crashed alien space vehicles, etc. that the militaries of many governments have gone to extraordinary lengths to cover up and suppress the existence of UFOs and the ETI presence. These cover ups and the secrecy surrounding them may have been at first sight necessary to prevent extraterrestrial technology from falling into the hands of enemy or hostile nations. On the surface this may seem like a rational explanation, except that over the decades, since the end of World War Two, the secrecy of this subject has been subverted and redirected into a global agenda of financial and military domination.

A high percentage of the public know that UFOs are real and that an extraterrestrial presence orbits above us and even walks among us. It now finds itself in an undesirable position of information catch up in which it has a right to an honest disclosure from its elected officials.

Interestingly, not everyone within the government or in the military and intelligence communities are opposed to a public disclosure on this subject as many have come out and made supportive statements of the UFOs and ETI reality. In this sixth volume numerous quotes from all sectors of society, from US Presidents, national and International government leaders, from military officials, astronauts and renown thinkers and scientists have declared the legitimacy of the UFO phenomenon. Such bold statements are in complete opposition to the denials and lies that such extraterrestrial phenomenon does not exist!

The right to an individual's unfettered self- investigation of the truth is a God-given right and no one or any organization has the right to impede that process. Because there is no transparency or official disclosure from our elected officials only deception, lies, denials and coverups there is an increasing distrust to officialdom. This in turn has led to an awaking by the public to investigate truth for themselves to invention and innovation to develop new forms of disc shaped aircraft and alternative energy sources.

With the realization after decades of so many eyewitness reports and accounts globally that UFOs and ETI are real, it has become apparent that human initiated peaceful contact and communications with ETI can be established and maintained bypassing governments and militaries! Such citizen-initiated group like CSETI and Rama which have been operating for over 25 years proves that ET contact and communication is possible and sustainable.

It is now possible for the average citizen to be no longer a second-hand researcher or a passive observer, but they can become proactive researcher who can communicate directly with visiting ETI! This may be the quickest way for the public to be on par with military intelligence, but more than that, we have the opportunity to establish friendships with other intelligent species beyond our planet.

A spiritual and societal re-alignment in our behaviour and thinking is required from all of us and it may mean adopting new values and morals, particularly if our old worn-out spiritual beliefs do stand the test of time in which we live. All things have a season of birth, fruition, death and renewal, making way for the new in its stead!

Such a renewal is now, as the world finds itself becoming a global community gradually resolving its differences and prejudices. We must mature as a species and leave our adolescence behind. When we do, we will finally enter into a golden age of civilization the likes of which former ages have never seen. When we are unified living in peace and harmony as one global commonwealth, then we will be given an invitation to join in greater unity taking our rightful places among the myriad of denizens of the universe.

Thanks for coming along on the ride and learning that we were never alone in the universe!

Terry Tibando - September 2015

THE ROSETTA STONE OF ETI
CONTACT AND COMMUNICATIONS

CHAPTER 99
OFFICIAL GOVERNMENT DISCLOSURE VERSUS
THE PEOPLE'S DISCLOSURE

CHAPTER 100
US PRESIDENTIAL QUOTES AND GOVERNMENT COVER-UP 225

CHAPTER 101
QUOTES FROM U. S. MILITARY WITNESSES ABOUT UFOS............235

CHAPTER 102
QUOTES FROM INTERNATIONAL MILITARY WITNESSES ABOUT UFOS......246

CHAPTER 103
QUOTES FROM U. S. GOVERNMENT WITNESSES ABOUT UFOs........252

CHAPTER 104
INTERNATIONAL QUOTES FROM WORLD LEADERS
AND OFFICIALS ABOUT UFOS 256

CHAPTER 105
QUOTES FROM ASTRONAUTS AND COSMONAUTS ABOUT UFOS 267

CHAPTER 114
PREPARATION FOR CONTACT AND SOME "INALIENABLE TRUTHS" 400

CHAPTER 115
GLOBAL UNITY BEFORE GALACTIC UNITY 426

CHAPTER 116
GLOBAL UNITY - A PREREQUISITE FOR ADMISSION INTO GALACTIC
CIVILIZATION..438

CHAPTER 117
CAN YOU PUT THAT IN WRITING? XENOLINGUISTICS -
EXTRATERRESTRIAL SCRIPTS AND SYMBOLS... 453

CHAPTER 118
WHO BEST SPEAKS FOR THE PEOPLE OF EARTH WHEN CONTACT WITH
EXTRATERRESTRIAL INTELLIGENCES IS IMMINENT? 463

CHAPTER 122
SOME ANSWERED QUESTIONS ABOUT MILITARY COVER UPS, UFOS, ETI, ZERO POINT ENERGY

CHAPTER 123
A GLOBAL VISION - HUMANITY MATURES TO BECOME AN EXTRATERRESTRIAL CIVILIZATION

THE ROSETTA STONE OF ETI CONTACT AND COMMUNICATIONS

CHAPTER 93

WHAT DO SCIENTISTS REALLY KNOW ABOUT UFOS AND ETI?

If you had asked an astronomer or any scientist 70 years ago of what they thought about unidentified flying objects, you probably would get an answer like what is that? Is that like a comet or meteorite or maybe, a meteor or asteroid that zips through the atmosphere at night? Could that be some foreign country's secret aircraft that has illegally entered our airspace? Quite frankly, most scientists really wouldn't know what you were talking about but, that all that changed by the late 1940s and early '50s with the public becoming more aware of what was flying around in their skies, particularly after **Kenneth Arnold's** sighting of nine silvery objects that skipped above the Cascade Mountains of Washington State in 1947.

If you were to ask the same questions again to the same scientists and astronomers at that time, you would get an entirely different set of answers, such as, that has something to do with flying saucers and aliens, doesn't it? It's fascinating, that there might actually be other life in the universe and it just happens to be coming to our planet or I don't believe in stuff like that sort of nonsense, it's all pseudoscience!

By the mid-'50s to the current time, science has been divided on the reality of the subject matter of UFOs and ETI with the majority of scientists staying clear of any investigation into the matter. Once the military intelligence of post-war nations determined what it was, that Earth was being monitored by Extraterrestrial Intelligences and with the crash retrievals of several downed ET spacecraft and the recovery of alien bodies, *"the whole subject went deep black"* and a serious disinformation campaign was launched with severe penalties imposed upon military personnel and those in the scientific community for speaking out about the subject, all under nation's catchall policy of **"National Security"**.

Spin control was necessary to keep the mass populace in line with the newly developed **Military Industrial Complex's** agenda, terms like **"UFOs and Flying Saucers"** simply became misidentified flying objects, hoaxes, and public attention-seeking. A ridicule factor prompted by military officials was propagandized via the news media into the public domain whenever someone reported a strange object flying peculiarly in the sky. Little wonder then, that any scientist would even broach the subject matter for any serious consideration of research and investigation. Reputation and tenure were on the line and saving face among one's peers was a strong motive, let alone being able to work in one's chosen career field without the **"UFO *Sword of Damocles"*** hanging over one's head!

But these threats and intimidations were not enough to those few rare and curious scientists who recognize that a true anomaly is always worth investigating as the potential for a new discovery or breakthrough in science may be of benefit to humanity as a whole. These courageous individuals were not deterred by the new National Security policy. Perhaps, it was because they may have already achieved a certain measure of success in their careers that nothing could threaten that success or take away their achievements; they therefore, felt that UFOs and ETI was a noble pursuit of science into that mysterious realm of the unknown. This is after all, what science is all about!

Quotes from Mainstream Scientists

Here is the first chapter in this book of quotes about unidentified flying objects from notable public officials and scientists. We begin with some quotes from scientists on the subject that still mystifies and perplexes the hallow halls of mainstream science. Some are more open minded and positively responsive while others are in a dark abyss of denial.

Dr. Peter A. Sturrock, Professor of Space Science and Astrophysics and Deputy Director of the Center for Space Sciences and Astrophysics at Stanford University; Director of the Skylab Workshop on Solar Flares in 1977 wrote in the Journal of Scientific, a very insightful and acutely accurate assessment of the scientist dilemma that surrounds the UFO topic:

"In their public statements (but not necessarily in their private statements), scientists express a generally negative attitude towards the UFO problem, and it is interesting to try to understand this attitude. Most scientists have never had the occasion to confront evidence concerning the UFO phenomenon. To a scientist, the main source of hard information (other than his own experiments' observations) is provided by the scientific journals. With rare exceptions, scientific journals do not publish reports of UFO observations. The decision not to publish is made by the editor acting on the advice of reviewers. This process is self-reinforcing: the apparent lack of data confirms the view that there is nothing to the UFO phenomenon, and this view works against the presentation of relevant data." **Sturrock, Peter A., "An Analysis of the Condon Report on the Colorado UFO Project," Journal of Scientific Exploration, Vol. 1, No. 1, 1987** and http://www.ufoevidence.org/topics/science.htm

Sturrock asks why scientists "shudder" when topics such as parapsychology, ufology, and cryptozoology are brought up. Answer: because, in each of these areas,

(1) There is widespread, well-documented fraud, and

(2) there is a community, or, better, an industry, of "true believers" whose minds are CLOSED to the possible non-existence of the phenomenon. These two together are enough to make any rational person shudder and put on his "I'm from Missouri" hat.
http://www.ufoevidence.org/documents/doc780.htm

Bernhard Haisch, Ph.D., *"Cut through the ridicule and search for factual information in most of the skeptical commentary and one is usually left with nothing. This is not surprising. After all, how can one rationally object to a call for scientific examination of evidence?"*

Dr. Jacques Vallee, astrophysicist, computer scientist and world-renowned researcher and author on UFOs and paranormal phenomena. He worked closely with **Dr. J. Allen Hynek**. Commenting on the need for science "to search beyond the superficial appearances of reality":

16

"Be Skeptical of the Skeptics" "Skeptics, who flatly deny the existence of any unexplained phenomenon in the name of 'rationalism,' are among the primary contributors to the rejection of science by the public. People are not stupid, and they know very well when they have seen something out of the ordinary. When a so-called expert tells them, the object must have been the moon or a mirage, he is really teaching the public that science is impotent or unwilling to pursue the study of the unknown." **Vallee, J., Confrontations, New York: Ballantine Books, 1990** and http://www.ufoevidence.org/topics/SkepticsAnalysis.htm

But, as already stated, some scientists are not deterred and view these topics with interest, particularly when globally, so many of the public continually report strange flying objects in the skies or unusual encounters with strange humanoid beings on the ground not native to planet Earth. There is more here than just the publicity-seeking individuals wanting to get their 15 minutes of fame or notoriety, something that does not fall into the neat, tidy parameters of human thinking and science. Why?

This is the reason why some scientists become involved in a new area of research and investigation because there are similarities to conventionality but, a whole lot of differences that seem just beyond the current understanding of science.

Sturrock further states: *"... scientists need at least three modes of thought that I call "curious," "creative" and "critical." These requirements, though they may be quite general in their applicability, come sharply into focus when one deals with anomalies within mainstream science or with anomalous phenomena that seem to reside outside of science as we know it."*

Dr. Maurice Biot, aerodynamicist, and mathematical physicist *concluded:*

"The least improbable explanation is that these things are artificial and controlled...my opinion for some time has been that they have an extraterrestrial origin."

Werner Von Braun, rocket scientist who was instrumental in the development of Nazi Germany's V2 rocket and later, the American space program commented on the mysterious events during the re-entry phase of the Juno 2 rocket during a test flight in an article from the January 1959 issue of "News Europa":

"We find ourselves faced by powers which are far stronger than we had hitherto assumed, and whose base is at present unknown to us. More I cannot say at present. We are now engaged in entering into closer contact with those powers, and in six or nine months time it may be possible to speak with some precision on the matter."

Louis Breguet, French aircraft designer, and manufacturer:

"The discs use a means of propulsion different from ours. There is no other possible explanation. Flying saucers come from another world."

Dr. Richard F. Haines, retired NASA senior research scientist at Ames Research Center and the Research Institute for Advanced Computer Science where he worked on the **International**

Space Station; in the preface of his book CE-5:

"What I found (in doing research for the book Project Delta) was compelling evidence to claim that most of these aerial objects far exceeded the terrestrial technology of the era in which they were seen. I was forced to conclude that there is a great likelihood that Earth is being visited by highly advanced aerospace vehicles under highly "intelligent" control indeed." **CE-5: Close Encounters of the Fifth Kind by Richard F. Haines, Ph.D.; 1998; published by Sourcebooks Inc.; Naperville, Illinois, USA; ISBN 1-57071-427-4**

UFO researchers like Stanton T, Friedman, Leslie Kean and James E. MacDonald have independently concurred that regardless of the intense public interest in Unidentified Flying Objects (UFOs) and the Extraterrestrial presence coming to Earth, there is, however, still a failure to investigate this phenomenon by science and therefore, science is in default of its basic mission statement. The betrayal of a marginalized public by science has seen no further independent, federally financed scientific research conducted on these phenomena since the flawed and biased 1969 Condon report.

Dr. Frank Halstead of the Darling Observatory, Minnesota—1957:

"Many professional astronomers are convinced that saucers are interplanetary machines."

Once again, a statement like Halstead's is indicative of that time period but, in recent times when pressured to express a public opinion on the UFO subject, most scientists and astronomers are not as forthcoming as they would be privately, behind closed doors, among themselves. In public, however, these mainstream scientists offer nothing more than ridicule or scorn upon the topic of UFOs.

Dr. J. Allen Hynek, former Chairman of the Dept. of Astronomy at North Western University and scientific advisor to **Project Bluebook** from 1952-1969 made these following admissions:

"When I first got involved in this field, I was particularly skeptical of people who said they had seen UFOs on several occasions and totally incredulous about those who claimed to have been taken aboard one. But I've had to change my mind." --1972

"It reminds me of the days of Galileo when he was trying to get people to look at the sunspots. They would say that the sun is a symbol of God; God is perfect; therefore, the sun is perfect; therefore spots cannot exist: therefore there is no point in looking." *--Hynek in Newsweek, Nov. 21, 1977, p. 97*

"I was there at [Project] Bluebook and I know the job they had. They were told not to excite the public, not to rock the boat... Whenever a case happened that they could explain--which was quite a few--they made a point of that and let that out to the media... Cases that were very difficult to explain, they would jump handsprings to keep the media away from them. They had a job to do, rightfully or wrongfully, to keep the public from getting excited."
http://www.stargate-chronicles.com/site/

18

Lee Katchen, NASA atmospheric physicist in an announcement on June 7, 1968, stated that he believed, based on his examination of 7,000 reports, that UFOs have an extraterrestrial origin:

"UFO sightings are now so common, the military doesn't have time to worry about them, when a UFO appears, they simply ignore it. Unconventional targets are ignored because apparently, we are only interested in Russian targets, possibly enemy targets. Something that hovers in the air, then shoots off at 5,000 miles per hour, doesn't interest us because it can't be the enemy. UFOs are picked up by ground and air radar, and they have been photographed by gun camera all along. There are so many UFOs in the sky that the Air Force has had to employ special radar networks to screen them out."

Robert J. Low, project coordinator of the **Colorado University UFO Project** (a.k.a. **The Condon Committee**), in a memorandum of instruction from August 9, 1966 made this inadvertent comment which went public and set off a firestorm of controversy and conspiracy charges around the US Air Forces ultimate intentions and attitude around the UFO subject which have never been extinguished to this day. It is often quoted and gives the true goal of the Project: "to either get the thing out of the way without hurting any of the scientists' credibility or to comply with the rumored Air Force directive to produce a report showing UFOs to be unworthy of scientific consideration."

"The trick would be, I think, to describe the project so that, to the public, it would appear a totally objective study but, to the scientific community, would present the image of a group of non-believers trying their best to be objective, but having an almost zero expectation of finding a saucer." http://www.ufoevidence.org/documents/doc1744.htm

Cultural attitudes toward science are ever-changing in a rather predictable fashion as we bear witness to the outdated concepts and standards of youth as taught by our parents and the institutions of their time while as adults we perceive that reality forces us all down highways of rapid change that we are barely able to keep up with. In the forefront of rapid change is the unstoppable, irresistible force of science, one of the twin pillars of knowledge in society, the other being religion.

As **Richard Dolan** points out:

"Science, we were taught, is a bastion, indeed the foundation, of intellectual freedom in the world. It is an independent search for truth and the destroyer of social and religious myths."
"How independent is science? In whose interest is it practiced today? This is no idle question, for gone are the days of scientists following their intellectual passions in a search for truth."
Science, Secrecy, and Ufology by Richard M. Dolan; copyright ©2000 by Richard M. Dolan; Published in February/March issue of UFO Magazine

It has become a common practice and a rule of thumb that when dealing with military departments, intelligence agencies and government contractors one merely has to follow the money to find out what is really going on and what program may be operating. To this may also be added the science community. *"Science is an expensive business and you need sponsorship,"* says Dolan.

James Lovelock, a pioneer in environmental science back in 2000 according to Richard Dolan, told him the following:

"Nearly all scientists are employed by some large organization, such as a governmental department, a university, or a multinational company. Only rarely are they free to express their science as a personal view. They may think that they are free, but in reality, they are, nearly all of them, employees; they have traded freedom of thought for good working conditions, a steady income, tenure, and a pension."

Reflect for a moment that wars, whether we like or not are the biggest and primary contributors to scientific breakthroughs! Why, because as exemplified by the First and Second World Wars and all subsequent smaller wars and conflicts, "the military has been *by far* the biggest sponsor of scientific work."

This was publicly acknowledged by the former **US President Dwight D. Eisenhower** in his farewell address (1961) when he saw that the labour force behind the US Military during WWII was greatly needed and necessary in winning and ending the war but, warned that the continued buildup of military armaments in a time of peace was due to the partnership of private industry and the military working together. In that now renowned speech Eisenhower warned of increasing power of the military-industrial complex and further stated:

". . . we must guard against the acquisition of unwarranted influence, whether sought or unsought, by the military-industrial complex. The potential for the disastrous rise of misplaced power exists and will persist.

We must never let the weight of this combination endanger our liberties or democratic processes. We should take nothing for granted. Only an alert and knowledgeable citizenry can compel the proper meshing of the huge industrial and military machinery of defense with our peaceful methods and goals so that security and liberty may prosper together."

To this growing military industrial machination, it has historically come to light in the past years through the release of British UFO documents that **President Eisenhower** and **Prime Minister Churchill** both were made aware from air pilot debriefing reports that the world was being monitored by intelligent beings from other worlds. It could be said that Allied Forces had an order to shoot down these craft even in the unlikelihood that the alien spacecraft may be a new type of enemy weapon. From a least, a couple of encounters with something unknown particularly the **1942 Los Angeles UFO Incident**, the military and intelligence community were put on high alert to this UFO phenomenon and before the Second World War ended there were extreme levels of high interest and classification that encompassed the subject.

In order to better understand the phenomenon, the only logical recourse for the military and the various intelligent branches was to classify the whole subject above top secret, even higher than the atomic bomb and sponsor the best scientific minds to work on the problem for as many years as was necessary. We know of some of the most brilliant minds who were involved in this phenomenon in some way, most noteworthy were **Edward Teller, Detlev Bronk, Lloyd**

20

Berkner, Vannevar Bush, David Sarnoff, Thornton Page, H. P. Robertson, J. Allen Hynek, and Lincoln La Paz.

"These men were some of the elite power scientists in the world, and intimately connected with the American defense establishment. And yet, we find them looking at UFO reports. Of course, let us not forget Harvard astronomer and UFO debunker extraordinaire, **Donald Menzel***, who, unbeknownst to the world, was deeply involved with the American intelligence community, in particular, the super-secret* **National Security Agency."** Science, Secrecy, and Ufology by Richard M. Dolan; copyright ©2000 by Richard M. Dolan; Published in February/March issue of UFO Magazine

To keep a tight lid on the subject scientists were offered financial inducements and swore an oath of secrecy and National Security policies were passed into law making it a crime to reveal inside knowledge about the phenomenon. A carefully orchestrated infrastructure of lies, denials, as well as campaigns of misinformation and disinformation were perpetrated upon the inquiring public who went searching for answers to the UFO mystery and even mainstream scientists added insult to injury by offering nothing more than ridicule or scorn upon the topic of UFOs.

Scientist like most people enjoy the benefits of a great wage package and therefore, will always follow the money and in return whenever they are employed or contracted into highly secretive and compartmentalized projects and programs, they are told or warned not to publicly disclose positive statements on things that they are secretly working on.

Essentially, compartmentalization ensures that rogue scientists working on the inside never fully get to understand the projects which may be related to the UFO phenomenon thus, legitimate widespread ignorance is maintained among scientists with regard to the UFO subject. Their world becomes the black world of science and even among scientists who have figured out that they are involved in such things as reversed alien technology, national security oaths and threats keep them in line. Outside of this black world of science, mainstream scientists in non-classified fields of research and development are most likely in a world of ignorance when it comes to the UFO phenomenon, taking their lead from other scientists who scorn and ridicule the subject because that may actually be working on the inside and have a vested interest to keep it all secret and covered up.

Bruce S. Maccabee, Ph.D., 1986 MUFON International Symposium Proceedings:

"For nearly 40 [more than 50] years, the science establishment has ignored the UFO problem, relegating it to the domain of "true believers and mental incompetents" (a.k.a. "kooks and nuts"). Scientists have participated in a "self-cover-up" by refusing to look at the credible and well-reported data. Furthermore, some of those few scientists who have studied UFO data have published explanations which are unconvincing or just plain wrong and have "gotten away with it" because most of the rest of the scientific community has not cared enough to analyze these explanations. The general rejection of the scientific validity of UFO sightings has made it difficult to publish analyses of good sightings (in refereed journals of establishment science)." http://www.ufoevidence.org/topics/science.htm

Clark McClelland, Aerospace Engineer and Technical Assistant to the Apollo Program Manager during the Apollo moon landings, also assisted in almost six hundred launches at Cape Canaveral, and in addition to working in the Mercury and Gemini programs, Space Lab and the Space Station was heavily involved in the Space Shuttle program. These quotes come from Clark's website, The Stargate Chronicles.

"As the Gemini Capsule entered orbit, the RCA world tracking team began to realize that 'our' capsule was not alone as viewed through their incoming telemetry, visual theodolite, and other high-powered optical data. Our capsule had four 'visitors'. The RCA team was ordered to run a recheck of the situation to be certain ghost images were not the cause. The Titan II stages were also excluded as causing the images After much huddling and discussion, the intelligent determination was that we had other physical objects up there with our Gemini capsule The official NASA determination was that the objects were the torn particles or remains of the Titan upper stage that apparently entered orbit with the Gemini capsule. I was at the news conference and I nearly began to laugh. How could a broken stage overtake the capsule and stop slightly ahead of the capsule to accompany it an entire orbit around the earth? But I held my laugh to save my job." --Commenting on the April 9, 1964, unmanned launch of the Gemini-Titan 2.

"The day will arrive when the governments of earth will finally admit we are not alone, that humans have come face to face with other life forms from the cosmos."

Dr. James E. McDonald, Professor of Atmospheric physics, University of Arizona. 1967:

"I have absolutely no idea where the UFO's come from or how they are operated, but after ten years of research, I know they are something from outside our atmosphere."
http://www.stargate-chronicles.com/site/

Dr. J. C. MacKenzie, Chairman of the Canadian Atomic Energy Control Board and former president of the National Research Council.--January, 1952

"It seemed fantastic that there could be any such thing. At first, the temptation was to say it was all nonsense, a series of optical illusions. But there have been so many reports from responsible observers that they cannot be ignored. It seems hardly possible that all these reports could be due to optical illusions."

Dr. Harry Messel, Professor of Physics at Sydney University, Australia, in a 1965 statement:

"The facts about saucers were long tracked down and results have long been known in top secret defense circles of more countries than one."

C. B. Moore, General Mills Meteorologist, and expert on weather balloons, when asked whether he believed the Roswell Incident could be explained by a Mogul balloon:

"Based on the descriptions, I can definitely rule this out. There wasn't a balloon in 1947 or today that could account for this incident."

Dr. Herman Oberth, the father of modern rocketry:

"UFOs are conceived and directed by intelligent beings of a very high order and they are

propelled by distorting the gravitational field, converting gravity into useable energy. There is no doubt in my mind that these objects are interplanetary craft of some sort. I and my colleagues are confident that they do not originate in our solar system, but we feel that they may use Mars or some other body as sort of a way station. They probably do not originate in our solar system, perhaps not even in our galaxy." --This comment was apparently made sometime in 1954, I don't know the source, but it seems consistent with the following quote from The American Weekly of Oct. 24, 1954:

"It is my thesis that flying saucers are real and that they are space ships from another solar system."

"We cannot take the credit for our record advancement in certain scientific fields alone. We have been helped." (The reporter asks *"By who?") "The people of other worlds."* --From a statement to a group of reporters after his retirement in 1960. http://www.stargate-chronicles.com/site/

Dr. Walther Riedel, research director and chief designer at Germany's rocket center in Peenemunde, also worked on classified projects for the U.S. after WW2. From LIFE Magazine, April 7, 1952:

"I am completely convinced that [UFOs] have an out-of-world basis."

Dr. John Sathco, an Astronomer at the University of Southern California, 1973 (source: Real Audio from Lost Encounters of 1973)

"There are in excess of 200 reports of the type that we had from down in Louisiana, from people claiming that they have had direct contact with a spacecraft full of aliens. I mean 200 reports from witnesses who are as reliable or more so than these people. I'm not counting the reports from the obvious crackpots that have an axe to grind...If you accept them at face value then you're forced to accept that we have been visited.

Wilbert Smith, Electrical engineer who convinced the Canadian government to establish **Project Magnet** to study the UFO phenomenon and later served as engineer-in-charge of the project.

"The matter is the most highly classified subject in the United States Government, rating higher even than the H-bomb. Flying saucers exist. Their modus operandi is unknown but concentrated effort is being made by a small group headed by Dr. Vannevar Bush." --From a declassified Canadian government memorandum dated Nov. 21, 1950.

"...it soon became apparent that there was a very real and quite large gap between this alien science and the science in which I had been trained. Certain crucial experiments were suggested

and carried out, and in each case, the results confirmed the validity of the alien science. Beyond this point, the alien science just seemed to be incomprehensible." --In a speech concerning experiments allegedly suggested by EBEs (Extraterrestrial Biological Entities); March 31, 1958, http://www.stargate-chronicles.com/site/

Clyde Tombaugh, the astronomer who discovered Pluto, in a letter dated September 10, 1957. The phenomenon was also witnessed by his wife:
"The illuminated rectangles I saw did maintain an exact fixed position with respect to each other, which would tend to support the impression of solidity. I doubt that the phenomenon was any terrestrial reflection....I do a great deal of observing (both telescopic and unaided eye) in the backyard and nothing of the kind has ever appeared before or since."

Dr. Weisberg--From a memo by the director of the Borderland Science Research Foundation, Layne Meade, in 1949 concerning a description given by Dr. Weisberg, a Canadian physics professor who apparently examined some retrieved discs for the U. S. Air Force at Edwards AFB.
"Like a turtle's back, with a cabin space some fifteen feet in diameter. The bodies of six occupants were seared and the interior of the disc had been badly damaged by intense heat."

Yale Scientific Magazine (Yale University) Volume XXXVII, Number 7, April 1963.:

"Based upon unreliable and unscientific surmises as data, the Air Force develops elaborate statistical findings which seem impressive to the uninitiated public unschooled in the fallacies of the statistical method. One must conclude that the highly publicized Air Force pronouncements based upon unsound statistics serve merely to misrepresent the true character of the UFO phenomena."

Zhang Zhousheng, an astronomer at the Yunnan Observatory in Chengdu City, China-- Zhousheng and others nearby watched a strange glowing, spiral object moving steadily across the sky for about five minutes on the evening of July 26, 1977.

"What was especially important was that, at a distance of 180 kilometers apart, the records about the direction of movement of the strange aerial body in space, made independently by at least two different observers was basically the same. To the present time, this strange phenomenon has not been satisfactorily explained, yet there were thousands of good observers who had seen it."
http://www.ufoevidence.org/documents/doc1744.htm

Black World of Science Versus White World (Mainstream) Science

The **Black World of Science (BWS)** and the **White World of Science (WWS)** are not well-known concepts and the American public might be forgiven if their perception followed along the typical racial lines that it was a special initiatives program setup for Black Americans to become more involved in the field of science. In one sense they would be right but, this is rather a peculiar situation to America and not the norm in most other countries. This perception as it relates to Ufology unless you are a Ufologist or a researcher of paranormal sciences, is unknown to most people, who wouldn't know what you were talking about. Military Industrialists and some government officials, on the other hand, do know and they are not talking publicly about it.

To be fair, there are no informative sources to be found and very few books written on the topic, even on the internet, if one were to **Google** or **Yahoo** the subject "Black World of Sciences", you would get websites that guide you to a record album, a rock or country band or a comic book or to Black American websites! These websites are not what the Black World of Science is all about! To get a basic understanding of **BWS** you must root the information out from the few available books on the subject and research the rest from other sources like the few scant documents obtained from the **FOIA** on black projects and programs but, even these will be of limited information.

It is apparent from the above quotes made by astronomers, physicists, and scientists that they are convinced that the UFOs or Extraterrestrial spacecraft and Extraterrestrial Intelligence exist beyond any shadow of a doubt. It is also obvious that the science community is divided into two polarized groups known as the **"Black World of Science"** and the **"White World of Science"**.

The **Black World of Science** are essentially sciences where covert R&D black projects and programs are undertaken in complete secrecy, hidden from public view and knowledge thriving upon an almost inexhaustible flow of financial capital, resources and manpower buttressed with an airtight security system all working in some remote or unknown location away from prying eyes. BWS can best be characterized as covert, beyond top secret, it is rogue in nature, it is corrupt and self-serving benefitting only a minority of wealthy and powerful elite, it is beyond government oversight and control, surrounded most times in an atmosphere of oppressiveness, jealously guarded under false ideals of national security, compartmentalized in every facet of its operations, "the right hand never knows what the left hand is doing", only a few on top know the whole "picture"; black projects and programs are highly restricted and divided into **SAPs (Special Access Programs)** and **USAPS (Unacknowledged Special Access Programs)** with an inclusive or a need-to-know, otherwise most people including high military brass, government officials, and even the US President are "out of the loop". Scientists are paid extremely well but, they are expected to perform and produce, they have to all outward appearances ***"sold their souls to the devil"***. Such deep black programs are never acknowledged publically and governmentally they are as transparent as a lead-lined brick wall within a mountain under the ocean! They have one other aspect that the White world of Science doesn't have, the ability to access through military means retrieved crash saucer wreckage and alien bodies!

The **White World of Science,** obverse to this Machiavellian clandestine world, is an invigorating breath of fresh air! It is the science we all know and appreciate in providing the marvels, comforts, and the cures to everyday living. It may be characterized as providing the most good for the most number of people, it is for the most part methodical following the scientific approach to all things, skeptical toward new things but, imminently practical in research and development, it is, unfortunately, ***two to five generations behind*** the Black World of Science in technological development and breakthroughs, there are some classified and top secret programs, communications with all branches of sciences is good, scientists in the **WWS** rarely, if ever become millionaires from their careers and generally discoveries and advancements have full disclosure and are open to public scrutiny.

Historically, early in the Twentieth Century up to the current times, war has been the primary reason for a secret, hidden world of science enabling one or more nations to gain power and dominance over enemy nations. This was been particularly true in the First and Second World Wars with Germany "blitzkrieging" its way across a technologically unprepared and bewildered Europe, which for the most part was still used to fighting in an1800s Napoleonic style. Secret military programs in armament development would be a deciding factor and initially, Europe was no match against this new lightening violence of aerial assaults and larger cannon bombardments, massive troop deployments and radical new concepts in weapons development.

War has been a quick learning curve for most nations to develop technological advancements in weapons and in aviation which re-enforced military brinkmanship among nations in massive arms build-up during periods of cold war or in peace time. No nation wanted to be unprepared should there be any future outbreaks of war. *"Necessity is the mother of all invention"* and ***"science could be considered as the father of invention!"***

Breakthroughs in science are sometimes out of necessity and sometimes occur as a result of an ***"oops moment"*** or as an accidental coincidence to some other direction of research which often become the cornerstones to whole new fields of science. The science of aviation in the early 20th Century brought about the development of engine-powered aircraft, air transportation, and military air forces. Once humanity tasted the liberating freedom of flight, mankind could barely keep its feet on the ground and war saw to it that the sciences of aviation, aeronautics and aerospace kept up to military demands as well as to civilian needs.

Manmade Flying Saucers in the White World of Science

Fixed wing aircraft dominate the skies but, radical new concepts in aircraft designs hinted at a possible future where single wing and disc-shaped craft powered by propellers, jets, and other propulsion systems would one day become the aircraft standard in the world. Even in the early 1900s, scientists, inventors, and aeronautical engineers were drafting saucer-shaped craft and constructing prototypes as proof of concept of their designs. A good example of man's ingenuity and creativeness in aeronautical engineering and design can be found at the **"Celtic Cowboy Co."** website where hundreds of original aircraft concepts actually flew, glided, or simply jumped or hopped into the air! http://celticowboy.com/Round%20Aircraft%20Designs.htm

One can see an evolution of aircraft design spanning more than a hundred years, since the turn of the 20th Century where many nations experimented in building round, single fixed-wing, bi-wing multi-wing aircraft, rotary wing or helicopters and all sorts of flying contraptions. Not all the aircraft designs shown below are round, some are truly odd and unflight worthy but, they are also fascinating and certainly entertaining.

As would be expected, it is only natural that aircraft designers and engineers would have based their aircraft concepts on the birds of the air and thus, fixed wing aircraft proved to be a reliable design but, round wing or disc shape craft also proved to be a reliable design as well. What was the motivation to design aircraft that did not follow the observable rules of nature? Was it the wheel shape found on wagons or automobiles or the ancient Greek throwing discus? It seemed that aeronautical engineers were obsessed with designing disc-shaped craft as if trying to recapture a forgotten memory from eons long ago or were they simply experimenting with different concepts?

The following gallery of photographs indicates the name of the aircraft and in some cases the year it first appeared. The first section of photos is the Disc/Saucer or "round-like" aircraft. The next section of photos is the "Blended Wing" Aircraft and a section termed "Single Fixed Wings Aircraft ". These aircraft are neither round, flying wings or blended wings, they are simply unique. Some are also airships incorporating current technology and are truly massive in size.

Disc or Saucer Shape Aircraft: Many aircraft are civilian built and others are definitely military planes but, were not classified at a top secret level except during or after the Second World War for a short period of time. Though many of the aircraft appeared to be too flimsy-looking to fly because most aircraft construction was made from canvas, wood, and wire with few metal parts, they actually worked quite well, however, strong breezes or winds might cause them to become uncontrollable.

The "Sky Car" worked similar to an umbrella opening and closing in a pumping action providing its pilot and passenger with the first known air-conditioned vehicle!
(Google Images)

The same or similar "Air Car" which never flew but hopped along in an amusing manner probably giving a rather bumpy ride!
(Google Images)

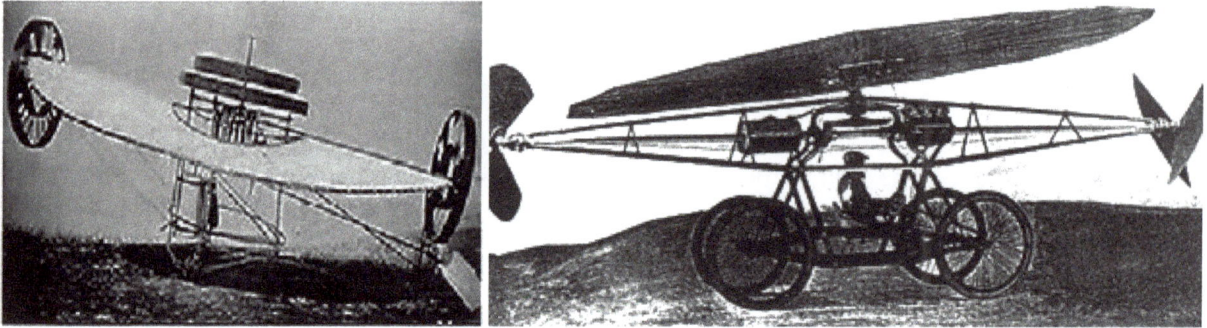

**Botts, 1903 (left) and Lataste Aeroplane Gyroscopique (right)
are both gyroscopic designed aircraft**
https://www.pinterest.com/vrocampo/aircraft-research-and-experimental-early-years/?lp=true
and http://www.theaerodrome.com/forum/showthread.php?p=702880

Lipkowski Helocopter, 1905 (left) and Villiard (right)
https://disciplesofflight.com/nemeth-parasol-strange-aircraft/ and
https://www.pinterest.com/vrocampo/aircraft-research-and-experimental-early-years/?lp=true

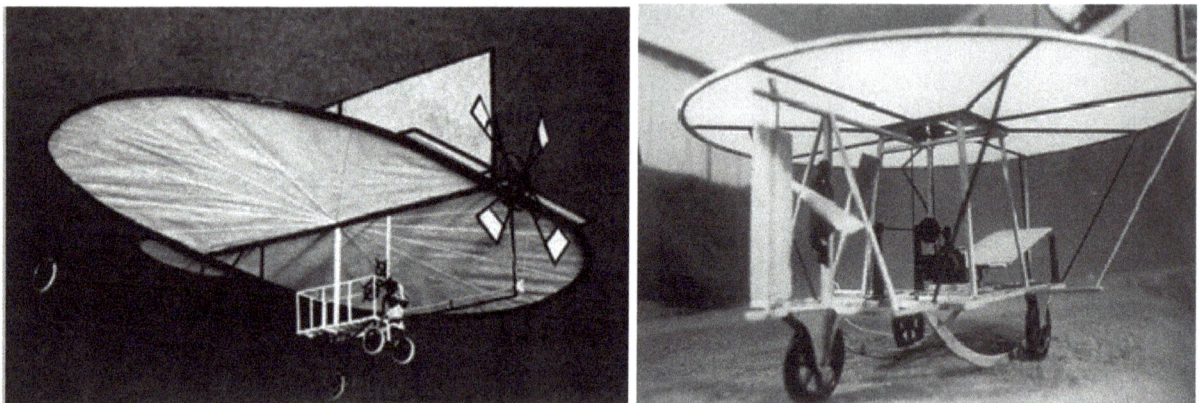

Underwood, 1907 (Replica) (left) and Donovan Monoplane, 1909 (right)
https://www.pinterest.com/vrocampo/aircraft-research-and-experimental-early-years/?lp=true

Thiersch, 1910 (top) and Davidson Gyro-Copter, 1911 (bottom)

Italian Design which in which the disc wing appears balloon-like (left) and Kitchens Annualar, 1911(right) built by John George Aulsebrook Kitchen, Scotforth, Lancashire

Lee-Richards Annualar (left) and Lee - Richards Biplane Annualar (Replica) (right)

The McCormick-Romme "Umbrella Plane" also called the "Cycloplane" 1910 on the ground and in flight at the Cicero Flying Field, August 30, 1911
https://www.pinterest.com/vrocampo/aircraft-research-and-experimental-early-years/?lp=true

© Frank Rezich
aerofiles.com

© Frank Rezich
aerofiles.com

Recently tested in Chicago, saucer plane made 135 m.p.h. and landed like a parachute.

Two Paul Nemeth designed "Umbrellaplanes" were also known as the "Roundwing", they were built in 1934
http://www.pprune.org/aviation-history-nostalgia/332082-silhouette-challenge-1436.html
and https://disciplesofflight.com/nemeth-parasol-strange-aircraft/

Before the start of the Second World War, many scientists and aeronautical engineers were still fixated on the saucer concept as being the ultimate aircraft design to fly the skies as it represented better stability and less drag over the entire airframe. The problem, however, was that unless the aircraft had ailerons or a tail rudder or rotated as it flew it was unstable in basic flight maneuvers. It would be a few more decades before computers and gyroscopic mechanisms would solve the stability problems.

The **Avrocar** built by **Avro Canada** was a proof of concept of the saucer design which flew low to the ground implementing the **Coanda Effect** but, it too suffered instability problems, as it

30

wobbled as it flew. The Avrocar was funded by the US Army for 12 million dollars and built at **RCAF base Malton**, Ontario. This "wobble" instability problem was eventually resolved; however, it came too late as the project was cancelled before production could begin. The US Army had both Avrocar prototypes sent Stateside, one to be displayed in the **Smithsonian Air Museum** and the other stored in a US air base hangar. Curiously, the US military was still highly interested in the saucer configuration for military aircraft and later designs out of **Skunk Works** and **Phantom Works** were engineered to full flight capability! The reader should also review the earlier section in this book which illustrated the disc-shaped aircraft and the single fixed wing aircraft built by both the German Nazis and the American militaries during the Second World War.

The world's first widely publicized "Flying Saucer" was the Canadian-built VZ-9 Avrocar that flew in 12 November 1959. It could only fly 3 Feet off the ground, it was essentially the world's first hovercraft vehicle
https://www.hexapolis.com/2014/08/10/14-peculiar-specimens-human-made-aircraft-look-like-ufos/2/

Astro V by Dynafan, 1964

Saucer-shaped helicopter gets lift from air pump

Air pumped from inside this inverted dish by the impeller on top creates an air flow over the upper surface that causes the craft to rise vertically. Reason, says the maker, Astro Kinetics, of Houston: less pressure in this air layer than that of normal air below. This eliminates need for rotor blades, while maneuverability and forward speed are said to be the same as for a helicopter. Motor: a 135-hp. Mercury outboard.

The inverted "dish" or saucer-shaped helicopter utilizes the Coanda Effect for lift

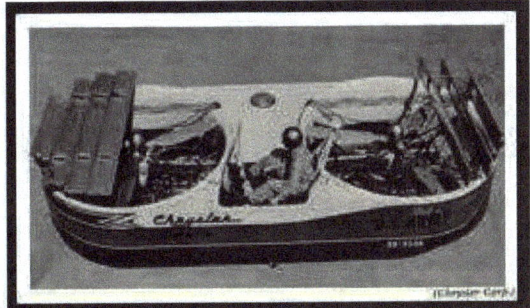

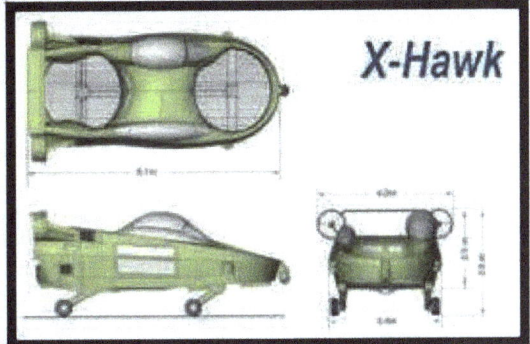

Various Air Cushioned Vehicles or "hover cars"

**This is either a brilliant homemade Chinese "flying saucer"
or a slice and dice accident waiting to happen.
Note the pilot left his shoes on the bench!**

Lighter-than-air Silver Sphere
https://en.wikipedia.org/wiki/PAGEOS

Sky Ship, UK 1975
http://www.abovetopsecret.com/forum/thread146492/pg1

Thermoplane by Aviastar in Russia
http://www.abovetopsecret.com/forum/thread420998/pg1

Saucer shape airship - Locomo Sky Thermoplan
http://www.roswellufomuseum.com/research/ufotopics/earthmadeflyingsaucers.html

Flying/Blended Wing Aircraft: the aircraft that follow are a combination of aero-form types incorporating straight fixed wings and disc or curve structures which in their strange configuration actually enable the aircraft to fly.

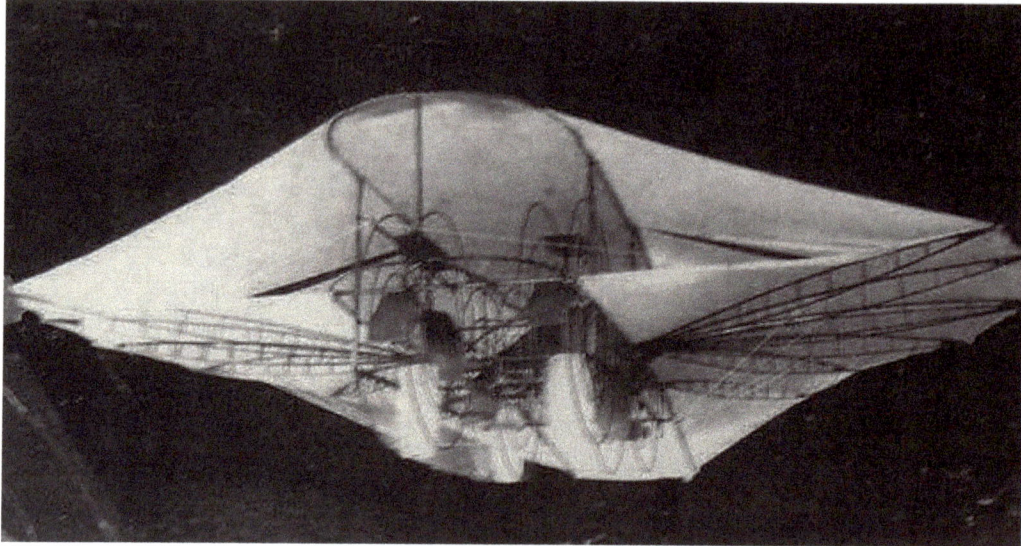

Ezekiel Airship
https://www.standeyo.com/stans.files/cannon_ezekiel.html

Gillespie, 1905
https://www.pinterest.com/pin/223350462748488389/

Williams, 1908
http://www.aerofiles.com/_wh.html

Call II
http://www.ctie.monash.edu.au/hargrave/call.html

Edwards Rhomboidal, 1910
http://flyingmachines.ru/Site2/Crafts/Craft28410.htm

Unknown
(Google Images)

Flapjack and Pancake Planes: all these planes follow the uniwing disc configuration which has now become a proven and stable wing shape for planes yet for some reason never became the accepted standard for aeronautical design and engineering.

JOY JX, 1935
http://www.shu-aero.com/AeroPhotos_Shu_Aero/Aircraft_J/Joy/index.html

Synder "Dirigiplane" (left) and the Snyder A2 (right), also called the Arup 1
http://svammelsurium.blogg.se/2011/july/pa-annorlunda-vingar.html

Arup – 5
http://svammelsurium.blogg.se/2011/july/pa-annorlunda-vingar.html

Hoffman Flying Wing (left) and Johnson Uniplane (right)
http://www.century-of-flight.net/Aviation%20history/flying%20wings/early%20US%20flying%20wings.htm
and http://www.aerofiles.com/_j.html

Russian K-12, 1936
http://www.aviastar.org/air/russia/kalinin_k-12.php

This aircraft was known as the Flying Flounder. It was designed and built by Cheston Eshelman and flew in 1942
https://www.youtube.com/watch?v=JpV68abt8uc

The US Navy's yellow "Flying Flapjack" Vought V-173
http://www.fiddlersgreen.net/models/aircraft/Vought-XF5U.html

40

**The "Flying Flapjack" Vought V-173(left) in flight and
the XF5U-1 Flying Flapjack V/STOL (right) which never flew**
https://forum.warthunder.com/index.php?/topic/185928-xf5u-flying-flapjack/

German fighter called Sack AS6 V-1
http://ufxufo.org/german/sack.htm

Straight Fixed-Wing Aircraft: these aircraft are typically a single fixed aircraft powered by either props or jets and may have flap and rudder stabilizers. These aircraft found their true expression during the Second World War when both the Germans and the Americans armed forces were experimenting with new and radical aircraft designs that would enable longer range missions, faster and more maneuverable aircraft with bigger and heavier payloads. Single fixed wing aircraft were the most researched design which decades later would evolve to delta wing aircraft

BICH aircraft
https://civilianmilitaryintelligencegroup.com/16079/the-chyeranovskii-bich-series-parabola-wing-aircraft

Horten Ho Vc
http://ufologie.patrickgross.org/aircraft/horten5.htm

Horten Ho-229
http://worldwartwo.filminspector.com/2014/06/super-weapons-of-luftwaffe.html

Hitler's Hortens
http://etepguerra.blogspot.ca/2012/11/producao-de-armamentos.html

Naranjero 1, a Horten built in Argentina after the war
http://www.aviastar.org/air/argentina/fma_iae-38.php

Lippisch DM1
https://airandspace.si.edu/collection-objects/lippisch-dm-1

Northrop, 1929 Flying Wing 1
http://www.century-of-flight.net/Aviation%20history/flying%20wings/early%20US%20flying%20wings.htm

44

Fiesler F-3 early 1930s (left) and Armstrong-Whitworth 52 (right)
https://picclick.com/Flying-Triangle-Fieseler-Wasp-Delta-Wing-Aircraft-352014852994.html and
https://www.quora.com/What-are-the-notable-flying-wing-aircraft-before-or-during-World-War-II

Northrop YB, 1949
https://www.revolvy.com/main/index.php?s=Horten%20Ho%20229&item_type=topic

Northrop B-35 Flying Wing
and Northrop P-61 Black Widow
USAF Museum Photo Archives

Northrop B-35
https://www.pinterest.com/pin/308285536962801990/

Northrop XP-79
https://www.revolvy.com/main/index.php?s=Horten%20Ho%20229&item_type=topic

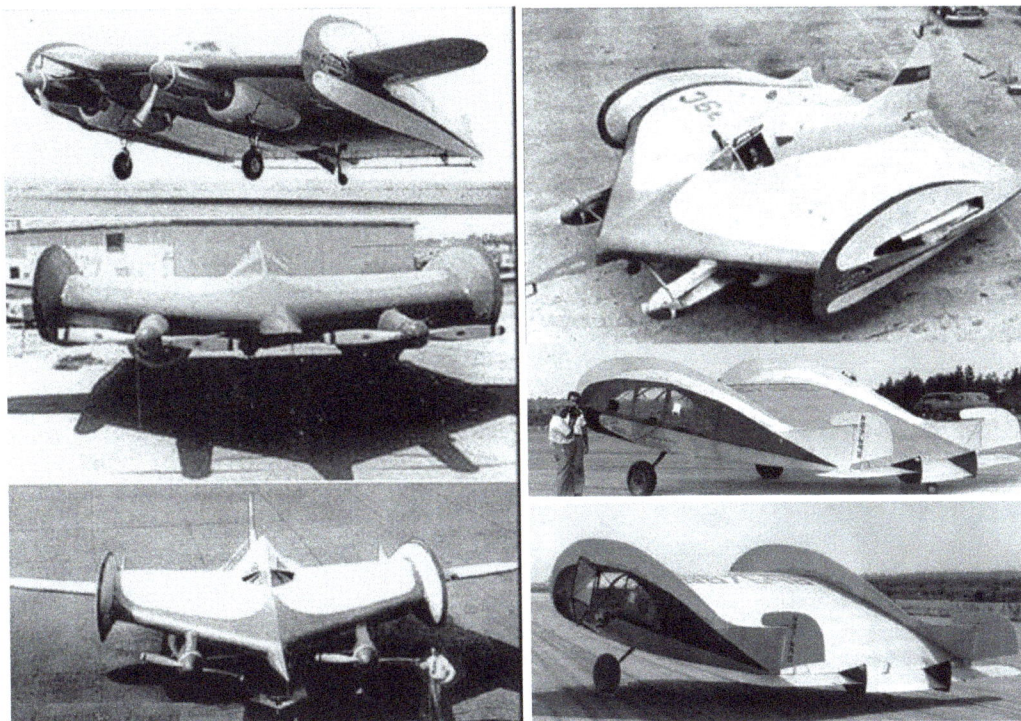

Built by William E. Horton of Santa Ana, California in 1951. Not truly wingless, but essentially a highly modified Cessna UC-78 with a more airfoil-shaped fuselage than wing.

https://www.pinterest.com/pin/558798266238137906/ and http://www.hitechweb.genezis.eu/liftingbodies2.htm

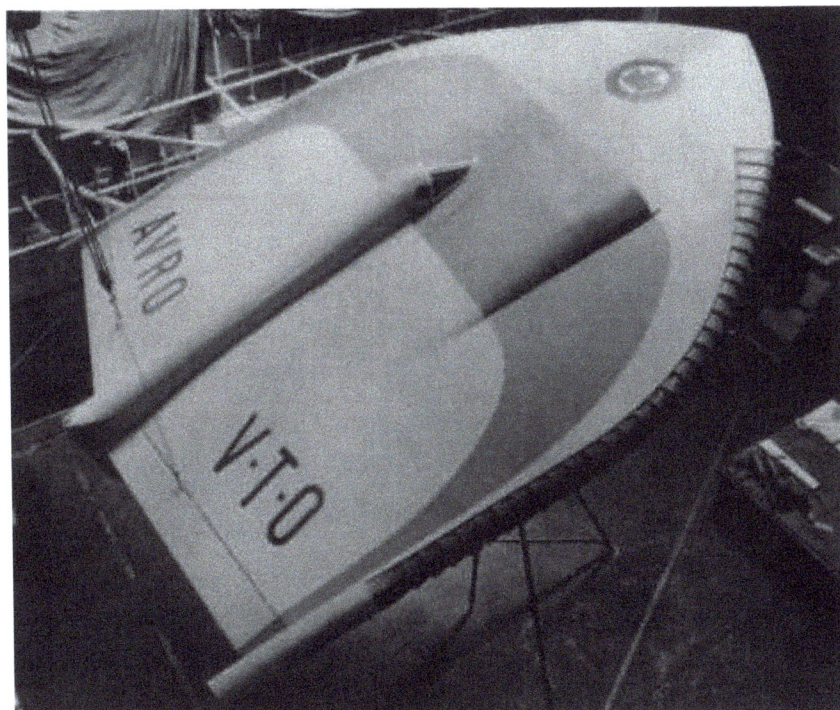

Avro Project Y-2: the "flat-riser"

http://ufo-joe.tripod.com/gov/avrocar.html#SPENCER

**This Geo Bat aircraft model flies exceptionally well and
there are plans to build a full-scale aircraft**
http://www.aerobataviation.com/gallery

Many of these aircraft reflect the evolution in aviation and in aeronautical engineering limited only by the imagination of the human mind but, unlimited by the human spirit to soar like a bird in the skies.

It should not come as a surprise to anyone from the above aircraft photos that in the White World of Science that progress in aviation is steady with occasional technological leaps forward particularly in aeronautical design. The saucer/disc configuration is a proven and viable concept in aircraft design but, major technological breakthroughs in the White World Science are yet to be made that will make this design feasible for space travel.

CHAPTER 94

SOME VISIONARY AERONAUTICAL ENGINEERS

Below, we will take a closer look at a few of the better known visionary aeronautical engineers who have made major contributions to aviation and have influenced or inspired other engineers and aircraft designers. Each of these exceptional aeronautical engineers strongly believed that the saucer or disc concept represented the future of aviation and aerospace. The fact that we are still building fixed-wing aircraft in the 21st Century instead of disc or saucer shaped aircraft is not because they are not a feasible alternative, but rather, that they represent a technology that is being deliberately suppressed and kept within the covert **Black World of Science**!

Henri Marie Coandă

Henri Marie Coandă; (7 June 1886 – 25 November 1972) was a Romanian inventor, a visionary aerodynamics pioneer, and builder of an experimental aircraft, the Coandă-1910 described by Coandă in the mid-1950s as the world's first jet, a controversial claim disputed by some and supported by others. He invented a great number of devices, designed a "flying saucer" and discovered the **Coandă Effect** of fluid dynamics.
http://en.wikipedia.org/wiki/Henri_Coand%C4%83

Henri Coandă
http://www.patentdesign.ro/famous-inventors-and-inventions/henri-coanda/?lang=en

Henri Coandă attended Military High School in Romania and graduated as a sergeant then, he enrolled in a technical institute in Berlin, Germany becoming an artillery officer. His passion, however, was in aeronautics so he headed to Paris, France and rolled at *École Nationale Superieure d'Ingenieurs en Construction Aéronautique* where he graduated as an aeronautics engineer.

In 1910, he designed and built an aircraft known as the Coandă-1910, which he displayed publicly at the second International Aeronautic Salon in Paris that year. The plane used a 4-cylinder piston engine to power a rotary compressor which was intended to propel the craft by a combination of suction at the front and airflow out the rear instead of using a propeller. Though the Coandă-1910 was contested by contemporary sources of his time as not being a true motor jet or alleged that it never flew or it that had crashed on test flight. None of these rumours or assertions were true. Coandă's colleague at Huyck Corporation, G. Harry Stine, a rocket scientist, author and "the father of American model rocketry" and much later, Rolf Sonnemann and Klaus Krug from the University of Technology of Dresden, all had acknowledged that the **Coandă-1910** was the world's first jet.

The world's first jet, the Coandă-1910
https://en.wikipedia.org/wiki/Henri_Coand%C4%83

Coandă continued to build other aircraft in both Britain between 1911to 1914 and then in France during the First World War. In 1934 Coandă was granted a French patent related to the **Coandă Effect** (the tendency of a fluid jet to be attracted to a nearby surface), probably the most famous of discoveries which became his chief contribution to aviation. During 1935 he used the same principle as the basis for the design of a disc-shaped aircraft called *Aerodina Lenticulara*, a "flying saucer" that used an unspecified source of high-pressure gases to flow through a ring-shaped vent system. This design was similar to the flying saucers later developed by Avro Canada before being bought by USAF and becoming a classified project. This effect has also been utilized in many aeronautical inventions such as the British designed hovercraft of the 1960s http://en.wikipedia.org/wiki/Coanda_Effect#Applications and http://en.wikipedia.org/wiki/Henri_Coand%C4%83

50

It was Coandă's assertion that *"These airplanes we have today are no more than a perfection of a toy made of paper children use to play with. My opinion is **we should search for a completely different flying machine, based on other flying principles**. I consider the aircraft of the future that which will take off vertically, fly as usual and land vertically. **This flying machine should have no parts in movement. The idea came from the huge power of the cyclones."***

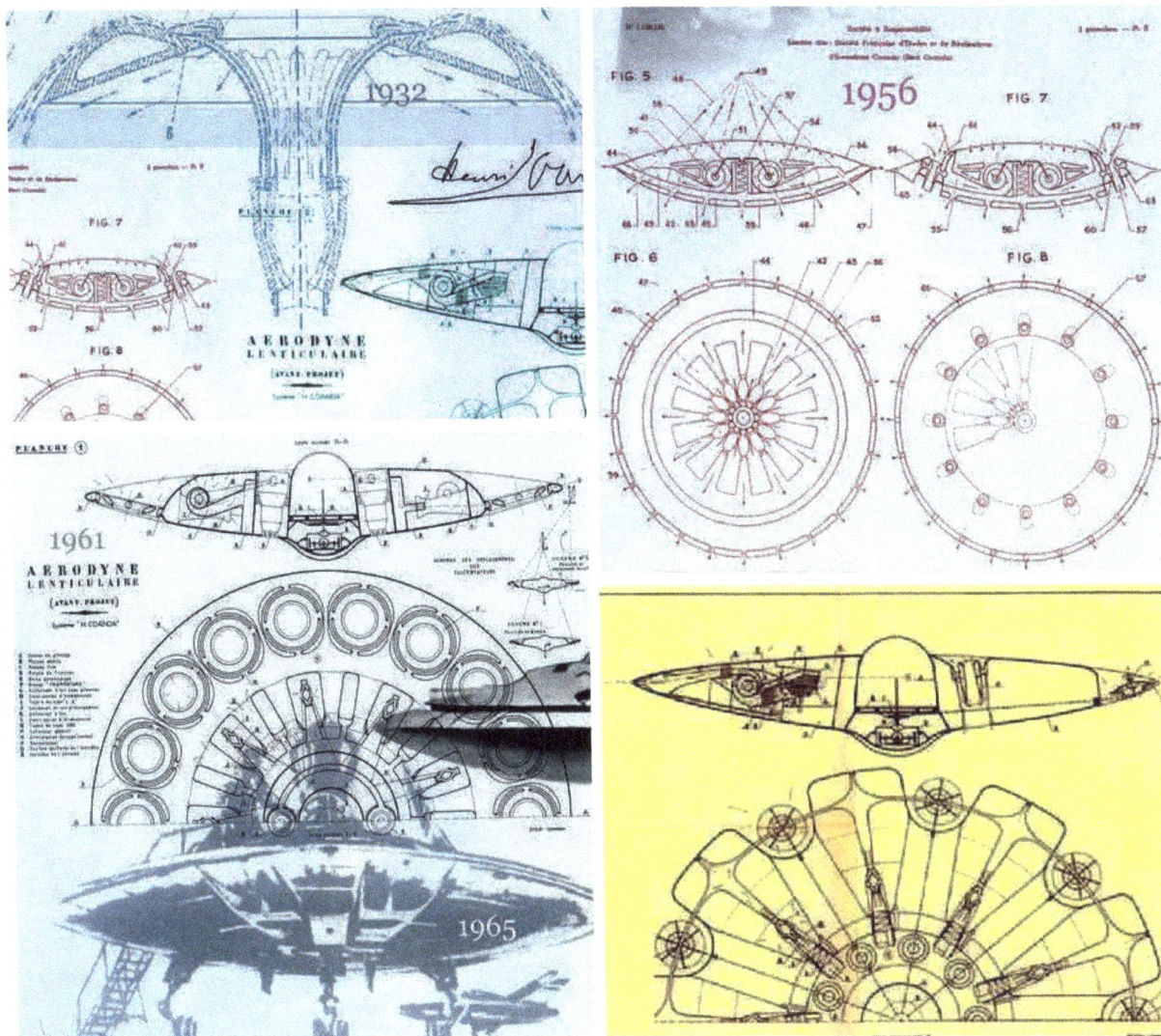

Coanda's 1934 patent drawings of the "Aerodina Lenticulara" *(Flying Saucer)*
http://petrumihaisacu.blogspot.com/2013/02/prima-farfurie-zburatoare-din-istoria.html

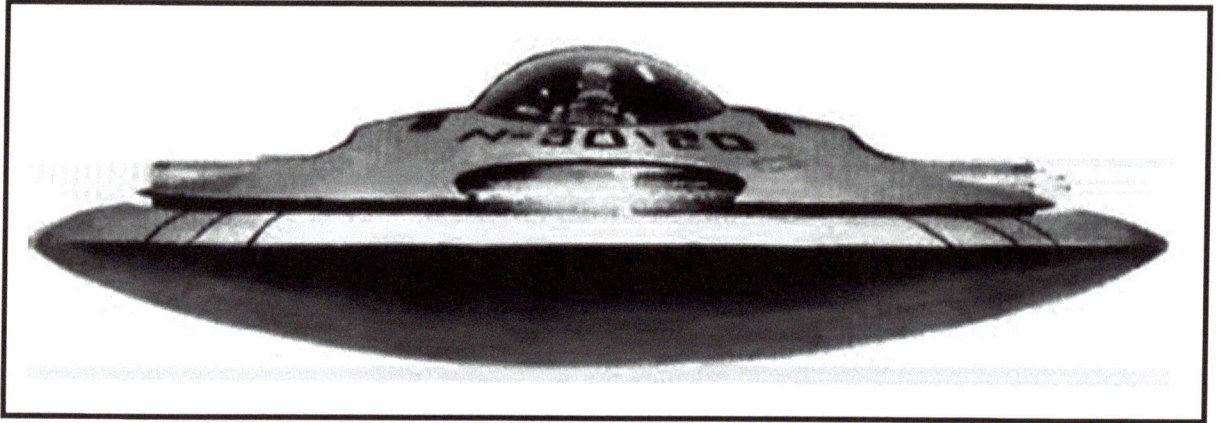

Henri Coanda's Flying Saucer GETOL aircraft prototype

Rene Couzinet

René Couzinet (born 20 July 1904, Saint-Martin-des-Noyers, Vendée, died 16 December 1956) was a French aeronautics engineer and aircraft manufacturer. The Société des Avions René Couzinet manufactured a range of Couzinet aircraft during the 1920s and 30s.
http://en.wikipedia.org/wiki/Ren%C3%A9_Couzinet

Rene Couzinet a French visionary aeronautics engineer and aircraft manufacturer

Couzinet designed three-engine passenger aircraft, transport planes and fighter aircraft for the French Air Force during WWII against the German Air Force as well as hydroplanes for water lake and river use.

An area investigated by Couzinet was the ***multiple wing aircraft* or *RC 360***, a machine that had the shape of a flying saucer. Back in France, and without support from the State for eventual airplanes of his conception, Couzinet became interested with the problems of vertical takeoff (a serious aviation concern back in the fifties). In his factory of l'Ile de la Grande-Jatte at Levallois-Perret, he built a wood 3/5-scale model of the *RC 360*, which was introduced as the future of aviation. But, René Couzinet's flying saucer never flew, because the administration had estimated that investing in such "far out" project would be foolhardy. http://aerostories.free.fr/constructeurs/couzinet/page3.html

He released photographs in 1952 of a full-scale model of his Flying Saucer, the Couzinet RC-360, a **Vertical Takeoff Aircraft (VTOL)** (see photos above) that used two contra-rotating discs powered by three engines. A second model was designed with six Lycoming engines (180 hp each) and one Marcel Dassault Viper turbojet.
http://www.laesieworks.com/ifo/lib/VTOLdiscs.html

According to a declassified US Air Force report, the modified version incorporates a principle of operation similar to that used in **Couzinet's** original proposal, that is, two contra-rotating discs superimposed to annul gyroscopic effect. The discs are supported by a fixed central section in which the cockpit, the engines, and landing gear are located. There are now 50 adjustable vanes around the periphery of each disc instead of the 48 in the earlier proposal.
http://www.cufon.org/cufon/couzinet.htm

Couzinet builds a wood and steel 3/5 scale model, ***not intended to fly***, of "Aerodyne with multiple wings RC 360", in his workshop on the island of La Jatte in Levallois-Perret. The craft in the shape of flying saucer would decades later, inspire "wannabe" Ufologists to make computer altered versions to post on their internet websites. Couzinet displays the model in his La Jatte workshop, and in aeronautical meetings from July 1955 on, including the international aeronautical fair of Brighton in England.

The craft was supposed to be able to ***take off vertically and then fly like a jet plane***. It was thus planned to be fitted, around ***a motionless cockpit***, with six 180 HP Lycoming engines in three pairs that were supposed to make t***wo contra-rotating crowns of 96 small wings ROTATE, at the circumference of the craft***. A 135 HP Marcel Dassault Viper with 745 kg ***was mounted below the cockpit for horizontal propulsion***.
http://ufologie.patrickgross.org/aircraft/couzinet.htm#sv1

According to many news articles and websites, it was claimed that Couzinet's "flying saucer" obviously ***never flew, it was never built***. The administration of aeronautics had rightly estimated that the unrealizable project would be only a waste of taxpayer's money. It did not deter the sensationalist press of 1955-1956 by making headlines with "the French Flying saucer".

Moreover, three years earlier, the Anglo-Canadian aeronautic engineers had tried a similar project, more seriously funded, **Robert Frost's Avrocar.** *It went far beyond building a **wood model**.* But all the efforts remained in vain. Far from taking off and darting away in the sky at the expected supersonic speed, Avrocar could hardly do better than clumsily hover less than one meter above the ground at 45 km/h, getting damaged as soon as it was "flown" over a non-flat terrain. Avrocar did not behave like a plane at all, but merely like a very ill-conceived hovercraft, missing the "skirt" that true hovercrafts later used to trap an "air cushion" that enabled them to hover without much contact above the ground or water. And although, some Ufologists or *conspiratorialists* believe or want to make believe that this project was a big secret then, it a secret only for a few months, and the aeronautical authorities in France were not unaware of the project's failure and could only see that **René Couzinet** had no new solution at all.
http://ufologie.patrickgross.org/aircraft/couzinet.htm#sv1

But the media, of course, fell in love with the new saucer.

The Libération newspaper of July 2-3, 1955, who saw the pictures of the saucer model in the workshop of Levallois-Perret, writes that it is "a true revolution in aeronautical history." Paris-Match, in 1956, shows the saucer model and other smaller Couzinet saucer models as the "*French flying saucer*", a "*success*" needing only a few additional months of development to fly. Not only "*the flying saucer exists*", but it will "*soon will replace airplanes*", and "*a model will fly before the end of the year. The final saucer, ten times larger, is planned for 2 pilots and 6 people. It's impressive wooden model is ready. Couzinet, who received no help so far, will build it in nine months.*" And in his introduction, the writer claims that the magazine shows the **Aerodyne RC 360**" for the first time as if unveiling a big secret, although the official and public presentation took place one year earlier.

A photograph of a circular one-metre diameter model surmounted by smoke is published as proof that "*the saucer tends to go up*" because its "*higher rotor... sucks smoke*". Decades later a writer on the web, although being a Couzinet admirer, comments with cruel but truthful irony that his **"vacuum cleaner"** will also fly soon, for it also sucks smoke...

The new sensation reached the foreign Press as soon as 1955, with for example: the newspaper *Philadelphia Inquirer* for July 5, 1955, in the USA, or *Popular Science* in the USA, headlining "Giant Pie Cooked Up By Frenchman is Latest Flying Saucer", or *Aeroplane And Commercial Aviation News*, who sees in 1955 a "*promising*" side in Couzinet's saucer. It is shown in *Flying* magazine in 1955, in *Naval Aviation News* in 1956; *Life* magazine for June 18, 1956, publishes a sketch of the saucer in flight captioned "*It takes off vertically*", without specifying that it is merely *hoped* to take off vertically *if it is ever built successfully*. So much for a so-called "secret" of the French saucer...

Expectedly, it reaches the Ufological literature, with an article by the editor of the British *Flying Saucer Review*, Derek Dempster, in September 1955. But Couzinet is drowning in debts and dramatically disillusioned as his lasted inventions convinced nobody.

On Sunday, December 16, 1956, he gets a revolver, assassinates his wife Gilberte Chazotte, widow of Jean Mermoz, then commits suicide. René Couzinet rests in the cemetery of Bagneux,

in Montrouge, in the Pas-de-Calais, France.
http://ufologie.patrickgross.org/aircraft/couzinet.htm#sv1

There is a mystery here which seems to indicate that Couzinet and perhaps many of the French and American press along with various aviation and science magazines may have known something about the French flying disc that appears to have been overlooked in later articles and missing from many web and blog sites.

However, be that it may, *the mystery is evident in the three photographs below and the answer to the mystery is right before your eyes in the pictures. Take a look at the images below and examine them carefully, compare the three photographs particularly the top image which is claimed in most news press articles to be a "wood model" and "non-functional", in that it could not fly" then, compare it to the two lower photo images, you'll notice something curious!*

Question: How is it possible for this wooden model's contra-rotating vanes at the outer circumference rings, both top and bottom **APPEAR TO BE ROTATING AT AN INCREDIBLE SPEED??? Is the Couzinet craft being engine and wing tested?!!!**

It is this author's assessment that this is not a non-functional "wood model" but a fully or partially functional metal flying saucer!!!

Notice that both vane rings are rotating to the point of being nothing more than a blur and appear to have a strobe effect comparable to a phonograph record player wherein the static top photo, there is no movement and the vanes top and bottom can be clearly seen at rest!!!

Now, this is either a very clever hoax for the time period in which the photographs were taken or it has been computer altered in recent times to add further confusion and disinformation to an already difficult subject in proving reliable authenticity. These photos can be found on almost any UFO website and they are all the same photo images, so is this a real functional flying saucer? The author thinks so!

From the **minijets.org** website, one of the few websites to display the Aerodyne craft **"being engine tested"**:

"In 1955 Couzinet investigated in the multiple wing aircraft or RC 360. This bizarre machine had the shape of a flying saucer. Without support from the State for eventual airplanes of his conception, Couzinet became interested with the problems of vertical takeoff (it was the fad in the fifties).

In his factory of l'Ile de la Grande-Jatte at Levallois-Perret, he built a wood 3/5-scale model of the RC 360, which was introduced as the future of aviation. But, René Couzinet's flying saucer never flew, because the administration had estimated that investing in such "far out" project would be foolhardy." *So we are told and science marches on!*
http://www.minijets.org/index.php?id=169

Designed by Rene Couzinet, this engineless model of the French aerodyne has a diameter of almost 27 feet. It will be powered by three 135-horsepower engines and the turbojet reactor visible on the underside in the lower view. Philadelphia Inquirer, July 5, 1955. (World Wide Photo)

These three photos are rather interesting as it appears that Couzinet's Aerodyne is being engine tested; note the blur of the contra-rotating vanes of the outer rings (middle and bottom images) which means this is not a wooden model but a full metal construction craft

http://www.laesieworks.com/ifo/lib/VTOLdiscs.html

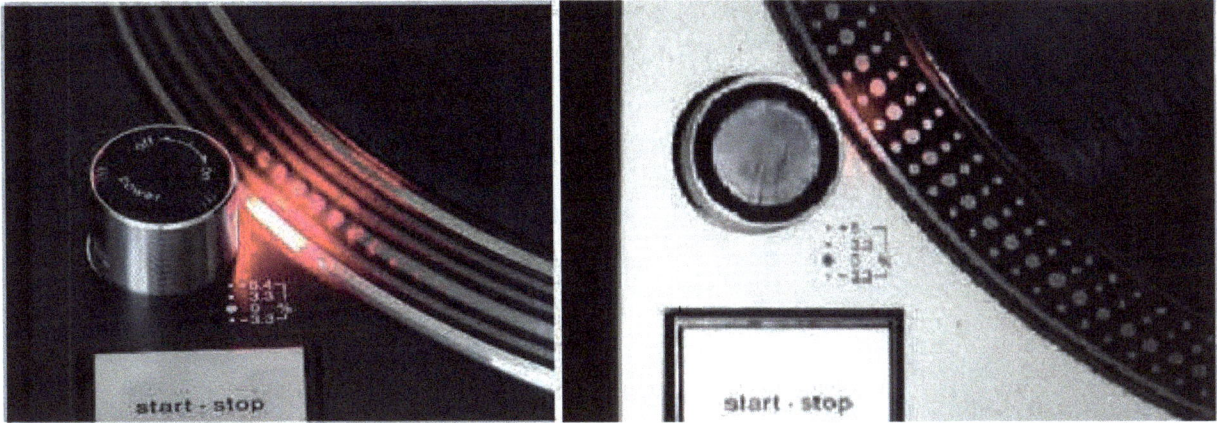

Strobe effect caused by a light shining on the equally spaced dots of a phonograph turntable. Compare this to the strobe effect or blur of Couzinet Flying Saucer
http://talk.ltn.com.tw/article/breakingnews/1839389

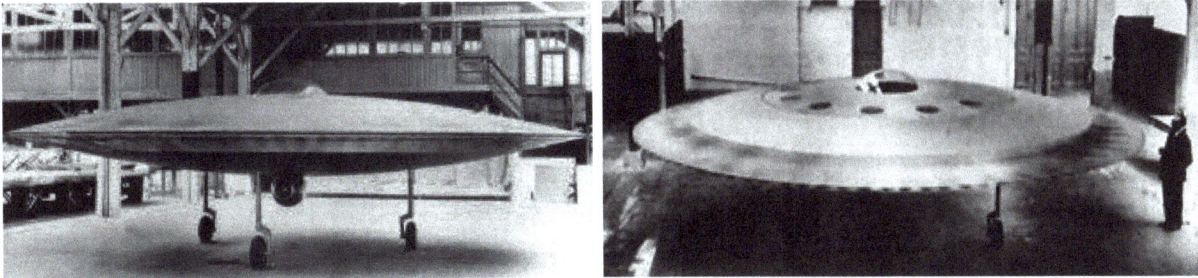

The photograph above, taken during an engine running test, show the Aerodyne as it will look just before vertical take-off with its contra-rotating vanes merging into a blur.
http://www.laesieworks.com/ifo/lib/VTOLdiscs.html

Couzinet Flying saucer at rest
http://www.laesieworks.com/ifo/lib/VTOLdiscs.html

Thomas Townsend Brown

Thomas Townsend Brown (March 18, 1905 – October 22, 1985, Zanesville, Ohio, USA) was an American physicist who developed theories concerning the link between electromagnetic and gravitational fields as postulated by **Dr. Albert Einstein**. His brilliance allowed him to transform theories and concepts into demonstrable applications involving disc-shaped apparatus which created and utilized temporary, localized gravitational fields.

Townsend Brown was the celebrated prodigy of electrical science during the 1920s, was respected as an innovative physicist throughout the mid-century, and is generally regarded as the seminal pioneer in the field of anti-gravity research and technology.
http://www.minijets.org/index.php?id=169

Thomas Townsend Brown, genius and prodigy of electrical science
https://pulsoelectromagnetico.wordpress.com/2010/12/01/electrogravedad-townsend-brown-biefeld/

In 1921, Brown discovered what was later called the **Biefeld-Brown Effect** while experimenting as a teenager, with a Coolidge X-ray tube. This is a vacuum tube with two asymmetrical electrodes. Brown noticed that there was a force exerted by the tube when it was connected to a high-voltage source. This force was not caused by the X-rays but instead was related to ionized particles created at the small (sharp) electrode and moving to the large (flatter) electrode. Later, in 1923, he collaborated with **Paul Alfred Biefeld** at Denison University, Granville, Ohio. He started a military career afterwards and was involved in a number of science programs.

In 1930 he joined the United States Navy and conducted fundamental research in *electromagnetism, radiation, field physics, spectroscopy, gravity and other topics*. He later worked for Glenn L. Martin and, still later, for the **National Defense Research Committee (NDRC)** and the **Office of Scientific Research and Development**, headed at that time by Dr. **Vannevar Bush**. After 1944 he worked as a consultant to the **Lockheed-Vega Aircraft Corporation**.

Brown's work became very controversial due to the similarity between his work and what is believed to be the propulsion method of some observed UFO's. This has made it the effect to become something of a cause célèbre in the UFO world, where it is seen as an example of something much more exotic than **Electrokinetics**. His name is also often mentioned in the same breath as the so-called **"Philadelphia Experiment,"** as a possible candidate along with **Nikola Tesla, A.L. Kitselman,** and **Dr. Einstein. Charles Berlitz** devoted an entire chapter of his book The Philadelphia Experiment to a retelling of Brown's early work with the effect, implying he had discovered some new electrogravitic effect being used by UFOs.
http://en.wikipedia.org/wiki/Thomas_Townsend_Brown and
http://www.antigravitytechnology.net/thomas_townsend_brown.html

There is much mystery around Brown's life as pulling out information about his work his like "pulling teeth or root-canal work", the truth is often difficult to get at, even his own family never knew what he was working on except perhaps, either the government, the military or the CIA. Thus, little is really known about him and outside of the government, his family and some investigating Ufologists, no one seems to have known or heard about him!

Much of what we know about him from the UFO literature and some **Freedom of Information Act (FOIA)** documents seems always "to trail off down some dark corridor of intrigue or run into a brick wall of officially *"classified"* material."

Expanding on his original discoveries through more than four decades, Brown built numerous wingless devices that flew without any conventional means of propulsion – the precursors to a future generation of gravity-defying **"electric spacecraft."**
http://www.antigravitytechnology.net/thomas_townsend_brown.html

In 1955, Brown went to England, and then France where he worked for La Société Nationale de Construction Aéronautique du Sud-Ouest (SNCASO) on secret research called Projet Montgolfier, a study of the **Biefeld-Brown Effect**. The Biefeld-Brown Effect is now widely referred to as **Electrohydrodynamics (EHD)** or sometimes **electro-fluid-dynamics**, a counterpart to the well-known **magneto-hydrodynamics**. Small models lifted by this effect are sometimes called **"lifters"**.

In 1956, the aviation trade publication "Interavia" reported that *Brown had made substantial progress in antigravity or electrogravitic propulsion research*. US scientists felt earlier that Brown's work with anti-gravity and electro-gravitic propulsion was nothing more than a demonstration of electric or ion wind but after his work in France proving the Biefeld-Brown Effect worked in a vacuum, American scientists were forced to reconsider their initial findings of his earlier work.

Later, top U.S. aerospace companies had also become involved in such research (see United States gravity control propulsion research (1955 - 1974)) which became a classified subject by 1957. Others contend Brown's research simply reached a dead end and lost support… *at least this is what we are told as a way to throw other scientists and alternative energy-seekers off the scent of real breakthrough science.*

The effect relies on corona discharge, which allows air molecules to become ionized near sharp points and edges. Usually, two electrodes are used with a high voltage between them, ranging from a few kilovolts and up to megavolt levels, where one electrode is small or sharp, and the other larger and smoother. The most effective distance between electrodes occurs at an electric field gradient of about 10 kV/cm, which is just below the nominal breakdown voltage of air between two sharp points, at a current density level usually referred to as the saturated corona current condition. This creates a high field gradient around the smaller, positively charged electrode. Around this electrode, ionization occurs that is, electrons are stripped off the atoms in the surrounding medium and they are literally pulled right off by the electrode's charge.

This leaves a cloud of positively charged ions in the medium, which are attracted to the negative smooth electrode, where they are neutralized again. In the process, thousands of impacts occur between these charged ions and the neutral air molecules in the air gap, causing a transfer in momentum between the two, which creates a net directional force on the electrode setup. This effect can be used for propulsion (see EHD thruster), fluid pumps and recently also in EHD cooling systems. (See the sketch by **Paul LaViolette** below and also **B-2 Bomber** photos)
http://www.antigravitytechnology.net/thomas_townsend_brown.html

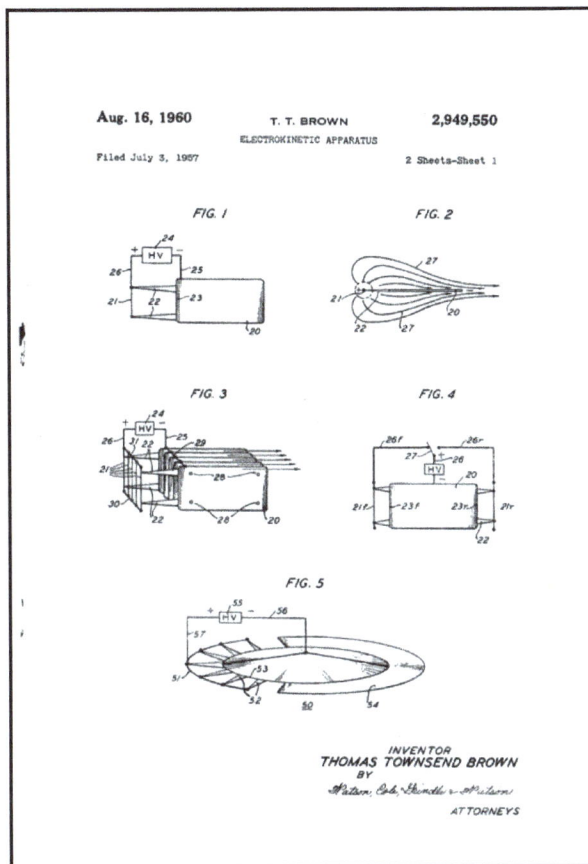

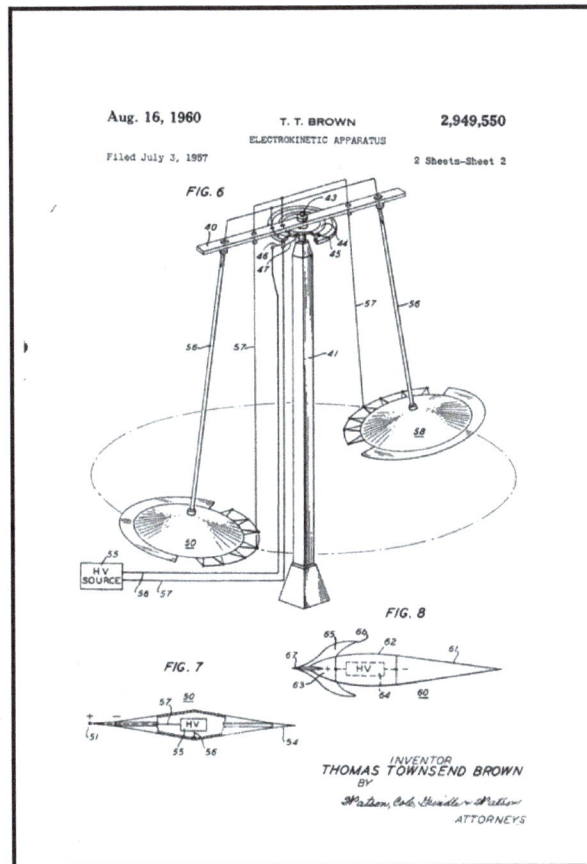

**Patent of Thomas Brown's Electrokinetic saucers and apparatus
demonstrating flight by means of the Biefeld-Brown Effect**
http://www.google.ca/patents/US2949550

Though the effect he discovered has been proven to exist by many others, Brown's work was controversial, because he and others even believed that this effect could explain the existence and operation of **Unidentified Flying Objects (UFOs).**

Brown was an early investigator of UFOs and in 1956 helped found the **National Investigations Committee on Aerial Phenomena (NICAP).** Though Townsend resigned not long after NICAP was founded, NICAP was an influential force in civilian UFO research through 1970. The organization's activities drew the attention of the **Central Intelligence Agency (CIA),** several high-level officers of which joined the group. Brown's research has since become something of a popular pursuit around the world, with amateur experimenters replicating his early experiments in the form of **"lifters"** *powered by high voltage.*
http://en.wikipedia.org/wiki/Thomas_Townsend_Brown

There is some speculation by some people that the **Tesla Coil** may be related to the **Biefeld-Brown Effect** point to the fact that upon Tesla's arrival to the USA, he was carrying plans for a *"flying machine".* However, it is also pointed out that the only common denominator of a Tesla Coil and the Biefeld-Brown Effect is that each of them generates a high voltage as a requirement of their vital role in their respective experiments. High field gradients between electrode plates can be produced by an AC circuit powered by Tesla coils.

Brown examining one of the Electrokinetic discs before attaching it to the "Carrousel"

Townsend Brown tests his Electrokinetic flying saucers in flight

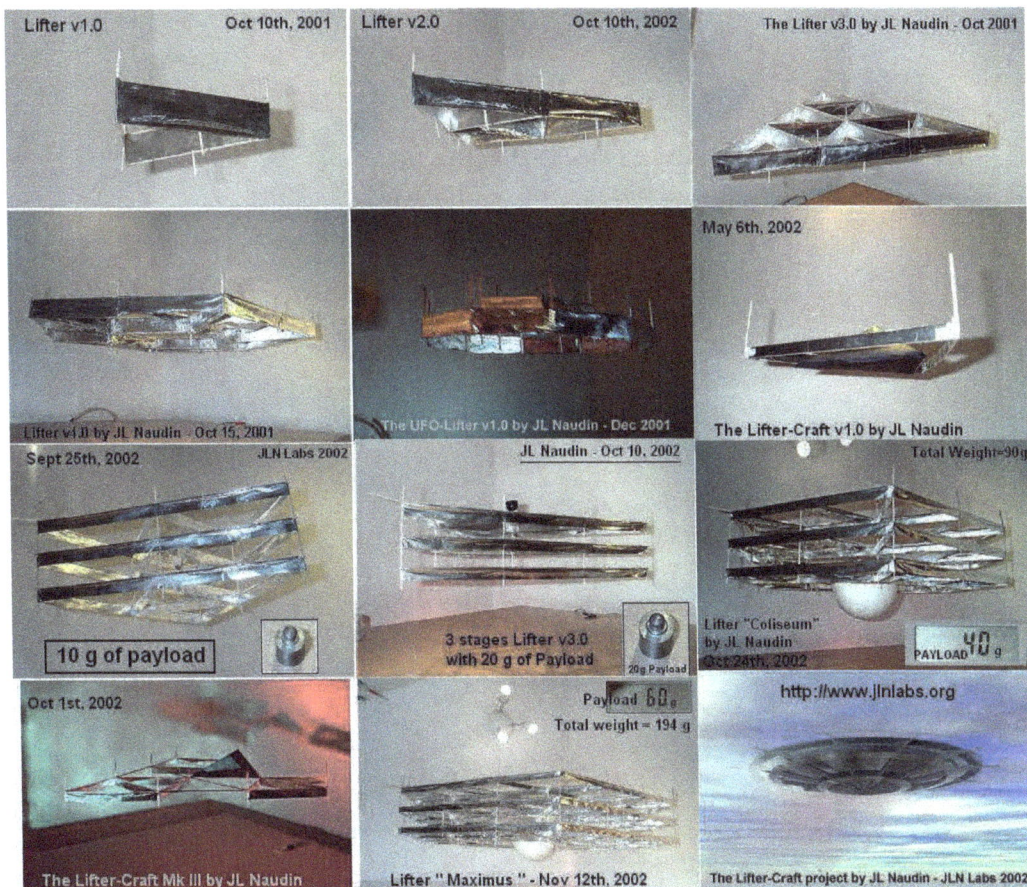

Various modern day "Lifters" from JLN Labs demonstrating the Biefeld-Brown Effect

63

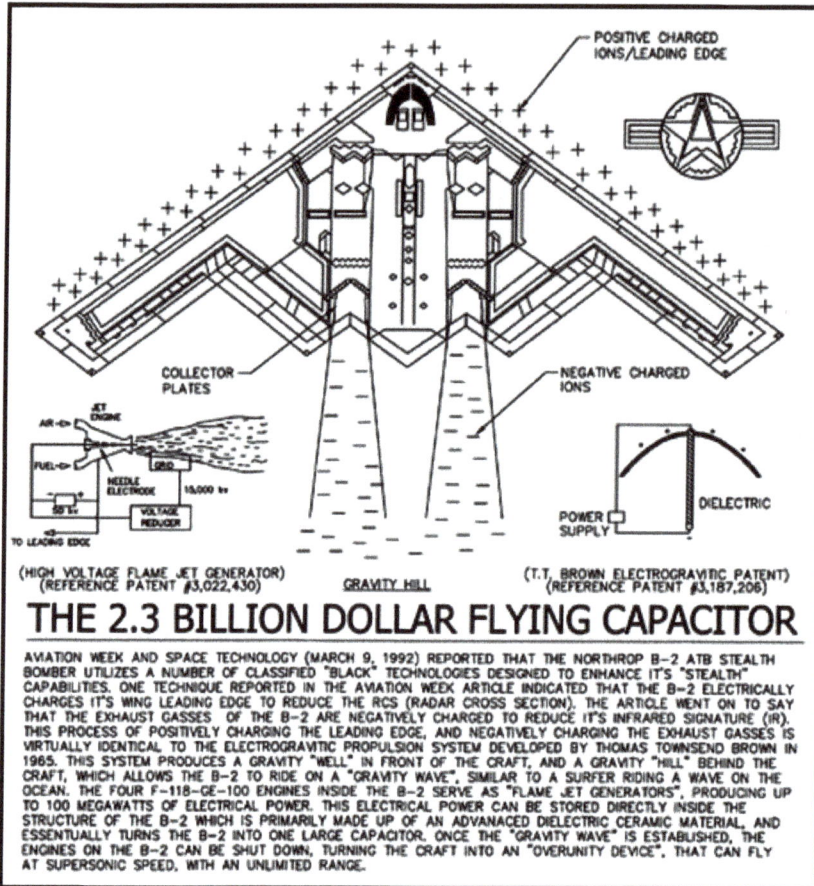

THE 2.3 BILLION DOLLAR FLYING CAPACITOR

AVIATION WEEK AND SPACE TECHNOLOGY (MARCH 9, 1992) REPORTED THAT THE NORTHROP B-2 ATB STEALTH BOMBER UTILIZES A NUMBER OF CLASSIFIED "BLACK" TECHNOLOGIES DESIGNED TO ENHANCE IT'S "STEALTH" CAPABILITIES. ONE TECHNIQUE REPORTED IN THE AVIATION WEEK ARTICLE INDICATED THAT THE B-2 ELECTRICALLY CHARGES IT'S WING LEADING EDGE TO REDUCE THE RCS (RADAR CROSS SECTION). THE ARTICLE WENT ON TO SAY THAT THE EXHAUST GASSES OF THE B-2 ARE NEGATIVELY CHARGED TO REDUCE IT'S INFRARED SIGNATURE (IR). THIS PROCESS OF POSITIVELY CHARGING THE LEADING EDGE, AND NEGATIVELY CHARGING THE EXHAUST GASSES IS VIRTUALLY IDENTICAL TO THE ELECTROGRAVITIC PROPULSION SYSTEM DEVELOPED BY THOMAS TOWNSEND BROWN IN 1965. THIS SYSTEM PRODUCES A GRAVITY "WELL" IN FRONT OF THE CRAFT, AND A GRAVITY "HILL" BEHIND THE CRAFT, WHICH ALLOWS THE B-2 TO RIDE ON A "GRAVITY WAVE", SIMILAR TO A SURFER RIDING A WAVE ON THE OCEAN. THE FOUR F-118-GE-100 ENGINES INSIDE THE B-2 SERVE AS "FLAME JET GENERATORS", PRODUCING UP TO 100 MEGAWATTS OF ELECTRICAL POWER. THIS ELECTRICAL POWER CAN BE STORED DIRECTLY INSIDE THE STRUCTURE OF THE B-2 WHICH IS PRIMARILY MADE UP OF AN ADVANCED DIELECTRIC CERAMIC MATERIAL, AND ESSENTUALLY TURNS THE B-2 INTO ONE LARGE CAPACITOR. ONCE THE "GRAVITY WAVE" IS ESTABLISHED, THE ENGINES ON THE B-2 CAN BE SHUT DOWN, TURNING THE CRAFT INTO AN "OVERUNITY DEVICE", THAT CAN FLY AT SUPERSONIC SPEED, WITH AN UNLIMITED RANGE.

Paul A. LaViolette cracks the classified "deep black" technology behind how the USAF B-2 Bomber flies. It is an Over Unity Device
http://www.viewzone.com/event66.html

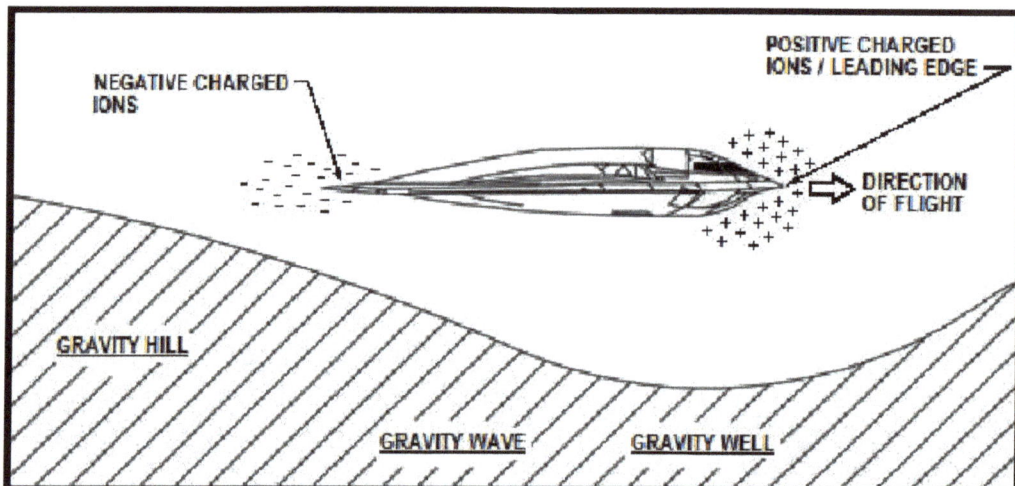

The B-2 Bomber rides gravity waves as it flies much like Flying Saucers
http://www.viewzone.com/event66.html

The B-2 Advanced Technology Bomber. In 1993 Paul LaViolette demonstrated that it is propelled by T. Townsend Brown's electrokinetic technology

http://www.thelivingmoon.com/forum/index.php?topic=776.0

Is this air turbulence or ionization of the atmosphere by the B-2 Bomber?

http://www.bugimus.com/stealth/stealth.html

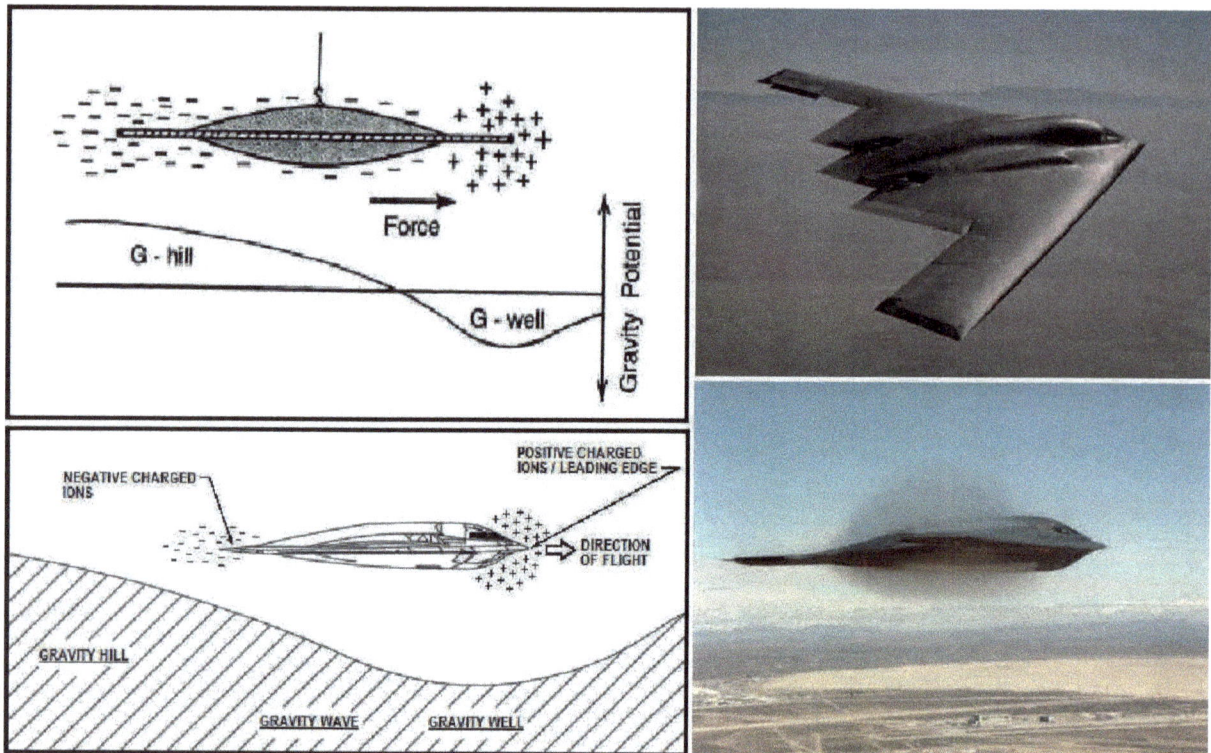

The B-2 bomber utilizes the Biefeld-Brown Effect to fly as can be seen for comparison in the drawings and in the actual photos of the aircraft

Flying Car Concepts

When we look at the various attempts to build a personal flying vehicle for everyone to have in their own garage, one can't help but recall in earlier times that **Henry Ford** in 1926 had made a similar claim to have a **"flying car"** or **"sky flivver"** (small experimental single-seat aeroplane).This, however, never materialized and so, the public was left wondering for some time when they would own and fly their own personal sky car.

Some progress in that direction was made in 1956 when the **Ford Advanced Design** studio built the **Volante Tri-Athodyne,** a 3/8 scale concept car model. It was designed to have three ducted fans, each with their own motor, that would lift it off the ground and move it through the air. In public relation release, Ford noted that *"the day where there will be an aero-car in every garage is still some time off"*, but added that *"the Volante indicates one direction that the styling of such a vehicle would take"*.

In the same year, the **US Army's Transportation Research Command** began an investigation into **"flying jeeps",** (see previous photos images above) ducted-fan-based aircraft that were envisioned to be smaller and easier to fly than helicopters. In 1957, **Chrysler, Curtiss-Wright**, and **Piasecki** were assigned contracts for building and delivery of prototypes. They all delivered their prototypes; however, Piasecki's VZ-8 was the most successful of the three. While it would normally operate close to the ground, it was capable of flying to *several thousand feet*, proving to be stable in flight. Nonetheless, the Army decided *that the "Flying Jeep concept was unsuitable for the modern battlefield",* and concentrated on the development of conventional helicopters. In addition to the army contract, Piasecki was developing the **Sky Car**, a modified version of its VZ-8 for civilian use.

The Piasecki PA-59K "Flying geep", "Sky car", or "Airgeep" (top and bottom) was designed for military use and never intended for civilian use or enjoyment
http://www.piasecki.com/geeps_pa59k.php

http://www.piasecki.com/geeps_pa59k.php

67

The **Piasecki PA-59K Airgeep** was designed to research this concept. Vertical lift, propulsion, and control were derived from two ducted horizontal rotors in tandem. The PA-59 could fly up to 75 mph close to the terrain and was not dependent on ground effect, permitting flight between trees, buildings and other obstacles. It could also hover, land and travel as a ground vehicle on its three wheels.

The pilot was positioned on the starboard side of the vehicle to keep his collective pitch control lever away from the open side of the machine. This also permitted the pilot to look down over his right arm giving him precise clues of the machine's motion relative to near obstructions. The enclosed rotors made flight close to ground personnel feasible without danger of injury. The downwash was surprisingly different than the helicopter and gave the pilot clear local visibility in flying over sand, water, and snow, unlike the blinding recirculation characteristics of a helicopter rotor. http://www.piasecki.com/geeps_pa59k.php#

A **flying car,** also to refer to roadable aircraft and hovercar, is a hypothetical personal aircraft that provides door-to-door aerial transportation (*e.g.,* from home to work or to the supermarket) as conveniently as a car and without the requirement for roads, runways or other specially-prepared operating areas. Such aircraft lack any visible means of propulsion (unlike fixed-wing aircraft or helicopters) so they can be operated in urban areas, close to buildings, people, and other obstructions.

The flying car has been depicted in fantasy and science fiction works such as *Chitty Chitty Bang Bang*, *Harry Potter and the Chamber of Secrets*, *The Jetsons*, *Star Wars*, *Blade Runner*, *Back to the Future Part II* and *The Fifth Element* as well as articles in the American magazines *Popular Science*, *Popular Mechanics*, and *Mechanix Illustrated*.
http://en.wikipedia.org/wiki/Flying_car_%28aircraft%29

The Piasecki Airgeep was an amazing feat of technical aviation but, for the US Army to state *that the "Flying Jeep concept was unsuitable for the modern battlefield"* makes no sense. There has to be more to this story but okay, fine, be that it may. What about public use? It certainly would have fulfilled the public's needs! After all, the public was still waiting and asking, "Where's my flying car?"

The flying car was a common feature of science fiction and futuristic conceptions of the future, including imagined near futures such as those of the 21st century. For instance, less than a month before the turn of the millennium, the journalist Gail Collins noted:

"Here we are, less than a month until the turn of the millennium, and what I want to know is, what happened to the flying cars? We're about to become Americans of the 21st century. People have been predicting what we'd be like for more than 100 years, and our accouterments don't entirely live up to expectations."... "Our failure to produce flying cars seems like a particular betrayal since it was so central to our image."

As a result, flying cars have become a running joke; the question "Where is my flying car?" is emblematic of the supposed failure of modern technology to match futuristic visions that were promoted in earlier decades. http://en.wikipedia.org/wiki/Flying_car_%28aircraft%29

68

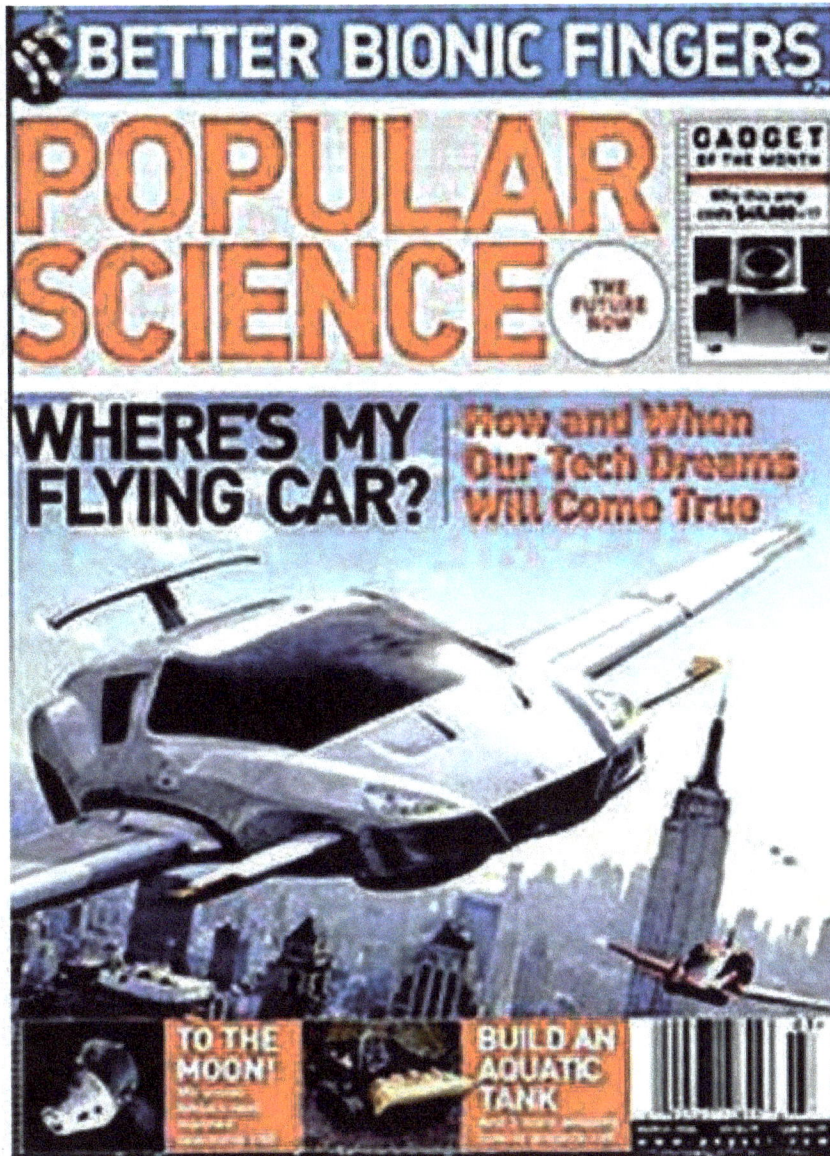

"Where's my flying car?" On the March 2008, cover of *Popular Science* reported on flying cars and related futuristic aircraft throughout the 20th century

Even Popular Science has routinely done articles on potential concepts and actual prototypes but as yet, apart from tormenting the public with promises of an amazing future in personal flying, there is yet to be a company that can deliver on the promise! Could Canadian- born **Paul Moller** fulfill the promise and deliver the goods to an exasperated and waiting public?

Paul Sandner Moller

Paul Sandner Moller (December 11, 1936, Fruitvale, British Columbia Canada) is another brilliant engineer who has spent the past forty years developing the **Moller Skycar** personal **vertical takeoff and landing (VTOL)** vehicle. Moller is a professor emeritus at the University of California, Davis and lives in Davis. He was featured in *Popular Science's* January 2005 issue and recently appeared on the **Coast To Coast AM.**

Paul Moller and the Skycar 400
http://wikicars.org/en/Paul_Moller

The **M200 Neuera** (formerly, the **M200G Volantor**) is a prototype of a flying saucer-style hovercraft, designed by aeronautics engineer Paul Moller. The vehicle is envisioned as a precursor to the **Moller Skycar M400**. The M200G Volantor uses a system of eight computer-controlled fans to hover up to 10 feet (3 m) above the ground. **Volantor** is a term coined by Moller meaning a "vertical takeoff and landing aircraft."

**The Moeller XM4-3 (top left) flew (tethered) in 1966 and the Moeller 200 X
with canopy parked (top right) flew (tethered) in 1968.**
http://www.the-big-picture.org.uk/wordpress/?page_id=9132 and
http://listas.20minutos.es/lista/que-auto-te-gustaria-tener-para-pasear-los-domingos-57535/

The engine technology developed for the **Skycar** has also been adapted as a UAV platform called the aerobot. The rotapower engine itself has been spun off to a separate Moller company, **Freedom Motors**.

In 1972, Moller founded Supertrapp Industries to market his invention of an engine silencing system. Moller sold Supertrapp in 1988 in order to fund development of his Skycar and its rotapower engine

A **Skycar** is not piloted like a traditional fixed wing airplane and has only two hand-operated controls, which the pilot uses to inform the computer control system of the desired flight maneuvers. All eight engines operate independently and, as demonstrated in during a tethered flight, will allow for a vertical controlled landing should anyone fail. The Skycar's ducted fans deflect air vertically for takeoff and horizontally for forward flight. The ducted fans also encase the propellers, which prevents bystanders from being exposed to moving blades as well as improving aerodynamic efficiency at low speeds. http://en.wikipedia.org/wiki/Moller_Skycar

The Moller Skycar M400 is a prototype personal VTOL (vertical take-off and landing) aircraft which is powered by four engine nacelles, each with two computer-controlled Rotapower engines (Wankel rotary engines) and is approaching the problems of satellite-navigation, incorporated in the proposed **Small Aircraft Transportation System**. Moller also advises that, currently, *the Skycar would only be allowed to fly from airports & heliports.* Moller has been developing VTOL craft since the late 1960s, but no Moller vehicle has ever achieved free flight out of ground effect. The proposed Autovolantor model has an all-electric version powered by Altairnano batteries.

Moller's disc craft and his Skycar are based on the Coanda concept of 1934 and the Couzinet turbojet saucer design RC 360 of 1956 and therefore, really doesn't represent new propulsion technology other than new material compositions. Below are some of the variants that Moller Industries have developed:

Moller M150 Skycar
> The initial single seat technology demonstrator, incorporating the fuselage of a Bede BD-5 with two of Moller's ducted fan propulsor units. Prototype only; never flown.

Moller M400 Skycar
> The prototype version powered by four Moller propulsors incorporating Rotapower 500 Wankel rotary engines; has flown several times to date without a pilot but tethered via slack safety line to an overhead crane.

Moller 400 Skycar
> Production version; unbuilt.

Moller 100LS and 200LS
> Proposed 1-and-2 seat volantor air vehicles, similar to the 400 Skycar.

Moller Neuera
> Flying Saucer-type volantor with 2 seats; has flown several times with a pilot but tethered via slack safety line to an overhead crane. This volantor is meant to operate in ground effect only.

Close up of the Moeller 200 X without canopy parked on runway flew (tethered) in 1968
http://psipunk.com/worlds-first-flying-car-enters-production/

Some of the Moeller "toys" that come with a hefty price tag for anyone with money in their pocket to burn and yes they do fly but getting past government flight regulations may be a big hurdle to fly over cities or "into the sky blue yonder!"
http://www.bizjournals.com/sacramento/news/2013/11/05/moller-raising-money-skycar-crowdfunding.html

In 2003, the Securities and Exchange Commission sued Moller for civil fraud *(Securities And Exchange Commission v. Moller International, Inc., and Paul S. Moller, Defendants)* in connection with the value of shares after the initial public offering of stock, and for making unsubstantiated claims about the performance of the Skycar. Moller settled this lawsuit without admitting guilt by agreeing to a permanent injunction against claiming projected worth of Moller International stock and paying $50,000. The shareholders of Moller International - collectively known as SOMI ("Shareholders of Moller International) banded together on a website to tell the Moller-side of the SEC issue.

In 2007, Moller announced that the M200G Volantor, a successor to the Moller Skycar, would hopefully be on the market in the United States by early 2008. His proposed Autovolantor model includes an all-electric version powered by Altairnano batteries.

Moller's credibility has been questioned in recent years because of the vaporware nature of his creations. In April 2009, the *National Post* characterized the Moller M400 Skycar as a 'failure', and described the Moller Company as "no longer believable enough to gain investors".

On May 18, 2009, Dr. Moller filed for personal protection under the Chapter 11 reorganization provisions of the federal bankruptcy law; however, **Moller International** (corporation) did not file for bankruptcy and continues to do business as of this writing.
http://en.wikipedia.org/wiki/Paul_Moller and http://moller.com/dev/

The future of the Moller Skycar and the other disc shape craft has yet to be written but, at this current time, a big dark cloud hangs over any further developments unless, there is a major breakthrough in vehicle power and in-flight performance…***untethered and providing there are no aviation regulation and patent hurdles to overcome!***

CHAPTER 95

TRICKS, TRAPS, THREATS AND LEGAL HURDLES FROM THE US PATENT OFFICE IN THE NAME OF "NATIONAL SECURITY"

This is the reality in the white world of science, where if you fulfill all of the patent requirements there are always agencies and people who will see to it that you do not succeed or your invention or device will never see the light of day in the public domain. This is particularly true when radical concepts and demonstrable inventions intrude into the covert world of black science. The **BWS** does not like competition and will find a way to remove the competition from existence.

Every time an inventor, engineer or a scientist has a great idea that comes off the drawing board and into the cold reality of a physical prototype which is then publicly announced to the world as an invention that will change the world, someone from the government or an intelligence agency is watching and listening and they may decide to pay you a visit. You must either prove your claims or find yourself quickly discredited and a possible lawsuit filed against you, as Dr. Paul Moller has discovered. You may find yourself having a cease and desist order filed against you and your efforts to become the next "garage inventor" to make a difference in people's everyday lives comes to a grinding halt or hit the proverbial brick wall.

Should your invention live up to its claims and you try to patent your invention, the **US Patent Office** may not give you a patent number for your prototype for mass production of as it may violate National Security regulations on any number of accounts.

If your invention is all that it's cracked up to be and more, then the men in the black suits will come and confiscate your all hard work, all your papers, documents, equipment and the prototypes claiming some nonsense that it is dangerous, it breaks many laws and national security risks, disrupts the status quo or simply say nothing and if you try to stop them, you may find yourself, threatened or physically harmed or arrested and thrown into jail or even, murdered!!!

Tom Valone is a former Patent Examiner who was fired about six years ago for producing a conference in Washington DC on these new energy technologies. Valone recently won a lawsuit against the US Patent Office and was awarded reinstatement and six years of back pay. In a 2001 email to **Gary Vesperman**, Valone wrote in part:

"As a former Patent Examiner, I can tell you that the number of **"secretized"** patents in the vault at the Patent Office (Park 5 Bldg.) is closer to 4000 or more. They never receive a patent number, and the inventor is rarely, if ever, compensated by the government for use of the invention." http://www.apparentlyapparel.com/free-energy.html

An understandable reason for suppressing certain types of energy inventions is that the knowledge behind them is also capable of producing tremendously destructive advanced electromagnetic weapons such as the "death ray" apparently invented by Nikola Tesla. Hence many such new energy technologies, particularly those using this kind of knowledge of advanced electromagnetic principles, are considered "dual use" technologies that are among the 4,000 un-numbered patent applications confiscated in a vault at the US Patent and Trademark Office because of their military potential and the need to keep that knowledge from America's enemies. http://www.theorionproject.org/en/suppressed.html

The **U.S. Patent Office** has a nine-member committee that screens patents for national security implications. A hidden purpose of this committee is to also screen energy-related patents which could threaten the power and fossil fuel companies, etc.

Canada's patent office doesn't have a similar screening committee. It is recommended that energy patents possibly in danger of being classified should be first applied for in Canada. Once granted, up to one year is allowed to apply for the same patent in the U.S. Patent Office. Now the patent cannot be classified because it is already out in the public domain, courtesy of Canada. Maybe, the lyrics in the national anthem of Canada are correct after all… *"**The True North, strong and free!**"* (Bold italics added by author for emphasis). http://www.theorionproject.org/en/suppressed.html

Inventors for the most are pure-hearted and desire to make the world a better place by improving the lives and the well-being of their fellow inhabitants, but unfortunately, they face such perils as poverty, slander, ridicule, and neglect. Inventors of new energy devices and energy generation systems that are revolutionary in their concepts and designs have frequently been bullied by mega energy-related corporations and their allied lackeys within the US Government, who seek to maintain their energy enslavement of the people by ruthlessly suppressing development of energy inventions. Implementation of legal as well as illegal tactics of suppression has included imprisonment on false charges, IRS harassment, burglaries, bribery with huge sums of money, and *even murder if the inventor was too stubborn to heed warnings or undeterred by lesser actions.* http://peswiki.com/index.php/Directory:Suppression#Overview_Documents

Sometimes, apparently to con investors out of money, it seems (or is made to appear) that the supposed energy inventions have been science-fiction props or incorrectly measured devices or were more often simply *"grounded out"* (short-circuited) in some manner to cause the energy device to be ineffective, or that the alleged inventors may have tried to cover up the scam by claiming to be conspiracy victims. and https://www.youtube.com/watch?v=48AkxqT16gk

Text of Generic Patent Secrecy Order

SECRECY ORDER

(Title 35, United States Code (1952), sections 181-188)

NOTICE: To the applicant above named, his heirs, and any and all of his assignees, attorneys and agents, hereinafter designated principals:

You are hereby notified that your application as above identified has been found to contain subject matter, the unauthorized disclosure of which might be detrimental to the national security, and you are ordered in nowise to publish or disclose the invention or any material information with respect thereto, including hitherto unpublished details of the subject matter of said application, in any way to any person not cognizant of the invention prior to the date of the order, including any employee of the principals, but to keep the same secret except by written consent first obtained of the Commissioner of Patents, under the penalties of 35 U.S.C. (1952) 182, 186.

Any other application already filed or hereafter filed which contains any significant part of the subject matter of the above-identified application falls within the scope of this order. If such other application does not stand under a security order, it and the common subject matter should be brought to the attention of the Security Group, Licensing, and Review, Patent Office.

If prior to the issuance of the secrecy order, any significant part of the subject matter has been revealed to any person, the principals shall promptly inform such person of the secrecy order and the penalties for improper disclosure. However, if such part of the subject matter was disclosed to any person in a foreign country or foreign national in the U.S., the principals shall not inform such person of the secrecy order, but instead shall promptly furnish to the Commissioner of Patents the following information to the extent not already furnished: date of disclosure; name and address of the disclosee; identification of such part; and any authorization by a U.S. government agency to export such part. If the subject matter is included in any foreign patent application or patent, this should be identified. The principals shall comply with any related instructions of the Commissioner.

This order should not be construed in any way to mean that the Government has adopted or contemplates adoption of the alleged invention disclosed in this application; nor is it any indication of the value of such invention.

(The harsh punishment for a violation of this secrecy order, should an inventor exploits or even simply discusses his or her invention which is classified by a patent secrecy order, is 20 years in federal prison. In effect, the US Government brutally and suddenly orders unlucky energy inventors to keep absolutely quiet and not do any more work on their inventions – without compensation for their well-meaning efforts. Thus a shocked, intellectually shackled and frustrated inventor would end up losing everything he or she had invested in his or her invention. The public is also ruthlessly denied any benefits from the invention.)
http://www.theorionproject.org/en/suppressed.html

Suppression of energy technology occurs when an individual or group which is more powerful than the new developer(s) or inventor(s) tries to directly or indirectly censor, persecute or otherwise oppress the innovations and technology, rather than engage with and constructively develop or accommodate the new technology.

Government or industry may often act in this way. Suppressed inventions are taken into the realm of business, rather than strict politics. **Nikola Tesla** has been the object of several

76

conspiracy theories, with claims relating to revolutionary energy generation and distribution technologies which may or may not have been utilized by **"HAARP"**, an American military-funded research program. Claims that **Wilhelm Reich's "Orgone Energy"** was suppressed by the establishment do rest on historical facts: his incarceration (and death while in prison) as well as the FDA-instigated burning of his books. What remains unclear is whether prosecuting him for alleged medical infringement had any relationship to the possibility that Orgone could be a power source. **http://peswiki.com/index.php/Directory:Suppression#Overview_Documents**

Case Histories of Inventors Whose Inventions Were Hijacked by the U.S. Patent Office

The following case histories (which can been hyperlinked) were last updated in late Aug. 2007 and indicate the inventors and their projects which they were working on and seeking a patent. Their inventions were usually suppressed, and the inventor was told to remain silent about the nature of his invention with a legal non-disclosure gag order placed upon him, his family, his friends, and associates. Should the inventor breach this order, he would be fined and/or imprisoned and if necessary, he and his work would be discredited, or he would be threatened, and in some cases harmed or kidnapped, even murdered:

- Adam Trombley: Trombly-Farnsworth Solid State Oscillating Electromagnetic System
- Al Wordsworth: Generator and Carburetor
- Bill Williams: Joe Cell-Powered Truck
- Bob Aldrich (Reporter): Suppression of Vibrating Energy Source used by Farmers mid- 20th Century
- Bob Boyce VeriChipped & Given Cancer
- Bob Lantz Energy Invention Suppression
- Bruce DePalma: N Machine
- Charles N. Pogue: High Mileage Carburetor
- Dave Wetzel: Veggie Oil back-taxes
- David G. Yurth, Ph.D. (Reporter): Remediating Nuclear Waste Materials
- Dean Warwick: Ampliflaire efficient wood-burning stoves
- Dennis Lee: Freon Engines (note: there is some skepticism to the claims of Dennis Lee)
- Designex Inc. in Toronto, Canada
- Edwin V. Gray: Free Energy EVGRAY Generator and Free Energy Engine
- Electric Mass Transit
- ERR Fluxgenerator
- EV1 (GM Electric Vehicle)
- Fish/Kendig: Variable Venturi Carburetor
- Frank Richardson: Magnet-Based Electrical Generator and Bladeless Steam Turbine
- Fred Bell (Conspiracy Theorist)
- Gary Vesperman (Reporter): US versus Japanese Support of Cold Fusion
- Gary Vesperman (Reporter): CIA Agents at 1996 Tesla Convention
- George Arlington Moore: Carburetion
- Grant Hudlow: Method of Converting Garbage and Tires to Gasoline and Fertilizer
- Henry T. Moray: Free Energy Generator

- Hitachi Magnetics Corporation: Magnet Motor
- Honda (Turbine Generator)
- Howard R. Johnson: Johnson Free Energy Motor
- Gerald Schaflander: Hy-Fuel - solar-produced hydrogen turned into liquid fuel
- Gianni A. Dotto: Anti-Aging and Anti-Gravity Thermionic Couple
- Joel McClain and Norman Wooten: Magnetic Resonance Amplifier
- John Andrews: Water-to-Gasoline Additive
- John Bedini: Electromagnetic Overunity
- John Searl's UFO and Generator Technology Joseph C. Yater: Heat-to-Electricity Converter
- Joseph Newman: Energy Machine
- Ken Rasmussen: Water-to-Energy Electrolysis Process
- Keshe Foundation (counter-gravity, propulsion, and free energy generator technologies) (M.T. Keshe was 'kidnapped' in Canada over false nuclear weapons concerns)
- Lester Hendershot: Free Energy Generator
- M. DeGeus: Infinite Battery
- Mike Brady: Perendev 300 kW Self-Running Electromagnetic Generator
- Neil Schmidt: Hydraulic Wind Turbine
- Orion Project (Dr. Steven Greer's initiative to bring scientists and inventors together in a peaceful type of Manhattan Project for energy development)
- Otis T. Carr: OTC-X1 Flying Saucer
- Paul Pantone: GEET system for fuel/plasma technology
- Richard Diggs: Liquid Electricity
- Robert Golka: High-Powered Tesla-Type Energy Tower
- ?????? Mixed Chemical Stone
- Robert Stewart: Stewart Cycle Heat Engine
- RomeroUK Dynamo ("romerouk dynamo")
- Ron Brandt: 90 MPG Carburetor
- Ron Hatton: Gadgetman Groove
- Russ George: Plankton Seeding
- The Silicon Mine of the Netherlands
- Stewart Harris: Theory of Magnetic Instability
- Tim Thrapp: Energy Invention Suppression
- Welton Myers: Myers' Efficient Carburetor
- United Nuclear DIY Hydrogen Fuel System Kit Blocked by the CPSC
- Viktor Schauberger: Jet-Turbine (Schladming Group development suppression in 1980s)
- WaveReaper (ocean wave energy technology)
- Wilhelm Reich (Orgone Energy)

http://peswiki.com/index.php/Directory:Suppression#Overview_Documents

These are more scientists that have experienced suppression of their inventions as well as personal attacks and threats on their lives because they dare to bring alternative energy systems out into the public domain:

Adam Trombly (Interview): The Truth about Zero Point Technology

Adam Trombly (Speech): Climate Change Factors, Ozone Layer Crisis, and Zero Point Energy Technologies

Adam Trombly: Trombly-Kahn Closed-Path Homopolar Generator
Andrew Leech (Reporter): Suspicious Deaths of Inventors in Australia
Andrija Puharich: Method and Apparatus for Splitting Water Molecules
Allen Caggiano: 100+ MPG Fuel Implosion Vaporization System
Bill Jenkins (Reporter): Free Energy Machine
Bob Dratch: Thorium Powerpack
Bob Lantz: Lantz Water and Power System
Brazil: Ethanol Produced from Sugar Cane
Canadian Scientist: Standalone Water-Based Electricity Generator
Charles N. Pogue: 200+ MPG Carburetor
Christopher Bird/Walter (Reporter): Energy
Daniel Dingel: Converts More than 100 Cars to Run on Water
David Crockett Williams (Reporter): Non-Drug Industrial Hemp as Bio-Fuel
David G. Yurth (Reporter): Remediating Nuclear Waste Materials
Dick Belland: 100 MPG Carburetor that Runs on Gasoline Fumes
Dr. Timothy Trapp: 127 Energy Technologies
Em-Tech Technologies: Advanced Solar Photo-Voltaic Crystal Lattice Cells
Floyd Sweet: Vacuum Triode Amplifier
Frank Roberts: Water Car
Gary Vesperman (Reporter): Energy Inventors are Buzzed by Black Helicopters
Gary Vesperman (Reporter): Shielding Over-Unity Power Converters
Gary Vesperman (Reporter): Six CIA Agents at 1996 Tesla Society Symposium.
Gary Vesperman (Reporter): US versus Japanese Support of Cold Fusion
Gary Vesperman: My Car was Fire Bombed July 3, 2006
Gerald Schaflander: Solar-Produced Hydrogen Turned into Liquid Hy-Fuel
George Wiseman: Fuel Savers
Honda: 60 MPG 1992-1994 Honda Civic VX
Howard Rory Johnson: Magnatron – Light-Activated Cold Fusion Magnetic Motor
Idaho Inventor: Advanced Zero-Point Energy Device
IPMS-Chernovitsky: Super Ceramics
IPMS: Energy Storage/Battery Devices
IPMS: High-Temperature Gas Plasma Detonator
IPMS-Kiev and Arzamas-16: Super Magnets
IPMS: Micro-Channels and Filters
IPMS: Thermal Electric Cooling Devices
IPMS: Thorium-227 Electricity Generator
Ira Einhorn: Free Energy and Mind Control Researcher
James Watson: 8-Kilowatt Battery-Popper Motor
Jim Powell (Reporter): Flywheel/Dual Hydraulic Cylinder
John Bedini: 'School Girl' Motor and Battery Energizer
John Richardson: 90+ MPG Carburetor; Atomic Isotope Generator
Joseph C. Yater: Heat-to-Electricity Converter
Mitchell Swartz: U.S. Patent Office Blocks Cold Fusion Patents
Nikola Tesla: Wireless Power and Free Energy from Ambient
Paul Brown: Hyper-Cap E-Converter
Paul M. Lewis: Airmobile

Ph.D. Electrical Engineer: Advanced Form of Plasma-Discharge Energy

Phil Stone: Engine Runs on Water

Remy Chevalier (Reporter): NiMH Batteries; Solid-State Lithium-Ion Batteries

Richard Diggs: Liquid Electricity Engine

Robert Bass: Low-Energy Nuclear Transmutation

Shell Oil Company: Achieves 376.59 MPG with a Modified 1959 Opel in 1973

Stanley Pons and Martin Fleischman: Cold Fusion

Stanley A. Meyer: Water Fuel Cell-Powered Car

Stefan Marinov: Magnetic Vortex Hyper-Ionization Device

Teruo Kawai: Motive Power Generating Device

Thomas E. Bearden, Ph.D.: Motionless Electromagnetic Generator

Thomas E. Bearden, Ph.D. (Reporter): J.P. Morgan Emasculated Electrical Engineering Theory

Thomas Henry Moray: Radiant Energy Pump/Electricity Generator

Tom Ogle: 100+ MPG Oglemobile

Viktor Schauberger: Jet-Turbine

Volcheck: Engine Powered by Gas with Unusual Expansion Properties

Walter Rosenthal (Reporter): mall Electrical Power Converter

Wilhelm Riech: Orgone Energy Motor

William Bolon: Automobile Steam Engine

Yasunori Takahashi: Magnetic Wankel Motor

http://www.theorionproject.org/en/documents/Gary_V.pdf

Energy Invention Suppression Case Statistics

Number of Energy Invention Suppression Incidents **95**
Number of Dead, Missing, or Injured Energy Inventors, Activists, and Associates **20**
Number of Energy Inventors and Associates Threatened with Death **32**
Number of Energy Researchers and Associates Imprisoned or Falsely Charged........................ **5**
Number of Inventions Classified Secret by U.S. Patent Office **5000**
Number of Incidents Involving Oil Companion **9**

Number of **Incidents of Energy Invention Suppression** by the **United States Government, Patent Office, Central Intelligence Agency, Federal Bureau of Investigation, U.S. Marshals, Army, Air Force, Navy, Bureau of Alcohol, Tobacco, and Firearms, Defense Intelligence Agency, S.W.A.T. Teams, National Security Agency, U.S. Postal Service, Department of Energy, Department of State, Securities and Exchange Commission, Food and Drug Administration, Department of Defense, Department of Homeland Security, Internal Revenue Service, Rural Electrification Administration, White House, Consumer Product Safety Commission, Small Business Administration, and Canada's "Royal Canadian Mounted Police"!** **59**
http://peswiki.com/index.php/Directory:Suppression#Statistics

Names of Companies, Banks, State Agencies, Private Groups, and Universities Involved with **Energy Invention Suppression – Standard Oil, Zapata Petroleum, Atlantic Richfield, Exxon-Mobile, Shell Oil Company, General Electric Company, Yakuza, California Air Resources Board, Organization of Petroleum Exporting Countries, Wells Fargo Bank, Ford Motor Company, General**

Motors Corporation, Massachusetts Institute of Technology, Queen of England, Kollmorgan, World Bank, Rockefellers, Carlyle Group, and Bush Family.
http://www.theorionproject.org/en/documents/Gary_V.pdf and
http://www.apparentlyapparel.com/free-energy.html

Invention Secrecy Still Going Strong

The **Federation of American Scientists** has published an extremely important article written by **Steven Aftergood** is an electrical engineer by training (B.Sc., UCLA, 1977), entitled "Invention Secrecy Still Going Strong'. It comes from a mainstream organization and corroborates information **The Orion Project** has presented. We are constantly asked, *"If better energy systems exist, why are they not available for public use?"* The following article addresses one reason: The systematic suppression of energy inventions by abuse of the national security provisions of U.S. law. This means that thousands of inventions have been suppressed- and more than that number through national security orders not issued via the patent process. This is why the has a specific strategy to develop and bring out to the public such energy inventions: One that stands up to these abuses. With your help, we can do it! http://fuel-efficient-vehicles.org/energy-news/?page_id=983

There were 5,135 inventions that were under secrecy orders at the end of Fiscal Year 2010, the **U.S. Patent and Trademark Office** told Secrecy News last week. It's a 1% rise over the year before, and the highest total in more than a decade.

Under the **Invention Secrecy Act of 1951,** patent applications on new inventions can be subject to secrecy orders restricting their publication if government agencies believe that disclosure would be "detrimental to the national security."

The current list of technology areas that is used to screen patent applications for possible restriction under the Invention Secrecy Act is not publicly available and has been denied under the Freedom of Information Act. (An appeal is pending.) But a previous list dated 1971 and obtained by researcher **Michael Ravnitzky** is available.

Most of the listed technology areas are closely related to military applications. But some of them range more widely.

Thus, the 1971 list indicates that patents for solar photovoltaic generators were subject to review and possible restriction if the photovoltaics were more than 20% efficient. Energy conversion systems were likewise subject to review and possible restriction if they offered conversion efficiencies "in excess of 70-80%."

One may fairly ask if disclosure of such technologies could really have been "detrimental to the national security," or whether the opposite would be closer to the truth. One may further ask what comparable advances in technology may be subject to restriction and non-disclosure today. But no answers are forthcoming, and the invention secrecy system persists with no discernible external review. http://blogs.fas.org/secrecy/2010/10/invention_secrecy_2010/ and
http://www.fas.org/sgp/othergov/invention/35usc17.html

On May 13, 2010, **Dr. Steven Greer** updated his supporters via his internet website, **THE ORION PROJECT** with important news regarding how his efforts to bring scientists together to work on alternative energy generation devices and systems was being sabotaged from a very high-level group with the Intelligence Community. Part of that account is below is one more example that **"the powers that be...Majestic"** do not want advanced energy systems out in the public domain simply because they will no longer will have control over the minds of the populace!

For the past two years, The Orion Project has worked to raise funds to build a facility where we can bring scientists and inventors together in a peaceful type of **Manhattan Project** for energy – to develop new sources of energy that will get us off the fossil fuel economy. We have also worked to identify scientists and inventors capable of this work. For the past hundred years, scientists such as Nicola Tesla have worked on such devices. The fact that many have tried, and we are still using predominantly fossil fuels – the same fuels used in the 1800s – can be illustrated by our recent experience.

We reported last December that we had under contract a very talented scientist who wanted to work with us. Because he still is doing work for a private company that is linked to the **Intelligence Community**, he preferred working with his identity concealed. A side note is that he had agreed to work with us in the fall of 2008. At that time, he met with the Board of Directors and took a contract home to read and sign but called us within three days saying that he was being deployed abroad for fifteen months. We heard nothing else from him until a year later – precisely after Dr. Greer publicly disclosed the energy briefing, he had put together for President Obama.

He resurfaced and said that he wanted to work with us to develop new energy technologies. He met with the Board of Directors again. He assured us that he was cleared by his **"shepherds"** in the Intelligence Community to work with us to build advanced energy systems, but he was not allowed to work on advanced propulsion systems – that is, he could build systems to power our houses and businesses, but *he would not be allowed* to reproduce the advanced propulsion systems he has developed in the past. *[Oh, Really?! Everyone, contact Obama - and your senator! - Thugs are still in control in the CIA.]*

We spent a long weekend with him and agreed upon a system he described that would be akin to a Tesla-type high-voltage system that would extract energy out of the Earth's energy field. It could run twenty-four hours a day, independent of solar energy or wind, and be more cost-effective, as well as non-polluting. This scientist also described three more technologies he could and would develop for us. He and his wife agreed that they wanted to work with us and would move to Virginia when we could fund a facility. They presented a timeline for completion of the first device, with work on the other devices to follow after that. Given his credentials and independent high-level confirmation from scientists who had reproduced his work, we were excited to move forward with this particular scientist.

Things went well and were very positive for the first couple of months. He contacted Dr. Bravo several times a week to say he was working on the project and was quite excited about it, although he apologized that the work was going more slowly than he had hoped.

In mid-March, there was a sudden change. He called **Dr. Jan Bravo** to say that he was receiving threats from "foreign nationals" that had to do with his other job obligations. He said that phone calls with the **Orion Project Board** had been recorded and "played back" to him. He said there was no problem with us (**The Orion Project**) – in fact, several times over the past couple of months, he reiterated that his handlers in the Intelligence Community had checked us out and okayed our working together. He confirmed to Dr. Bravo that he planned to finish the device he was working on and come to work with us in the future. But then, over a course of six days, he changed his stance from "Yes, I'm anxious and excited about working with you guys" to "No, I can't work with you."

Dr. Bravo had multiple conversations with him during that period. He said several people who worked in "security" visited him and they gave him disturbing information. One of these he identified (and named) was a past Director of the Central Intelligence Agency. Apparently, he was getting disinformation – because he had previously mentioned multiple times that we had been thoroughly checked out by the Intelligence Community before he was allowed to sign a contract with us.

Nothing had changed about us in the past few months. Hence, someone, somewhere must have decided that he indeed was capable of producing a paradigm-changing device, and The Orion Project was the group that could get it out to the world – and wanted that work to stop. This is called oppositional research or **"oppo research"**, and it consists of distorting truth or *telling absolute lies meant to deceive.* One absurd statement he made is that he was told that we were teaching terrorists how to explode weapons of mass destruction by using their minds. http://fuel-efficient-vehicles.org/energy-news/?page_id=983

Just after this occurred, *three more scientists* who had agreed to work with us – with impeccable credentials and experience in the aerospace industry, a major government laboratory and military projects – were all provided with disinformation or **threatened** in some way to keep them from working with us. This is an astonishing example of how a group who wishes to retain power – this group called **"Majestic"** – keeps its grip on the world, maintaining the fossil fuel economy with its multi-trillion dollar derivatives value. Coal-mine shafts collapse worldwide and kill many, vast amounts of oil continue to spew into the Gulf of Mexico, while many in the world continue without basic energy resources.

We must have been on the right track for these beasts to circle their wagons with illegal wire-tapping of our conversations, interference with a legal contract, and bringing in a powerful former CIA director to intimidate and kill this initiative.

Further, since we started The Orion Project, Dr. Greer has received **multiple death threats and repeatedly been stalked.** This has accelerated in the last year. There are FBI files open on this as well as local law enforcement files. Dr. Greer's computer was **hacked** into by one of the most lethal programs a criminal can use. One of our chief engineers who is volunteering with The Orion Project had his email **hacked** into. A specific email from an inventor with a sensitive document that involved a Department of Defense memo and the inventor's contact information was **deleted** from his inbox – these were the only things tampered with.

So, what can we do in the face of these highly orchestrated suppression efforts?

We have, in the last couple of weeks, created a very extensive file (including both audio and paper documents) that contains information relating to these events. In the event that anything happens to any of us (injury, false legal charges, disappearance, death), and we cannot go forward, copies of this information will be released to the public. The file includes every bit of information about the scientists who can do this work and are not allowed to as well as very reliable evidence and information about those involved in the suppression – names, addresses, phone numbers, and specific details of actions.

We have contacts who can get it to the President, as well as the highest people in the media and Hollywood. And it will be released to you. We feel this is the best strategic step we can take because we cannot enter into protracted legal battles with these individuals with limitless money and power. Our one weapon – the best one we could have – is the truth, and the documentation we have of the truth.

Our goal is not to get into endless fights with these corrupt individuals; we just want to give humanity a chance to move forward with earth-saving, life-saving technologies. http://fuel-efficient-vehicles.org/energy-news/?page_id=983

John Hutchison

Consider the case of Canadian **John Hutchinson**, "home/garage inventor extraordinaire" from New Westminster in British Columbia found himself the victim of "free energy" suppression and the confiscation of his equipment, not once but twice! In 1979, Hutchison accidentally discovered a remarkable phenomenon while experimenting with longitudinal waves - waves that another inventor, **Nikola Tesla** had experimented with. As a result of the very powerful radio wave interferences in his experiments, the term: **The Hutchison Effect** was given for the strange things that would occur with different objects. Heavy objects - even non-magnetic, non-metal objects would levitate or fly into the air. Objects of metal, porcelain, wood and rubber were also affected, and hard alloy metals would become soft and pliable because the Hutchison Effect.

When rumours of this man in the Greater Vancouver region had found a way to tap into a new energy source to levitate objects and fluids or to transmutate and recombine organic material with metal or to cause metals to become "plastic" at a molecular or atomic level, it immediate caught the public's attention and set alarm bells ringing in the military and intelligence offices throughout North America. People were showing up at his New Westminster apartment unexpectedly wanting to see what Hutchison was up to. In fact, the world, in general, was beating a path of curiosity to his door. News media, documentary film companies, Ufologists and new age energy investigators have come calling to film and see demonstrations of his new energy source.

According to the book "The Hutchison Effect - An Explanation", John Hutchison even performed his experiments for scientists from Los Alamos Laboratory. The effect has been videotaped many times and even broadcast on network television. A complete understanding of the phenomenon has yet to be found, but the implications of its potential seem mind-boggling.

At some point between Canadian and American intelligence co-operation, much of his equipment was confiscated by the **RCMP (Royal Canadian Mounted Police)** from his apartment home and when he relocated his newly replaced equipment to a garage shed, he returned one day to find almost everything stolen a second time from his new laboratory shed.

The intelligence communities from both countries were sending a clear message to Hutchison and any other would-be inventors of alternative energy generation systems, especially if they employ any zero point energy, quantum or space energy systems which could be used as a propulsion system. Cease and desist and keep your mouth shut or find yourself in a life threaten situation!

His work on the levitation of objects used old WWII equipment such as large army and navy generators, radar equipment and **Tesla** induction coils believed to employ some aspect of microwave resonance as a **"beam projector"** *where any object or liquid* was made to levitate as opposed to having any internal source causing an anti-gravity effect. Hutchison has since *"thrown in the towel"* on doing any more alternative free energy research for the simple fact that the Canadian or the American governments cannot be trusted not to interfere with his private research work particularly in matters of free energy.

This is yet, another creative and innovative mind whose dream for a better world has become a living nightmare because the **"powers that be"** do not share his world vision of providing the public with free energy systems to make their lives easier.

It is now believed that much of the **Hutchison Effects** that **John Hutchison** created was similar to Tesla's work on frequency, vibration and resonance in which he unknowingly created using standing waves *(Two opposing waves combine to form a standing wave. This phenomenon can occur because the medium is moving in the opposite direction to the wave, or it can arise in a stationary medium as a result of interference between two **waves** traveling in opposite directions)* that emanated from a beam-gun emitter which surrounds objects and nullified the effects of gravity upon them.

It is now believed that much of the **Hutchison Effects** that **John Hutchison** created was similar to Tesla's work on frequency, vibration and resonance in which he unknowingly created using standing waves *(Two opposing waves combine to form a standing wave. This phenomenon can occur because the medium is moving in the opposite direction to the wave, or it can arise in a stationary medium as a result of interference between two **waves** traveling in opposite directions)* that emanated from a beam-gun emitter which surrounds objects and nullified the effects of gravity upon them.

We know from leaked accounts that major breakthroughs pertaining to alien saucers were made in the mid "50s that indicated a basic understanding of anti-gravity, electromagnetism, and electrogravitics. These breakthroughs were due largely in part to the above mentioned scientists; one of these scientists stands out above the rest, a genius ahead of his time: **Nikola Tesla.**

(Top to bottom and left to right) John Hutchison's New Westminster apartment balcony; inside his over-crowded apt. suite; adjusting his levitation "beam projector"; a friend holding pieces of a disintegrated metal bar; disintegrated metal bars, a kitchen knife fused into another piece of metal and a molecularly pulverized metal cylinder

https://www.flickr.com/photos/timventura/6221811101 and http://www.antigravitytechnology.net/john_hutchison.html
http://www.abovetopsecret.com/forum/thread552859/pg1 and http://www.hutchisoneffect.com/Hutchison%20Effect%20Samples.php

Nikola Tesla

Nikola Tesla (July 10, 1856–January 7, 1943) arrived in the United States in 1884 as a poor Serbian immigrant hoping to change his fortunes and to provide his new country with some inventions from the brilliance of his futile mind. In fact, he provided America and the world with over 1200 new inventions that would still are still being used in current times. Chief among all Nikola Tesla's inventions, three great inventions stand out from the rest:

1. Rotating magnetic field… which electrified the entire world!!
2. Wireless or radio broadcasts.
3. Wireless transmission of electricity to any point on the earth.

Nikola Tesla

Nikola Tesla made some major discoveries at the turn of the 20th Century which proved that energy could be tapped from the Earth itself as well as from the ambient flux of quantum space which he called **cosmic** or **radiant energy** but, what we now know as the **Zero Point Energy** or **Quantum Energy.** In 1901 Tesla had already registered the use of cosmic energy in the Patent Office in New York, patent no. 685,957; this was over a hundred plus years ago. He proved that electrical energy flowed better as **Alternating Current (AC)** as opposed to **Thomas Edison's Direct Current (DC)**, which he incorporated into the world's first hydroelectric dam at Niagara Falls. http://www.apparentlyapparel.com/2/post/2012/05/nikola-teslas-wireless-electric-automobile-explained.html

Tesla also demonstrated that electricity could also be wirelessly transmitted to another receiving station hundreds of miles away or to individual homes or to automobiles, aircraft, and ships.

There would be no need of wires to connect homes or buildings to an electrical power generating plant as energy could be radiated out from central transmission towers or power plants.

But, the most amazing electrical feat from the incredible mind of Nikola Tesla was his demonstration in 1931to a small crowd of people that it is possible to power our vehicles without a drop of fossil fuel. He took a **Pierce Arrow automobile** and removed the gasoline engine, replacing it with a small 80-horsepower alternating-current (AC) electric motor drove for hours, at speeds as high as 90 mph. The Pierce Arrow ran without *gasoline,* or *electrical batteries, or any external power sources!*

In 1901 Tesla had already registered the use of cosmic energy in the Patent Office in New York, patent no. 685,957. In 1931 Tesla proved that cosmic energy or quantum energy could power and run a Pierce Arrow (above) without gasoline or a single drop of oil!
http://fuel-efficient-vehicles.org/energy-news/?page_id=952

This claim of running an automobile on the ambient energy flux of quantum space is hotly contested by skeptics and debunkers, even going so far as to discredit the relatives of Tesla and the few witnesses to the demonstration. It is truly amazing that there are people who want to hinder progress or side track it while helping those in power to propagate an alternative destiny for mankind that benefits a minority of people. However, when an idea's time has come, no power on Earth can stop its irresistible march forward. It is destined to be!

At a local radio supply shop he bought 12 vacuum tubes, some wires, and assorted resistors, and assembled them in a circuit box 24 inches long, 12 inches wide and 6 inches high, with a pair of 3-inch rods sticking out. Getting into the car with the circuit box in the front seat beside him, he pushed the rods in, announced, *"We now have power,"* and proceeded to test drive the car for a full week, often at speeds of up to 90 mph without any recharge. So, where did the power come from?

Careful research reveals that Tesla's *"black box"* of circuits and tubes (also called a valve amplifier) were acting much like radio wave receiver/amplifier where the small rods acted like an "on/off" switch when pushed into the box and the antenna received *the radio waves from the air* or *the ambient energy from the quantum vacuum flux of space* which was amplified by the tubes in the *"black box"*. http://www.apparentlyapparel.com/2/post/2012/05/nikola-teslas-wireless-electric-automobile-explained.html

Today, over 80 years later, it is still possible to convert any gasoline engine vehicle into an all-electric vehicle and it will operate for hours – without having to stop and recharge. Not a drop of oil, gasoline, hydrogen fuel, natural gas or water! No combustion engine! No exhaust system! No pollution! Ever!!!

This is a modern Valve Amplifier, though not the same as the one use by Tesla to power his 1931 Pierce Arrow automobile. Newer valve amplifiers use transistors not tubes
http://www.pinsdaddy.com/valve-amplifier_YDoCIK1Z9kxPuwS0d827V8rNctPT7QT4n1kk*quuxOw/

That's right! **NO FOSSIL FUELS** and **NO ELECTRICAL BATTERIES,** nothing but a few metal rods protruding from the engine block that were tapping into and collecting **radiant energy** aka. **Zero Point Energy** aka. **Quantum Energy**!!!

Now, if you re-read the last few paragraphs again and you may be asking yourself, why you've never heard about this amazing automotive demonstration by Tesla that occurred over eighty years ago. The answer is quite simple.

The big burgeoning oil companies in America owned by people like **J. P. Morgan** and **Rockefeller, et al** killed the patent and suppressed the revolutionary technology which later became *militarized!* They were manipulating other big businesses that were siding with new alternative energy forms with threats, intimidation and to maintain the status quo and to further their own agendas and wallets!

That means that for over 100 years we have never needed fossil fuels like environmentally polluting oil, gas, and coal to power our automobiles or our homes or our factories and plants or our airplanes, trains and ocean ships. For the last 80 plus years, we did **NOT NEED, ONE DROP OF OIL** to run our transportation systems or cities, absolutely no pollution need have contaminated our rivers, lakes and oceans, our atmosphere or the global environment as a whole. We did not need to be on the oil grid system, nor did we need to be forced to pay ever-escalating prices at the fuel pumps to run our cars and our transportation infrastructure.

WE HAVE ALL BEEN DECEIVED! We have been duped into thinking that there was no alternative to fossil fuels! We have been forced to go blindly along with a corrupt system that only benefitted the wealthy families and the corporate elite while robbing us daily and forcing the middle-income wage earners into a downward spiral toward poverty. The money you pay out of your pockets daily over the last one hundred plus years was stolen from you!!! If you feel like this author does, you should be absolutely **pissed off**!!! You should be *outraged to the Nth degree*. As consumers, it's time to **DEMAND JUSTICE** be served against these oligarchal, mega-wealthy oil baron Nazis!!!!!

How much longer are we to endure this corruption within our society? How much will we keep paying to these mega-oil robber-barons before we say, enough is enough!? It's time to wake up people and take back your power that you have blindly given away to a few corrupt individuals who no longer represent you!!!

Not long after Tesla remarkable inventions came to light then, other scientists and inventors brought out similar energy generating devices which either powered automobiles without gasoline or demonstrated over unity energy devices. Each scientist, however, inevitably ran into the "buzz-saw" of suppression and cover-up from the elite corporations of status quo.

C. Earl Ammann

A young 28 year old inventor, **C. Earl Ammann**, demonstrated his invention on August 8, 1921, by attaching it to an old automobile and running it about Denver city streets with fuel or batteries. **K. H. Isselstein,** a painter and decorator of Spokane, Washington and **Dr. Keith E. Kenyon** of Van Nuys, California both gave signed witness testimony that the invention by Ammann was revolutionary in its concept.

Ammann had shown Isselstein some of his home electrical brilliance in his attic by placing placed some steel bars on a worktable and picked up a coil which looked like a loose coupler. After placing the coils on the steel rods, he touched the opposite terminal. The bell rang with great force, and there was quite a spark, too. Upon inspection by Isselstein, he saw that there was no battery inside the device and the wire was iron. https://fuel-efficient-vehicles.org/energy-news/?page_id=971

In Ammann's basement was the **Activator Transformer** the size of two fists, which had to be within 10 miles of the radius of the generator coils. The activator was not in contact with any visible wires or appliances. It was activated by the electric currents which surge around the earth and activate the compass needle. By cutting into these currents, earl said, we can obtain unlimited power.

A year later Earl demonstrated h**is Cosmo Electric Generator** in Denver. He had placed two copper spheres on the front fenders of his car in place of the headlights. From these copper spheres he obtained enough power to drive that old jalopy all over Denver, (see Tesla's Electric Car).

C. Earl Ammann (left) stands beside his invention the Cosmic (Atmospheric) Electric Generator (two copper balls) attached to his old vehicle.
https://lifeboat.com/blog/2020/02/this-is-a-newspaper-story-about-c

While Earl was demonstrating his invention all over the streets of Denver, the power had been cut off in the foothills. ***In spite of this, when he went to Washington DC shortly afterward to try to obtain a patent on his Cosmo Electric Generator, he found that charges had been filed against him claiming he had a device to steal power from the power lines.*** https://fuel-efficient-vehicles.org/energy-news/?page_id=971 Bold italics added for emphasis.

The automobile which Ammann used for his demonstration Monday was the body and chassis of an electric vehicle. There are said to be no batteries in the car. It propelled itself with remarkable speed at the touch of the foot, climbed hills and glided through a maze of traffic under easy control.

Some skeptical people wonder if he had a storage battery concealed inside of the power cylinder, Ammann said:

"As badly as I would like to show the inside of my invention, I can't, for I have not yet obtained the patent rights. It would be exposing the result of seven years of work to open the cylinder. I leave for Washington this week to obtain the patent rights. When I return, I will gladly show everything and I can only say, wait until then and time will tell.

"I have bucked every law of the textbooks to perfect the invention. It appears on the order of the wireless telephone, but it is decidedly different, except that the electricity is derived from the air. It will run anywhere except under water.

The automobile is only a simple test. The generator will light buildings, do away with steam turbines, and, in fact, propel any kind of engine motor".

J. **N. Davis**, the proprietor of the Davis Electric Garage company, at 921 East 14th Avenue, and one of the oldest electrical men in Denver, made a thorough study of the generator.

"I believe that Mr. Ammann has at least made the invention which will revolutionize power", Mr. Davis said. *"Of course, we don't know what is inside of the generator and the inventor would be foolish to show us. We have long known that certain minerals exist, which if properly arranged together, would furnish power. That, in substance, according to the blueprints of the invention, is the basis of the whole thing.*

"If the generator has been perfected to the extent that it will propel an automobile, the rest of its work is assured. It will be the greatest invention of the age. The electricity obtained from the air, first passing through the generator, would be available for any use". https://fuel-efficient-vehicles.org/energy-news/?page_id=971

Lester Hendershot

In 1960, **Lester Hendershot**'s device (called a **'magnatronic generator'**) was researched by the U.S. Navy's Office of Naval Research. The generator was reported to have lit a 100-watt lamp by **'induced radio frequency energy.'** The project ended when Hendershot committed "suicide".

"He is reported to have accepted an offer he couldn't refuse being paid never to work on his device again."

Author's Rant: This is a common practice to keep the inventor silent by being bought off! Big money speaks and most people listen and then comply!

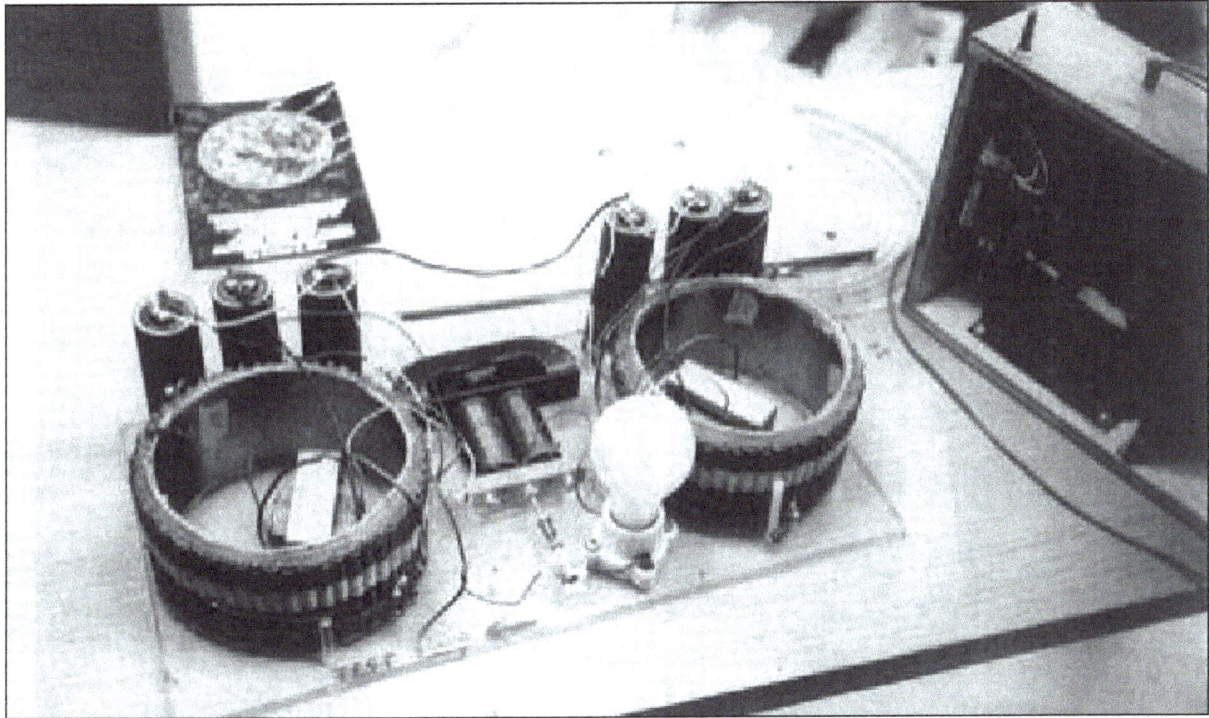

The Hendershot Magnatronic Generator
https://content.instructables.com/ORIG/FC2/3PPI/HSHDUKZ8/FC23PPIHSHDUKZ8.pdf

Alfred Hubbard

In 1920, at the age of 19, **Alfred Hubbard** built a coil and motor based on a **Radio Principle** – Armature winding new invention that ran his boat 10 knots an hour on Portage Bay in Seattle. He called this *'fuel-less'* unit an *'atmospheric power generator.'* Hubbard claimed it could operate for years; drive a large car; light an office building, and fly a plane around the world nonstop. Little is known about the **Hubbard Coil**. He was forced into obscurity because of what his motor would do to the present industries. "Since Alfred Hubbard worked with **Tesla** for a short period, it seems likely that his transformer is based on (information from) Tesla." http://fuel-efficient-vehicles.org/energy-news/?page_id=952

T. Henry Moray

A Doctor of Electrical Engineering, **T. Henry Moray** wrote the book 'The Sea of Energy in Which the Earth Floats' in 1960. It contains results of a 50-year study with another type of **'atmospheric energy collector.'** During FDR's reign, Moray became the chief engineer of the western branch of the Rural Electrification Agency. He built a device that weighed 55 pounds and produced up to 50,000 watts. An ignorant co-worker was so enraged by the strange principles Moray tested, that he destroyed the machine. By today's standards, the loss would be over a million dollars. Moray wrote that *'frequencies maybe developed which will balance the force of gravity to a point of neutralization.'*

Moray: *"It was during the Christmas holidays of 1911 that I began to realize the fact that the energy I was working with was not of a static nature but of an oscillating nature, and that the energy was not coming out of the Earth but that it rather was coming into the Earth from some outside source."*

Ed Gray

Edwin Gray Sr., 48, has fashioned working devices that could:

Power every auto, train, truck, boat and plane that moves in this land — perpetually.
Warm, cool and service every American home — without erecting a single transmission line.
Feed limitless energy into the nation's mighty industrial system — forever.
And do it all without creating a single iota of pollution.

"He flicked a switch and the tiny battery sent a charge into the capacitors. He then plugged in six 15-watt electric bulbs on individual cords — and a 110-volt portable television set and two radios. The bulbs burned brightly, the television played, and both radios blared — and yet, the small battery was not discharging."
(Gray had a neighbor, an engineer, who had met Tesla and is considered the source of Gray's motor). http://fuel-efficient-vehicles.org/energy-news/?page_id=952

In the **Victorian Age** of the late 1880's, newspapers, various trade journals in the sciences and public demonstrations of new science wonders were predicting a coming age of "free electricity" and "free energy" in the near future. The future for mankind looked hopeful and literally "bright" with new expectations and wonders for all to behold. The US Patent Office could barely keep up with the flood of new inventions that were occurring every day; there was something new and marvelous to change the fortunes of the common people.

The common people for the first time in history could envision the concept of a utopian future, a good abundant life filled with the marvels of modern transportation and communication, employment for all, everyone with a modern home and plenty of food on the table. The eventual abolishment of all known diseases and poverty would be a distant memory of the past. It would appear at long last that a golden age of civilization was about to dawn on mankind and everyone would get **"a piece of the pie."** Then, almost out of nowhere the harsh cold reality of corruption and greed set in, control and disclaimers on certain technology limited this technological explosion and the energy breakthroughs that were highly anticipated evaporated and disappeared almost overnight. What happened to the promise of a brighter, hopeful future? Where was the "free electricity", the "free energy" which had just appeared before the beginning of the 20th Century? Could the promises of science been wrong or was someone else controlling our future?

The problem wasn't that fantastic energy technologies weren't being developed along with the other science breakthroughs, the real problem was that none of these breakthroughs, cost saving energy technologies were getting out into the consumer market or for commercial use. Here is a short list of some of the "free energy" technologies that did make their way into the hands of the

94

consumer which are proven beyond all reasonable doubt as overunity devices. The common denominator in all of these discoveries is that they use a small amount of one energy form to control or release a large amount of a different kind of energy. Many of them tap the underlying **Aether** or **Zero Point Energy** field in some way; a source of energy conveniently ignored by "modern" science and deliberately kept off the public radar.
http://peswiki.com/index.php/Directory:Aether

1. **Radiant Energy/Cold Electricity**. Nikola Tesla's Magnifying Transmitter, T. Henry Moray's **Radiant Energy Device**, Edwin Gray's **EMA Motor**, and **Paul Baumann**'s **Testatika Machine** all run on **Radiant Energy** (aka the quantum ambient energy flux of space). This natural energy form can be gathered directly from the environment (mistakenly called "static" electricity) or extracted from ordinary electricity by the method called "fractionation." Radiant Energy can perform the same wonders as ordinary electricity, at less than 1% of the cost. It does not behave exactly like electricity, however, which has contributed to the scientific community's misunderstanding of it. The **Methernitha Community** in Switzerland currently has 5 or 6 working models of fuelless, self-running devices that tap this energy.

2. **Permanent Magnets**. **Dr. Robert Adams** (NZ) has developed astounding designs of electric motors, generators, and heaters that run on permanent magnets. One such device draws 100 watts of electricity from the source, generates 100 watts to recharge the source, and produces over 140 BTU's of heat in two minutes! **Dr. Tom Bearden** (USA) has two working models of a permanent magnet powered electrical transformer. It uses a 6-watt electrical input to control the path of a magnetic field coming out of a permanent magnet. By channeling the magnetic field, first to one output coil and then a second output coil, and by doing this repeatedly and rapidly in a "Ping-Pong" fashion, the device can produce a 96-watt electrical output with no moving parts. Bearden calls his device a **Motionless Electromagnetic Generator (MEG).** **Jean-Louis Naudin** has duplicated Bearden's device in France. The principles for this type of device were first disclosed by **Frank Richardson** (USA) in 1978. **Troy Reed** (USA) has working models of a special magnetized fan that heats up as it spins. It takes exactly the same amount of energy to spin the fan whether it is generating heat or not. Beyond these developments, multiple inventors have identified working mechanisms that produce motor torque from permanent magnets alone.

3. **Mechanical Heaters**. There are two classes of machines that transform a small amount of mechanical energy into a large amount of heat. The best of these purely mechanical designs are the rotating cylinder systems designed by Frenette (USA) and Perkins (USA). In these machines, one cylinder is rotated within another cylinder with about an eighth of an inch of clearance between them. The space between the cylinders is filled with a liquid such as water or oil, and it is this "working fluid" that heats up as the inner cylinder spins. Another method uses magnets mounted on a wheel to produce large eddy currents in a plate of aluminum, causing the aluminum to heat up rapidly. These **magnetic heaters** have been demonstrated by **Muller (Canada), Adams (NZ) and Reed (USA).** All of these systems can produce more heat than standard methods using the same energy input.
http://www.apparentlyapparel.com/free-energy.html

4. **Super-Efficient Electrolysis**. Water can be broken into Hydrogen and Oxygen using electricity. Standard chemistry books claim that this process requires more energy than can be recovered when the gases are recombined. This is true only under the worst-case scenario. When water is hit with its own molecular resonant frequency, using a system developed by **Stan Meyers** (USA) and again recently by **Xogen Power, Inc.,** it collapses into Hydrogen and Oxygen gas with very little electrical input. Also, using different electrolytes (additives that make the water conduct electricity better) changes the efficiency of the process dramatically. It is also known that certain geometric structures and surface textures work better than others do. The implication is that unlimited amounts of **Hydrogen fuel** can be made to drive engines (like in your car) for the cost of water. Even more amazing is the fact that a special metal alloy was patented by **Freedman (USA)** in 1957 that spontaneously breaks water into Hydrogen and Oxygen with no outside electrical input and without causing any chemical changes in the metal itself. This means that this special metal alloy can make Hydrogen from water for free, forever.

5. **Implosion/Vortex**. All major industrial engines use the release of heat to cause expansion and pressure to produce work, like in your car engine. Nature uses the opposite process of cooling to cause suction and vacuum to produce work, like in a tornado. **Viktor Schauberger** (Austria) was the first to build working models of Implosion Engines in the 1930's and 1940's. Since that time, **Callum Coats** has published extensively on Schauberger's work in his book "Living Energies" and subsequently, a number of researchers have built working models of **Implosion Turbine Engines**. These are fuel-less engines that produce mechanical work from energy accessed from a vacuum. There are also much simpler designs that use vortex motions to tap a combination of gravity and centrifugal force to produce a continuous motion in fluids. http://www.apparentlyapparel.com/free-energy.html

6. **Cold Fusion**. In March 1989, two Chemists, **Pons and Fleishman** from the University of Utah (USA) announced that they had produced atomic fusion reactions in a simple tabletop device. Their claims were "debunked" within 6 months and the public lost interest. Nevertheless, **Cold Fusion** is very real. Not only has excess heat production been repeatedly documented, but also *low energy atomic element transmutation has been catalogued*, involving dozens of different reactions! This technology definitely can produce low-cost energy and scores of other important industrial processes.

7. **Heat Pumps**. Everyone has a "free energy device" in their home and they are probably not even aware that they own one. It's your refrigerator! It is probably the only machine that they own that is an "overunity" device which electrically operates as a heat pump. You may also be fortunate enough to own two **overunity**, the other "free energy" or "overunity machine" which may have been built into your home at your specific request is your furnace, which also operates as a heat pump. It uses one unit of energy (electricity) to move three units of energy (heat) in the case of your refrigerator or four units in the case of your furnace. This gives it a **"coefficient of performance" (COP)** of about 3 and 4or more, respectively. Your refrigerator uses one amount of electricity to pump three amounts of heat from the inside of the refrigerator to the outside of the refrigerator or one unit of electricity to four or more units of heat out from your furnace which is desired. In the case of the furnace this is a fair use of the technology but, in the case of the refrigerator, it too puts out heat which is proportional to the room temperature of the kitchen, instead of more cold thus, it's not a practical use of this

96

amazing technology. This is its typical use, but it is the worst possible way to use the technology. Here's why. A heat pump pumps heat from the "source" of heat to the "sink" or place that absorbs the heat. The "source" of heat should obviously be HOT and the "sink" for heat should obviously be COLD for this process to work the best. In your refrigerator, it's exactly the opposite. The "source" of heat is inside the box, which is COLD, and the "sink" for heat is the room temperature air of your kitchen, which is warmer than the source. This is why the COP remains low for your kitchen refrigerator.

http://www.apparentlyapparel.com/free-energy.html

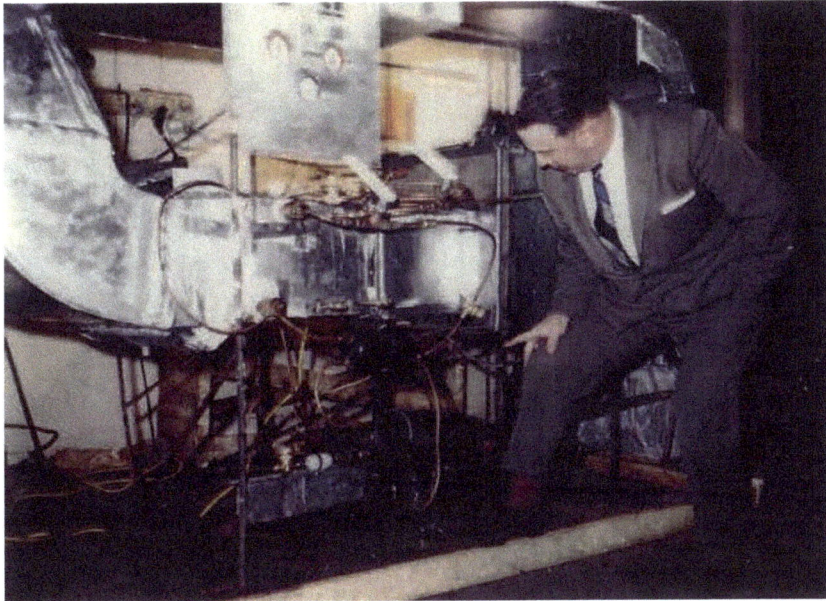

A Heat Pump installed and operating in a Toronto business building
(Google Images)

These are some of the overunity devices that have been release into the public domain because of their relative harmlessness and very low threat to national security but, if you have a device that will take us all off the oil and utilities grid infrastructure then, expect the government or some arm of the government to come down very hard on you.

You can say goodbye to that potential billion dollar money maker and you can count yourself lucky because people (inventors) and their families have been known to be killed for trying to create a better world through science.

Even if you play and abide by the rules of **LAW** as a good dutiful citizen, you will find that neither the government nor does **BIG BUSINESS** abide by those same Laws. The laws of the land are there to contain and control the public, but Big Business, the government, the military and various intelligence agencies are given free-reign to do whatever they please as long as they don't get caught.

It may be better for you to choose an altruistic approach and not focus on the almighty dollar, in the beginning, better to look for long-term benefits down the road, the wealth will eventually come, because people will believe in your invention and will want to buy it. Make multiple prototypes and sequester them away in different locations around the country or have them patented in other countries like Canada where the law is less limiting and national security issues are not a problem as they are in the US.

The U.S. Patent Office has a nine-member committee that screens patents in order to protect "national security".

An understandable reason for suppressing certain types of energy inventions is that the knowledge behind them is also capable of producing tremendously destructive advanced electromagnetic weapons such as the *"death ray"* apparently invented by **Nikola Tesla.** Hence many such new energy technologies, particularly those using this kind of knowledge of advanced electromagnetic principles, are considered "dual use" technologies that are among the 4,000 un-numbered patent applications confiscated in a vault at the **US Patent and Trademark Office** because of their military potential and the need to keep that knowledge from America's enemies.

A hidden purpose of this committee is to also ***find and remove from public access energy-related patents which could threaten the fossil fuel and power monopolies.***

Canada's patent office doesn't have a similar screening committee. It is recommended that energy patents possibly in danger of being classified should be first applied for in Canada. Once granted, up to one year is allowed to apply for the same patent in the U.S. Patent Office. Now the patent cannot be classified because it is already out in the public domain, courtesy of Canada. http://www.theorionproject.org/en/suppressed.html

"'O Canada! ... The True North, strong and free!" Yeah, right! We are quickly becoming the 53rd state of America as we seem to emulate far too much what the US does, good and bad!

It's time to start thinking independently and decide once and for all where our true destiny lies!

Getting your invention out to benefit the world requires out thinking the powers that be and sometimes the best thing to do is get as much publicity as is possible, surround yourself with like-minded people and by-pass the patent office altogether and go straight into mass production. By the time the government figures it out, it will be too late, as hopefully thousands or millions of people will already have their hands on your invention.

Always keep in mind that the government, after all, works for the people not for themselves. Let the US government always remember who they work for and should act only upon the will of people and not upon their own will. The will of the people is enshrined in the . Above all other words in the Constitution, boldly stated with clear prominence are three important concise words: **"We the people"!** Patent Offices serve the people, not the government or a small elitist group of people!

Every year in the US Patent Office many inventors file for patents hoping to get that all important patent number. Many inventions never get accepted because they don't live up to the claims of the inventor, however, those that do receive patents are still questionable and it really makes one wonder why the US Patent Office even gave them a patent number as there are no prototypes to demonstrate proof of concept, only a drawing with statements of intent.

The technology for building saucer-shaped aircraft in the white world of science are nowhere near as advanced as the fully functional black world flying saucers which demonstrate some of the amazing flight characteristics similar to alien spacecraft, but nevertheless, these patents are available from the US patent office. Because these white world saucers don't present a national security risk nor represent advanced concepts or technology, therefore, the US Military has no use for these public domain saucers; in other words, *they are not advanced alien technology*!

Below are just a few of the thousands of flying saucer patents, disc shape craft designed to transport people either in the air or through space. They do not operate on alternative energy generation or propulsion systems but traditional early 1900s technology. You be the judge if these patents are realistic as per their claims.
https://www.youtube.com/watch?v=48AkxqT16gk

CHAPTER 96

US PATENTS: DISC AIRCRAFT

Propeller & Jet-Driven Wingless Aerodynes, Lenticular Aircraft, Discoid Aircraft, & c.

In this chapter we examine the patented schematics of flying discs and saucers. (This file does not include force field propulsion & space drives)

Acknowledgements: Visit Man-Made Flying Saucers (http://www.aspubs.com/mmsaucers/) and World Space Drive Archives for a thorough treatment of this subject, plus: anti-gravity and space drive (unidirectional motion rectifiers) patents from around the world, and lots of photos, documentation, &c... http://www.rexresearch.com/wingless/wingless.htm

USP # 6,270,036 (8-7-01) ~ Blown-Air Lift Generating Rotating Airfoil Aircraft Lowe, Charles S., Jr.

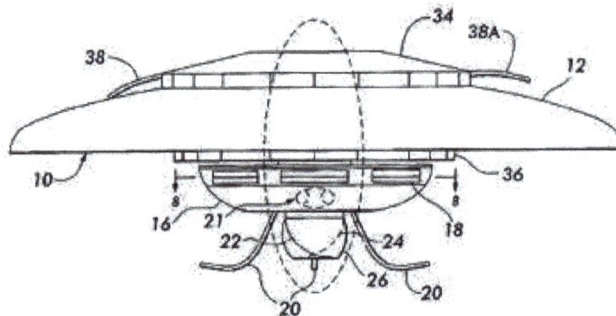

**USP # 6,254,032 (7-3-01) ~ Aircraft, &c...
Bucher, Franz**

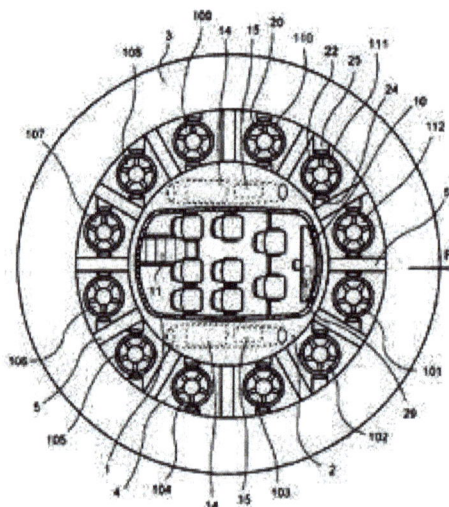

**USP # 6,179,247 (1-30-01) ~ Personal Air Transport
Milde, Jr., Karl F.**

100

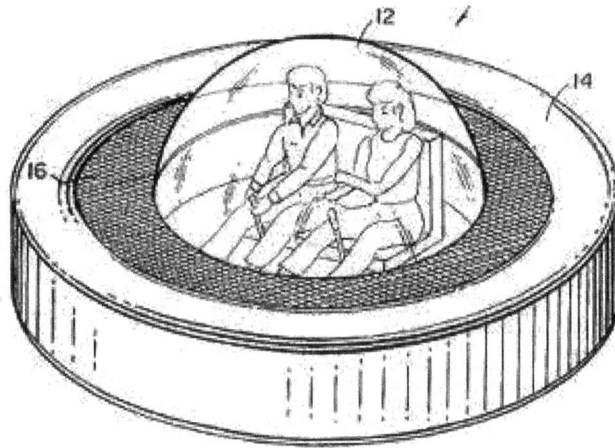

USP # 6,113,029 (9-5-00) ~ Aircraft Capable of Hovering Flight
Salinas, Luis A.

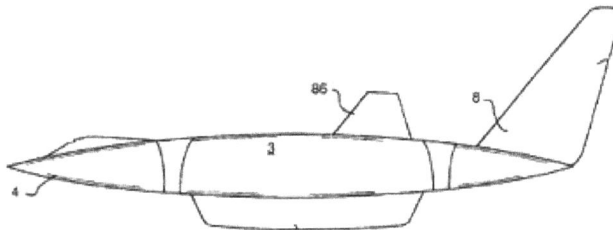

USP # 6,082,478 (7-4-00) ~ Lift-Augmented Ground Effect Platform
Walter, William. C., *et al.*

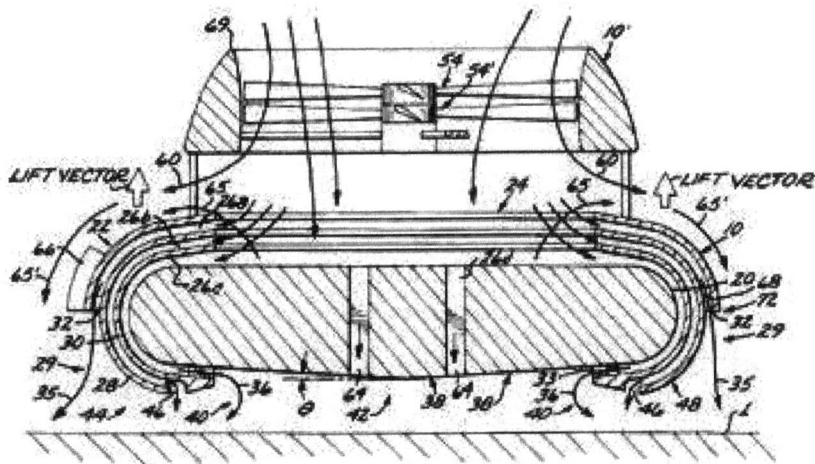

USP # 6,073,881 (6-13-00) ~ Aerodynamic Lift Apparatus
Chen, Chung-Ching

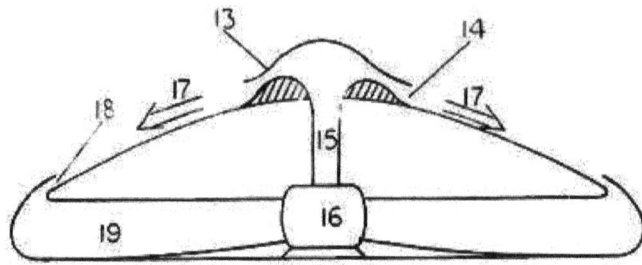

USP # 6,068,219 (5-30-00) ~ Single-Surface Multi-Axis Aircraft Control
Arata, Allen A.

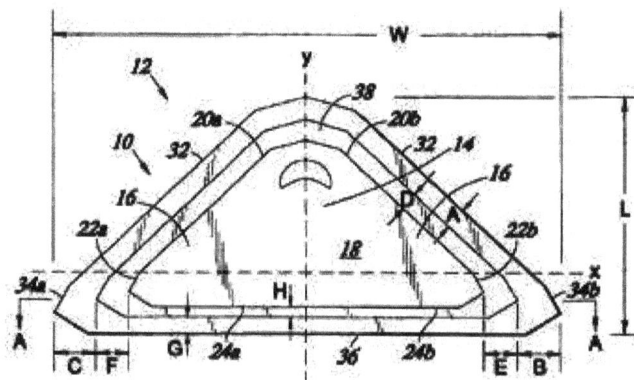

USP # 6,053,451 (4-25-00) ~ Remote-Control Flight Vehicle Structure
Yu, Shia-Giow

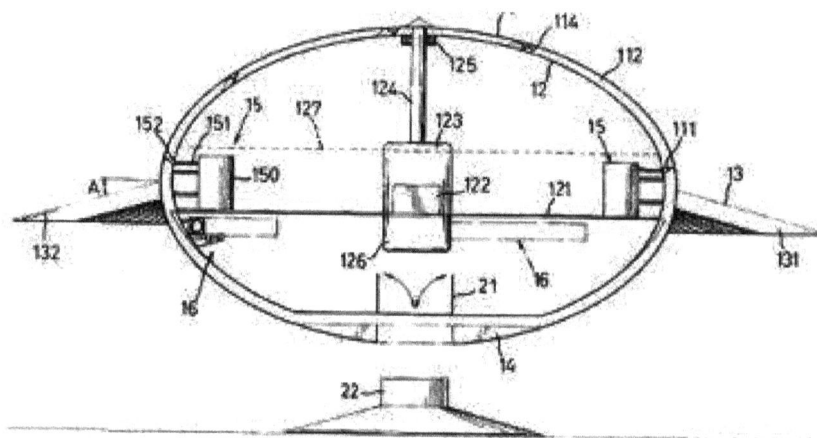

USP # 6,050,520 (4-18-00) ~ VTOL Aircraft
Kirla, Stanley J.

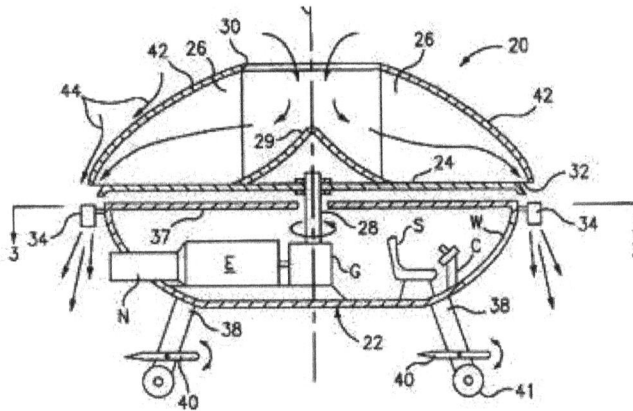

USP # 6,016,991 (1-25-00) ~ Evacuated Rotating Envelope Aircraft
Lowe, Jr., Charles S.

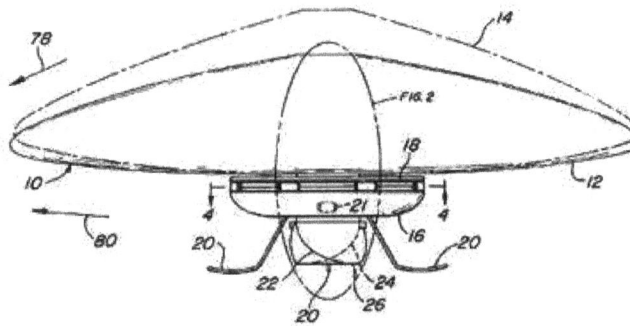

USP # 5,971,321 (10-26-99) ~ Body-Lift Airplane Assembly
Libengood, Ronald L.

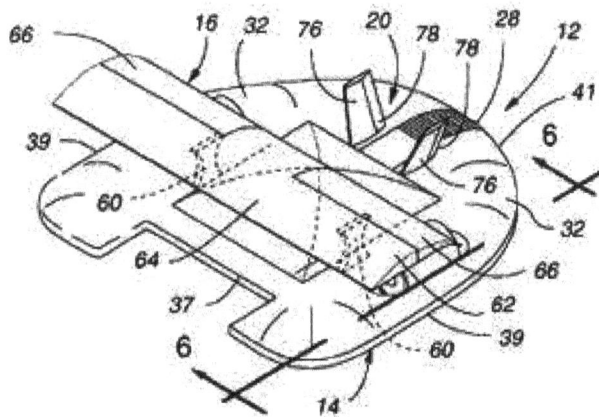

USP # 5,895,011 (4-20-99) ~ Turbine Airfoil Lifting Device
Gubin, Daniel

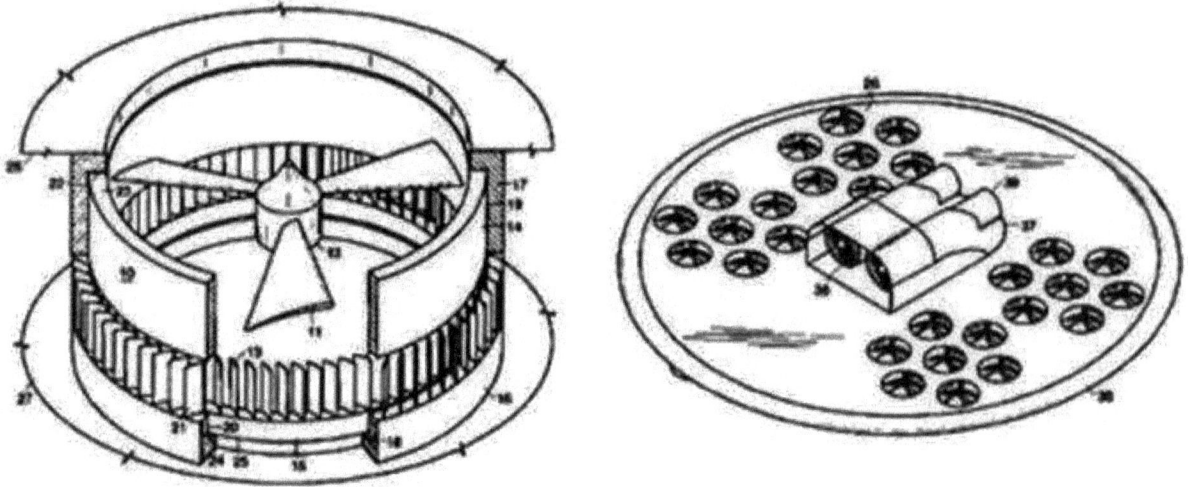

USP # 5,881,970 (3-16-99) ~ Levity Aircraft Design
Whitesides, Carl W.

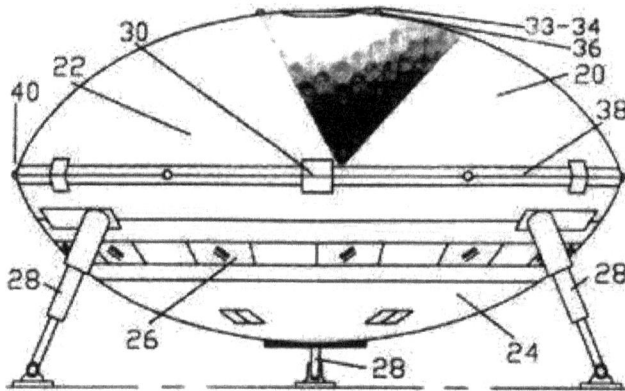

USP # 5,836,543 (11-17-98) ~ Discus-Shaped Aerodyne Vehicle...
Kunkel, Klaus

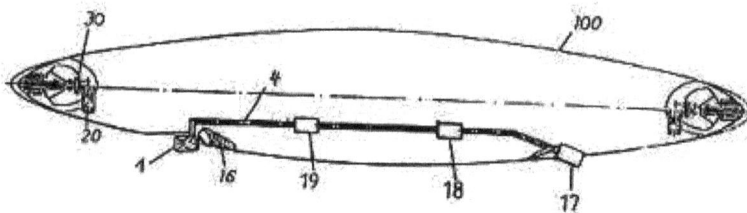

USP # 5,836,542 (11-17-98) ~ Flying Craft &c...
Burns, David J.

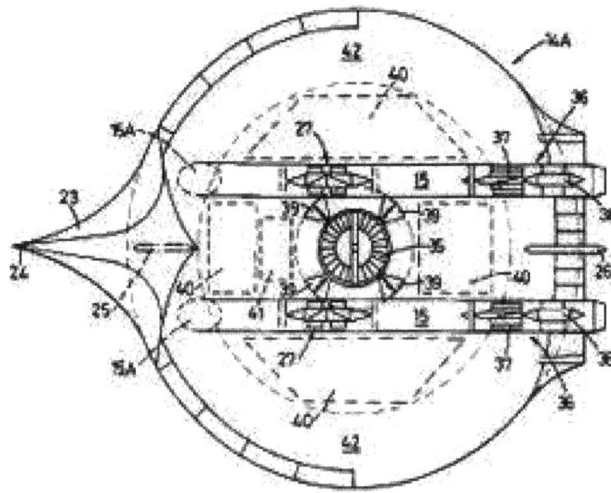

USP # 5,803,199 (9-8-98) ~ Lift-Augmented Ground Effect Platform
Walter, William C.

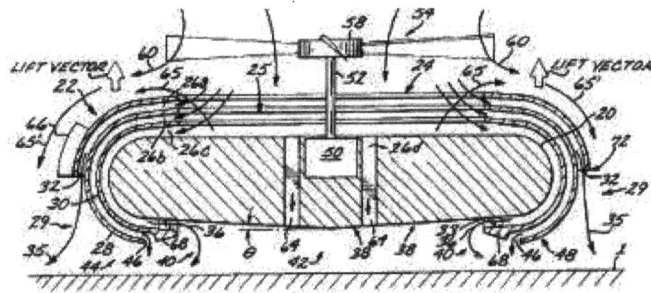

USP # 5,730,391 (3-24-98) ~ Universal Fluid-Dynamic Body for Aircraft & Watercraft
Miller, Jr., John A., *et al.*

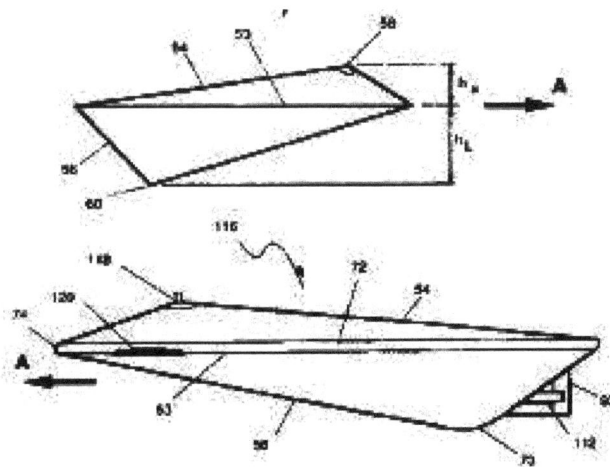

USP # 5,730,390 (3-24-98) ~ Reusable Spacecraft
Plichta, Peter & Buttner, Walter

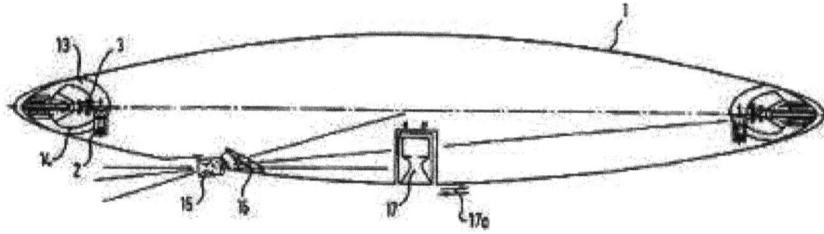

USP # 5,653,404 (8-5-97) ~ Disc-Shaped Submersible Aircraft
 Ploskin, Gennady

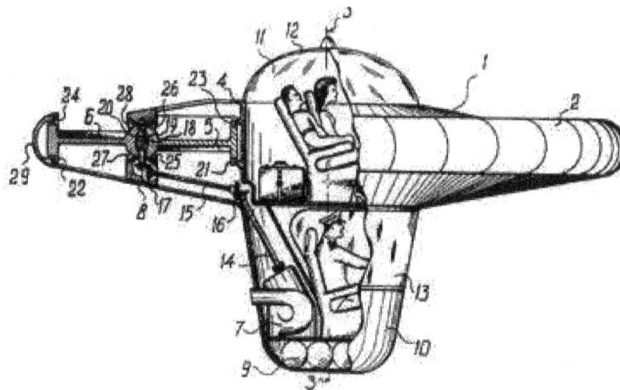

USP # 5,520,355 (5-28-96) ~ Three-Wing Circular Planform Body
Jones, Jack M.

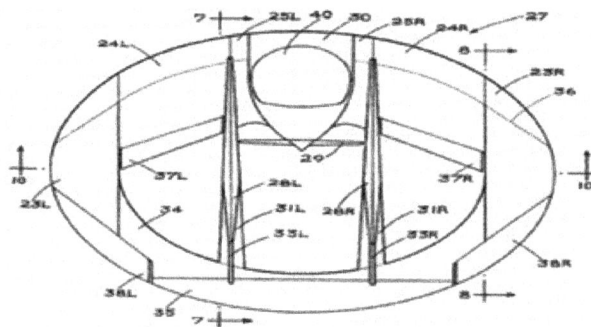

USP # 5,351,911 (10-4-94) ~ VTOL Flying Disc
Neumayr, George A.

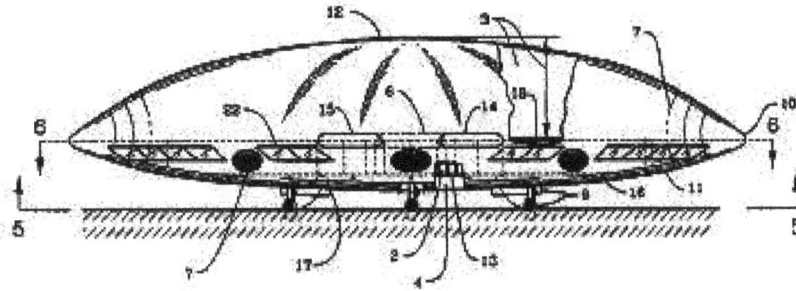

USP # 5,344,100 (9-6-94) ~ Vertical Lift Aircraft
Jaikaran, Allan

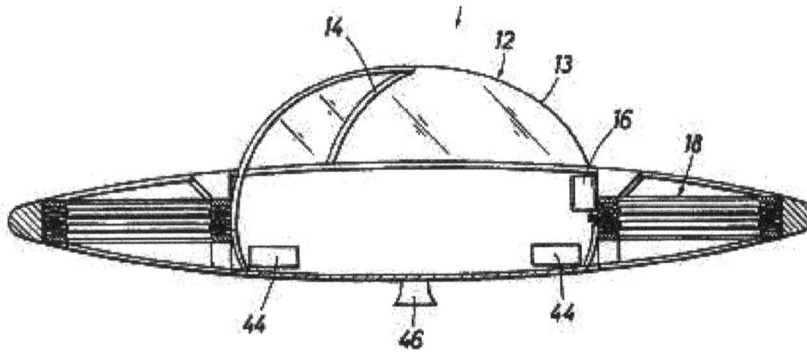

USP # 5,318,248 (6-7-94) ~ Vertical Lift Aircraft
Zielonka, Richard H.

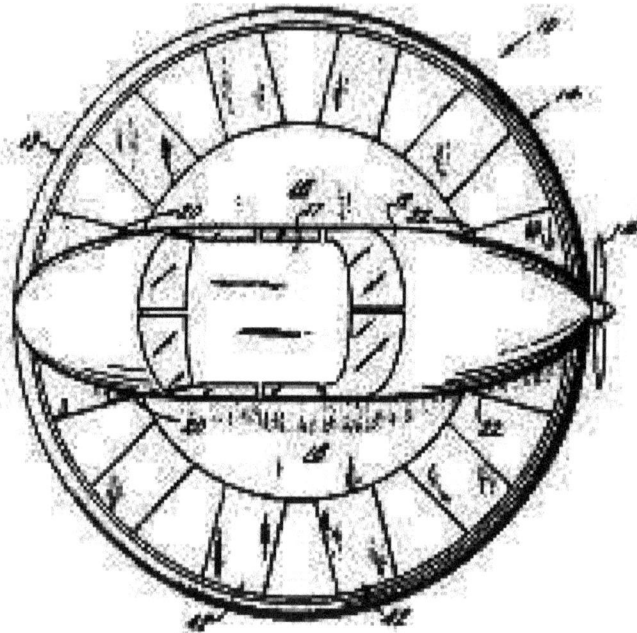

USP # 5,303,879 (4-19-94) ~ Aircraft with a Ducted Fan in a Circular Wing
Bucher, Franz

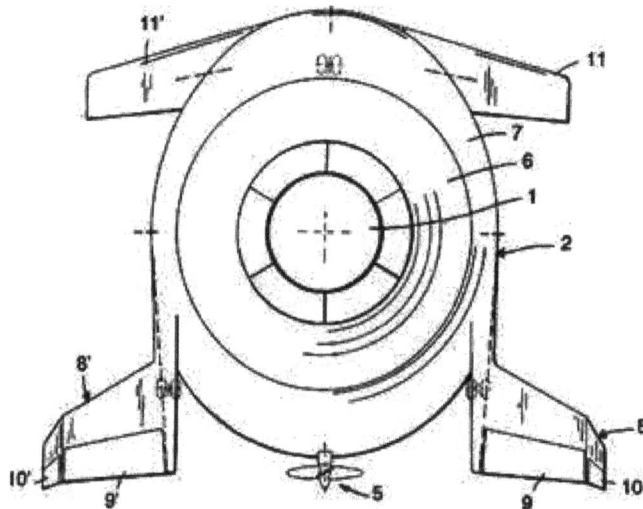

USP # 5,295,571 (11-9-93) ~ Aircraft with Gyroscopic Stabilization System
Blazquez, Jose M.

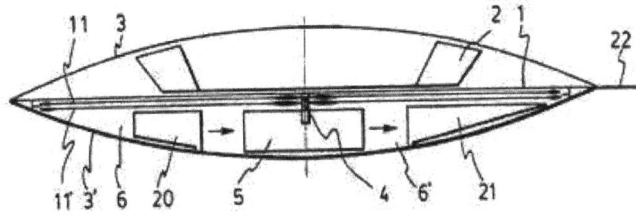

USP # 5,213,284 (5-25-93) ~ Disc Planform having Vertical Flight Capability
Webster, Stephen N.

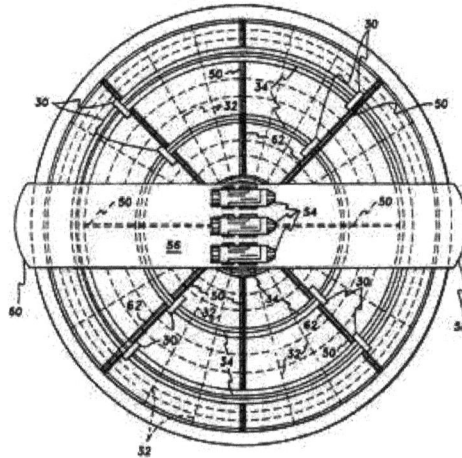

USP # 5,203,521 (4-20-93) ~ Annular Body Aircraft
Day, Terence

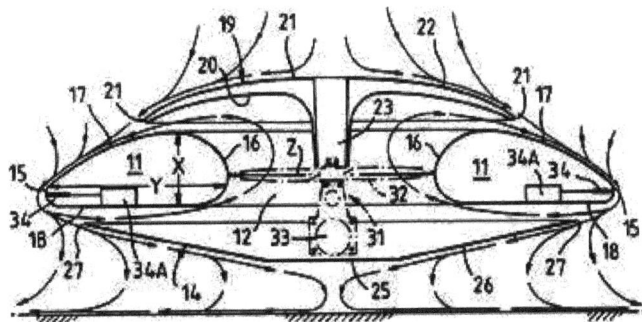

USP # 5,178,344 (1-12-93) ~ VTOL Aircraft
Dlouhy, Vaclav

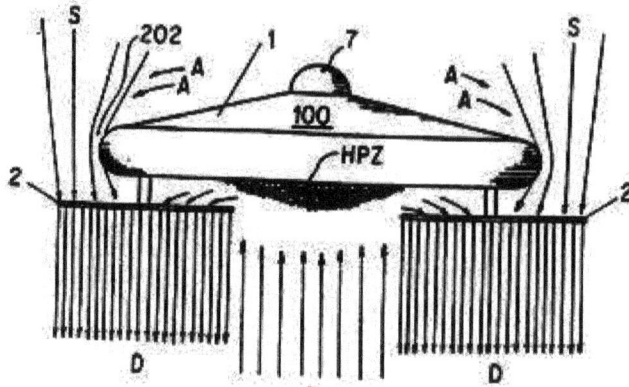

USP # 5,170,963 (12-15-92) ~ VTOL Aircraft
Beck, Jr., August H.

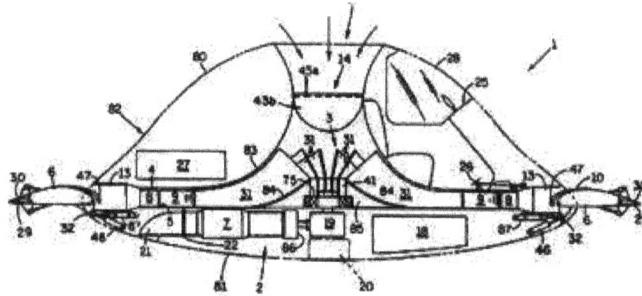

USP # 5,149,012 (9-22-92) ~ Turbocraft
Valverde, Rene L.

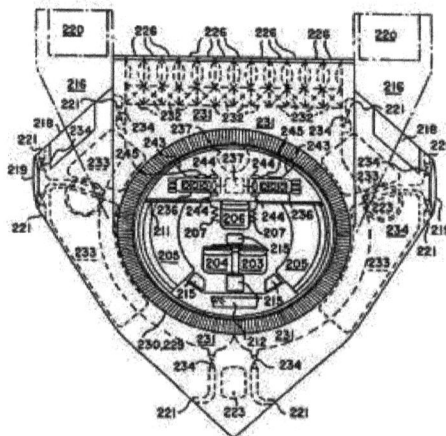

USP # 5,115,996 (5-26-92) ~ VTOL Aircraft
Moller, Paul S.

110

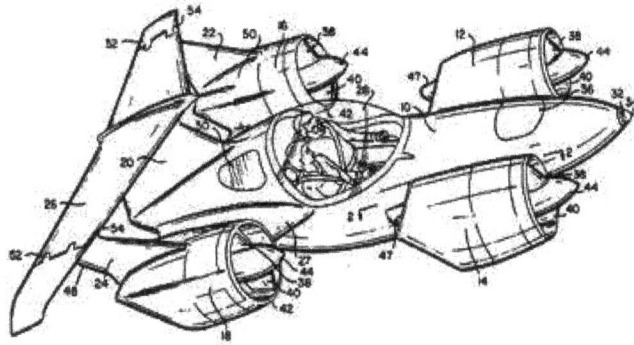

USP # 5,102,066 (4-7-92) ~ VTOL Aircraft
Daniel, William H.

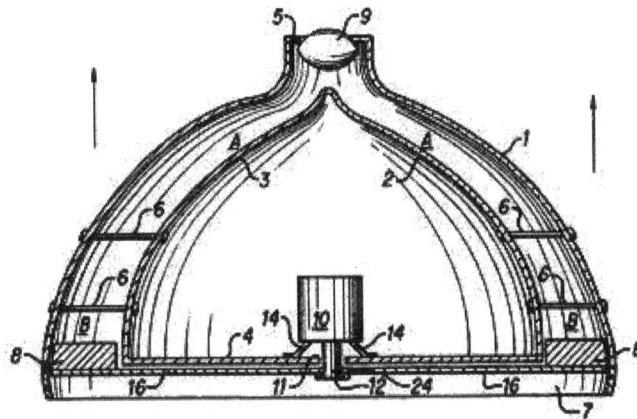

USP # 5,054,713 (10-8-91) ~ Circular Airplane
Langley, Lawrence W., *et al.*

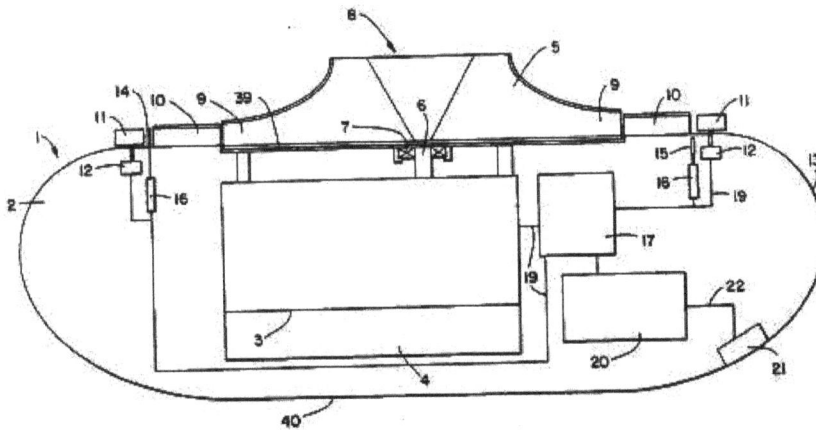

USP # 5,064,143 (11-12-91) ~ Aircraft Having a Pair of Counter Rotating Rotors
Bucher, Franz

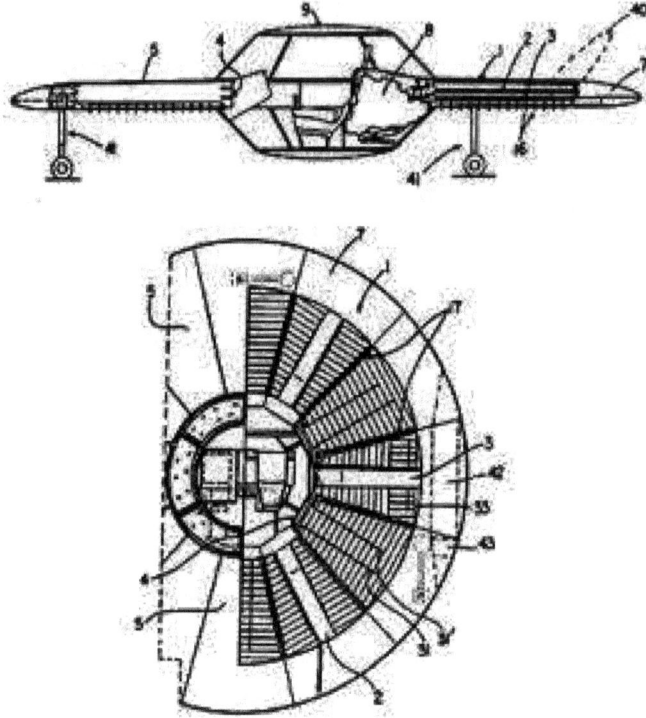

USP # 5,046,685 (9-10-91) ~ Fixed Circular Wing Aircraft
Bose, Phillip R.

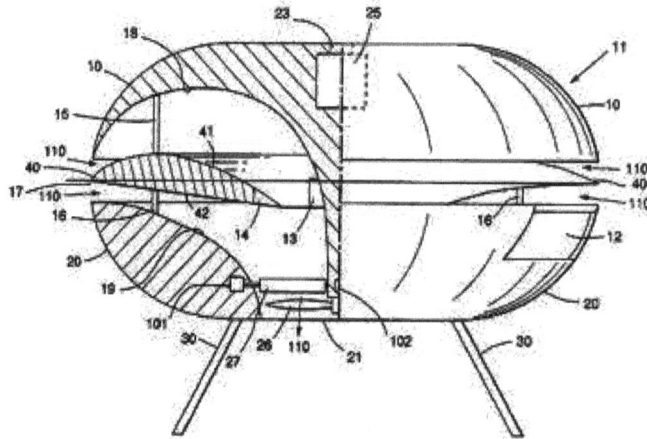

USP # 5,039,031 (8-13-91) ~ Turbocraft
Valverde, Rene L.

112

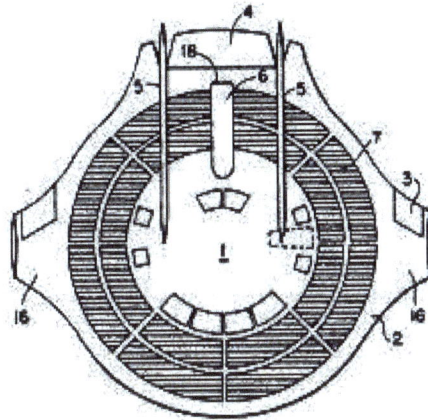

USP # 4,976,395 (12-11-99) ~ Heavier-Than-Air Disk-Type Aircraft
von Kozierowski, Joachim

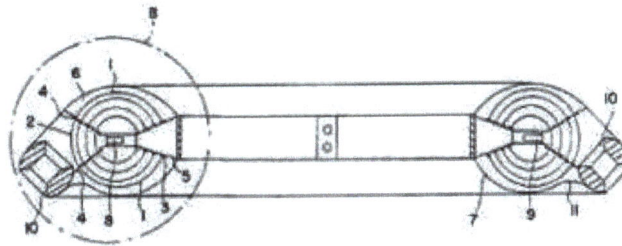

USP # 4,955,962 (9-11-99) ~ Remote Control Flying Saucer
Mell, Christian

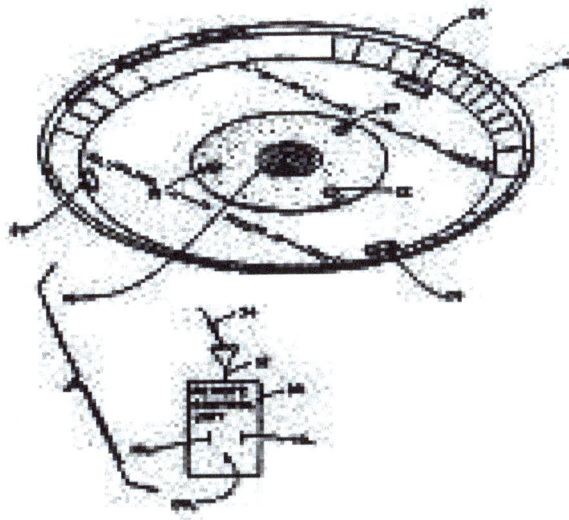

USP # 4,941,628 (7-19-90) ~ Lift Generating Apparatus, &c.
Sakamoto, Yujiro, *et al.*

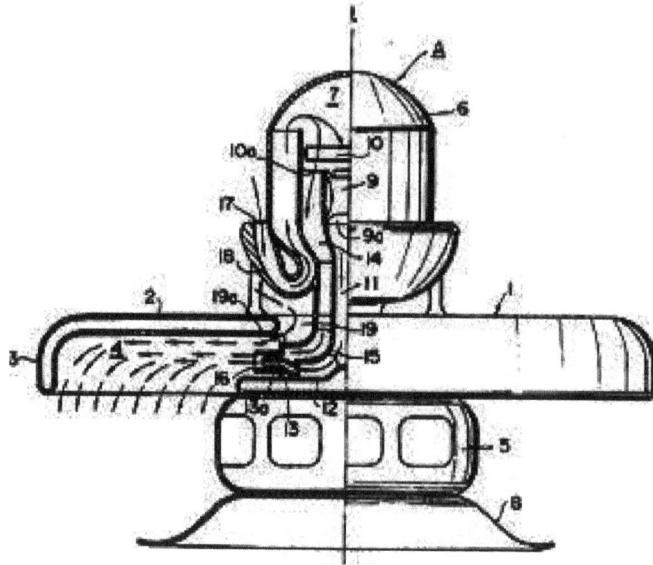

USP # 4,824,048 (4-25-89) ~ Induction Lift Flying Saucer
Kim, Kyusik

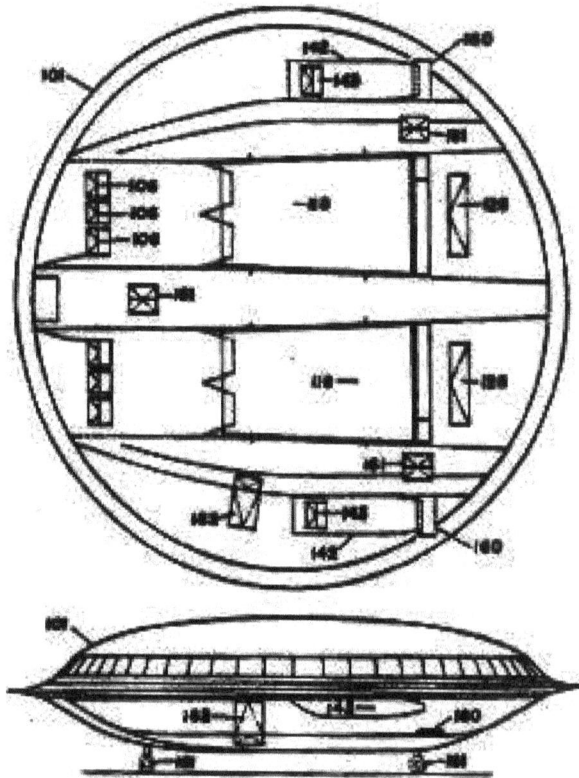

USP # 4,804,156 (2-14-89) ~ Circular Aircraft
Harmon, Rodney D.

114

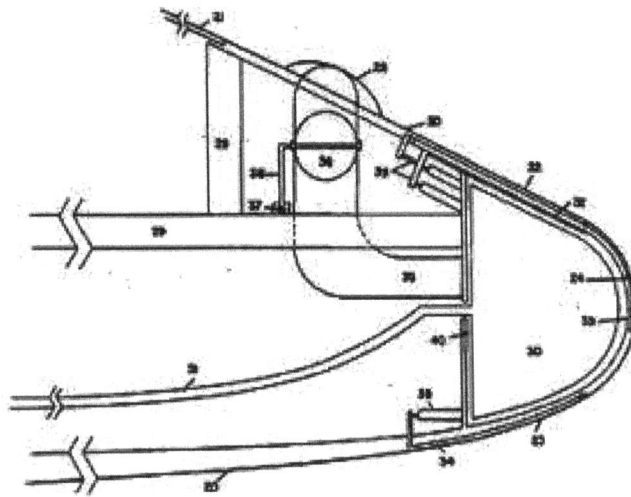

USP # 4,796,836 (1-10-89) ~ Lifting Engine for VTOL Aircrafts
Buchelt, Benno

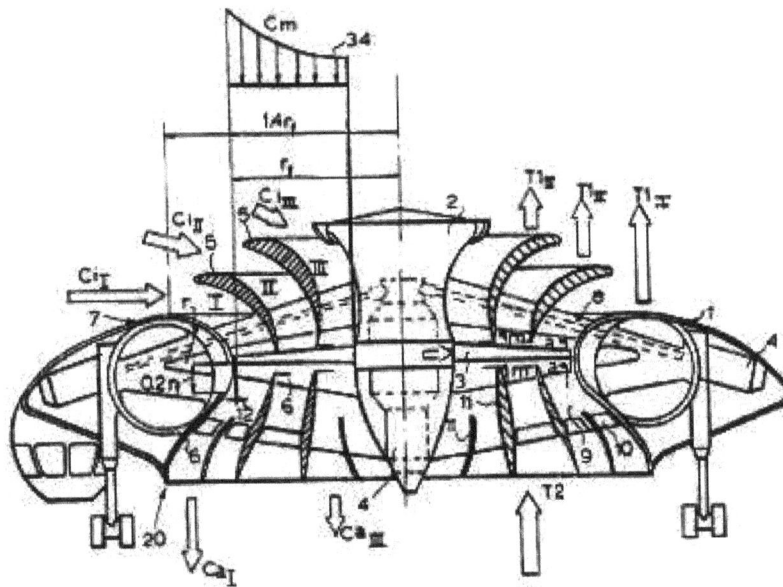

USP # 4,795,111 (1-3-89) ~ Robotic or Remotely Controlled Flying Platform
Moller, Paul S.

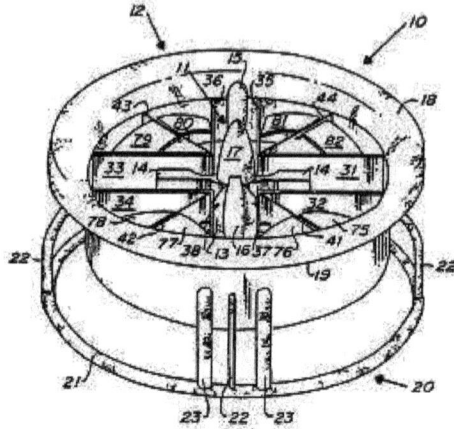

USP # 4,214,720 (7-29-80) ~ Flying Disc
DeSautel, Edwin R.

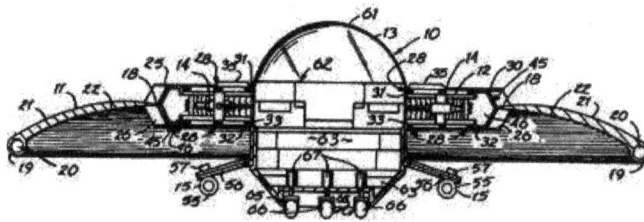

USP # 4,196,877 (4-8-80) ~ Aircraft
Mutrux, Jean L.

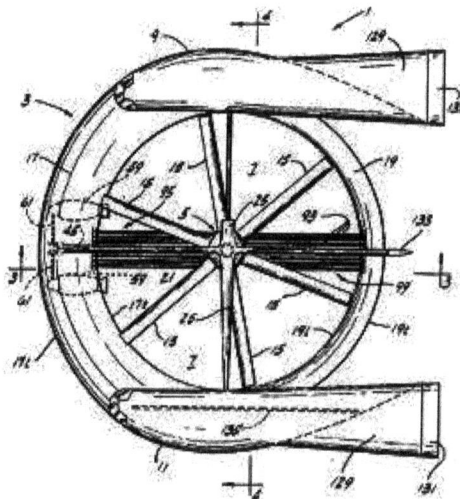

USP # 4,193,568 (3-18-80) ~ Disc-Type Airborne Vehicle
Heuvel, Norman L.

116

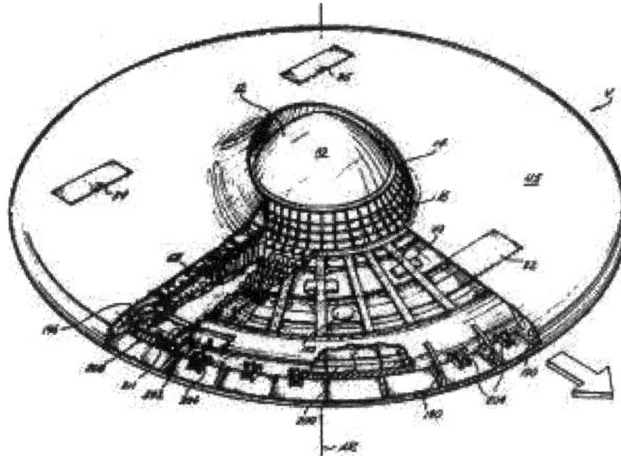

USP # 4,165,848 (8-28-79) ~ Rotary Thrust Device...
Bizzarri, Alfredo

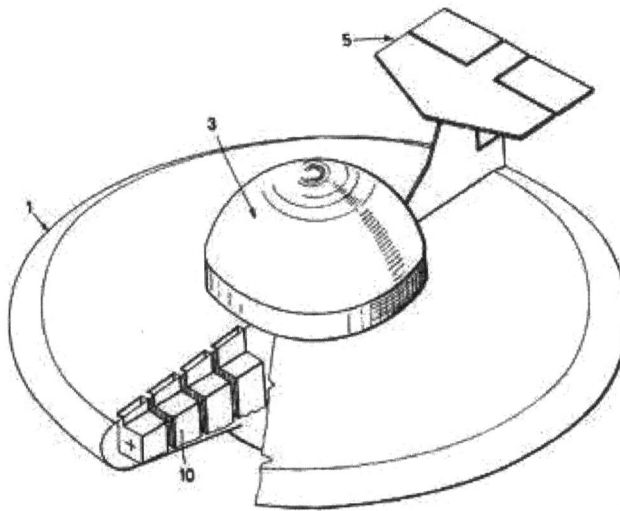

USP # 4,117,992 (10-3-78) ~ Vertical Lift Device
Vrana, Charles K.

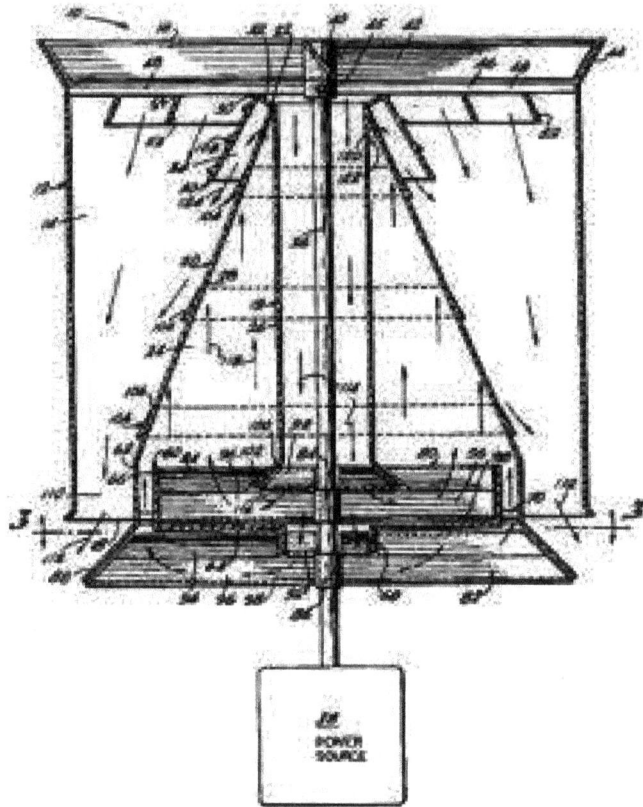

USP # 4,050,652 (9-27-77) ~ Gyro Foil
DeToia, Vincent D.

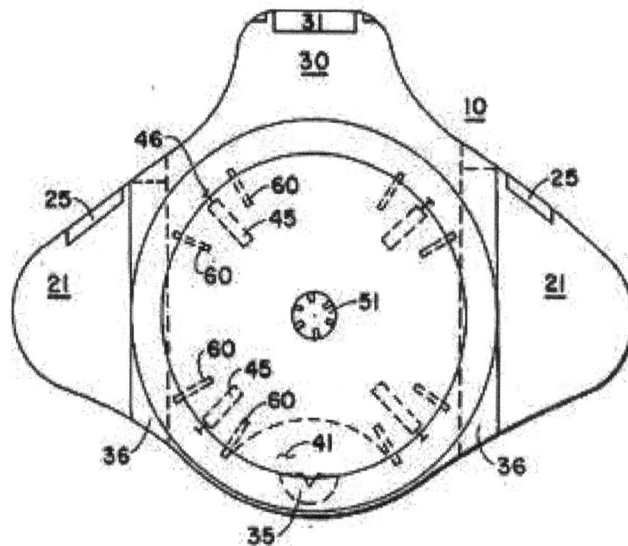

USP # 4,023,751 (5-17-77) ~ Flying Ship
Richard, Walter A.

118

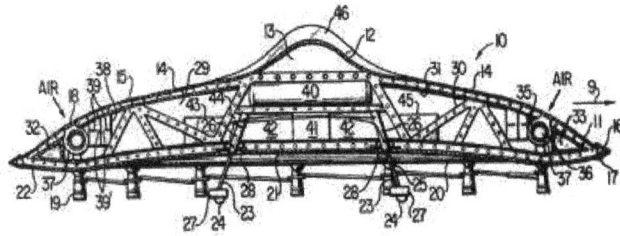

USP # 4,014,483 (3-29-77) ~ Lighter-Than-Air Craft
MacNeil, Roderick M.

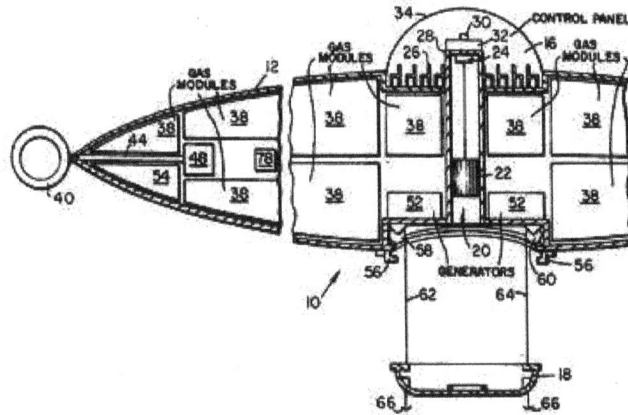

USP # 3,933,325 (1-20-76) ~ Disc-Shaped Aerospacecraft
Kaelin, Joseph R.

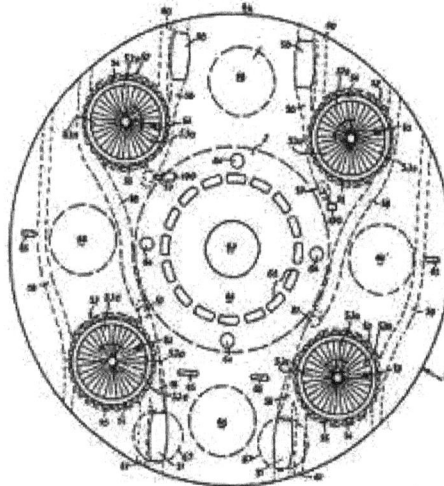

USP # 3,871,602 (3-18-75) ~ Circular Wing Aircraft
Kissinger, Curtis D.

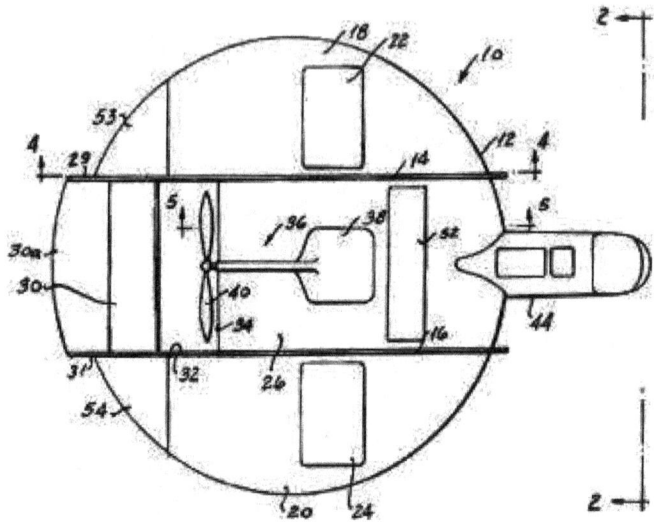

USP # 3,774,865 (11-27-73) ~ Flying Saucer
Pinto, Olympio F.

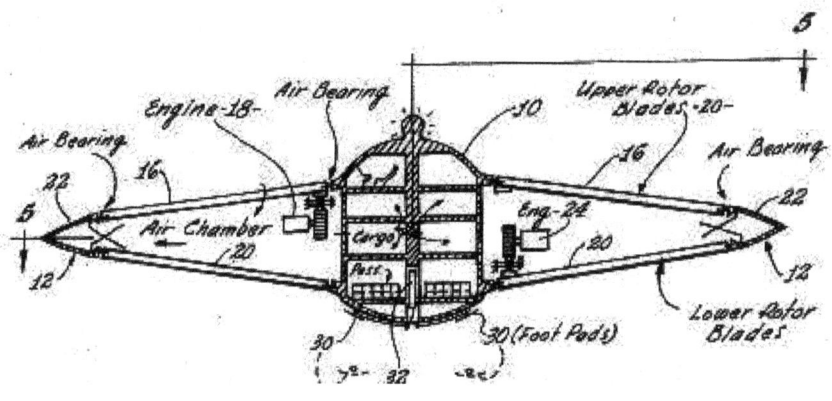

USP # 3,750,980 (8-7-73) ~ Aircraft with VTOL Capability
Edwards, Samuel L.

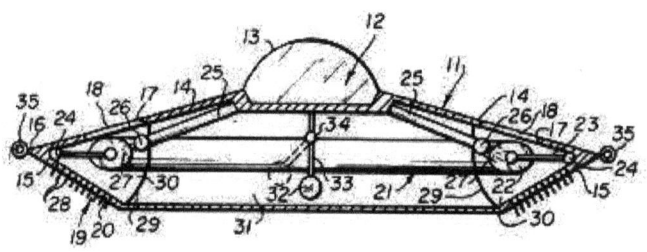

USP # 3,697,020 (10-10-72) ~ Vertical Lift Machine
Thompson, Raymond V.

120

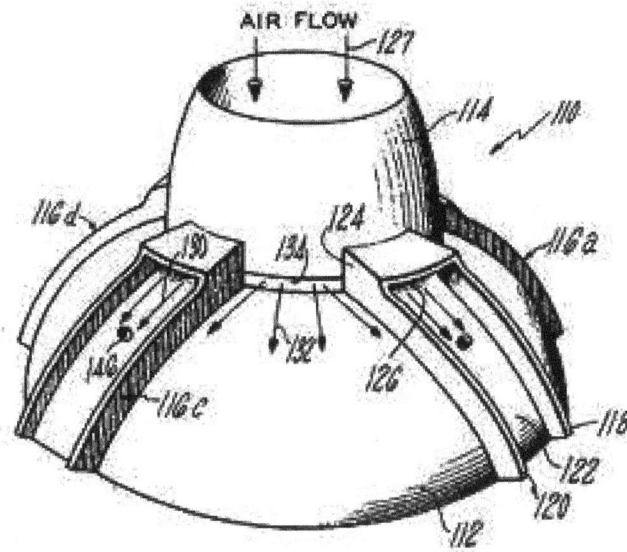

USP # 3,690,597 (9-12-72) ~ VTOL Aircraft...
Di Martino, Renato

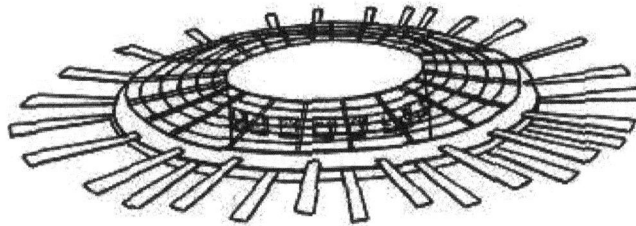

USP # 3,640,489 (2-8-72) ~ VTOL Aircraft
Jaeger, Karl

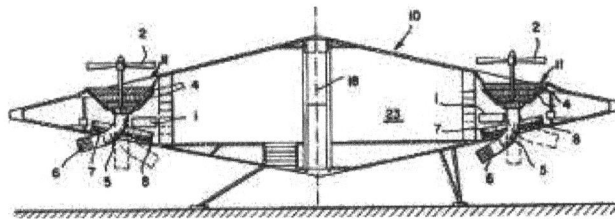

USP # 3,630,470 (12-28-71) ~ VTOL Vehicle
Elliot, Frederick T.

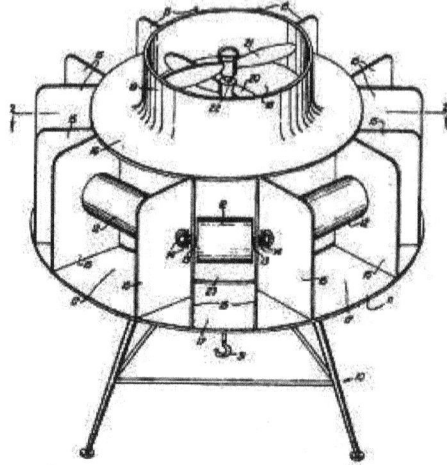

USP # 3,614,030 (10-19-71) ~ Aircraft
Moller, Paul S.

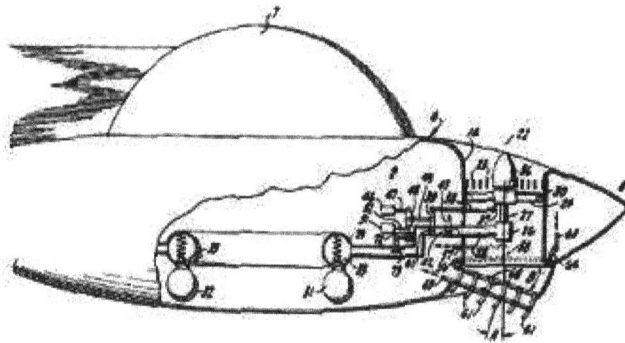

USP # 3,612,445 (10-12-71) ~ Lift Actuator Disc
Phillips, Duan A.

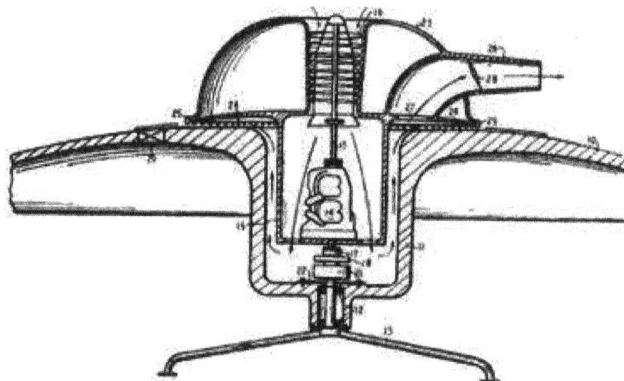

USP # 3,599,902 (8-17-71) ~ Aircraft
Thomley, John W.

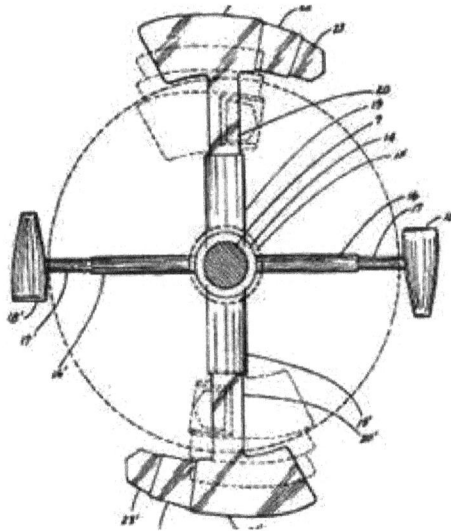

USP # 3,537,669 (11-3-70) ~ Manned Disc-Shaped Flying Craft
Modesti, James N.

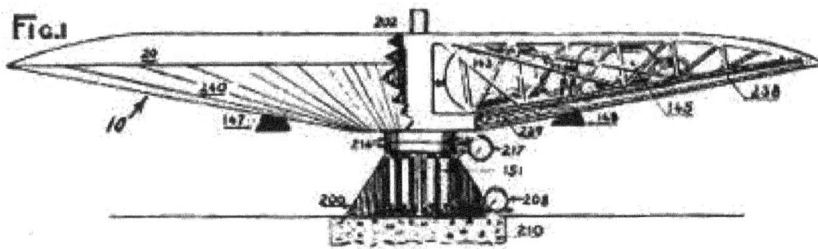

USP # 3,503,573 (3-31-70) ~ Disk Flying Craft
Modesti, James N.

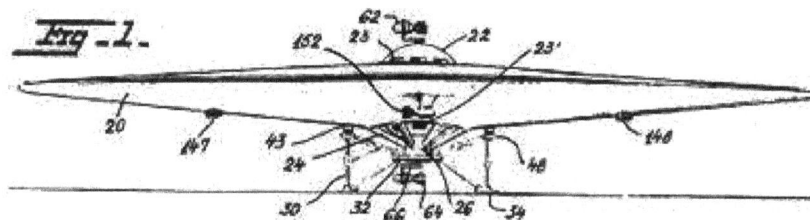

USP # 3,469,802 (9-30-69) ~ Transport
Roberts, J. R., *et al.*

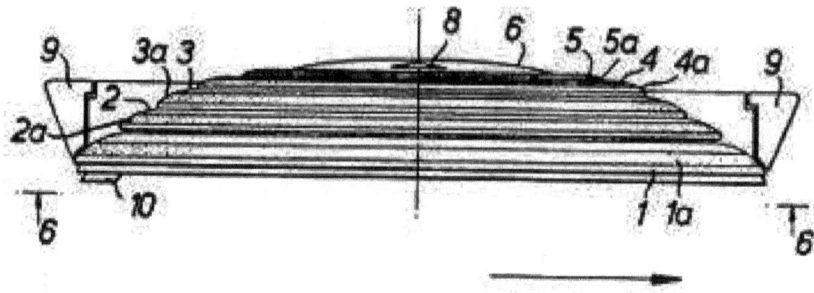

USP # 3,437,290 (4-8-69) ~ Vertical Lift Aircraft
Norman, Francis A.

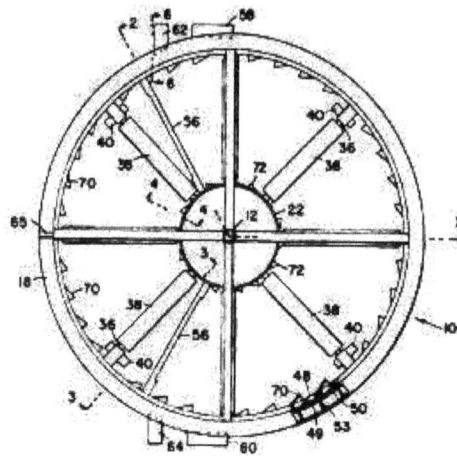

USP # 3,432,120 (4-11-69) ~ Aircraft
Guerrero, E.

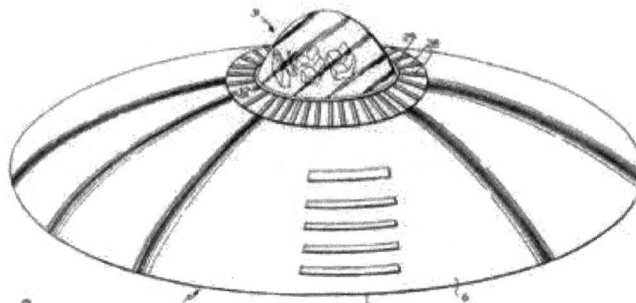

USP # 3,410,507 (11-12-68) ~ Aircraft
Moller, Paul S.

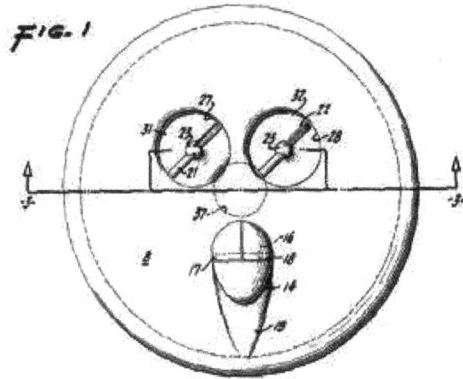

Fig. 1

USP # 3,397,853 (8-20-68) ~ Fluid-Sustained Vehicle
Richardson, William. B.

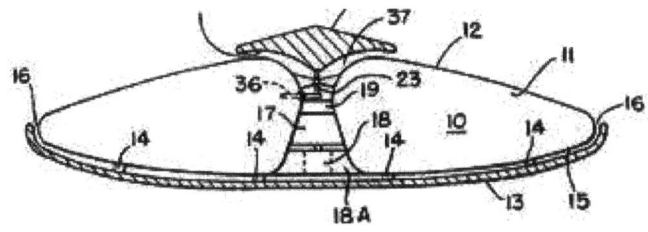

USP # 3,395,876 (8-6-68) ~ Aircraft with Housed Counter-Rotating Propellers
Green, Jacob B.

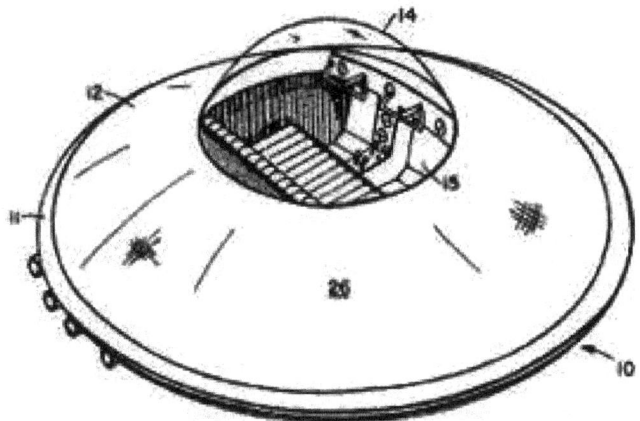

USP # 3,387,801 (6-11-68) ~ VTOL Aircraft
Kelsey, C. W.

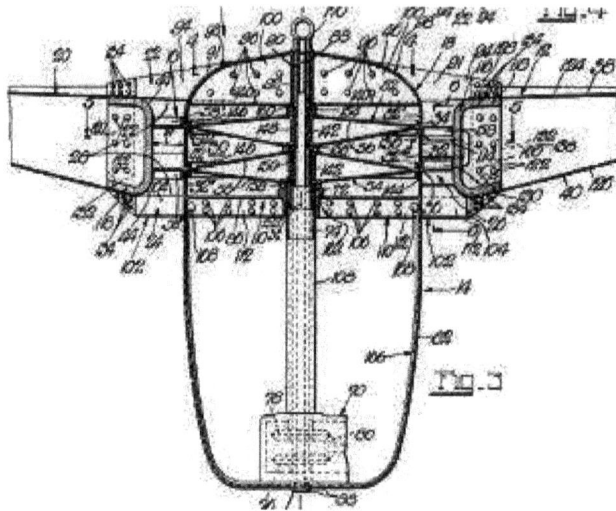

USP # 3,321,156 (5-23-67) ~ Universally Maneuverable Aircraft
McMasters, Douglas Q.

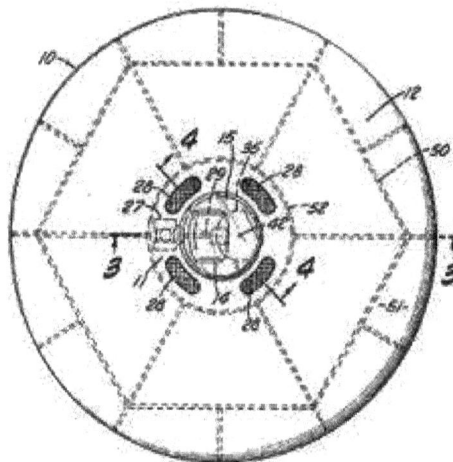

USP # 3,312,425 (4-4-67) ~ Aircraft
Le nnon, C. D., *et al.*

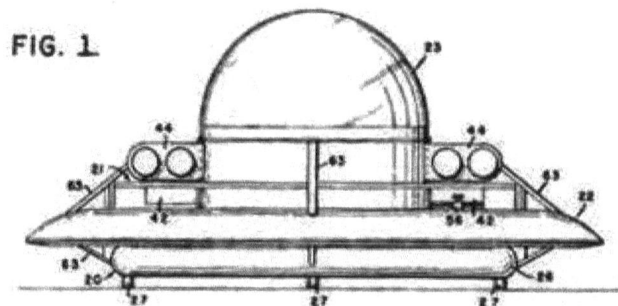

USP # 3,243,146 (3-29-66) ~ VTOL Aircraft
Clover, P. B.

126

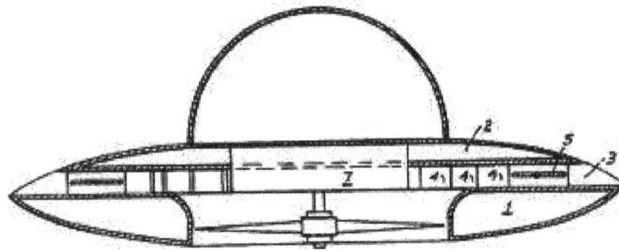

USP # 3,237,888 (3-1-66) ~ Aircraft
Willis, William M.

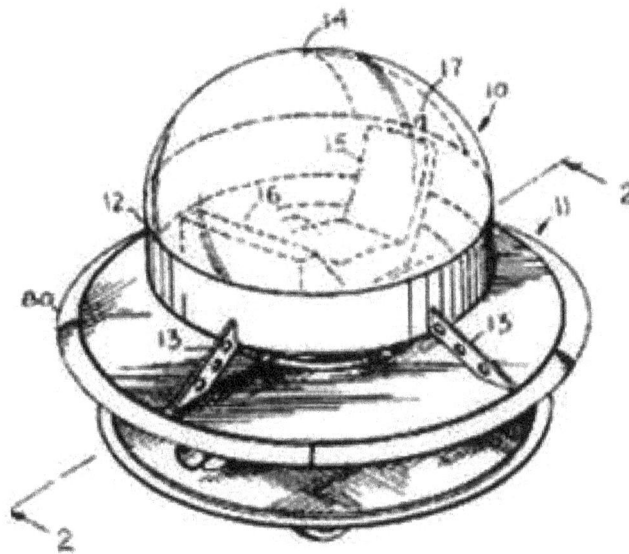

USP # 3,199,809 (8-10-65) ~ Circular Wing Flying Craft
Modesti, James N.

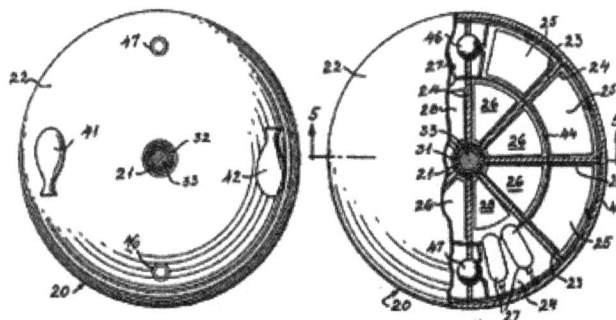

USP # 3,182,929 (5-11-65) ~ VTOL Aircraft
Lemberger, Robert A.

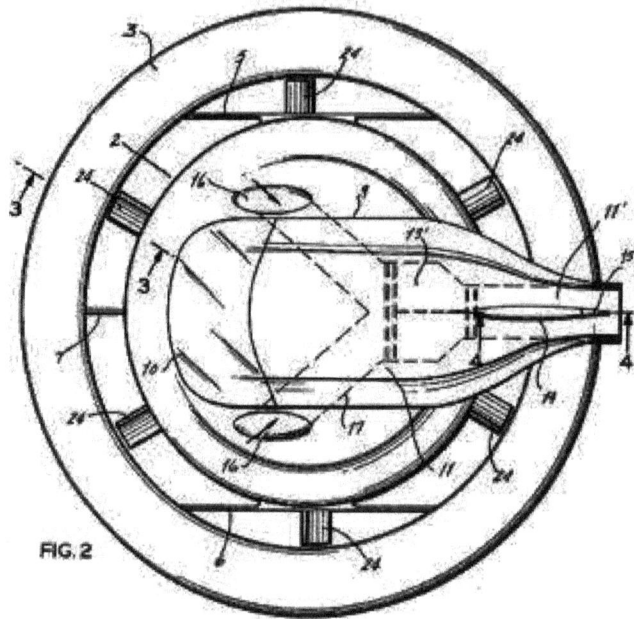

FIG. 2

USP # 3,124,323 (3-10-64) ~ Aircraft Propulsion & Control
Frost, John C. M.

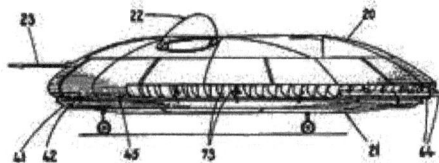

FIG.1

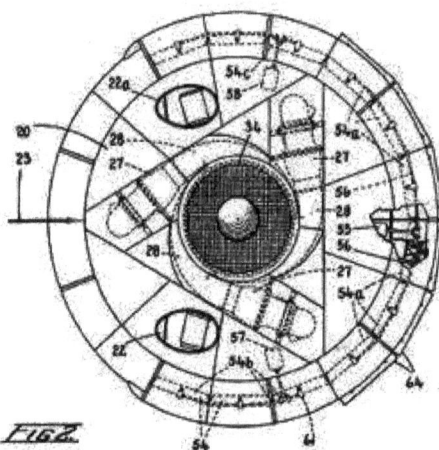

FIG.2

USP # 3,123,320 (4-3-64) ~ Vertical Rise Aircraft
Slaughter, E. E.

128

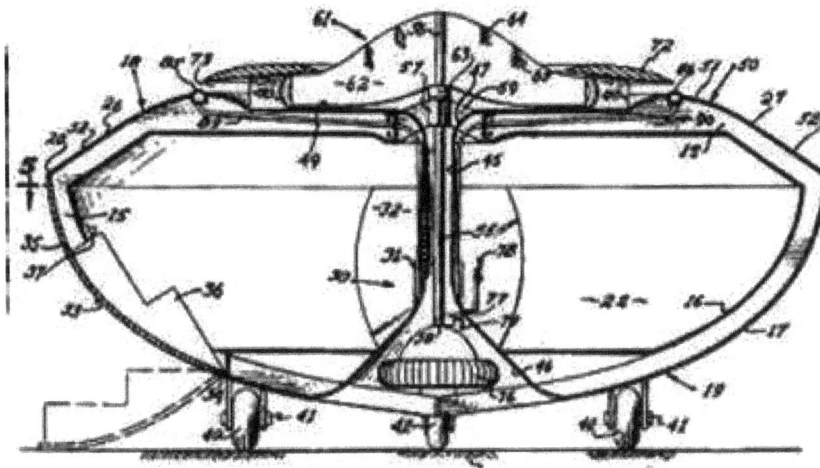

USP # 3,073,551 (1-15-63) ~ Vertical Lift Aircraft
Bowersox, Joseph W.

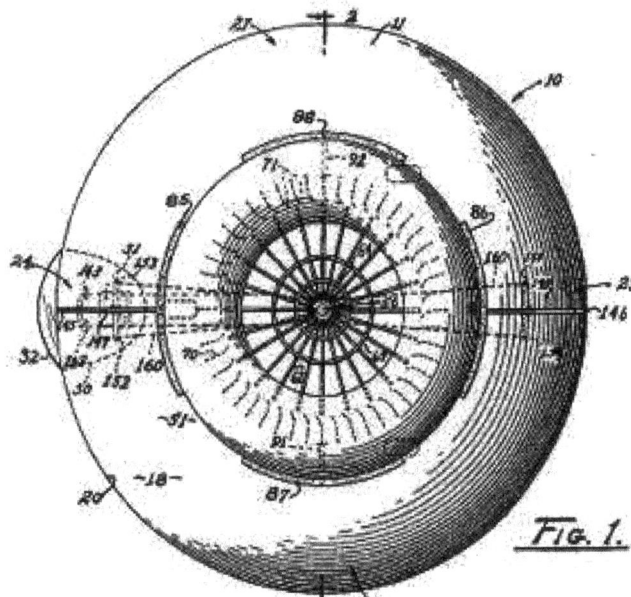

Fig. 1.

USP # 3,072,366 (1-8-63) ~ Fluid-Sustained Aircraft
Freeland, Leonor Z.

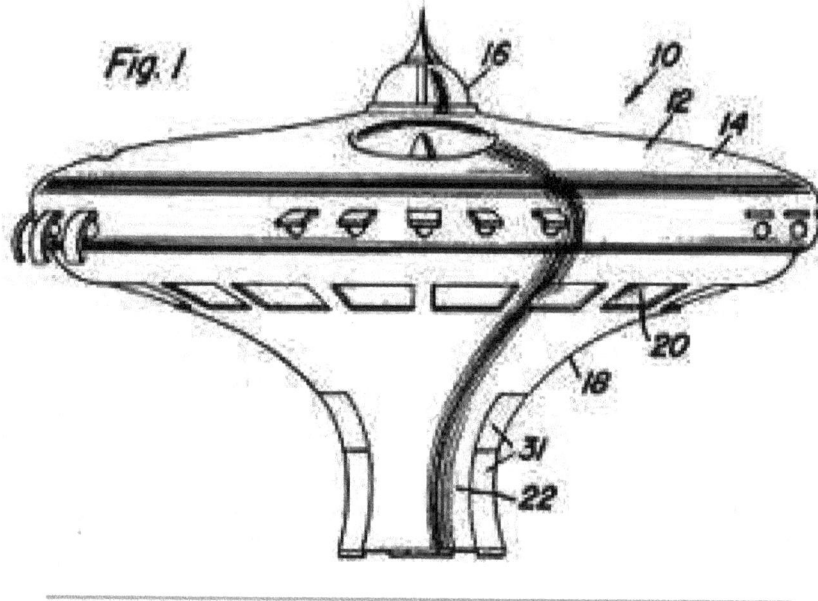

USP # 3,067,967 (12-11-62) ~ Flying Machine
Barr, I. R.

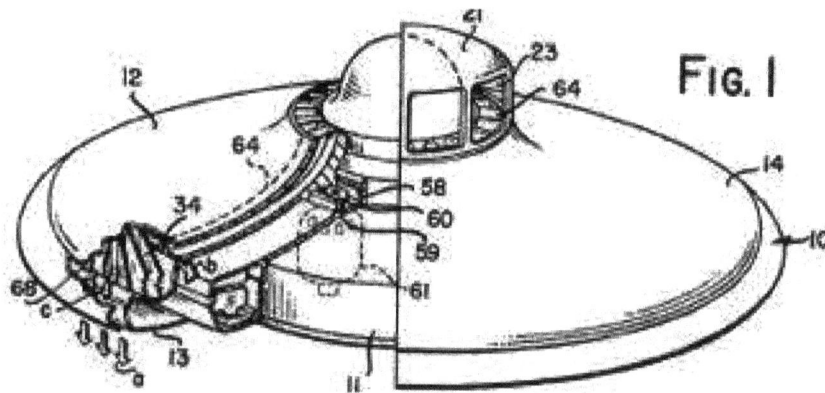

USP # 3,066,890 (12-4-62) ~ Supersonic Aircraft
Price, Nathan C.

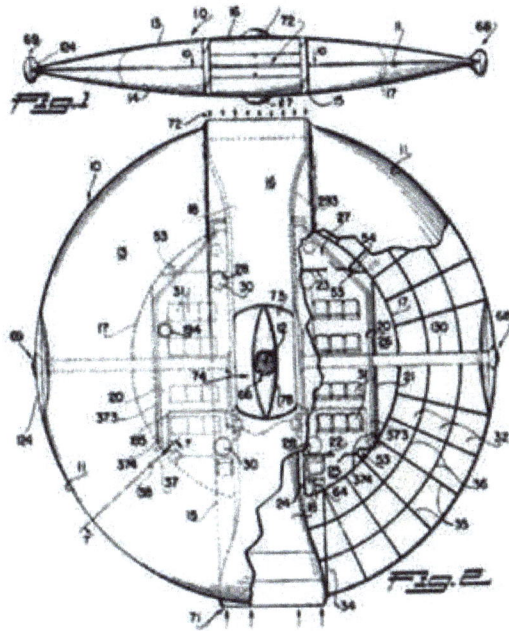

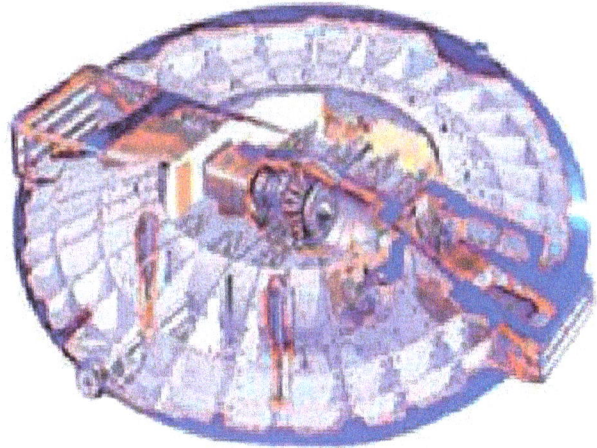

USP # 3,065,935 (11-27-62) ~ VTOL Aircraft
Dubbury, J., *et al.*

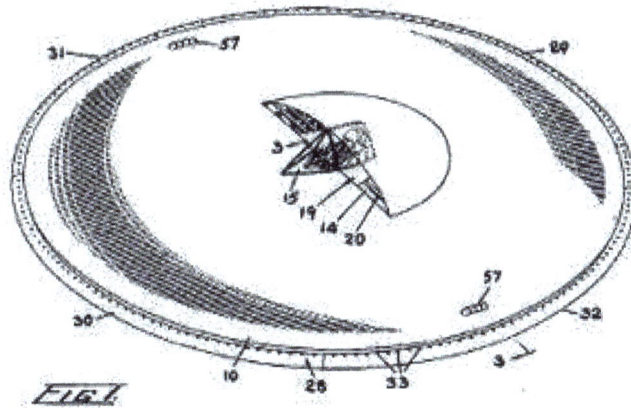

Frost, John C. M., *et al.*

USP # 3,051,415 (8-28-62) ~ Fluid-Sustained Aircraft
Frost, John C. M.

USP # 3,051,414 (8-28-62) ~ Aircraft with Jet Fluid Control Ring
Frost, John C. M.

USP # 3,024,966 (3-13-62) ~ Radial Flow Gas Turbine Engine Rotor Bearing
Frost, John C. M.

USP # 3,022,963 (2-27-62) ~ Disc-type Aircraft...
Frost. John C. M., *et al.*

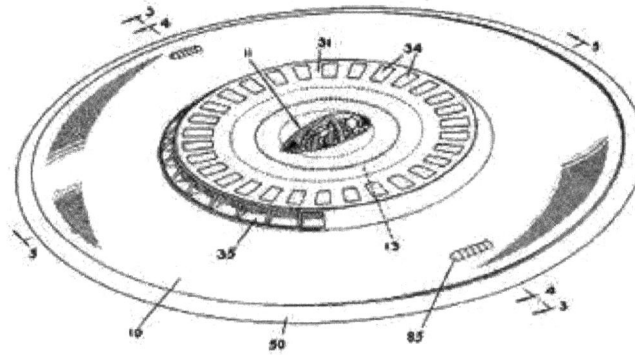

USP # 3,020,003 (2-6-62) ~ Disc Aircraft...and USP # 3,018,068 (1-23-62) ~ Disc Aircraft...
Frost, John C. M., *et al*

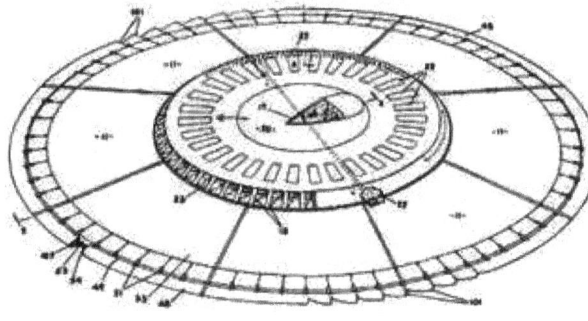

USP # 3,020,002 (2-6-62) ~ VTOL Control
Frost, John C. M.

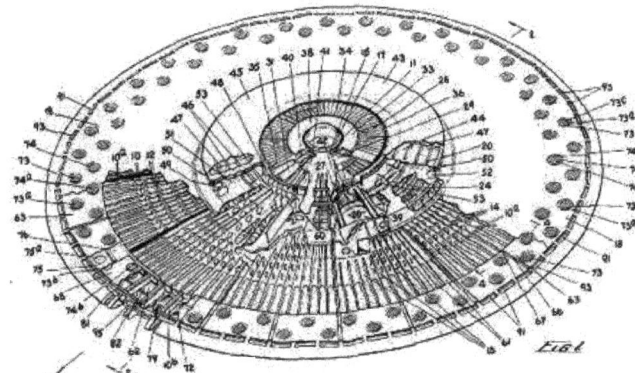

USP # 2,997,254 (8-22-61) ~ Gyro-Stabilized Vertical Rising Vehicle (Discoid)
Mulgrave, Thomas P., *et al.*

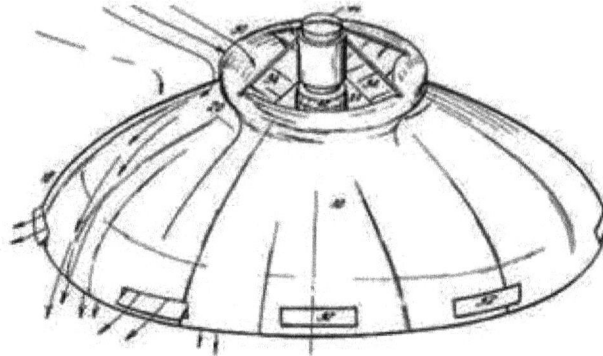

USP # 2,988,303 (6-13-61) ~ Jet-Sustained Aircraft
Coanda, Henri

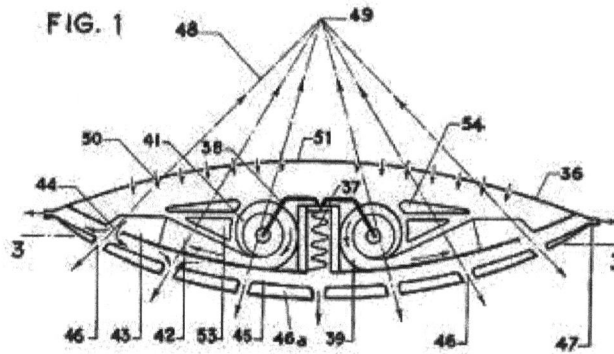

USP # 2,953,320 (9-20-60) ~ Aircraft with Ducted Lifting Fan
Parry, Robert D.

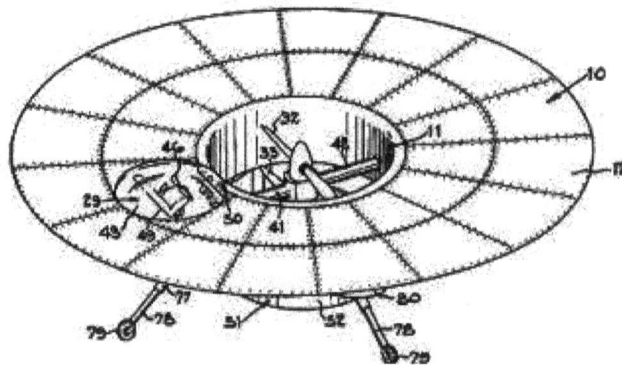

USP # 2,944,762 (7-12-60) ~ Aircraft
Lane, Thomas R.

USP # 2,939,648 (6-7-60) ~ Rotating Jet Aircraft with Lifting Disc Wing...
Fleissner, H.

USP # 2,937,492 (5-24-60) ~ Rotary Reaction Engine
Lehberger, Arthur N.

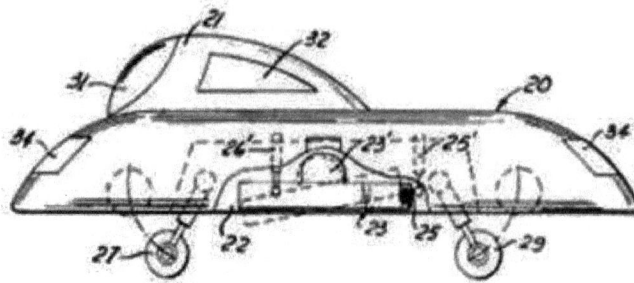

USP # 2,935,275 (5-3-60) ~ Disc-Shaped Aircraft
Grayson, Leonard W.

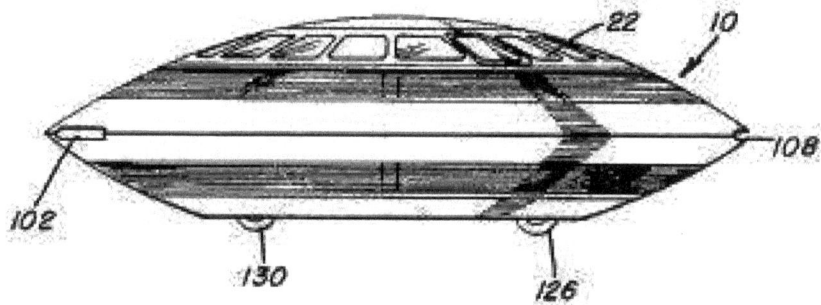

USP # 2,927,746 (3-8-60) ~ Toroidal Aircraft
Mellen, Walter R.

136

USP # 2,918,230 (12-22-59) ~ Fluid-Sustained & Fluid-Propelled Aircraft
Lippisch, Alexander M.

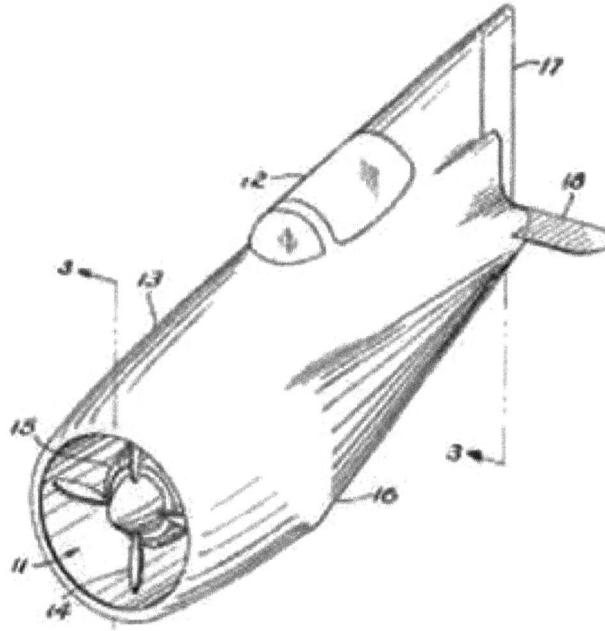

USP # 2,876,965 (3-10-59) ~ Circular Wing Aircraft...
Streib, Homer F.

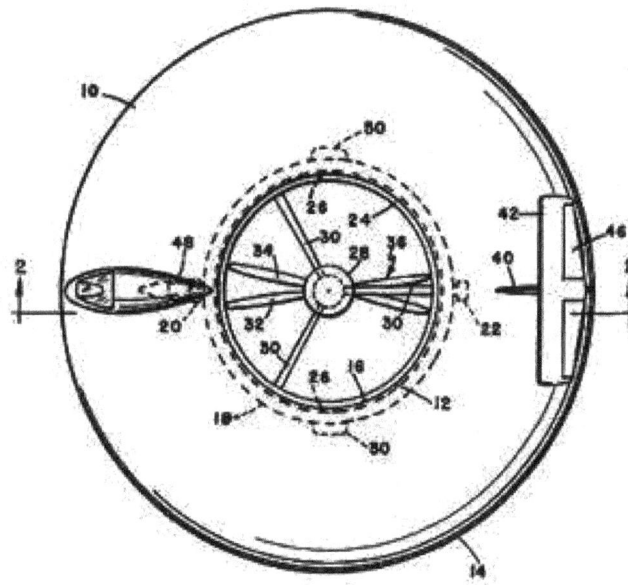

USP # 2,863,621 (12-9-58) ~ Vertical & Horizontal Flight Aircraft
Davis, John W.

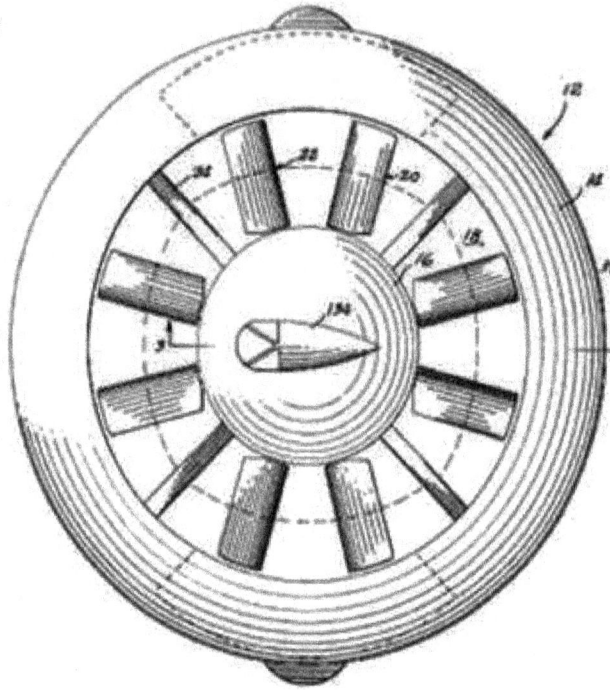

USP # 2,801,058 (7-30-57) ~ Saucer-Shaped Aircraft
Lent, Constantin P.

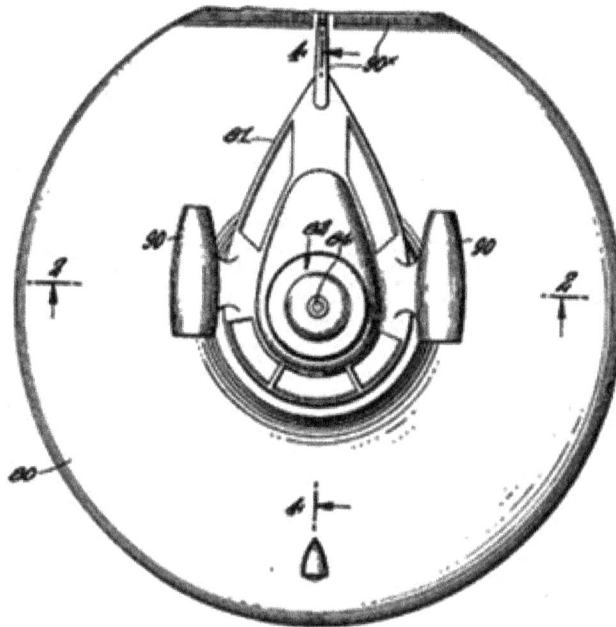

138

USP # 2,777,649 (1-15-57) ~ Fluid-Sustained Aircraft
Williams, Samuel B.

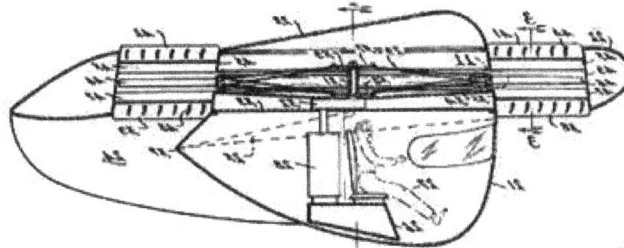

USP # 2,772,057 (11-27-56) ~ Circular Aircraft &c...
Fischer, John C.

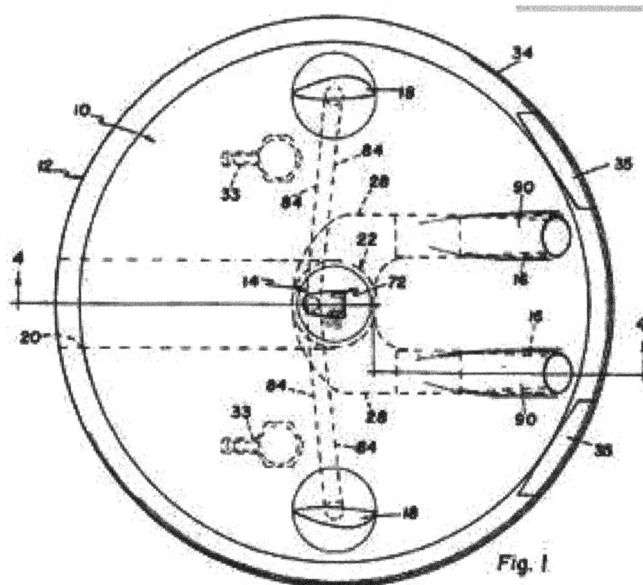

Fig. I

USP # 2,736,514 (2-28-56) ~ Convertible Aircraft
Ross, Robert S.

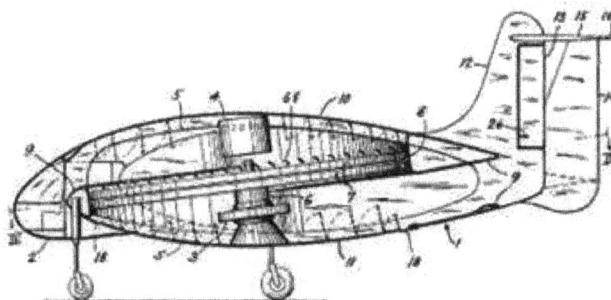

USP # 2,730,311 (1-10-56) ~ Impeller Propelled Aerodynamic Body
Doak, Edmond R.

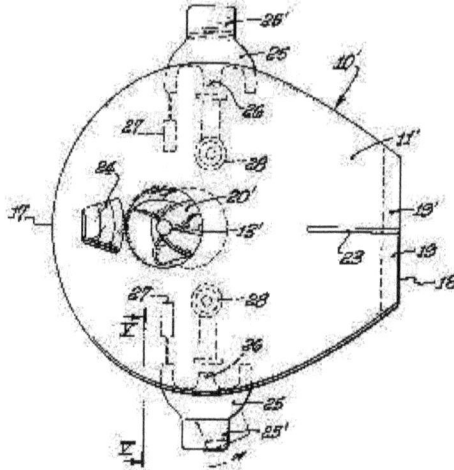

USP # 2,718,364 (9-20-55) ~ Fluid-Sustained & Propelled Aircraft...
Crabtree, E.L.

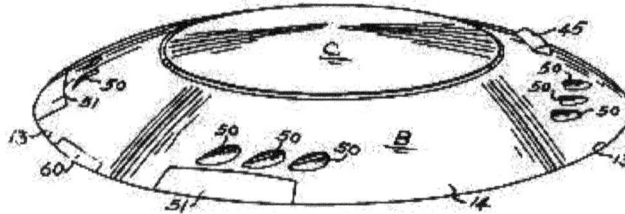

USP # 2,619,302 (11-25-52) ~ Low Aspect Ratio Aircraft
Loedding, Alfred C.

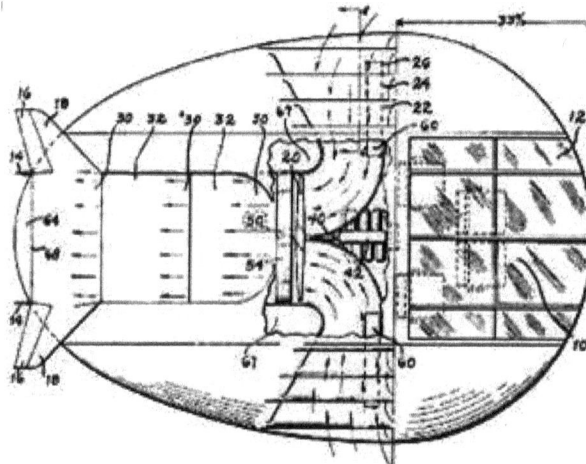

USP # 2,567,392 (9-11-51) ~ Fluid-Sustained Aircraft
Naught, Harold

USP # 2,431,293 (11-18-47) ~ Airplane of Low Aspect Ratio
Zimmermann, Charles H.

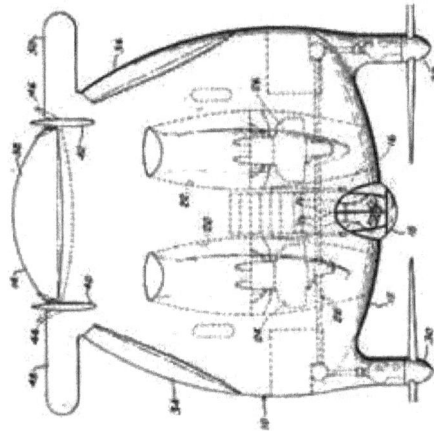

USP # 2,377,835 (6-5-45) ~ Discopter
Weygers, Alexander G.

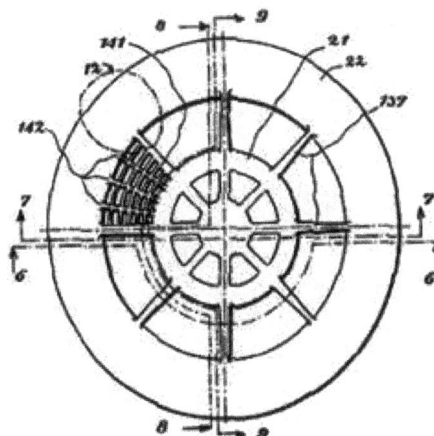

USP # 1,887,411 (11-8-32) ~ Aircraft Construction
Johnson, R. B.

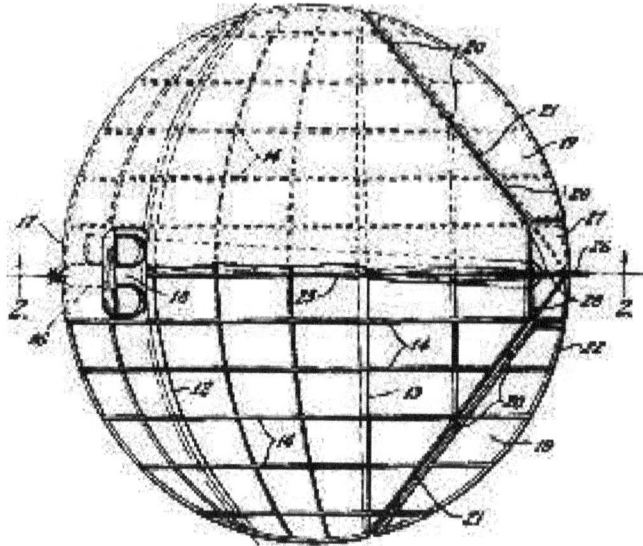

USP #5653404 A and USP #5653404

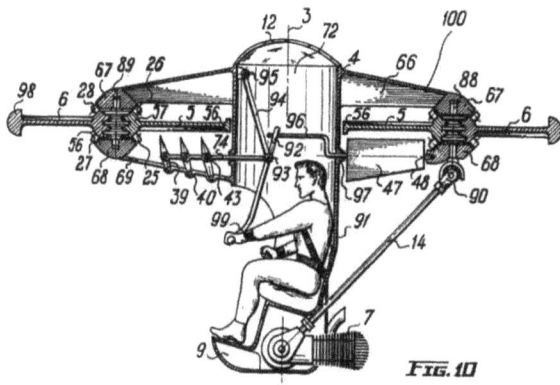

FIG. 10

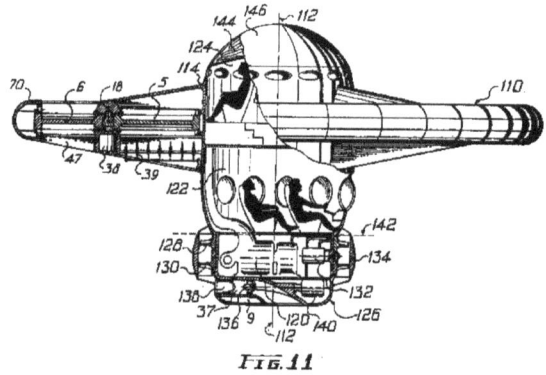

FIG. 11

USP # 3062482 ~ Gas Turbine Engined Aircraft
J. C. M. Frost

142

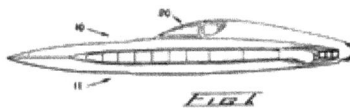

Nov. 6, 1962   J. C. M. FROST   3,062,482

GAS TURBINE ENGINED AIRCRAFT

Filed Aug. 25, 1953   8 Sheets-Sheet 1

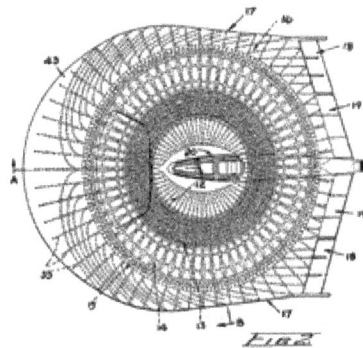

INVENTOR.
J.C.M.FROST
BY
Maybee & Legris
ATTORNEYS.

USP #4193568

U.S. Patent   Mar. 18, 1980   Sheet 3 of 5   4,193,568

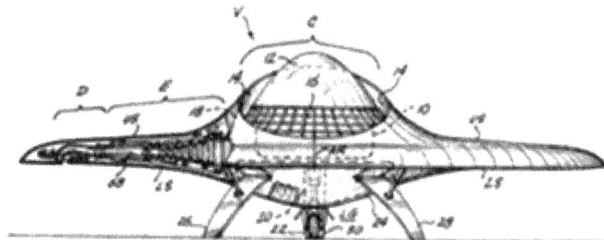

Fig. 3.

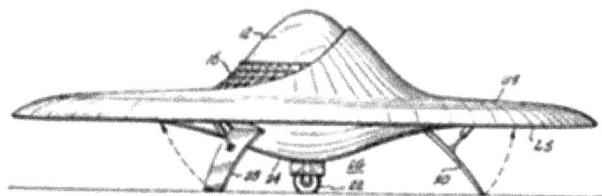

Fig. 4.

USP# 4,663,932

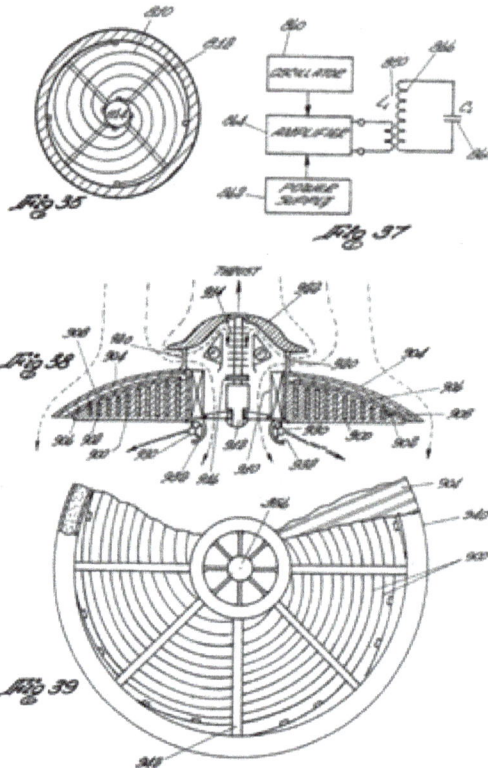

Designed by Frank G Young of New York, in 1972, the engine (5) of this vertical take-off "air and space craft" folded down, enabling ascent or descent "similar to... a falling maple seed".

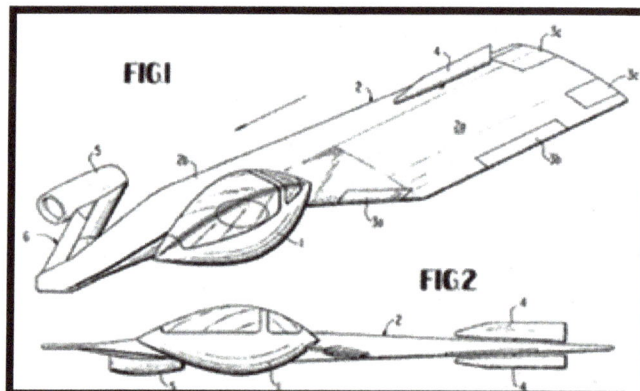

**A truly bizarre single wing jet aircraft with a cockpit pod carried on the side.
It probably flew in helicopter-type fashion much like a falling Maple or Box Elder seed**

All of the above patents can be found at: http://www.rexresearch.com/wingless/wingless.htm

144

Here are some of the types of propulsion systems and the basic operation of some of the saucer shape craft from the patents above to give a flavour of the inventiveness of the designs as well as a look into the mindset of their inventors.

Induction Lift Flying Saucer

This invention USP # 4,824,048 (4-25-89) by Kyusik Kim, (from above patents) relates to the propulsion system of an aircraft. It utilizes a liquid fuel pre-vaporization and back burning induction jet oval thrust nozzle which is fitted onto the exit nozzle of a conventional turbojet engine having a ram constriction air inlet plenum-engine pod located forward of the aerodynamic generating channel. The aerodynamic generating channel is located forward and above a vacuum cell induction lift wing and below recycling air inductor vanes.

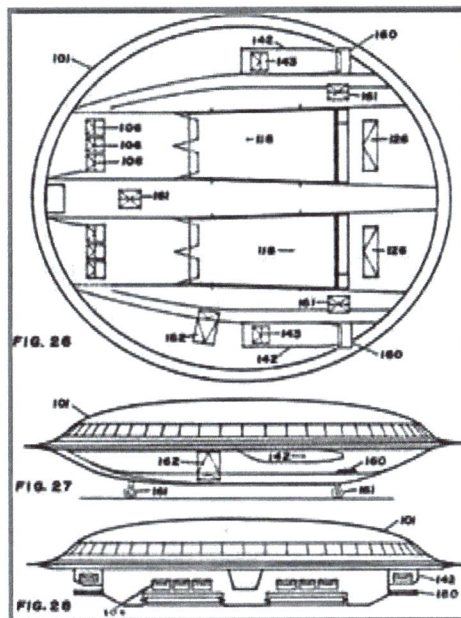

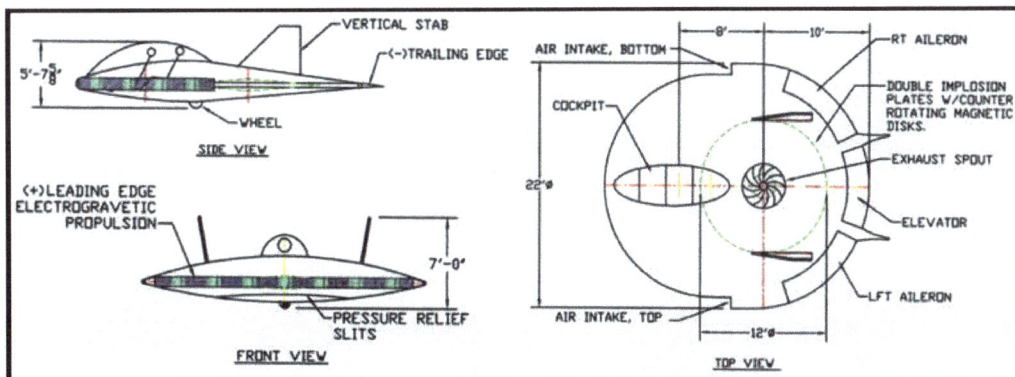

A saucer shape design with vertical stabilizers, aileron, and elevator with cockpit near the leading edge and exhaust spout off center toward the trailing edge

https://www.google.com/patents/US4824048

Circular Wing Flying Saucer

Inventor: **Brunner, Ashton Frank (Bush, LA)**
The ornamental design for a circular wing flying saucer, as shown and described. U.S. design patent D487, 048 invented by Brunner.

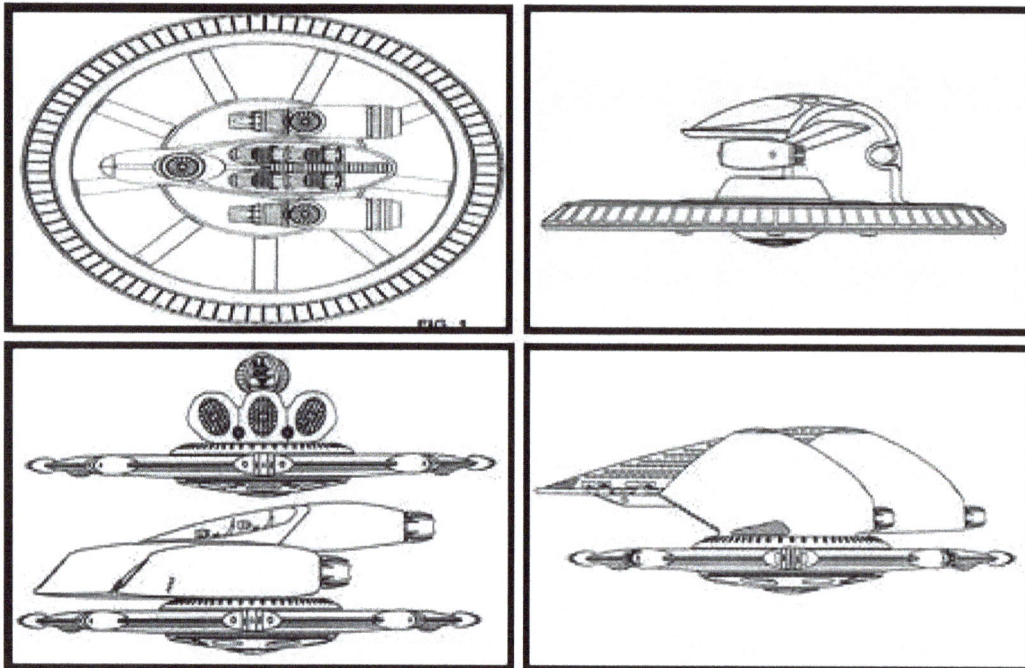

http://www.wikiwand.com/id/Klasifikasi_wahana_UFO and http://www.google.co.zm/patents/USD541206 and http://www.google.mu/patents/USD471507 and http://www.google.mu/patents/USD484452

Gyro Stabilized Flying Saucer Model

A model having a flying saucer shaped body which is provided with lift by means of a thrust producing device such as a reciprocating Wankel or turbine engine and a propeller. The body is prevented from rotating by means of counter rotational fins, and stability in the horizontal plane is provided by means of an internal gyro rotor actuated by the airflow from the propeller.

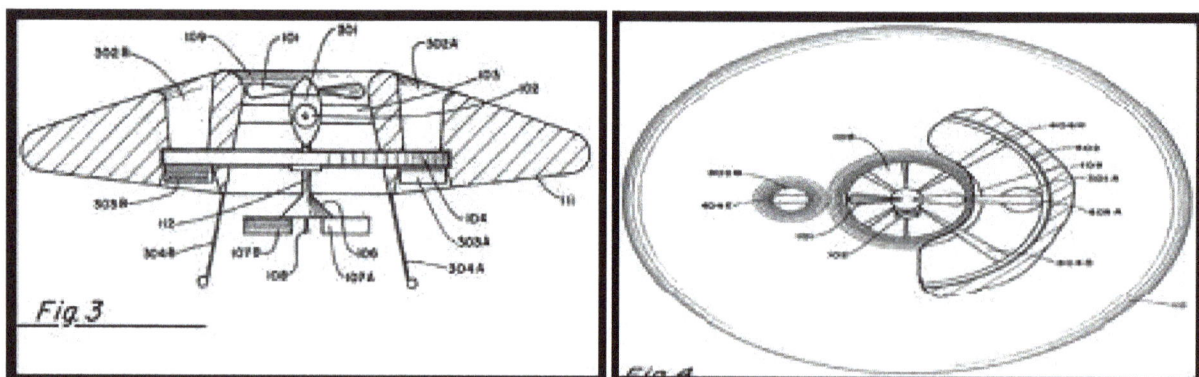

https://www.google.com/patents/US4461436

British Rail Flying Saucer from Flying Saucers to Vacuum-Sealed Trains

British futurist inventors back in the early 1970s foresaw an era where as these patents from the **European Patent Office** show, there were grand plans for the future of transportation for both a flying saucer and as a part of the **British Rail system** patent in 1973 .

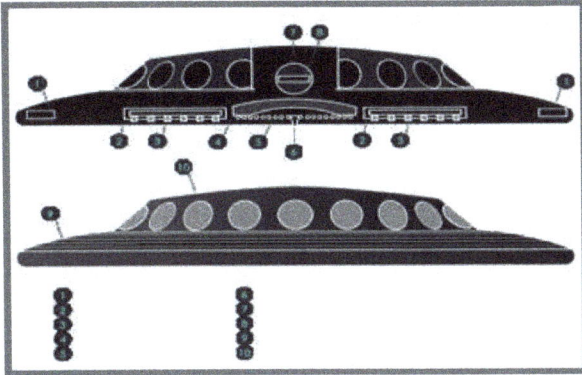

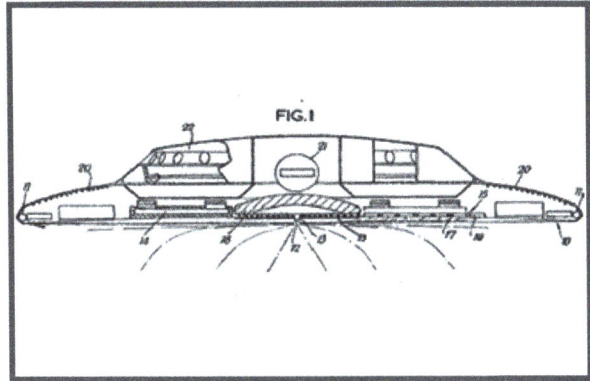

Vertical Take-off Flying Platform

Patent No. 2,953,321 - The **Vertical Take-off Flying Platform** invented by Robertson, Stuart and Wagner was part of US Army attempt for single personnel transport system which works exceedingly well but never put into production. (See below).

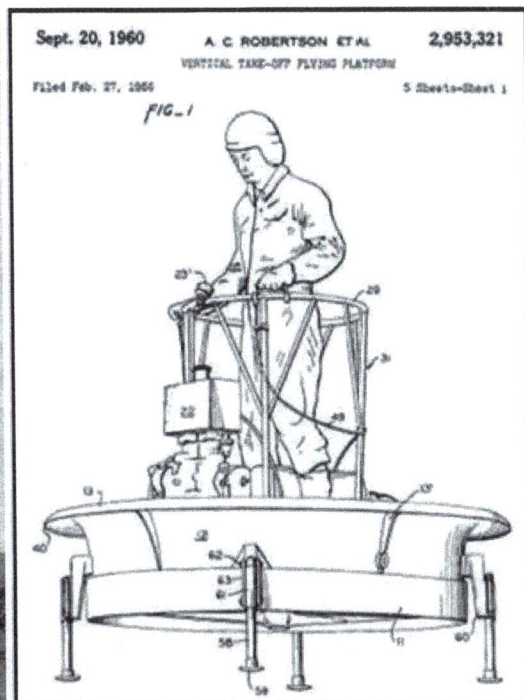

CHAPTER 97

MANMADE FLYING SAUCERS FROM
THE BLACK WORLD OF SCIENCE

With the voluminous sightings worldwide of **UFOs (unidentified flying objects),** any researcher investigating this subject must give serious consideration that not all saucer shape craft may be of Extraterrestrial in origin. From the many patents above and the many photos taken of disc shape craft, a good percentage of reports may, in fact, be describing manmade flying saucers. "Patents from the US patent office and other nations throughout the world show that people have at least been attempting to build craft that match the description of these so-called "flying saucers." http://alienufoparanormal.aliencasebook.com/2008/07/20/flying-saucer-patents--have-you-designed-yours-yet.aspx

What will become apparent as the reader goes through this subsection is the fewness in numbers of photographic evidence of any black science flying saucers? The reason is the very deep covert and secret world in which the evidence is hidden and that what little is known in the public domain of the everyday world has been leaked by insiders or whistleblowers. Getting this information out of the labyrinthian black vaults and off the black shelves into an official disclosure venue will take an act of God!

The **Avro Car**, although, less of a black world of science project did have various aspects of the program that were initially black and kept from the public limelight. The Avro Car was developed as a saucer shaped craft that the US Army could use as an airborne geep carrying two or more passengers utilizing a ducted fan to achieve its lift by means of internal jet turbines. It was designed to take off and land in a vertical manner and travel at high velocities, however, it was limited by its own ground effect (the **Coanda Effect**, whereby the air pushed down creates a cushion below the craft that it sits on) thus, the research project was discontinued.

The Avro Car
http://www.diseno-art.com/encyclopedia/strange_vehicles/avrocar.html

While the above patent utilizes known technology, the craft would still have the appearance and maneuverability of that ascribed to UFOs. However, there is more than just a single patent utilizing known technology. There are patents for things like plasma propulsion and magneto-hydrodynamic propulsion. It operates by sending high frequency, high voltage electricity from the top of the craft to the sides or the bottom of the craft. This approach was utilized in Nazi saucers during WWII. http://alienufoparanormal.aliencasebook.com/2008/07/20/flying-saucer-patents--have-you-designed-yours-yet.aspx

It should also be mentioned that Nazi scientists utilized many concepts and methods to develop the perfect saucer aircraft and most of the designs were an evolutionary process incorporating all the concepts and theories of propulsion and aerodynamics into one synergistic ultimate flying saucer craft. These became the first real black programs and projects developed for wartime use.

It is now accepted that Nazi scientists actually built and flew a disc-shaped craft before the end of WWII and if so, where is the prototype craft, today?
http://discaircraft.greyfalcon.us/HAUNEBU.htm

"One thing must be considered and that is the two ways a patent can be validated. The first is with a working model and the second is it is logically provable using mainstream science. The technology powering the craft that are seen doing amazing things has been patented and is mostly within the public domain now. The craft exists and the technology does as well. It is only a matter of time before it is utilized and mankind moves into the space-age."
http://alienufoparanormal.aliencasebook.com/2008/07/20/flying-saucer-patents--have-you-designed-yours-yet.aspx

Any attempt of a search of black projects related to saucer or disc shape craft will more than likely come up short as the covert Black world of Science does not publish or seek patents for their secret aircraft or spacecraft. These types of projects are **unacknowledged** or **special access projects** known only to the military and a few contractors. Whatever has come into the public domain has been from **Freedom of Information** requests which usually has significant amounts of the documents "blacked-out" or the information has been leaked from an inside source who felt that the public should know what is going on in the hidden world of black science. What we do know in Ufology has also distilled down through history as much as five or six decades later and in some cases, there are still wartime secrets from WWII that have not yet been revealed.

We know that German scientists were operating in a feverish state of research and development in aviation to fulfill Hitler's request for **Wunder Waffen** *("Wonder Weapons")* to bring ultimate victory for Germany over its enemies. **Henry Stevens** has written on this subject in his book, "Hitler's Flying Saucers" as has **Joseph Farrell** in his book "The SS Brotherhood of the Bell", as well as many other Ufologist historians providing names of German Nazi scientists, their saucer projects, their secret or hidden locations of development as well as some sketches, plans and photographs of the completed prototype flying discs. The reader should review the section in this book on Nazi saucer projects discussed earlier.

Italian researcher **Renato Vesco**, in his classic and well-researched book, "Intercept – But Don't Shoot: The True Story of the Flying Saucers" asserts that the Nazis were working on many advanced propulsion systems and rudimentary anti-gravity devices to power their disc-shaped, or lenticular, aircraft. Intercept UFO by Renato Vesco (Originally published as a Zebra Book ib1971 under the title "Intercept – But Don't Shoot"); 1974; published by Zebra Publications, Inc., ISBN 0-8468-0010-1

When word reached the Führer in 1937 that a crash saucer had been found in the German countryside only the best military and scientific minds were considered for research and development that would make the breakthroughs required from back engineering of the saucer.

Although vastly influenced by the capture of an actual disc, research into circular aircraft had been going on as far back as the experiments in Italy before 1920 with crude jets attached to aerodynamic discs. These Nazi saucer projects were deep black programs that were highly compartmentalized and rigidly guarded. Hitler's Flying Saucers by Henry Stevens; 2003; published by Adventures Unlimited Press; Kempton, Illinois, USA; ISBN 1-931882-13-4

The ET craft served to advance propulsion, electronics and provided clues to workable designs, but the scientists were at a loss regarding the metallurgy as the craft was composed of materials unknown to them and impossible to duplicate in the lab.

Some of Germany's most advanced theoretical engineers and physicists were personally tasked by Hitler to get the technology working and make it available for the war effort. In that regard, geniuses like the **Horton brothers**, **Ballenzo**, **Habermohl**, **Miethe,** and **Schriever** were recruited.

Later, the expertise of **Viktor Schauberger** was tapped. He was the inventor of the revolutionary imploder motor that created an imploding vortex. That motor may have been the basis for later S.S. experiments in Poland with the notorious **Glocke (Bell) device** that reportedly created inter-dimensional rifts in space-time. The S.S. Brotherhood of the Bell: The Nazis' Incredible Secret Technology by Joseph P. Farrell; 2006; published by Adventures Unlimited Press; Kempton, Illinois, USA; ISBN 1-931882-61-4

150

Artist's interpretation of first Nazi disc built by Schriever
http://www.roswellufomuseum.com/research/ufotopics/naziufocrash.html

Documents captured by the Allies after the war ended indicate that Schriever was the first to have some success with the disc technology. His own research, augmented by what had by then been deduced from the retrieved saucer, enabled him to build a working craft powered by specially designed jet engines. The craft, however, was unstable. The first two versions ended in disastrous crashes killing the test pilots. **Man-Made UFOs, 1944-1994: Fifty Years of Suppression by Renato Vesco and David Hatcher Childress; 1994; published by Adventures Unlimited Press; Stelle, Illinois, USA; ISBN 0-932813-23-2** and http://www.roswellufomuseum.com/research/ufotopics/naziufocrash.html

As the years advanced and the Allies began counter-attacking Germany, some Nazi officials began pressuring Schauberger to adapt a version of his **Repulsine imploder** to serve as the motive force for a different type of disc craft using the vortex propulsion motor.

Although the Nazi engineers and scientists built their first experimental saucers from light steel or heavy aluminum, the steel still proved too heavy and the aluminum too soft and not tough enough to withstand the stress subjected to it when molded into a lens-shaped craft.

Frustrated scientists finally gave up trying to unravel the mystery of the alien metal and succeeded in creating their own version with an alloy of aluminum and magnesium in 1944. The new alloy was exactly what they needed to design bigger and better craft. The metal was light, yet more durable than simple aluminum.

According to documents and the testimony of surviving Polish slave laborers, the Germans actually created an assembly line to manufacture a flying disc weapon called the Kugelblitz (ball lightening).

The Allied pilots called them **Foo-fighters.**

The **Kugelblitz** was the first real flying saucer ever manufactured by any nation on Earth. Several of the tests were secretly attended by Hitler and his S.S. chief, **Heinrich Himmler**. http://www.roswellufomuseum.com/research/ufotopics/naziufocrash.html

151

According to captured documents, the radio-controlled craft were made in the underground factories at Thuringia and the craft varied in size from 10 to 15-feet across. Reportedly they were amazingly maneuverable and were able to achieve speeds of more than 1,250 miles per hour. They emitted a strong electrostatic field designed to disrupt the electrical circuits of conventional aircraft causing enemy planes to falter, dive, and crash.

Prototypes of other saucers were designed and engineered mock-ups were made. Some made it as far as wind tunnel tests. A few disc craft other than Schriever's early attempts were constructed and flown.

S.S. Reichsführer Himmler recommended **S. S. Grupenführer Hans Kammler** to head the saucer projects and save the Fatherland. Hitler agreed. Later Kammler is said to have overseen the Glocke project as well.

Undoubtedly, Germany's super weapons—specifically those they were perfecting in the closing months of the war—would have defeated the Allies and helped Germany conquer and rule Western and Eastern Europe, suppress the U.S.S.R. and keep America neutered—even with the new atomic weapon the U.S. had developed.

But the Germans ran out of time.
http://www.roswellufomuseum.com/research/ufotopics/naziufocrash.html

Historically, this was the world's first look into the operations of the black world of science initiated by Nazi Germany in the late "30s and early '40s and later was subsumed into America's own military-industrial complex and into a few other post-war allied countries due largely in part by the capture of 1500 Nazi scientists from Germany through **Operation Paperclip.**

Jim Marrs author of such books as "Alien Agenda" and the "Fourth Reich" has stated that through Operation Paperclip, *"Germany may have lost the war, but Nazism won!"* as it still survives today in small radical and subversive groups within the government of the United States and a few other countries.

Perhaps, no better example of the secret world of black science in America can be found than in the late **Colonel Phillip Corso's** book, "The Day After Roswell". Col. Corso "took over the **Foreign Technology** desk at R&D and was asked by his commanding officer General Arthur Trudeau to use the army's ongoing weapons development and research program as a way to filter the Roswell technology into the mainstream of industrial development through the military defense contracting program."

Essentially, Corso "seeded" alien technology from the **1947 Roswell saucer crash/retrieval** to some of America's largest technology and industrial corporations which help the development of such things as integrated circuitry, fiber optics networks, lasers, accelerated particle beam devices, night vision electronics, Kevlar material in bulletproof vests and stealth capabilities.

Col. Corso even admits that there was evidence of the same or similar technology that was being used by the Nazis toward the end of the Second World War. The army did not know if the alien

152

presence over military missile bases and a few mutilated cattle represented a threat from Extraterrestrials but, the military mindset was not to be caught with their pants down should things develop into the worse-case scenario.

False assumptions and worst-case scenarios like these, however, only help to spin a negative direction of information regarding the UFO/Extraterrestrial presence and agenda on Earth which eventually distills down into today's UFO community. At some point during or after Corso's retirement from the army, everything pertaining to the UFO and ETI matter went deep black and even Eisenhower, the current president of America at that time admitted he had lost control of the UFO situation.

The whole UFO/ETI question had developed a life of its own, fostered and controlled by the military industrial complex or Majestic 12 or MJ 12 or PI 40 or whatever it is called at this time. The very people who were voted into office to oversee such matters, like the US President were now cut out of the loop. Rogue elements with the M.I.C. inspired by radical Nazi/fascist doctrines and eschatological beliefs were secretly steering the government away from the US constitutional system of oversights, checks, and balances. The US government, its people, and their spiritual destiny had been hijacked away from their true hopeful future toward some darker future of gloom and uncertainty.

CHAPTER 98

WELCOME TO THE PLANET EARTH – WHAT'S THE SECRET OF YOUR PROPULSION SYSTEM?

Knowing how alien technology works and in particular, what energy source powers and propels flying saucers and any other alien spacecraft has been the *"Holy Grail"* to the whole UFO mystery. It is the primary raison d'étre of the US military, the military industrial complex and the various intelligence agencies of the government. The acquisition of alien propulsion systems like the quest for the *"Holy Grail"* or the *"Philosopher's Stone"* would allow scientists to understand and implement this advanced technology into their own terrestrial based technology thus, giving that nation a greater edge of superiority both technically and militarily over other nations.

This then becomes the primary motivation among other military powers for an aggressive program to track, target and shoot down any and all Extraterrestrial spacecraft in orbit or in the atmosphere above their nation. Military brinkmanship has developed to a whole new level because of alien technology; along with nuclear weapons it has become just one of many weapons in a nation's weapons armament.

If scientists particularly those from the world of black science are helping to retrieve targeted, flying saucers to reverse engineer the propulsion systems of these spacecraft, it stands to reason that the big question on their minds is **"How are they getting here?"**

The standing joke among the UFO investigators that best sums up the motivations of the Military Industrial Complex and is at the heart of the UFO situation namely, how they are getting here is: ***"Welcome to the planet Earth! What planet do you come from and what's the secret of your propulsion system?"*** This amusing perception is really what has spurred the M.I.C. all these years since WWII, to find out all that they can, in whatever manner possible, about ETs and their spacecraft visiting the Earth.

It isn't just enough to design and build new disc shape aircraft to provide better stability and performance, as these pre-'50s manmade craft are simply not capable of duplicating the high-speed maneuvers that witnesses report when sighting flying saucers, they just don't have the same propulsion energy generating systems as UFOs.

Alien propulsion technology is directly connected to the development of alternative energy generation systems on this planet in both the white and black world of science. We know that it has something to do with how manmade flying saucers function from our examination of the R&D programs that the Nazis were involved in during the war. Later in post-war America particularly in the southwest, after the **Roswell Saucer Crash of 1947**, American scientists like **Einstein, Oppenheimer, Teller, T. T. Brown, Tesla and La Paz** and many other brilliant scientists were pulled from their research projects to be secretly brought in on the research and examination of the **Roswell ET craft**, the **Kingman ET craft**, the **Aztec ET craft** and other ET craft that were shot down between the late '40s and early "50s.

It's all about energy generation systems and the endless production of energy to power everything we can imagine and are capable of building. It has been the primary reason for the US Military to sponsor false investigation programs like **Projects Sign, Grudge and Project Blue Book** to placate the general public but to mislead the public in what the US Air Force was finding. It is also the reason behind the false assessments and conclusions reached by the USAF sponsored the University of Colorado's pseudo **Scientific Study into Unidentified Flying Objects.** The USAF knew long before this study, also known as the **Condon Committee Report** was even conceived or before it reached its final conclusions that not only were UFOs were in fact alien spacecraft, but that they knew that UFOs didn't represent a threat to national security as they weren't hostile and more importantly they knew that there was new and advanced technology to be gained from the research and reverse engineering of downed Et spacecraft!

Every conclusion that came out from this report was a **Bold-Face Lie** designed to diffuse the public media's reporting of UFO sightings by adding ridicule to such reports, disarming any further scientific investigation into the matter by white world scientists by referring to it as pseudo-science and hopefully painting the public's interest in this phenomenon by describing public witnesses as cranks, publicity seekers and not professionally trained observers.. Everything we thought we knew about UFOs since pre-World War Two up and until the current time has been a massive campaign of deceptions, lies and denials from the US military and the various intelligence agencies (known also as the **Military Industrial Complex**) to cover up the true reality of alien spacecraft.

The question of how Extraterrestrials were getting here from other star systems had already been asked and answered back in the early "50s, however, keeping that knowledge from the white world of science and from the public was also, of primary importance and concern because ETI don't play by Earthmen rules or more precisely by the rules of the US Military!!!

It was extremely difficult to damn near impossible to control the aerial maneuvers and mission sorties of Unidentified Flying Objects that flew over any nation or encroached upon that nation's military bases. To level out the playing field particularly from the standpoint of the US Military, covert *"Manhattan-style"* reverse engineering programs were implemented to crack the propulsion code of ET spacecraft. The success of such programs was matched only by the final working prototypes which later went into production under a newly established **US Space Force!**

Everything that we know about **UFOs** (ETI spacecraft) has been leaked out by "insiders" from within these covert black projects and programs or from leaked documents. No scientist from the white world of science has made any significant breakthrough in the technology of advanced propulsion systems other than theories, simply because they have, we the public would be utilizing it today. Even, the simple concept of a mass-produced flying car has never materialized over the last 100 years so, do we still need to hold our breath any longer? Most scientists have been told that the subject is either, taboo or pseudo-science and no scientist wants to ruin his reputation or tenure or lose a lucrative science contract. Threats by intimidation have kept most scientists under the ever-watchful eye and control by the M.I.C.

Yet, with all that has been said about this phenomenon, the public, for the most part, knows that it exists and can no longer be dissuaded from that conclusion. Strange alien objects continue to fly in our skies around the planet, often being pursued by slower flying military jet aircraft. Some of these strange craft are Extraterrestrial in nature and some are manmade technology indicating that new advanced propulsion systems have been engineered into exotic air/spacecraft that are now, almost routinely being observed but the general public.

**The Provo, Utah ARV (Alien Reproduction Vehicle) prototype,
a true manmade flying saucer or "Flux liner"**
http://ufologie.patrickgross.org/htm/usafprovo66.htm

The Bob Lazar - Area 51 "Sports Model" which is an actual ET spacecraft
http://www.zamandayolculuk.com/sayfa3.HTM

**The USAF Nuclear Powered Flying Triangle – TR-3B hovers and then completely
disappears in a ball of light when it powers up its propulsion system**

http://asian-defence-news.blogspot.ca/2015/08/strange-lights-over-el-paso-tx-separate.html and
https://www.youtube.com/watch?v=RJ_AwiUMOHA

Extraterrestrial Symbiotic Technology (EST) – Self-Aware ET Engines

Given the aggressive "shoot down" programs of Extraterrestrial spacecraft by the US military and the other military superpowers of the world and the many different ETI species that have by witnessed or captured by the military, we must conclude that not all ET spacecraft are the same nor do they all operate on the same propulsion principles or energy generation systems. This would explain the highly active, highly aggressive track, target and shot down of alien spacecraft. Therefore, to what extent do these alien technologies differ from each other or is there a sharing of technologies among Extraterrestrial intelligences, who work together in cooperative interstellar programs?

If we accept the testimony of **Bob Lazar** from his brief employment in the super secret AREA 51 military base that he saw about nine different types of ET saucers and spacecraft then, we must conclude, some if not all were from different ET civilizations or that each spacecraft has a unique function for operating in the Earth's atmosphere and environment based upon the appearance of it exterior hull. If we also accept David Adair's testimony (see below), at least one type of alien craft has a propulsion system that is self-aware which enables the engine to interface in a **symbiotic relationship** with the consciousness of its pilot!

David Adair

David Adair is an internationally recognized expert in space technology spinoff applications for industry and commercial use. As a child prodigy at age 11, David Adair became a top rocket scientist who built his first of hundreds of rockets which he designed and test-flew. At 17, he won "The Most Outstanding in the Field of Engineering Sciences" award from the US Air Force. At 19, he designed and fabricated a state-of-the-art mechanical system for changing jet turbine engines for the US Navy that set world-record turnaround times that still stand today. Adair is the president of Intersect, Inc., he lectures and provides consulting services to companies and organizations that want to know how to use the latest cutting-edge technological advances.

Testimony from **David Adair**, one of **Dr. Steven Greer's Disclosure Witnesses** reveals some unique aspects of ET technology that the USAF requested that he examine to figure out how it works. He soon had progressed to the point that he was drawing attention to his exploits by people such as **General Curtis LeMay** and **Wernher Von Braun.**

His complicated mathematical formulas in high school found their way to the eminent scientist, **Dr. Stephen Hawking**, who at that time had just received his Ph.D. in Theoretical Astrophysics and was at the beginning of his own career. When they met and David was asked for the source of his formulas, he sheepishly replied that many came to him in dreams. To that, Stephen Hawking replied, *"I get a lot of my ideas through dreams also. We dream on the same wavelength; therefore, that makes us brothers."*

As a young man of 17 David attracted the attention of the local news newspapers because of his rocket building, as well as the Air Force who showed up when he started building his second rocket which incorporated a new advanced rocket engine, an **Electromagnetic Fusion Containment Engine.**

158

David Adair is a brilliant rocket engineer who as a teenager was given the opportunity by the USAF to examine up close an alien propulsion system

https://www.youtube.com/watch?v=jqhLO4RZgdo

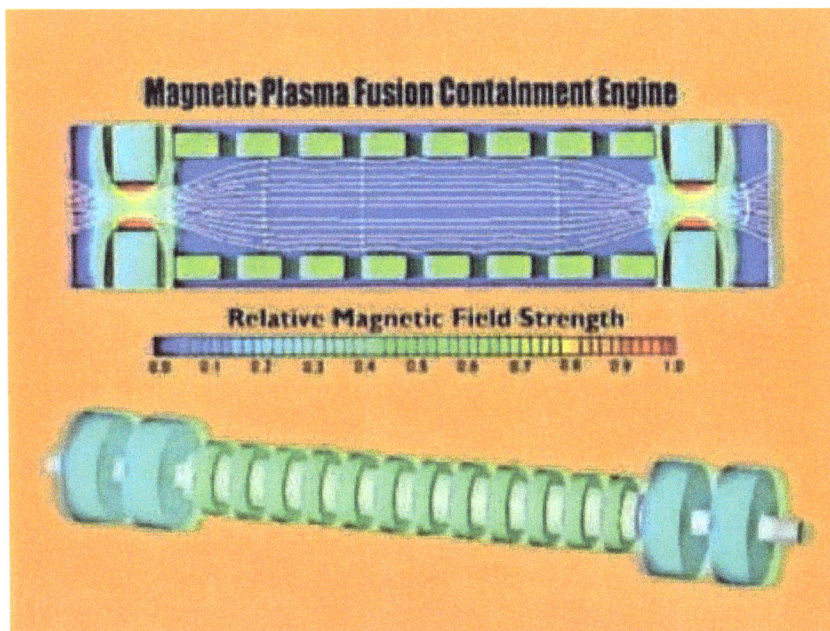

Seventeen year old Adair's Magnetic Plasma Containment Engine

https://www.youtube.com/watch?v=jqhLO4RZgdo

The Air Force knew Adair was on to something, so they funded him through the **National Science Foundation (NSF)** and Gen. Curtis LeMay, a friend and neighbour of the Adair family became David's "buddy" acting as his mentor and project manager.

Initially, Adair had difficulty "without the right electronics and the right formulas to compress and scale down his fusion engine" which would only fit a 30 story **ICBM Titan III** rocket housing at that time, however, more information-based dreams helped resolve the problems of reconfiguration of the fusion engine down to a workable size for a 12-foot-tall rocket housing.

Eventually, on June 20, 1971, the rocket was ready for launch from **White Sands Proving Grounds** and coordinates were plugged into the navigation system for it trajectory and landing point then, in front of Air Force officials and some *"men in black suits with mirrored sunglasses"*, the rocket was launched. Adair said, *"This thing took off so fast. It went from zero mph to **8,754 mph in about 4.6 seconds.** It was so fast that you couldn't even see it. It would be like trying to watch a bullet leave a rifle barrel. So everyone else at the launch site thought it blew up!"*

AREA 51 ("Dreamland Resort"), a US Military Base that allegedly does not exist!
At 17, David Adair was taken to Groom Lake by the US Military to retrieve his rocket
http://vortexhunter.com/area-51/

It landed 456 miles northwest of White Sands in an area called **Groom Lake**, in Nevada better known as **AREA 51!** Adair and the military officials caught a black DC 9 plane out of **White**

160

Sands to AREA 51to retrieve the rocket for inspection. Adair back in the early 1970s was rather naive about most things in life and didn't know about AREA 51 at the time and thought it was just a dry lake bed. Now this is where Adair's story becomes really interesting as his rocket launch was a precursor to even more advanced propulsion systems than his which were not of this world and that were kept in an area, said not to even exist or even acknowledged on any map!

Upon the landing, instead of going to the crash site of Adair's fusion rocket to see how it survived the parachute landing, he and Gen. LeMay and other military brass went drove toward some large hangars near the airstrip in go-cart like electric vehicles. They drove directly inside the central hangar whereupon the large hangar doors closed and the whole floor of the massive hangar began to descend some 200 feet below the ground.

Part of AREA 51 where David Adair was taken to an underground complex 200feet below a large central hangar, could this be the hangar in the composite images above?

http://area51phx.freeservers.com/mainbase.html

When they finally came to the bottom floor, Adair was impressed by the sheer size of the underground complex: *"It had a huge arched ceiling, but it went so far that you couldn't see the end of it. It just went forever. And I thought, "My God! You could park a hundred 747s in here and they wouldn't even be in the way!" At that point, I asked, "What in God's name did you do with all the dirt?" And they just looked really strangely at me. I guess they didn't expect me to try and figure things like that out. The walls were at least 30 feet high, and all along them were different workshops and laboratories and periodically there were big, huge, work bays. So we kept driving down past all kinds of aircraft that I had never seen. Some of them I had seen, like the **XB-70 (Valkyrie)**."* (Bold italics added by author). From an interview by Robert M. Stanley extracted from "Nexus Magazine, Volume 9, Number 5 (Aug-Sept 2002)" and *http://www.greatdreams.com/david-adair.htm*

Adair was also amazed by ceramic material that seem to cover all the walls everywhere and the lighting in the complex as everything was well lit with absolutely no shadows anywhere , yet there was no light source from which the lighting came from. *"The most interesting thing about this to me still is how well lit the underground area was. There were no shadows, anywhere. And there were no light fixtures, anywhere. I was wondering how they generated that much light. It didn't look like the walls were glowing, or the floor or the ceiling. But every square inch of this place was lit, and yet there was no visible source of light".*

"And after we had been driving for a while and we had passed a lot of different aircraft, we took a road to the left that took us away from a lot of the other activities. I could see a lot of people working on stuff. These aircraft appeared to be operational. Some of them I have never seen before or since. They were shaped like a reverse teardrop. And there were others that looked similar to the flying wing. One aircraft, the XB-70 (Valkyrie), was a delta-wing bomber built in 1959". http://www.greatdreams.com/david-adair.htm

Finally, the military officials and Adair arrive at some big steel doors, one of the officers get out and places his hand on a security scanner, there is a light flash at his head as his retina is also scanned whereupon, the doors open. To Adair, this was sophisticated technology being employed back in 1971as there were yet, *"no laptops, no modems, no fax, no VCR, no cell phones; we didn't even have handheld calculators. Texas Instruments developed those about five years later. So where in the hell did these guys get all this technology?"*

As soon as the military party entered the room, there is a noticeable temperature drop in this room, cooler than the large areas they had just come from and again, the same mysterious lighting was evident in this room as in the large complex areas, again, there were no shadows being cast, anywhere.

In the middle of the room sitting on this huge steel platform was a giant **electromagnetic fusion containment engine**! Adair immediately recognized it, because its configuration was similar to his but it was the size of a Greyhound bus. Adair's was about the size of a large watermelon! The similarities were unmistakable, as they appear to operate on the same principles but, with some unbelievable performance differences between the two.

Adair explains: "It was the same situation with my little engine and this thing they had stored underground. They both ran on the same principle, the same configuration, but the level of sophistication is like that of the Model A compared to the Viper engine. This thing they had was so powerful. There were so many design features that I didn't recognize, for reasons that became clear." http://www.greatdreams.com/david-adair.htm

"…They asked me if I liked what I saw. I said, **"Well, yeah, but I'm confused. I thought I was the first one to build one of these engines."**

And this is where things really started getting odd. The colonel that was with Dr Rudolph said, **"Son, you want to help us with this design here since yours is very similar to it. You do want to help your country, don't you?"**

…So at first I agreed with the colonel that I wanted to help. However, I was very curious and asked, **"Where are your people that built this engine?"** *He paused for a moment, then told me,* **"Well, they are on vacation right now. You're off on summer vacation, right?"** *And I said,* **"Okay! That's good. Did they leave any notes on their work that I can look at?"** *Then I was told,* **"Well, they took them with them as homework. You get homework."** *And I was thinking,* **"You know, this is really condescending. I am 17 years old."** *But that's how they treated 17-year-olds back then. So I thought,* **"Okay; I will play along with this asshole."**

162

*"I agreed to help them, but told them that I needed to get a closer look at the engine. And they agreed, at which point I walked up and got onto the platform. And the closer I got to it, the more I realized that these people had no idea what this engine was; they were still trying to figure it out. I could tell that it didn't belong to us. And when I was about three feet away, **the first thing I noticed was a perfect shadow of myself on the engine**. And what did I tell you earlier?"* Remember, there were **no shadows anywhere!** http://www.greatdreams.com/david-adair.htm

David wondered how his shadow was showing up on this thing and the fact that **his shadow moved about a half a second behind me**. That really got David's attention and he thought, **"If this is what I think it is a heat sensitive recognition alloy"** He then realized that humans don't have any known material that could do that. David Adair wanted to get a closer look up at the engine and got permission to climb to the top it to examine the damaged area. There was a four foot diameter hole in the side of it, and this area interested the young Adair.

Knowing his own engine design, he assumes that this strange engine has suffered "some kind of breach in the electromagnetic flux field that acts as the containment wall that harnesses the power of the reactor engine. These engines basically function like a magnetic bottle or sphere, and inside you have contained the power of the Sun or a hydrogen bomb continuously detonating. It's not impossible to figure out how this works, because it occurs all the time out in space. Black holes can suck an entire galaxy full of suns into their point of singularity. Obviously a black hole has no problem containing that fusion energy."
http://www.greatdreams.com/david-adair.htm

Like a mathematical model of a black hole there is a point of stabilization based on a figure-eight design, once it has stabilized it will always implode and consume itself without pulling everything around it in. But this engine at Area 51 had lost its stabilization in the figure eight or sweet spot, and that's why Adair was so curious about the hole.

*"The way this engine was built was really cool. There wasn't a single screw or rivet or weld seam anywhere on this entire device from end to end. It looked like it was grown rather than assembled. And I thought, **"Man, whoever built this really has some incredible manufacturing techniques."*** (Bold italics added by author for emphasis). http://www.greatdreams.com/david-adair.htm

This engine and no doubt the craft as well were created out of the "thought and ideation matrix" by the conscious will of the creators, in this case from the probable combined wills of the Extraterrestrial Intelligences or engineers. This means a 3D construct can be created into a perfect assembly unit using the atomic forces in the universe without, rivets, bolt or screws or even seams, it is in essence a unibody construction engineered atom by atom, molecule by molecule incorporating if necessary intelligence into the physical construction so that the engine and ship are a living entity! Sound or resonance is another way to construct certain geometric solids but, achieving specific forms and body types like flying saucers are still beyond human technology and capability. A highly evolved consciousness is the missing essential element to specific forms of physical creation and of interfacing intelligence into the creation!

Adair explains how he replicated this process to some extent in an experiment that he built which flew onboard one of the 1993 Space Shuttle missions. *"It was part of the GAS (Get Away Special) program. That's where you rent space in a 55-gallon drum for your project. The first thing I did was melt alloys together, and when you spin them in a weightless environment you can create any type of dimension you want, because I figured out a way to control this. There was always a question about how you shape liquid metals in a weightless environment. It's a containerless process. It's a real phenomenon."* **(Think of sand on a metal or glass plate with a wire or some type of contact to a violin, where sound from a violin resonates at a certain frequency through the wire to the plate, causing the movement of the sand particles to move around into particular geometrical patterns)** (Bold italics added by author for emphasis). http://www.greatdreams.co_m/david-adair.htm

Adair figured that by taking a glob of fluid floating in this weightless environment of space, it could be manipulated into a geometric shape and dimension to a corresponding sound wave from a Moog synthesizer, simply by playing notes. This machine generates interlocking standing sound waves that vibrate, even in space, and which allows the liquid metal to be shaped.

Suspicions arose in Adair's mind when he first saw the engine at Area 51 back in 1971 that whoever built this engine may have used this process. This aspect now raised an even larger question in Adair's mind as to who could have built an engine of this size in space. *"It would have to be deep space. Like intergalactic deep space, away from any planets or stars".*

Realize of course, that this is an educated assumption on the part of Adair who thought the engine was clearly built in a weightless environment. There is more than one way to accomplish this process, even on a planet; space is not required for this type of engineering!

"Anyway, when I placed my hands on the engine to pull myself up, I began climbing up the exterior of the engine, which was designed with an exoskeletal structure." http://www.greatdreams.com/david-adair.htm

Adair reasons with further tactile examination of the engine: *"It was warm, which didn't make any sense at all. It was so cold in that hangar; you could almost see your breath. I looked around on the floor and saw no power lines. And I asked myself,* **"How in the world could this alloy be staying warm?"** *And it was really hard. It was the hardest material I have ever touched. It didn't give anywhere. The surface cohesion-tension on it felt more like a baby's skin. It was supple but hard and warm. I was thinking,* **"What the heck is going on?"** *And as I was crawling up everywhere,* **I touched the surface and it reacted.** *When I turned and looked at the Air Force guys, all their mouths were hanging open. And so I assumed that the reaction they were seeing hadn't happened for them, because wherever I touched it there were these really amazing blue and white swirls moving down through the hull of this thing. It looked like wavelengths that you see on an oscilloscope. When I pulled my hands off, it stopped. And I said,* **"Wow! This thing is reacting!"**

164

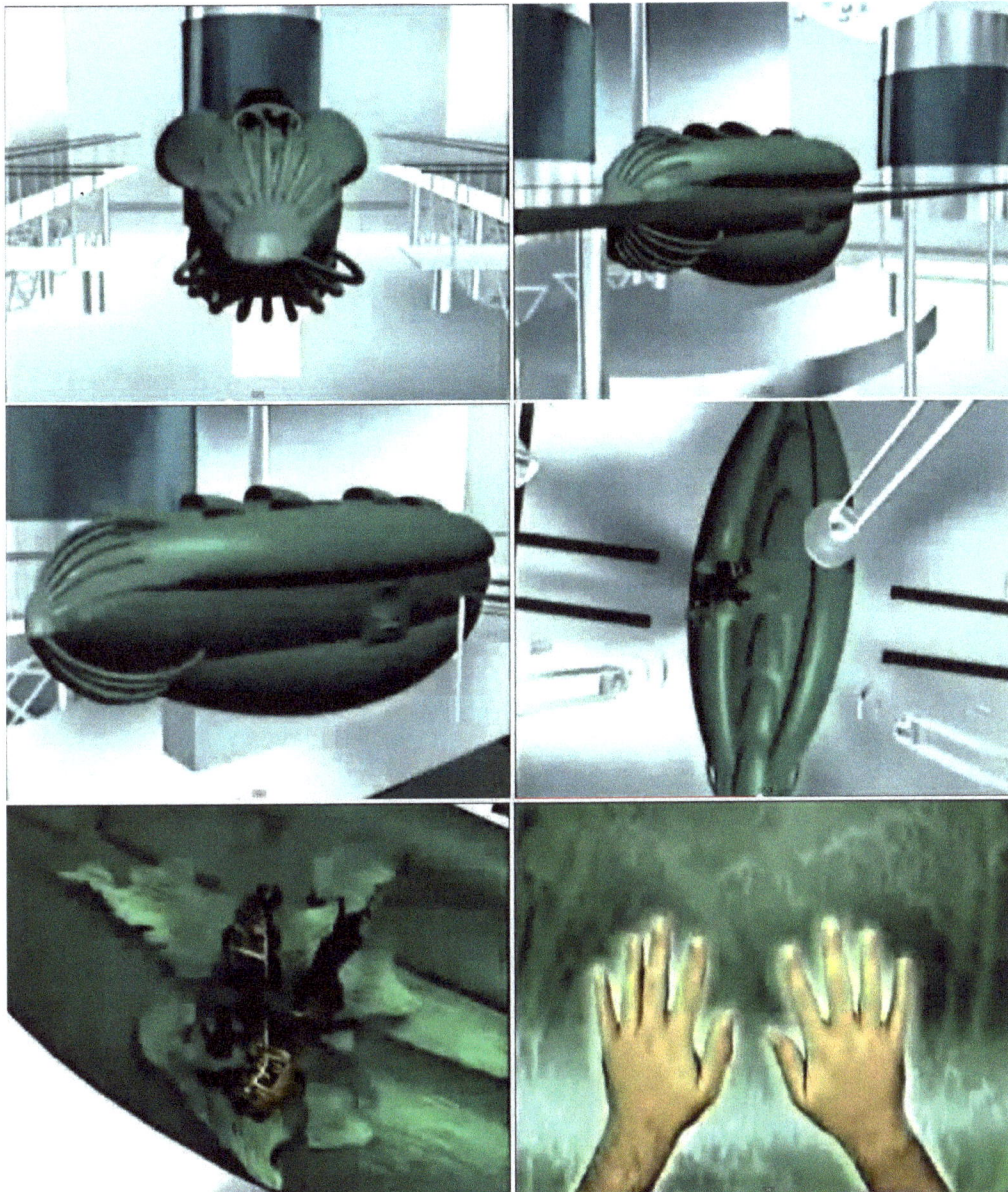

Artist's conception of the Alien Symbiotic Engine that David Adair closely examined in AREA 51 back in 1971. There were no shadows in this well-lit room and the engine seem to have neuron synaptic pathways built into its overall design providing the engine a reactive sentience to its immediate surroundings

https://ericsamsung.wordpress.com/2014/05/20/what-is-a-symbiotic-engine/

So I continued to climb up until I reached the centre area. It had these vertebrae that branched off, cascading, fiber-like. They looked almost like fiber optic cables filled with some kind of fluid. They were very small tubes the size of angel hair pasta. There were millions of these things cascading over the hull of this engine. And I thought, ***"Boy, these patterns look familiar."*** *Then it dawned on me:* ***they looked like neural synaptic firing patterns.*** *There were millions of them going out everywhere on this thing. So I thought that* ***maybe the engine was designed with an exoskeletal brain.*** *And at that point, I reached out and grabbed some of the fibers and found that*

*they were really tough and that **there was fluid in them. And wherever I touched, no matter what I touched, there would be a reaction to it like a tremor of visual lights.***
http://www.greatdreams.com/david-adair.htm

David also describes other features of the interior of the engine after he asked permission to inspect the damaged area inside of the reactor where it had blown open. Hesitantly they gave permission.

"Before I came out of that damaged area, totally pissed off because when I got down in this thing, they told me to make it brief. So I got down and looked in the area. Man, there was some incredible-looking technology up and down this engine. And I couldn't get more than three feet into it before I came up to a wall. And this wall, it was like the iris/shutter on a camera lens. It had lots of interlocking fans that contract or expand - and I've always thought that would make the coolest door. Well, there was this little round pod-thing there, and I just put my hand on it; and when I did, the wall just shuttered open. It made a slight noise.

"...I got to look deeper into the engine. And what I saw in there was fascinating. It was such a trip being there because whenever I worked on my fusion engines, everything was so small; some parts I even had to machine under a microscope. Now, here was a replication of my basic design that was big enough to walk through. But man, this thing what I had manufactured to achieve a certain function in my engine, this thing would have something else in its place. And this something else would be stuff I couldn't begin to recognize. There were these crystals that were facing each other. They were fabulous-looking crystals. And they were integrated into this plasma duct type thing.

And in my engine, I had such a hard time getting a cyclotron to curve the blast waves I needed for propulsion. This thing had some kind of venting system that allowed them to flush their plasma out through an area that looked like the gills of a shark. The whole thing was so organic looking. It looked like a living machine - both organic and inorganic incorporated together. It was an oxymoron. How do you explain something like that? So anyway, I just got to see a lot of stuff in there that I couldn't believe. *http://www.greatdreams.com/david-adair.htm*

Again, David Adair is describing a living mechanical device which is way beyond human technical capability; everything is integrated together and designed to work as a living whole. In fact, the engine was trying to heal itself but the damage to it, on its side, was like a gaping wound and part of the body had been destroyed completely beyond the ability of the engine to heal itself properly!

As David walked down into the damaged area of this thing and in an inquisitional cross-examination, he says to the Air Force officials**, *"You know, this thing is a power plant. It is more than a propulsion system. It is a power plant. It obviously came out of a big vehicle craft of some kind. Where is that craft located?"*** Now they were not happy with me, but I continued. ***"A craft like this must have had a crew. What did you do with those people? This is clearly not American or Soviet technology, is it, boys? This is some kind of extraterrestrial entity. How old is it? Did you dig it up? Is it millions of years old or did you guys shoot it down?"*** And man,

166

they got really upset. They told the MPs to take me down off the engine. As I was coming down, I was really pissed off. I was so pissed off because I had had enough.

At this point, I knew where I was. I knew that this engine was from somewhere other than Earth. I didn't know where it had come from or how long they had had it, but it was obvious that my whole world was coming undone in that moment. **I grew up in a world where the government would never lie.** *We had just landed on the Moon the year before. And here the Air Force had this technology and they weren't saying anything, which made me furious.*
http://www.greatdreams.com/david-adair.htm

Blessed with a photographic memory, David Adair is one of those unique individuals who was absorbing it all in when he left the compartment in the reactor he remembers that he didn't touch that little pod on the other and as soon as he passed that area, the door closed behind him. He never told the Air Force brass that he went into that part of the engine. Adair believes that they didn't even know there was another compartment in the interior that they could enter.

Adair believes *that* **the engine didn't allow them access. There was a presence about this engine,** *just like you have a presence of a person or an entity;* **it just had its own sentience. It was self-aware!** Adair came out of the engine and was totally pissed off because he knew there was no way the military could have built it. It was using some kind of crystal containment field power that they couldn't even imagine. Adair felt he would have to work on it for a long time to figure out how they were doing the fractions. **Where he was using the plasma in a linear mode, this thing was designed to go in any direction it wanted with its plasma flows. With our understanding of science, that would be impossible.**

David says: "This thing could do anything. And I really wondered who in the hell built it. So, as I started coming down the outside of the engine. After we got into a big argument, I noticed that now, wherever I touched the engine, it was no longer reacting with the nice blue and white swirls of energy. They had changed to a reddish-orange flame-looking pattern. And as I calmed down to try and figure out what that was, it changed back to the bluish white, more tranquil-looking pattern.

That's when I realized that the engine is not just heat sensitive; **it reacts to mental waves. It is symbiotic and will lock on to how you think and feel. This allows it to interface with you. And that means this thing was aware. And it knew it was there. And I knew that it knew I was there.** *http://www.greatdreams.com/david-adair.htm*

Imagine, a conscious, sentient living machine that reacts and interacts with its creator or controller in a symbiotic relationship which is probably, mutually beneficial permitting the operator to travel almost anywhere in the universe almost instantaneously or with a brief period of time to any point in the universe and back again.

This is real alien technology and in the words of **Carl Sagan** (paraphrased), *"Any highly advanced alien civilization visiting the Earth, their technology would appear as magic to humans! "***This author contends this as close to magic as it comes (at least, it is a science for which we do not as yet, understand) and thus, it is indeed Extraterrestrial in nature!**

Obviously, David Adair because of his brilliance was afforded the opportunity to build advanced rocket propulsion systems, to go where few people have gone in the black, hidden military world of science and not only see but, physically interact with Extraterrestrial technology and be able to return to the public domain to tell the story.

Adair was escorted off the base by Gen. LeMay but before boarding the plane back to Ohio, David managed to obtain a handful of graphite grease from the hangars doors and conceal it while pretending to be emotionally unglued and wanted to see and inspect his rocket on the dry salt lake bed. He was driven to his rocket and where he opened a panel on the rocket housing and subtly placed his handful of graphite into the fusion containment of the engine which set up a chain reaction that caused the rocket to explode. Adair suddenly came running back the military officials yelling for them to get out of the area as the rocket was about to blow, which it did! There was absolutely nothing left of the rocket or the engine, therefore, Adair's sabotaged was successful and undetected by the Air Force.

His knowledge of his **Magnetic Plasma Fusion Containment Engine, as well as his photographic memory of the alien engine that he examined,** are, for now, securely contained in his head and out of the hands of the US Military!

Dr. Steven Greer was going to have David Adair testify at the **Disclosure Project** event along with other witnesses but at the last moment had to have him sit it out as his "loose cannon" testimony was considered too risky and Greer felt the public wasn't ready for it at this time plus the fact that Adair was under age at 17 to have sworn an oath of secrecy which contravened US constitutional law There were two major differences that surfaced in the pre-screening between myself and the other witnesses. The first difference was that I had hardware contact inside a top-secret Air Force base and the other that I was minor at an age when they took me there. Big mistake on the Air Force's part, the Constitution of the United States prohibits signing a minor to a **National Security Oath**. As one high-ranking official refers to me in the Congressional Briefings was that *"I was the worst loose cannon on the deck."*

"When Dr. Greer sequestered me away from everyone, I believe he did not have any choice in the matter. It did not make sense for him to do that when he had worked so hard to get me there.

*What did make sense was this was **Standard Operating Procedure** of the covert community by dividing the force that is coming at them and having it fight amongst its own, thereby keeping the focus off of them. They were hoping when Dr. Greer sequestered me that I would get mad at him and then start a disruption which would have damaged the proceedings. I believe that the covert community is working with Dr. Greer and is using him and **CSETI** to get out a disclosure of information, but they control the content and the rate of flow. If you notice the vast majority of witnesses have seen lights in the sky or blips on a radar screen. We have not seen one reverse engineering scientist or technician come forward on tape or in person with their story. The reason according to Dr. Greer why we have not seen this type of witness is because they must have legal immunity from their signing of a **National Security Oath** This is where we can apply pressure onto members of Congress to set forth a Congressional Hearing granting full immunity to all who testify."*

http://web.archive.org/web/20040420185409/http://www.flyingsaucers.com/adair3.htm

So at the end of the day what are we to believe about David Adair and his testimony? His story is definitely intriguing and he has the evidence to back up what he says. As a Disclosure Witness and a **"hardware contact,"** he was concerned that he would be alone among the 12 Disclosure Witnesses that were **"soft contacts"** when a final piece of corroboratory evidence came from fellow Disclosure Witnesses particularly, **Stephen Lovekin.**

The **Disclosure Project** website indicates that **Brigadier General Stephen Lovekin** entered the military in 1958 and joined the **White House Army Signaling Agency** in May 1959, serving briefly in both the Eisenhower and Kennedy administrations until August of 1961. He had worked as a trained cryptologist with a Top Secret security clearance in the Pentagon during President Eisenhower's White House term during the '50s and was the military aide who regularly briefed **President Eisenhower** on UFO evidence and developments. Lovekin is currently a practicing attorney in North Carolina and a Brigadier General in the Army National Guard with the JAG division (Judge Advocate General).
http://ufologie.patrickgross.org/rw/w/stephenlovekin.htm

Adair recalls in a conference meeting just before the **National Press Club** Disclosure event that Lovekin stated that he worked as a cryptologist with a Top Secret security clearance in the Pentagon breaking codes. He said his superior office came into his office and laid down three pieces of metal, one about a foot long with strange writings or hieroglyphics on it. They tried to decipher it but couldn't and while Lovekin was telling his story Adair was busy writing some stuff down on paper and then handed to Steven Greer. Lovekin remembered 12 emblems and when he and Greer looked at the paper from Adair, they were shocked to see that these were the same symbols that Lovekin had seen and remembered from so many decades ago. When asked to explain this corroborating evidence, Adair stated simply he was standing on the engine where that piece metal came from back in 1971 and that there were 98 different emblems in all and because of Adair's photographic memory, he had remembered them all and he showed Greer and Lovekin the paper.

Adair said that they were grouped together in clusters on every piece of metal and device of the engine. Adair had figured out what they were, they weren't some sort of cosmic message but serial numbers for parts! That is how ETs keep track of parts as there were repeating symbols, one symbol being similar to our use of **Pi ($\pi\pi$)** which would show up every time there was *a curved radius!* Other symbols would *show up reversed* wherever, they were on the opposite side of the engine, ergo, *Left and Right*! (See image below).

This is very similar to how the Navy or Air Force would group out its numbers on aircraft engine parts, etc. and Adair declares that there is enough there to give you the ET alphabet and understand their language!!!

This is a truly amazing story by David Adair that has many implications for humanity on many different levels such as technology, cryptology, languages, propulsion, engineering and design, aviation and aerospace development, to name just a few.

What is truly remarkable is the similarity of the alien symbol for **Pi ($\pi\pi$)** which means that when an inventor, astronomer, scientist, etc. on Earth makes a discovery or builds something, there is always one or two or more people who do the same thing simultaneously at the same time, else on Earth. Therefore, it is hypothetically possible that an alien culture on another planet elsewhere in the universe might also have done the same thing in language as develop a similar symbol for **Pi (ππ).** It is a startling implication which means that there is an intelligent connection and greater design between humans and life everywhere in the universe! As in the Hollywood blockbuster movie "Contact" where Dr. Arroway (**Jodi Foster**) suggests that *"Math may be the universal language"…*this state statement may actually have a basis in fact, rather than fiction!

**ET Writing or Numbers that Stephen Lovekin had seen on strange metal parts and David Adair had seen on an alien symbiotic engine in AREA 51.
Note the Alien Pi (ππ) like symbol in the middle bottom of
the page compared to the Terrestrial Pi symbol!**
https://www.youtube.com/watch?v=jqhLO4RZgdo

Weird Science and Freakin' Magic

Most people first became aware of the term **WSFM (Weird science and Freakin Magic)** when they heard **Dr. Steven Greer** explain how he was told by an agent within the **(CIA) Central Intelligence Agency** who was in charge of the **WSFM** at the "Agency" *"which describes all the technologies and all the things that deal with transdimensional physics, things that go faster than the speed of light! These are things that resonate at resonate frequencies, beyond matter and beyond the electron… beyond the speed of the photon!"* And not surprisingly, but perhaps new to most people is that *"most of the things in the cosmos resonates beyond the speed of light!"* From a lecture "The Final Sequence" given by Dr. Greer in Barcelona, Spain in 2009

These are technologies that are so far advanced of the white world of science that they border on magic! Such top secret programs included devices built for dematerializing things, teleporting and space-time alteration. These are interdimensional technologies which interface with *thought and consciousness! "We're talking really advanced stuff"* explains Dr. Greer.

A priori: *All interstellar technologies are transdimensional and the heart of that science is consciousness!*

"Because they have technologies which interface directly with thought, coherent thought, much like a laser which is coherent light! These are technologies which are WSFM", as Dr. Greer explains further, *"they are technologies which are transdimensional that cross right over into consciousness and awareness!"*

Dr. Steven Greer and **Dr. Ted Loder** had at one point had procured a top secret inventor who had been released from his earlier military seclusion that had worked on some of these advanced secret programs as mentioned above and now was contracted to the good doctors to help them bring Tesla-related free energy technology to the world.

On Christmas day, 2009 as *"a Christmas present to the world,"* Steven Greer, M.D., and his science advisor, Ted Loder, Ph.D., from **The Orion Project**, announced on the **World Puja Network** that the "top secret" inventor they've talked about over the past year has been released early from military seclusion, and has contracted with them to begin bringing various technologies he has developed (in some cases at taxpayer expense) to the world, beginning with a free energy technology in the tradition of **Nikola Tesla**.

They are hoping to have a sample prototype to begin to disclose to the public this spring. However, much research and development, beta testing and durability testing will be needed before a completed device of several kilowatt output is ready for distribution to the public. Greer said that **The Orion Project** signed a contract with this top secret scientist some ten days prior to the Dec. 25 announcement. http://www.theorionproject.org and http://worldpuja.org/archives.php?list=host&value=steven&rnd=12576

Unfortunately, at the time of the writing of this textbook, Dr. Greer had this inventor pull from him almost at the last moment, again, by the military who assigned him to some new project out of the country just as he was prepared to work on one of the Tesla-like projects for the Orion Project.

Both **Doctors Greer** and **Loder** have been involved for years in combing the world for a technology like this that could break our addiction to polluting fossil fuels and the control paradigm associated with them; providing instead a cheap, clean, reliable energy technology capable of being scaled to power everything in a distributed manner, from cars to homes and individual appliances.

"This scientist, by far, is the most knowledgeable and skilled -- and genius -- inventor and physicist that I've ever met; and I've met a lot of interesting people", said Greer. The inventor, who will be unnamed for security purposes, has top clearances in the U.S. military.

Loder said, *"This is just what we've been searching for all these years. What he's bringing to the table is his thorough mathematical understanding of electronics and physics; [adding] modern technologies to optimize the process."*

Their purpose in making this announcement was both to bring in additional support for the project, as well as to increase the education of people to the alternatives and the hope that they bring for a better world coming. They are counting on the shield of protection provided from the community through this exposure so that this technology can make it to the world without being sequestered. They will need financial resources and moral support.

They plan to build a research laboratory to support this top-secret inventor so that he can have both the tools and personnel needed to accelerate the development of the technology for public use. During this phase, they plan to bring in half a dozen scientists and engineers. To date, this still remains Greer's objective and needless to say, if some benefactors with high altruistic motives would come forward and generously support the Orion Project, free energy would finally be within the grasp of every home and automobile owner in very short order.

Loder said the energy technology is based on concepts and prototypes and patents first put forth more than 100 years ago by Nikola Tesla and implemented by many scientists and inventors since then to various extents.

Greer said that some of his technologies have been replicated and verified by a senior scientist in the U.S. Department of Defense who has direct access to the Secretary of Defense. *"Independent replication is the sin quin non of science",* said Greer. http://www.theorionproject.org and http://worldpuja.org/archives.php?list=host&value=steven&rnd=12576

Apparently, there's a lot more where this came from, and once that energy technology has been delivered to the public, the inventor has other technologies that the Orion Project will begin to develop and bring forward. Greer said, *"We're dealing with someone who has a track record of thirty years of building these technologies, and being under contract with top secret programs; and has actually built things to the level of things dematerializing, teleporting, and space-time alteration. We're talking really advanced stuff."* Greer says that in the military secret ops, such projects are referred to as **WSFM** which stands for **"Weird Science and Freakin' Magic."**

172

The purpose of starting with the Tesla-based technology would be prove to the world that free energy can be had from the environment 24/7/365 via a device that is affordable even in the developing world, subsidized at first from sales of the device in the developed world.

"Let's walk before we levitate – get our civilization out of the dust bin of coal, and oil and sludge and mess. As civilization stabilizes; and we put in the means to ensure that these technologies are only used for peaceful means and are not militarized or weaponized; then these more advanced concepts can come forward." – Greer

More than a year ago, the Orion Project had thought they were going to be able to move forward with this inventor, but then he was spirited away into a top secret military operation and was incommunicado. They were told that it would be at least a year and a half before he would be available again.

Greer speculates that the many people petitioning the government for his release may have been instrumental in his early release so that they could resume their business relationship with him, which was cemented on around Dec. 15 but, as stated earlier, this inventor has been sequester away from Greer once more to prevent his helping Greer's quest for free energy for everyone.

Last January, Greer chose to compose a briefing for **President Barrack Obama** about various free energy technologies in general and about this inventor, in particular, asking that the Orion Project be allowed to work with the inventor to bring his technologies forward in a sort of "peaceful Manhattan Project". That briefing was then published on TheOrionProject.org website on Oct. 24.

Greer points out that some of the technologies developed by this top secret scientist were done at taxpayer expense, *"yet you never hear the Department of Energy talking about them. It's about time these technologies got into the hands of the public who paid for them"*, Greer said. Note that the initial technology this scientist is working for The Orion Project was not developed at taxpayer expense. http://www.theorionproject.org and
http://worldpuja.org/archives.php?list=host&value=steven&rnd=12576

Now the technology is being taken forward by a private organization, though apparently not necessarily with the full cooperation of the U.S. government. Greer posits that if the 100s of billions and even trillions of government spending for energy solutions were put into this technology, each person could have their own generator at home.

About a year and a half ago, the Orion Project launched a fund drive to raise $3 million for the purpose of developing and/or finding such a technology. They've not yet raised that amount and are asking their supporters to chip in to help raise the additional several $100k they'll need to build their research facility in Charlottesville, Virginian, not far from the historic Thomas Jefferson Monticello property. *"We'll have to create a whole new scientific process,"* said Greer.

Speaking of history-making, Greer said:

"This is one of the most significant undertakings in the history of our country, as we try to bring forward, very boldly, and decisively, the sciences and technologies that would get our country off of oil and gas and coal and nuclear power; and into clean, free energy; so that every home and business and facility in America and the world may be running on this type of energy, without any costs, once you have the device in place; and without any pollution or impact on the environment." - Greer-Loder announcement, Dec. 25, 2009, http://www.theorionproject.org and http://worldpuja.org/archives.php?list=host&value=steven&rnd=12576

Other **WSFM** technologies involve stealth and invisibility for the individual soldier on the battlefield, for tanks, battleships and aircraft and it isn't just on the drawing board or in some simplified prototype but, the technology according to some military historians and journalists already exists and is currently used in some of the battlefields around the world. https://www.youtube.com/watch?v=Rqi3jpBSvCc and https://www.youtube.com/watch?v=7zKQe-1BUFQ

There are satellites currently in orbit (about 200 plus miles up0 from a number of the world's super powers that can see through heavy-laden clouds to watch and track your every movement in high definition, read newspapers or an open document, even tell you the brand of the cigarette that you are smoking, if you are a smoker. These spy satellites can even eavesdrop on your conversations using high definition audio, they can hear a whisper! If you are a person of interest, you will be watched by one of the satellites. This is a military use of these types of satellites.

Even you cell phone will help intelligent agencies like the FBI, CIA and the NSA to track and listen in on your conversations *even when you turn off your cell phone!*

"Cell phone users, beware! The FBI can listen to everything you say, even when the cell phone is turned off. A recent court ruling in a case against the Genovese crime family revealed that the FBI has the ability from a remote location to activate a cell phone and turn its microphone into a listening device that transmits to an FBI listening post, a method known as a **"roving bug."** Experts say the only way to defeat it is to remove the cell phone battery. *"The FBI can access cell phones and modify them remotely without ever having to physically handle them,"* James Atkinson, a counterintelligence security consultant, told ABC News. *"Any recently manufactured cell phone has a built-in tracking device, which can allow eavesdroppers to pinpoint someone's location to within just a few feet,"* he added."

"The courts have given law enforcement a blank check for surveillance," Richard Rehbock, attorney for defendant John Ardito, told ABC News. Judge Kaplan's ruling said otherwise. *"While a mobile device makes interception easier and less costly to accomplish than a stationary one, this does not mean that it implicated new or different privacy concerns."* He continued, *"It simply dispenses with the need for repeated installations and surreptitious entries into buildings. It does not invade zones of privacy that the government could not reach by more conventional means."* But Rehbock disagrees. ***"Big Brother is upon us…1984 happened a long time ago,"*** he said, referring to the **George Orwell** futuristic novel ***"1984,"*** which described a society whose members were closely watched by those in power and was published in 1949. http://abcnews.go.com/blogs/headlines/2006/12/can_you_hear_me/

174

We live in the dark shadow of Orwell's world of "1984" as revelations by **Edward Snowden**, whistleblower and former CIA operative now living under protective asylum in Russia has attested to in the world's major medias. Whether you view Snowden as the people's hero or patriot in the disclosure of the corruption and criminal activity by the various intelligence agencies of the American government or as a traitor to the American "ideals of democracy", the American government want to get their hands on him to give him some well-deserved "American justice" for his too outspoken public pronouncements. According to Snowden, the cell phone hacking of any and everyone's phone in America and anyone elsewhere in the world by the CIA and the NSA is now a matter of routine; they can listen to anyone's phone conversation anywhere in the world without their knowing about it!
http://www.washingtonpost.com/opinions/snowdens-hypocrisy-on-russia/2014/02/07/23c403c2-8f51-11e3-b227-12a45d109e03_story.html and
http://www.huffingtonpost.com/bob-burnett/questioning-authority-edw_b_4744956.html

As mentioned in the British newspaper, The Guardian, there are people who have even nominated Edward Snowden for a **Nobel Peace Prize!** Two Norwegian politicians say the NSA whistleblower's actions have led to a "more stable and peaceful world order". Two Norwegian politicians say they have jointly nominated the former National Security Agency contractor Edward Snowden for the 2014 Nobel peace prize.
http://www.theguardian.com/world/2014/jan/29/edward-snowden-nominated-nobel-peace-prize

Time will tell, if Snowden will eventually be recognized on the global stage as a Nobel Peace hero or be serving time in an American jail (Guantanamo?)

Another listening device that really borders on magic is the device described by John Lear and Dr. Steven Greer that the military intelligence use in specific areas of interest, where the conversations by people in that area, can be recorded….one or two weeks later! That's right!! Recording conversations that are two weeks old, as if you were present at the time when the conversation was being held!!!

These listening devices can boggle the mind unless you understand the science behind it. Think of a rock or stone being tossed into a lake which creates endless ripples across the lake. Sound waves of any kind, even from conversations ripple into the ether and across time to a point where the energy from the wave eventually diminishes to non-existence. There is a point where such sound waves from conversations can be captured and recorded, if you know the area where people of interest have spoken, particularly those conversations that might be of interest to an intelligence agency. It's almost a form of audible time travel!!!

Whether we like it or not, most new discoveries and breakthroughs in science are militarized and then eventually weaponized because that is the way of the national security mindset of the military industrial complex…develop it and use it first before some other nation uses it against you! What a real sad state of affairs we live that any new technology developed has come down to the baser level of militarization!

Philadelphia Experiment

One of the strangest accounts of weird science is the famous or infamous case of the **Philadelphia Experiment** in which copious books have been written and movies have been made about it. Contrary to the popular cultural myths that surround the "Philadelphia Experiment" back in October 1943, did NOT, in fact, involve the **U.S.S. Eldridge (DE-173)** as most people believe, but a ship named **minesweeper IX-97** or "**Martha's Vineyard**". The Eldridge was part of a government disinformation campaign that seems to have worked for over 60 years.

The logs of the **Eldridge** show that it was in New York harbor in October 1943, when the **Philadelphia Experiment** took place and that it was decommissioned in 1946 and transferred to Greece in 1951, where she sailed as the H.S. Leon. The ship was sold for scrap in 1999.

The minesweeper **IX97**, however, was commissioned in March 1943 with **Lt. William W. Boyton USNR** in command and was logged in the area where the Philadelphia Experiment took place in October 1943. The IX97 was involved in 'Test Operations' along the New England coast from 1943-1946 and also "Training Exercises" off Newport, RI.

The IX97 was stripped in April 1946 and returned to its "owner" in September 1946, although there is no further record of the vessel and therefore, no way to examine it, unlike the Eldridge/Leon which was easily tracked up until 1999. http://thelostgunmen.blogspot.ca/2008/03/levitation-teleportation-time-travel.html and http://www.stealthskater.com/Documents/Beckwith_02.pdf

While everyone has been studying and investigating the U.S.S. Eldridge which has become an urban legend of sorts, the real Philadelphia Experiment, the ship IX97 seems to have "evaded everyone's radar, without even, so much as a blip or a ping from their sonar scanners!

The mystery that surrounds the events of the true Philadelphia Experiment has links to the weird science of alien technology, levitation, teleportation and time travel. The former stories as portrayed in the movies and in books of the naval experiment gone awry raises serious questions of their historical accuracy in light of the revelation brought forth by **Dr. Steven Greer** back in 2000.

Dr. Greer cites **Bob Beckwith** and **Drew Craig's** book: "Hypothesis- Superatoms, Neutrinos, and Extraterrestrials", 'Chapter 15: The Philadelphia Experiment'

According to Beckwith's experience, the **Philadelphia Story** did happen during WW2 (in early 1944) but was related to *the battle of sonar underwater just off our coast.* It was *NOT related* to the *battle of radar over the skies of Europe* as told in the movie "The PHILADELPHIA STORY". Also, the boat involved was *the minesweeper IX97, NOT the USS Destroyer Eldridge* as in the 1984 science fiction version of the event. This may appear to be disinformation to mislead and misdirect investigators and UFO researchers from what really happened.

The US and Britain early in the war we were losing the radar war over Europe, as well as losing the sonar battle against German mines and submarines off US shores but, it didn't prevent troops getting to England in time, not did it impinge the development of the atomic bomb or it completion before the Germans' own atomic bomb.

Would it have made a difference in order to end the war, before the Americans drop the first two atomic bombs on Japan? History would never know and only speculation and debate is left for the historians to figure it out.

In the actual Philadelphia Experiment, the **minesweeper IX97** was moved in time from a position in the Philadelphia harbor outside the Navy shipyard to a position at the dock in Newport News CT where the IX97 had been two weeks previously. As the experimental power was raised, nothing happened until suddenly, wham! It happened and sailors died as a result.

Author **Col. Philip J. Corso** has confirmed **The Experiment** from work he did with **Admiral Burke** while Chairman and chief policy maker of **President Eisenhower's National Security Administration.** Corso tells Beckwith of Burke's knowledge of the experiment and his deep concern and regret over the loss of life that resulted. Corso tells Beckwith that his recall of the event has greater detail than any source that he knows of and that the project continued and has progressed greatly since then.
http://www.beckwithelectric.com/ber/downloads/Hypotheses.PDF and **Hypotheses: Superatoms, Neutrinos, and Extraterrestrials by Bob Beckwith and Drew Craig; 1996 - 1998; clearwater, Florida, USA; ISBN 0-9657178-0-2** also,
http://www.beckwithelectric.com/ber/downloads/Hypotheses.PDF

The Experiment used and confirmed the principle of "Chapter 13: **Divided Space".** As a result, we now know that once the threshold of divided space is crossed, levitation, teleportation and time travel all become possible. *(It has always been suspected that the science and the energy requirements behind levitation was essentially the same for teleportation, time travel, and invisibility).* **(Bold italics added by author for emphasis).**

The ability of "gifted" humans to levitate, teleport and travel in time can now be supported, duplicated, and studied in university Biology and Psychology laboratories. Concurrently the barrier to low energy levitation, teleportation, and time travel can be crossed in Engineering and Physics laboratories. Results from studies with humans and studies with machinery can then be compared in recognition of the same underlying principle: **Divided Space!**

*(**Note:** Divided Space is what Dr. Greer refers to as the space on the other side of the* **Crossing Point of Light***.)*

We cannot creep, however, using conservative scientific methods into the realm of levitation, teleportation, and time travel any more than we could have crept slowly into setting off the bomb. http://www.beckwithelectric.com/ber/ and
http://www.beckwithelectric.com/ber/downloads/Hypotheses.PDF

History leading up to this subject starts with Nikola Tesla in 1907. According to some reports he performed an experiment which was immediately classified. When he passed away in 1943, a custody battle evolved over ownership of his many notes. Some were taken to the Beograd N.T. Museum in Yugoslavia. Others were reported to have been seized and kept secret by our government. Many guesses have been made as to the nature of his experiment. So why not include ours as well? It is possible that Tesla put an object at point A on a table and then moved it to point B on the table. By applying power to an inventive circuit, he was able to move it back to point A. Turning off the power brought it back to point B. This much is pretty well understood. Our guess is that he put coils under the table creating a *three-phase magnetic field around the object operating at a frequency of about 7.5 Hertz.* When he applied power to the coils, the object appeared to move from point B to point A; and when power was removed, it moved back again to point B. What he did was use the fields to move the object in time back to when it was at point A, and then remove power and let it come back to the present at point B.

The principle of *divided time* needs to be expanded into theories which separates the phenomena of levitation, teleportation, and time travel. Without these theories very small scale experiments should be safely possible without knowing in advance just what to expect. There may be a source of information from synchronous electric power generators that is not totally enclosed so that the rotors are visible at least on one end. Hydro generators are generally of this type. Actually, we have heard stories of generator rotors becoming hard to see as the generators are brought up to speed with their fields applied. Perhaps the phenomena is so common that operators pay no attention. http://www.beckwithelectric.com/ber/downloads/Hypotheses.PDF

It is likely that Tesla was not using Einstein's principles as much as he was defying them. The government may have convinced him to keep time travel a secret and he went through the rest of his life frustrated in feeling people didn't listen to him but not able to tell them of his experiments. Or perhaps he did tell some who would not believe him and considered him somewhat crazy. The answers may be found in his notes if they are ever declassified. Tesla's experiment may have been the secret *starting point for the Philadelphia Experiment*.

It is known that Tesla was interested in the possibility of transmitting alternating current power using purely magnetic waves (i.e., no transmission wires). Our hypothesis of a six-dimensional space uses the fifth and sixth dimension for electro and magnetic fields respectively. The experiment described herein used very large currents producing huge magnetic fields with only very small voltages to drive the currents. The waves used here would have been almost purely on the sixth, magnetic plane. If the stories of his experiments in 1907 are true, it seems likely that Tesla used the same fields produced by what are now conventional three phase currents. http://www.beckwithelectric.com/ber/downloads/Hypotheses.PDF

In 1942 **Bob Beckwith** went to work with General Electric as an engineer and developed the highly successful **Frequency Shift Keyed (FSK)** transfer trip equipment which eliminated one high voltage circuit breaker at each substation where generator power was stepped up in voltage for sending over a distance. The FSK telemetry was also used in test firing captured German V2 rockets at White Sands and has since been used in other applications worldwide.

178

It is important to remember and understand this electronic development because it was the reason why Beckwith became involved in The Experiment. Radar, cryptology, and sonar were the top three problematic issues facing electronic engineers during WW2 but the one that involves Beckwith was sonar, even though both the Germans and the Japanese sonar technology was an old technology which couldn't match the Allies. However, the Germans were placing sea mines probably by submarines in and near the harbours or in the open eastern coastline of America, so there was an urgency to remove or detonate these mines in order to keep the ship lanes open for troop movement to Europe from the US. It was, therefore, hoped that FSK sonar would detect the mines at a sufficient distance to find and destroy them and thus, Beckwith was invited by **Bell Laboratory** to work on the problem.

An experiment was carried out early in 1943 in a very secret lake outside Boonton NJ and unfortunately found out that the FSK sonar didn't work. The sonar head ran along a track in the lake. Using the FSK sonar, the head would run up and touch a dummy mine without seeing it!

The lab at the lake was operated by Bell Laboratories with **Dr. Horton** as our technical director. I remember **Dr. Vannevar Bush** being mentioned as the 'big man in charge' with names of **Einstein** and **Tesla** mentioned as involved. 'The bomb' wasn't ready and it was feared that the German submarine/mine problem might cause us to lose the war before the bomb could be used.

After the failure of the FSK sonar, we received a continuing contract to develop frequency modulated sonar at 26 kHz for surface to surface and surface to submarine communications. This was potentially a secret means of communications since the Germans couldn't hear 26 kHz and might not know of its presence.
http://www.beckwithelectric.com/ber/downloads/Hypotheses.PDF

For our communications experiments, we used the **Sardonix,** a luxury yacht converted for sonar work and the **minesweeper IX97, *which became the first ship to time travel.*** The **IX97** had degaussing cables hanging over the port and starboard sides of the boat. A third cable was supported on masts over the top of the boat. The officers took their ships where we needed them to go to carry out our experiments and had little involvement in the experiments except to volunteer to judge their ability to talk with the communications noise and distortion that was always present.

At General Electric we developed equipment for FM voice modulation of 26 kHz sonar 'heads' for ship to ship to submarine voice communications. These were tested from June 16 to July 7, 1944, at the Underwater Sound Laboratories at the New London CT Naval base.

On June 27, Dr. Horton appeared with a Bell Lab single sideband transmitter and receiver converted for the 26 kHz we were using; saying "Today we are going to test SSB. "We did and it worked better than FM for a reason that surprised us both. On an oscilloscope, the distortion of the FM and SSB were about the same. The difference was that the SSB distortion was linear. Even when the voice changed between 'rain barrel' to dolphin squeak quality, the meaning of the messages was recognizable. With FM the distortion was non-linear and reduced the voice quality to unintelligible.

The movie *THE PHILADELPHIA EXPERIMENT* stated that the experiment was for the purpose of making a ship invisible to radar. Radar was not a major problem to the Navy, whereas submarines and the new type of German mine were.
http://www.beckwithelectric.com/ber/downloads/Hypotheses.PDF

It appears that the true Philadelphia Experiment was planned in part at least by Dr. Horton and others at Bell Lab and people at the underwater sound lab that we worked with in the communications experiment. The IX97must have been the minesweeper involved since what other reason would there have been for the third overhead 'degaussing' cable?

Now, this is the point where the accounts of this story take on the overtones of the Philadelphia Experiment that most people are familiar with from the movie and the books.

The amazing happening just a few months before our experiment was the constant scuttlebutt conversation during meals and idle time as our boats moved in and out of the New London harbor. The stories were about *the disappearance and movement of the IX97* and disturbing results that led the skipper and experimenters to quickly shut down the experiment when they suddenly found themselves at the dock in Newport News.

Navy operators and undoubtedly at least one civilian in charge of the experiment may have been completely enclosed in the inner ships cabin space and 'went along for the ride' with no ill effects. The IX97 may have stayed in Norfolk long enough for those frightened operators of the experiment enclosed in the space to see where they were and suddenly turned off the power bringing the ship back to Philadelphia yards. *Unfortunate sailors on the deck or the dock must have been partly moved in time giving them the mind disturbing space separation*.

One most unfortunate mate fell from the deck to a position where he was trapped in the steelwork. Part of his body was inside and part outside of a cowling just forward of a port side cabin door. There was fresh paint on the inside and outside of this curved cowling. **(Recall the work of inventor John Hutchison and the knife embedded in a bar of steel from earlier photographs in the section which seem to employ similar scientific principles that were used on the IX97).**

180

The USS Martha's Vineyard (IX97) became the first ship to time travel in 1943
http://www.navsource.org/archives/09/46/46097.htm

 The main cabin formed the research room housing our experimental gear. The bridge was above us. A small room towards the stern housed ***three special looking generators*** motor driven from ships power. Controls for these generators were in the back of the main cabin. The IX97was very cramped whereas the Sardonix was deluxe with a lounge, kitchen, officers mess, and guest bedrooms. http://www.beckwithelectric.com/ber/downloads/Hypotheses.PDF

Beckwith believes that three phase currents were placed through the wires at a low frequency. This frequency could have been at one of the Earth's resonances recognized by Tesla and first measured by Schulman. The frequencies were approximately 7.5, 14 and 21Hz. I believe the generators operated over this range of frequencies. Both in spaceships per HYPOTHESES and in the experiment, this twisting field must be necessary to break the field of strong force lines and create an inner space separated from universal space. Once separated, ***the inner space containing the ship apparently moved about two weeks back to a time when the ship was berthed at Norfolk VA.*** As they gradually eased the power up, the experimenters found that lower power levels had no noticeable effect. Suddenly a threshold was crossed and wham! ***they traveled back two weeks in time instead of the desired 15 minutes or so needed to get out of the way of a mine. Actually, they may have hoped just to levitate or teleport and did not recognize the possibility of time travel!***

The idea of a discontinuity in the power level vs. travel function fits perfectly in support of the **Divided Space Hypothesis (DSH)** of "Chapter 13". As the current level in the three cables is increased, force lines begin to tear between the ship and universal space. Only when the level reached the point that all lines were broken was the IX97 free from the pull of the earth and the

time of the present. Once free, the boat was free to suddenly go wherever it is that things within a divided space go. In the case of the IX97, *that place was two weeks back in time to the place where it was at berth in Newport News.* Turning the power off when the skipper saw what had happened *reconnected the divided space to universal space, jerking the boat back to the Philadelphia Navy yard harbor.* The return of the IX97 to Philadelphia was *within seconds of the time it left, NOT in 1984* as in the science fiction version released to movie houses in 1984.
http://www.beckwithelectric.com/ber/downloads/Hypotheses.PDF

It could be that the jump point uses less energy if the frequency of the rotating magnetic field is synchronous with the Earth's Schulman resonant frequency. Tesla may have known of the relationship of the Earth's resonance but, not to the depth of predicting the results of The Experiment. Besides, *Tesla passed away in 1943 and may not have been personally involved in the experiment.* The frequency may have been varied along with the power level. With the power up and nothing going on, a small change in frequency may have moved into a well of high sensitivity equivalent to, say, a tenfold increase in power. The 'Q' of the Earth's resonances may be known to some readers and may well be between 10 and 100.

The size of the three generators in the rear of theIX97 were about that of a 50-horsepower 60-Hz motor. Scaled down to 7.5 Hz, the power used may have been about 15kw. Let's assume a weight of 1500 tons for the 1X97 and one kW equivalent to one horsepower. This gives 15x103watts/3x106 pounds = 5 milliwatts per pound. This is exactly the order of magnitude one can assume for a human who can levitate and teletransport without 'frying their brain'! And so another piece of the puzzle of **HYPOTHESES** fits together!

It seems clear that the power used was far greater than needed IF the phenomena were only better understood at the time. All that was intended was to move the minesweeper a mile or so out of the way of the mine, corresponding to a time movement of a few minutes. It then could be kept there until the mine surfaced and was destroyed. Of course, those details were not worked out since first it was necessary to find out whether the ship could, indeed, be quickly moved. One must remember that at the time, we were losing the war and especially a lot of shipping of material and troops to England. There was only time for one quick experiment as the minesweeper was vitally needed using the older technology with the expected high fatality rate. Knowing what was known after "The Experiment", the sailors on the deck could have been either sent off the ship or back inside before removing power in Newport News. We didn't know, however. The sailors who were injured by the experiment had little choice which could have improved their chance of survival.
http://www.beckwithelectric.com/ber/downloads/Hypotheses.PDF

The Philadelphia Experiment was much like the atom bomb it could either explode with a very large bang or it could fizzle. Some very forward looking scientists starting with Fermi realized that. As a result, the Manhattan program was organized from the start on the grand scale required. The bomb program could not creep in the methodology still followed by conventional science from a small bang in a test tube and gradually building to a bang of the size desired. Likewise, neither can the jump into divided space be made by the conservative scientific approach. *The Philadelphia Experiment must send a message to the free scientific world that*

182

profess all progress as creeping carefully along a path a step at a time with careful peer approval of each step. ***They will never time travel that way!***

HYPOTHESIS: Movement of a space into a divided space may be accomplished by creating a rotating magnetic field within the space. Separation of the divided space will occur suddenly above a line which is a function of the strength and the frequency of the field. The shape of the line will be that of a simple resonance of a cavity which in this case is the resonance of the Earth's atmosphere. Once the space is divided, objects within the space may levitate, teleport, or move in time. Parameters controlling the mix of these effects are unknown to the authors at present.

The authors suggest that the effect of the rotating magnetic field is to create a vortex in the neutrino field. We suggest that ***"gifted" humans produce the vortex by causing the DNA molecules in the body cells to form a spiral configuration.*** The neutrinos follow this configuration in sufficient number to create the separated space.

Using **nanotechnology,** it may be possible to create a curved surface of tiny moveable heavy metal surfaces so as to create a divided space that can be switched on and off by control of the moveable surfaces. http://www.beckwithelectric.com/ber/downloads/Hypotheses.PDF

HYPOTHESIS: If two rotating magnetic fields are operated in synchronism with one moving clockwise and the other counter-clockwise objects within one such space can be moved to the other space when power is applied.

This would be the fundamental principle of operation of ***the teleporter stations said to be operational between Eglin AFB and a base in Australia.*** It may be that the bodies of the sailors on the deck of the IX97had become separated into a number of isolated spaces. If so, heat and nerve signals could not flow across the dividing boundaries. It is easy to visualize the disruption of the functioning of a body so divided.

QUESTION: Is the drastic effect experienced by sailors participating in The Philadelphia Experiment related to conditions of mental patients in general? Could something be learned from reports of the experiment if they could be declassified? Surely this 54 year old information could now be made available to the mental health industry if requested for possible use in the free portion of our scientific community.

HYPOTHESIS: If an effect can be produced magnetically, surely some physicist can find a way to re-couple all of the strong force lines within a human body. Intuitively this seems no more difficult than a cat scan. If this is found to be a factor in mental illness, the procedure would certainly be beneficial. A procedure should be possible that is essentially without risk; surely so as compared to electric shock therapy.
http://www.beckwithelectric.com/ber/downloads/Hypotheses.PDF

Beckwith clearly remembers the scuttlebutt among those of us working together in New London concerning the experiments and I had no reason not to believe the stories which included men trapped in steelwork and men with very serious mental disorders. The mealtime and free time conversations may or may not have included Dr. Horton, but I believe that some did. As to the validity of the scuttlebutt concerning the ship moving experiment, I can only say that jokes of this kind simply were not made up during the war. Besides, how could one hoax a story involving so many people?

The Experiment could well have been planned and carried out by civilian scientists with little involvement of the Navy, as with the work in which Beckwith was involved. There seem to be no rumors of the technique being developed, and used during WW2, however, the Philadelphia Experiment surely was not forgotten. Where has this knowledge lead in the 50 years since the end of WW2? The magnetic mine is long since obsolete or if not, highly refined. The effect noted on the minds of the sailors near the minesweeper during the experiment and as illustrated by Fig. 15.3 may well have been independently developed into an insidious offensive anti-personnel weapon leaving little or no trace of its use.

Note: there were actually **three USS "Eldridge's** -- however the vessel that was involved in The Experiment is still in storage in the US with the cable marks still on the sides, and one of the "Eldridge"s was indeed sold to the Greek Government as the myth has it. It is also strongly rumored that the results of this experiment will be declassified in 2002.
http://www.beckwithelectric.com/ber/downloads/Hypotheses.PDF

The Quest for Anti-Gravity

The quest for anti-gravity has become the modern-day version of the quest for the Holy Grail! It is a quest for the perfect infinite energy source that will bring about a new order or level in human civilization. It is the quest to understand the mysteries of hyperdimensional physics that will allow mankind to free itself from obsolescent energy generation systems. The world awaits with long overdue expectancy for a new energy source that is infinite, environmentally safe and pollution-free which will replace the dirty, environmentally polluting fossil fuels of oil and coal, the dangerous thermal nuclear power plants and their radioactive toxins and contaminants and all other utility infrastructures that perpetuate the reliance upon an outdated grid system of energy generation that enslaves mankind upon its dependence.

Since the time of the great **Nikola Tesla** (and before), a small sector of the scientific community has known that so-called "free energy" can be extracted from our surroundings, and that many other exotic forces and effects, such as "anti-gravity", are also waiting to be liberated at our beck and call.

In his 1899 Colorado Springs experiments, Tesla discovered the **electro-gravitational (scalar) wave**, which oscillate the energy density of the vacuum and hence oscillate the curvature of space-time. So, over a century ago, it appears that Tesla had already produced a unified field theory of gravitation and electromagnetics. His discoveries were so fundamental, and his intent to provide free energy for all humankind was so clear, that it was probably responsible for the withdrawal of his financial backing, his deliberate isolation, and the gradual removal of his name from the history books.

The **zero-point energy of a vacuum** is the lowest energy vacuum state, with fluctuations taken into account. Even at low energies, quantum fluctuations continually arise and result in an incessant, extremely rapid and violent "jittering" of the energy momentarily present. The minimum energy due to these quantum fluctuations is called the **zero-point energy**. The amount of this energy is **HUGE**. Some scientists have hypothesized that *one cubic centimeter of pure vacuum contains enough energy to condense into $10^{80} - 10^{120}$ grams of matter*! *Quantum mechanically, no system of interest* (including even space-time itself) **can have zero energy**. The so-called "free energy" is actually obtained by tapping into the above-described zero-point energy.

The bedrock of much of modern science is **Classical Electromagnetic Theory (CEM). James Clark Maxwell** developed this 136 years ago in an exotic algebra known as *quaternions*. In order to render it more assimilable for use by working electrodynamicists, it was deliberately re-written in much simpler language by **Oliver Heaviside** (and **Gibbs**) in 1903. *This simplification (and truncation) eliminated a whole subset of the equations including the scalar electromagnetics and the gravitational aspects that were contained in the original theory.* At last count, there are *at least thirty-four known flaws* in Clark Maxwell's hoary old theory, *which is what is still taught in today's classrooms*!!! Some of the world's leading scientists, such as **Wheeler, Feynman, Bunge, Margenau, Barrett, Cornille, Evans, Vigier, and Lehnert** have all written about CEM's deficiencies.

When this missing *"Heaviside subset" of Classical Electromagnetic Theory* **is restored**, and the brilliant 1903 and 1904 work of a Cambridge University mathematician, **E.T. Whittaker**, factored in, *all of a sudden one has the supposedly elusive Holy Grail of Science* – a true **Unified Field Theory** *that unites* **General Relativity, Quantum Mechanics, Mind** *and* **Subtle Energy Phenomena** *and* **Classical EM Theory.**

But in fact, for the advancement of humanity, the real Holy Grail could be argued to actually be *contained in this missing subset*, not in the more grandiose sounding Unified Field Theory.

For it is this *"scalar potential"* that stresses local space-time, i.e. the 3 spatial dimensions AND time, which allows the *"bleed-through" of additional electromagnetic energy to create overunity electromagnetic systems.* Indeed, the restoration of this missing subset also shows that **Einstein's Theory of General Relativity** was also *only a subset of the real theory* that he was trying to write. Even though Einstein has been lionized for his theory of General Relativity (Time Magazine "Man of the Century"), he himself is on record as saying that the so-called foundations of physics need constant review and that his *Relativity Theory was not necessarily cast in concrete.*

A further impediment to the theoretical extraction of "free" energy was also imposed on what was left of Maxwell's already diluted EM Theory *by* **H.A. Lorentz** around 1902. *He simply arbitrarily threw away the monstrous amount of current outside the circuit that was not intercepted by the circuit, and that he could not theoretically account for.* This he termed *"of no physical significance!"* – even though it is approximately 10^{13} times greater than the intercepted current in our everyday electrical circuits! He thus perpetually locked EM systems in a theoretical and figurative iron box that would never allow them to go overunity and bleed in and capture additional energy.

Author' Rant: This is probably going to go down in history as the crime of the century or the millennium!

"Energy cannot be created or destroyed" and examples abound of where systems (such as heat pumps or windmills) put out more energy than they take in, simply by transducing other energy sources. *This is called OVERUNITY when the outputted energy added to additional transduced energy from another source puts out more energy than is provided by the original primary source.* Conventional science "allows" this in every aspect of "conventional science" **WITH THE SOLE (ARBITRARY) EXCEPTION OF ELECTROMAGNETISM.**

But, by not *"allowing"* the curvature of space-time locally, which is a way to open the gate to the *"free"* vacuum energy, *one can well see why some entrenched economic interests have discouraged investigation of this physicist into this awesome source of energy.* Indeed, we have been told that **THE US PATENT OFFICE HAS STRICT INSTRUCTIONS NOT TO ALLOW ANY PATENT FOR A MEANINGFUL OVERUNITY ELECTROMAGNETIC SYSTEM OR ONE THAT WOULD APPEAR TO THREATEN THE STATUS QUO OF OUR PRESENT POWER SUPPLIERS.**
http://www.bibliotecapleyades.net/disclosure/briefing/disclosure18.htm

Occasionally, however, the veil does inadvertently get lifted on some of the wondrous suppressed overunity systems, only to quickly be dropped again. Here are a few examples:

Tesla's Self-Powered Automobile

This has already been discussed earlier but bears repeating in brevity. In 1931, in a very secret program, **Tesla** built an overunity, self-powered electrical power system, placed it in a Pierce-Arrow automobile, and ran the car around successfully. A relative who rode with him in the car confirmed it many years later. Some details are described in **Marc Seifer**'s biography of Tesla as follows:

"The car [was] a standard Pierce Arrow, with the engine removed and other components installed instead. The standard clutch, gearbox, and drive train remained... Under the hood,

there was a brushless electric motor, connected to [or in place of] the engine... Tesla would not divulge who made the motor."

"Set into the dash was a "power receiver" consisting of a box... containing 12 radio tubes... A vertical antenna, consisting of a 6 ft. rod, was installed and connected to the power receiver [which was] in turn, connected to the motor by two heavy, conspicuous cables... Tesla pushed these in before starting and said: 'We now have power."

This Tesla device seems to have been remarkably similar to the radiant energy amplifier of T. Henry Moray described below. Also, Tesla coined the term "radiant energy" with respect to natural media in two of his patents.

Further, we have Barrett's mathematical demonstration that Tesla's actual patented circuits, when viewed in a higher topology, did indeed freely shuttle energy in the circuit as desired. In short, it appears that Tesla knew how to make circuits that asymmetrically self-regauged themselves creating overunity systems and therefore self-powering systems.

This is an entirely different operation than just the present entropic transfer of voltage used by electrical engineers today. It is more akin to deliberately regauging desired sections of the circuit so that excess energy appears there from an external source. In short, it is akin to asymmetrical self-regauging, and also to the type of operation of circuits that Kron (discussed below) discovered so laboriously but never entirely revealed.

No technical details were ever released on how Tesla's self-powered automobile system worked. The Tesla papers eventually turned over to his native country did not contain the actual "critical" papers present in his room at the time of his death. Those "critical" papers were illegally removed from his room as if Tesla were an illegal alien (he was a naturalized U.S. citizen, so the entire action was blatantly illegal). If those "critical" papers are still in existence, then they are still highly classified and hidden from conventional scientists. Cheney in her biography on Tesla reports finding the location of those papers.

The free energy electrical automotive power system part of Tesla's work was financed by the same world financier who financed Adolph Hitler's rise to power as well as much of the early Communist takeover of Russia. It is speculated that because of this financial control, Tesla would not have been allowed to put that car into production by one of the U.S. automotive companies http://www.bibliotecapleyades.net/disclosure/briefing/disclosure18.htm

The Moray Radiant Energy Device

In the early 1900's, **Dr. T. Henry Moray** of Salt Lake City produced his first device to tap energy from the metafrequency oscillations of empty space itself. Eventually, he produced a free energy device weighing sixty pounds and producing 50,000 watts of electricity for several hours. Ironically, although he demonstrated his device repeatedly to scientists and engineers, he was unable to obtain funding to develop the device into a useable power station that would furnish electrical power on a mass scale.

In the 1920's and 1930's Moray steadily improved his devices, particularly his detector tube, the only real secret of the device according to Moray himself. In his book, *"The Sea of Energy in Which the Earth Floats",* Moray presents documented evidence that he invented the first transistor-type valve in 1925, far ahead of the officially recognized discovery of the transistor. In his free energy detector tube, Moray apparently used, inside the tube itself, a variation of this transistor idea — a small rounded pellet of a mixture of triboluminescent zinc, a semiconductor material, and a radioactive or fissile material. His patent application (for which a patent has never been granted) was filed on July 13, 1931, long before the advent of the Bell Laboratories' transistor.

In test after test, Moray demonstrated his radiant energy device to electrical engineering professors, congressmen, dignitaries, and a host of other visitors to his laboratory. Once he even took the device several miles out in the country, away from all power lines, to prove that he was not simply tuning into energy being clandestinely radiated from some other part of his laboratory. Several times he allowed independent investigators to completely disassemble his device and reassemble it, then reactivate it themselves. In all tests, he was successful in demonstrating that the device could produce energy output without any appreciable energy input. According to exhaustive documentation, no one was ever able to prove that the device was fraudulent or that Moray had not accomplished exactly what he claimed.

The records are full of signed statements from skeptical physicists, electrical engineers, and scientists who came to the Moray laboratory and left with the complete conviction that Moray had indeed succeeded in tapping a universal source of energy that could produce free electrical power.

But in the face of all of this, the U.S. Patent Office refused to grant Moray a patent, first, because his device used a cold cathode in the tubes *(the patent examiner asserted it was common knowledge that a heated cathode was necessary to obtain electrons)* and, second, because *he failed to identify the source of the energy.* **(Author's note: Talk about stupidity! How did this man get a job at the Patent Office, unless, he's a shill?)** All sorts of irrelevant patents and devices were also presented as being infringed upon or duplicated by Moray's work. Each of these objections was patiently answered and nullified by Moray; nonetheless, the patent has still not been issued to this day, although the Morays still keep the patent application current.

John Moray, who operates the Research Institute in Salt Lake City, has been trying to continue his father's work since the basic unit was destroyed by a Russian double agent. Dr. Moray himself died in May 1974.
http://www.bibliotecapleyades.net/disclosure/briefing/disclosure18.htm

Gabriel Kron and the Negative Resistor

At the time of his death, **Gabriel Kron** was considered by some the greatest "non-linear scientist ever produced by the United States.

A negative resistor is defined as any component or function or process that receives energy in unusable or disordered form and outputs that energy in usable, ordered form, where that is the *net* function performed. We specifically do not include "differential" negative resistors such as the tunnel diode, thyristor, and magnetron which dissipate and disorder more energy overall than they order in their "negative resistance" regimes.

It appears that the availability of the Heaviside energy component surrounding any portion of the circuit may be the long-sought secret to Gabriel Kron's "open path" that enabled him to

188

produce a true negative resistor in the 1930s, as the chief scientist for General Electric on the U.S. Navy contract for the **Network Analyzer** at Stanford University. Kron was never permitted to release how he made his negative resistor but did state that, when placed in the Network Analyzer, the generator could be disconnected because the negative resistor would power the circuit. This negative resistor, one might add, was developed at the expense of the U.S. Taxpayer.

Since a negative resistor converges, surrounding energy and diverges it into the circuit, it appears that **Kron's negative resistor** gathered energy from the Heaviside component of energy flow as an "open path" flow of energy — connecting together the local vicinities of any two separated circuit components — that had been discarded by previous electrodynamicists following Lorentz. Hence Kron referred to it as the "open path." Kron describes this as follows:

"...the missing concept of "open-paths" (the dual of "closed-paths") was discovered, in which currents could be made to flow in branches that lie between any set of two nodes. (Previously — following Maxwell — engineers tied all of their open paths to a single datum point, the 'ground'). That discovery of open-paths established a second rectangular transformation matrix... which created 'lamellar' currents..." "A network with the simultaneous presence of both closed and open paths was the answer to the author's years-long search."

A true negative resistor appears to have been developed by Kron. He described his apparent success in 1945 stating: "*When only positive and negative real numbers exist, it is customary to replace a positive resistance by an inductance and a negative resistance by a capacitor (since none or only a few negative resistances exist on practical network analyzers).*" Apparently, Kron was required to insert the words "none or" in that statement. He also wrote that: "*Although negative resistances are available for use with a network analyzer, ...*" suggesting in rather certain terms that negative resistors were available for use on the network analyzer.
http://www.bibliotecapleyades.net/disclosure/briefing/disclosure18.htm

University of Moscow Scientists tested Overunity devices in 1930s

In the 1930s Russian scientists (Mandelstam et al.) at the University of Moscow and supporting agencies developed and tested parametric oscillator generators exhibiting COP > 1.0. The theory, results, pictures, etc. are in both the Russian and French literature, with many references cited in this particular translation. Apparently, the work was never resurrected after WW II.

The Original Point-Contact Transistor

The original point-contact transistor often behaved in true negative resistor fashion but was never understood. The point-contact transistor was simply bypassed by advancing to other transistor types more easily manufactured and with less manufacturing variances. Point-contact transistors can easily be developed into true negative resistors enabling COP> 1.0 circuits.

Burford and **Verner** (p.281) state that: "*...the theory underlying their function is imperfectly understood even after almost a century... although the very nature of these units limits them to small power capabilities, the concept of small-signal behavior, in the sense of the term when* *applied to junction devices, is meaningless, since there is no region of operation wherein equilibrium or theoretical performance is observed. Point-contact devices may, therefore, be described as sharply nonlinear under all operating conditions."*

Overunity device installed in Minuteman Missile - patented by Westinghouse

A frequency converter using 64 transistor stages and similar sophisticated feedforward and feedback mechanisms was placed in the original Minuteman missile, then deliberately modified to stop its demonstrated COP > 1.0 performance. After much investigation, it was found that the units were putting out some 105% as much energy as they received. Some were exhibiting COP = 1.15. Very quietly, Westinghouse engineers then obtained several patents surrounding the technology, but no further mention of it appears in the literature although DeSantis et al. showed that feedback systems with a multi-power open loop chain can produce COP > 1.0 performance. http://www.bibliotecapleyades.net/disclosure/briefing/disclosure18.htm

The Astronauts' Magnetic Boots

In the original magnetic boots for astronauts developed by **Radus** *et al.* at Westinghouse, the magnetic fields themselves — from *permanent magnets* — were simply switched! ***The astronaut could pick up his foot by simply switching off the permanent magnetic fields easily. They switched on again when he placed the foot down.*** He did not have to carry a huge battery around with him, to furnish enormous current to do that. And the magnets had a memory. (So far as is known, even today no one tells you that in many virgin magnets fresh from the factory, their very first use conditions them with a memory! *That fact can be used, for example, to create magnets whose fields appear normal, but which deviate from the normal behavior of ordinary magnets, including produce anomalies in their magnetic fields.*

It can easily be seen that when one can switch a permanent magnet's fields easily, and the magnet also has a built-in memory as did the Radus magnets, then with a little ingenuity in switching one could use such switchable magnets to produce a self-switching, self-powered permanent magnet motor. The magnet, being a permanent dipole, is already a particular kind of "free energy generator", since it continuously gates magnetic energy directly from the vacuum due to its asymmetry in the energetic vacuum flux.

The entire subject of making permanent magnets with memories, and how to use such in operational systems, is still a largely unexplored, extremely obscure territory. In fact, most researchers are not even aware that the phenomenon exists. Hoagland also discusses the **Astronauts' Magnetic Boots** on his website: The Enterprise Mission

Hitachi Engineers Confirm Overunity Process

Applications by Japanese inventor **Teruo Kawai** of adroit self-switching of the magnetic path in magnetic motors results in approximately doubling the COP. Modification of an ordinary magnetic engine of COP < 0.5 will not produce COP > 1.0. However, modification of available high efficiency (COP = 0.6 to 0.8) engines to use the Kawai process does result in engines exhibiting COP = 1.2 to 1.6. Two Kawai-modified Hitachi engines were rigorously tested by

190

Hitachi engineers and produced COP = 1.4 and COP = 1.6 respectively. The Kawai process, which can be constructed directly from the Patent with appropriate switching (e.g. photon), and several other Japanese overunity systems appear to have been blocked from further development and marketing.

The Magnetic Wankel Engine

The **Magnetic Wankel engine** should also be capable of COP > 1.0 and closed-loop self-powering, but apparently, it has also been suppressed, as have all Japanese COP > 1.0 EM systems. http://www.bibliotecapleyades.net/disclosure/briefing/disclosure18.htm

Johnson's Motors

Howard Johnson has built many novel linear and rotary motors and at least one self-powering magnetic rotary device that was later stolen in a mysterious break-in at his laboratory. Johnson used a bidirectional "two particle" theory of magnetic flux lines that can be justified by Whittaker's earlier work showing the internal bidirectional energy flows in all potentials and fields. He also utilizes controlled spin-waves and self-initiated precise exchange forces, which are known to momentarily produce bursts of very strong force fields.

His approach is to use highly nonlinear assemblies of magnets that initiate the foregoing phenomena at very precise points in the rotation cycle. In short, he seeks to produce precisely located and directed sudden magnetic forces, using self-initiated nonlinear magnetic phenomena. This is analogous to what the Wankel engine did using the Lenz law effect by sharply interrupting a weak current in an external coil. The **Lenz law effect** and other very abrupt field changes momentarily produce not only an amplified *Poynting* **energy flow component** but also an amplified *Heaviside* **energy flow component** as well.

Floyd Sweet's Vacuum Triode Amplifier

Floyd Sweet's solid state vacuum triode used specially conditioned barium ferrite magnetics whose H-field was in self-oscillation. The device produced a COP = 1.2×10^6 (that's 1.2 million!), outputting some 500 watts for an input of only 33 milliwatts. Sweet never revealed his complete ELF self-oscillation conditioning procedure for the magnets. However, **in ferromagnets, self-oscillations of (1) magnetization, (2) spin-waves above spin-wave instability threshold, and (3) magnons are known at frequencies from about 1 kHz to 1 MHz. Under controlled conditions, the apparatus also exhibited anti-gravity properties, producing a weight reduction of 90% in one** experiment.

It may be of interest that Kron was a mentor of Sweet, who was his protégé. Sweet, who is now deceased, worked for the same company, but not on the Network Analyzer project. However, he almost certainly knew the secret of Kron's "open path" discovery and his negative resistor. http://www.bibliotecapleyades.net/disclosure/briefing/disclosure18.htm

Dr. Deborah Chung's Negative Resistor

Dr. Deborah D. L. Chung, professor of mechanical and aerospace engineering at University at Buffalo (UB), is the leading "smart materials" scientist in this country, and a scientist of international reputation. She holds the Niagara Mohawk Chair in Materials Research at UB and is internationally recognized for her work in smart materials and carbon composites. On July 9, 1998, in a keynote address at the Fifth International Conference on Composites Engineering in Las Vegas, she reported having observed apparent negative resistance in interfaces between layers of carbon fibers in a composite material. The negative resistance was observed in a direction perpendicular to the fiber layers.

Her team tested the negative resistance effect thoroughly, for a year in the laboratory. There is no question at all about it being a true negative resistor. Dr. Chung submitted a paper describing the research to a peer-reviewed journal, and the University filed a patent application. Several negative articles appeared quickly in the popular scientific press. Conventional scientists were quickly quoted as proclaiming that negative resistance was against the laws of physics and thermodynamics. Others thought perhaps the UB researchers had made a little battery and were unaware of it.

On the website for the University of Buffalo, it was announced that the invention would be offered for commercial licensing. A Technical Data Package was available for major companies interested in licensing and signing the proper non-disclosure agreements. *Shortly thereafter this was no longer true, the data package was no longer available, and there was an indefinite hold on licensing and commercialization. It is still on hold as of this writing.*
http://www.bibliotecapleyades.net/disclosure/briefing/disclosure18.htm

Dr. Randell Mills and Blacklight Power

In early 1989, **Dr. Randell Mills** discovered that the hydrogen atom could be collapsed below its ground state and give up significant amounts of energy. At first, it was thought that he had discovered a new form of cold fusion. However, he showed that his discovery was indeed a new form of energy from the collapse of the hydrogen atom (which he calls **hydrinos**). An early report showed as much as 1,000 times as much energy out as input energy. This excellent amount of thermal energy was attributed to the catalytic reactions that provide a receptor for the energy emitted when the hydrogen collapses. The newsletter Fusion Facts named Dr. Mills as Scientist of the year for his work.

U.S. Patent 6,024,935 was granted in 2000 to Dr. Randell Mills and his company, **Black Light Power, Inc.** The patent was unusually large with 60 pages and 499 claims. The patent is for **Lower-Energy Hydrogen Methods and Structure**. Information about this new field is on the Web site http://www.blacklightpower.com/index.html

Cold Fusion

The phenomenon of **Cold Fusion** was first reported by Utah researchers **Stanley Pons** and **Martin Fleischman** and has been successfully replicated in several hundred experiments in

192

laboratories around the world in spite of a huge disinformation campaign. The situation with cold fusion is akin to the early days of aspirin. It works, but "conventional science" does not know how or why – ergo it is a threat to not only entrenched economic interests such as the centralized power industry but also to the orthodox scientific community.

Nuclear engineer **Dr. Thomas Bearden** has proposed several novel reactions, including the formation of time-reversal zones, which do explain cold fusion's mechanism. This is the because these novel experimental anomalies are in fact caused by using time as an energy source by transducing time-energy into spatial energy. Transducing one microsecond per second of time into spatial EM energy yields nearly 10^{11} watts of power. It can thus be seen that time itself is potentially a huge source of power and could well be the energy source of choice for the latter part of this century.

An apparent variation on cold fusion, the **Patterson Power Cell** uses a thin film to achieve the same spectral results, a finding apparently confirmed by **Professor George Miley** at the University of Illinois. According to **Dr. Eugene Mallove**, a noted scientist, and cold fusion analyst, during a test conducted by Motorola, the giant US electronics manufacturer, on the Patterson Power Cell. *"One cell produced 20 watts continuously with zero input power. It is under non-disclosure, but I have that raw data and there is no doubt that in numerous such tests have seen these effects where a thermal difference of let's say 15 degrees centigrade persisted for eleven hours in a cell no bigger than my thumb. You had the functional equivalent of a twenty Watt light bulb heating water."*

COP's of over 1200 have been reported from the Patterson Power Cell.

The foregoing examples are but a few of the known legitimate overunity systems. Lack of space precludes discussion of other proven systems by pioneers such as **Bedini, Watson, Fogal, Nelson, Weigand, Lawandy, McKie** etc.
http://www.bibliotecapleyades.net/disclosure/briefing/disclosure18.htm

Many other relevant papers can be seen on the **Department of Energy** public Website and at Tom Bearden.

Though it will never power anything of substance, on December 31, 1996, U.S. Patent Number 5,590,031 was awarded to **Franklin B. Mead Jr.** and **Jack Nachamkin** for a system for converting electromagnetic radiation energy to electrical energy. This is actually a proof of principle (using the **Casimir Effect**) in that it definitely proves that there are ways to extract usable EM energy from the vacuum. ***"It only takes one white crow to prove that not all crows are black! And that is definitely a fine little white crow."***

But the most ubiquitous gate of all for "free energy" is the simple little source **dipole**. The dipole becomes a universal kind of negative resistor extracting EM energy from the vacuum. Specifically, it absorbs energy from the time-domain (complex plane) and emits the energy in real 3-space. Unfortunately, lack of space precludes further discussion of this.
http://www.bibliotecapleyades.net/disclosure/briefing/disclosure18.htm

The Implications of UFO and ETI Disclosure

We have been told that the release of such information would create chaos and panic throughout society in almost every country on the planet. People would anarchy against governments, institutions of authority, the long-held cherished belief systems of science and religion, even the Military Industrial Complex. And who would blame them for such reactions to a new paradigm in their thinking brought about by a new reality that has never been confronted before in everyday existence? Quite frankly the people who be pissed off to the nth degree, after all, they have been lied to and misinformed for decades by the very people in officialdom whom they elected to represent them. All trust for governments, their political leaders, the military and the intelligence communities as well as the twin pillars in society: mainstream science and religion will have evaporated into a wisp of smoke.

The fact is that these governments, institutions of science and religion and the private industrial sector of the Military Industrial have all been duplicitous in the 70 plus years of cover-up and suppression of the knowledge of UFOs and ETI visiting the Earth. The recovered and reverse engineered alien technology, the science of which align with some terrestrial sciences particularly in the area of free, clean infinite energy generation and propulsion systems have deliberately been kept from the public consciousness for over a century! This has force the general public to be reliant upon polluting, environmentally dangerous fossil fuels and its inherent grid system dependency which has in turn created historical, unparallel financial burden and duress upon the world's population. Such grievous action is criminal and reprehensible. It is a crime against humanity! Only in an international court of law will world's outrage and call for justice be satisfied!

Let us be clear at this point that the implications actually go way beyond most governments or world leaders' knowledge and their control of the UFO matter to implement the proper checks, balances, and oversights. **Dr. Steven Greer** states the situation succinctly:

"The infrastructure needed to maintain and expand the level of secrecy which can deceive presidents and CIA Directors and senior congressional leaders and European Prime Ministers and the like is substantial - and illegal. Let me be clear, the entity which controls the UFO matter and its related technologies has more power than any single government in the world or any single identified world leader. The current state-of-the-art in secrecy is a hybrid, quasi-government, quasi-privatized operation which is international - and functions outside of the purview of any single agency or any single government. 'The Government' - as you and I and Thomas Jefferson may think of it - is really quite outside the loop. Rather, a select, tightly controlled and compartmentalized 'black' or unacknowledged project controls these matters. Access is by inclusion alone and if you are not included, it does not matter if you are CIA Director, President, Chairman of Senate Foreign Relations or UN Secretary General, you simply will not know about or have access to these projects." Dr. Steven M. Greer, "Understanding UFO Secrecy" and http://www.ufoevidence.org/documents/doc789.htm

The longer this **"Truth Embargo"** on the UFO/ETI cover-up and suppression continues in the governmental halls of officialdom, the greater will be the disastrous fallout from the official disclosure which must eventually be presented on the world stage. If heads don't roll in officialdom, certainly long-held seats of power will be overthrown. A carefully worded disclosure document may be the saving action of those who pretend to represent the governed in their midst but, it may come too little and too late and the implications of could actually bring about a new world order founded on justice for all!

Many governments worldwide will find they are impotent to provide real leadership with only an admission that they have lost political control decades ago to the ruthless machinations of a politically usurping, financially draining military industrial complex. This same M.I.C. has for decades enforced a *"Truth Embargo"* on the cover-up and suppression of knowledge regarding the existence of UFOs and the presence of ET Intelligence coming to the Earth behind a veil of lies, denials, disinformation, misinformation, and obfuscation.

Scientists too, could easily find themselves painted with a bull's-eye target on their backs as well, particularly those rogue scientists who have been willing duplicitous in their involvement with the black world of science, black ops and anything to do with black projects and programs that aided and supported the Military Industrial Complex, regardless of nation or political affiliation or even religious persuasion.

The disclosure of such information would see the upheaval and overthrow of many well-established institutions of society that have been around for thousands of years. Many of the world's orthodox religions like Judaism, Christianity, Islam, Buddhism, etc would be profoundly affected. Many of their adherents will pull away from the religion of their parents to seek truth and spiritual guidance elsewhere, finding that their former religious scriptures do not provide clear and concise writings to deal with the presence of another intelligent species or interstellar civilization that is visiting the Earth. Church, mosque and synagogue leaders will state that their religions *do allow* for a bigger universe that permits other intelligences but, *these will be manmade interpretations of spiritual writings and not as original sources of divine guidance.* Very few religions actually speak with absolute divine pronouncement to the fact that life exists elsewhere in the universe, they merely allude to it behind subjective interpretation of scripture.

This then is the **BIG FEAR** and the **BIG LIE** that the **Brookings Report** warned of when NASA commissioned them to give a report on the implications of discovery of an Extraterrestrial presence in the Solar System or on Earth. It was not the potential for mass panic among the public, except among the few handfuls of fundamentalist thinkers and eschatological religious nut-bars but rather, the fact that many long held and supported institutions of society would be rocked to their very foundations. Those who were in trusted positions of power and authority within the government and the military or as leaders of religious institutions, the bright minds of the scientific community and even, those of the wealthy corporate industrial sector, who all failed to protect the public's trust would find their worlds turned upside down. These would be the institutions which would feel the wrath and condemnation of an outraged public who had been misled and lied to for generations.

Mark Rodeghier, director of the **J. Allen Hynek Center for UFO Studies** wrote an article in *"The International UFO Reporter"*, Spring 1998, Vol. 23, number 1 that suggests that some UFO researchers in the US began asking for the release of UFO files from various government intelligence agencies using the **FOIA,** which may have lead many governments around the world to start releasing their UFO files onto the internet. This certainly appears to be the case as other

UFO researchers in other countries started to besiege their own governments for UFO files using the **Freedom of information Act** of their country.

In the middle to late 1970s, several individuals, mostly associated with the group **Citizens Against UFO Secrecy (CAUS)**, used the **Freedom of Information Act (FOIA)** to compel various U.S. government agencies to release their files on the UFO phenomenon. The **FBI**, the **National Security Agency (NSA),** and the **CIA** eventually complied with the law and released documents relating to UFOs, although the NSA did so only after a lawsuit was filed by CAUS. In addition, with the help of FOIA, the air force was forced to make available UFO documents which it had collected or produced after Project Blue Book closed in late 1969.

These documents do not contain a smoking gun to prove that the U.S. government has a secret UFO project. They did show that many agencies had an ongoing interest in the UFO phenomenon that was often independent of the air force's UFO project and that the interest continued after it ended.

Many of these documents have been made available to the public by UFO groups, including **CUFOS** and the **Fund for UFO Research**. It was also possible to obtain these documents directly from the agencies involved, which undoubtedly has increased their workload on a subject they consider to be unimportant.

The growth of the Internet has now provided at least two agencies--the FBI and the NSA--with a resource which they can use to reduce their workload. In the past few months, both agencies have placed all their released UFO documents on their Web sites. I assume that, in the future, people who contact the FBI and NSA for copies of UFO documents will first be referred to the Web sites, thus placing the burden of retrieval on the public.

Nevertheless, this arrangement is an excellent opportunity for all those who wish to directly read FBI and NSA UFO documents, at only the cost of a phone call to your Internet provider. The address (or URL) of each site is listed at the end of this article.
www.cufos.org/UFO_Documents_internet.html

So, is the *"lid finally coming off"* the UFOs and Extraterrestrial phenomenon or is this another carefully choreographed campaign by many of the world's governments to gradually release voluminous but carefully sanitized documents and files on the subject into the public domain? Will there be any real evidence or official admission by governments in all these 100's of thousands of documents and the millions of pages generated over the years of collecting reports that will finally state the obvious, something that the general public has known all along, that "we are not alone and never have been"!?

The fact that many governments have released hundreds of thousands of UFO documents during a 10 – 15 year period have more to do with the public demanding that their governments make these reports available to the public domain. Secret government UFO files which have been covered-up for the last 50-60 years, while denying their existence is in some ways, an admission of sorts that the governments took such matters seriously and a small victory for all UFO researchers and investigators. But, is it enough? Is there any real substance in the released documents that will turn the tide in favour of Ufology? Have the "boys in censorship" sanitized all the real important and "juicy" facts out of these files leaving merely volumes of worthless paper that simply state that "so and so" had a nighttime sighting of a bright light or a daytime sighting of a silvery flashing something which he couldn't recognize as a conventional flying object?

Do these documents prove that we have been visited by creatures from outer space in flying saucers? And has that fact been covered up by the governments of the world?

If the answer to all of the above questions is an emphatic "no", then why were they closed from public scrutiny for so long? Why is the government releasing them now? And how can anyone be sure they contain the truth?

The excitement generated by these document releases by nations has been in reality very disappointing with very little advancing the understanding of the UFO/ETI phenomenon globally or nationally. There are no official government admissions or acknowledgements that UFOs do indeed exist or that intelligent alien life forms pilot these strange spacecraft and are currently engaging diverse sectors of humanity.

There are some files that recount how some commercial and military pilots have seen and occasionally chased objects in the sky but that is nothing new in the consciousness of the public, even this author has seen military jets chase after unidentified flying objects on several occasions. Reports of little men or strange alien beings are again, beginning to become passé. The frequency of such sightings and ET encounters has been a gradual learning curve and mental conditioning that has developed over the decades since the late '40s!

The only thing that has come out from the release of all these thousands upon thousands of documents is that the governments of nations know more than the average "John Q. Public" knows and their admission is essentially that, not so much in words of official disclosure but in their actions to conceal important information from the public supposedly for their own good. Actions which essentially have not deviated since governments or branches of the government like the military first set up investigation teams to discover what they were dealing with in their nation's skies.

Perhaps, the campaign to release UFO files by some nations in such volumes was ahead of their normal release date was to give the appearance greater openness or it was intended to keep Ufologist pre-occupied researching the information for the next 5 to 10 years allowing the governments and the militaries some breathing room to figure what their next course of action would be. Will it be that the next step is a full official government disclosure of truth with honesty or a carefully worded announcement that will justify the decades of secrecy and suppression of information while scientists working behind the scenes for various branches of the government, the military, and the intelligence agencies determine what they were up against? Needless to say, officialdom was dealing with a superior technology many hundreds to thousands of years or more, ahead of our own! No official disclosure if it ever happens is going to paint the government and the military in an unfavourable light as they must maintain the image or at least the illusion that the governments and the powers that be know what they are doing for the sake of the best interests of the public.

As we have stated before such official disclosure on the subject of UFOs and ETI may come too little, too late as the whole UFO question become axiomatic, as it seems very clear that the phenomenon has a life of its own and may not be playing by human rules of engagement. The longer nations put off the inevitable disclosure process, the worst it will be for them, as people have given them the power act on behalf of them and they can just as easily remove that power and those people in power if they don't do the right thing. Time will tell and it's running out quickly for the governments!

Below is a list of hyperlinks that the reader can click on to open to view each nation's UFO files, you can now see for yourself what constitutes the **"evidence"** and make up your own mind and as some journalists have reported, "This is freedom of information working to inform and educate the public." At least this is the intended perception but, what all these reports don't tell you is the assessments and conclusions reached by governments. These are the important details that the public and the Ufologists are after which would form an official disclosure!

The exception is Canada! *The important difference between the released Canadian UFO files with other country releases is the inclusion of departmental analyses rather than simply reports of UFO sightings.* According to **Victor Viggiani** from **Exopolitics Toronto** , who has been monitoring the Canadian Government UFO website since its inception, *"The Canadian files do not simply list UFO sightings; they describe actions, meetings and inter-departmental memoranda generated by Canadian officials that attempt to make sense of the considerable onslaught of UFO sightings as well as referencing American problems with keeping abreast of UFO sightings."*

For example, a September 1967 memo titled "Unidentified Flying Objects (UFOs) - Investigations", was released which stated: *"a number of investigations of the reports suggest the possibility of UFOs exhibiting some unique scientific information or advanced technology which could possibly contribute to scientific or technical research."*

Undoubtedly the most significant documents are those associated with a 1950-1952 classified investigation and analysis of UFOs by a **Department of Transportation** team led by Wilbert **Smith**, a senior radio engineer. In his **Project Magnet Report**, Smith commented extensively on the flight performance of UFOs that were far in advance to anything known at the time:

"… it is difficult to reconcile this performance with the capabilities of our technology, and unless the technology of some terrestrial nation is much more advanced than is generally known, we are forced to the conclusion that the vehicles are probably extra-terrestrial, in spite of our prejudices to the contrary." http://www.examiner.com/article/canada-releases-ufo-x-files-to-the-world

Now these conclusions fly in complete contradiction to the findings of the **Condon Committee Report** which was sponsored by the USAF that concluded that no new scientific knowledge or

198

advancement in technology would be gain by further research or investigation into this phenomenon.

Lest anyone think that the release of various governments UFO files is a waste of time in researching their content, there are glimpses into what governments knew about the phenomenon. In a recent batch of UFO files released by the British government indicated that Winston Churchill and President Eisenhower knew that early during WWII that many of the war theatres were being monitored by aerial vehicles, not from this world! They decided together at that time that the public didn't need to know this new development given the state of the public's pre-occupation of the war and its eventual outcome.

There is also a "treasure trove" of UFO files recently released by the American intelligence agency, the NSA released to the public domain formerly classified UFO X-files which had never received any prior media attention. Of particular interested in the NSA Technical Journal Vol. XIV No 1 with FOIA Case number 41472 which has been titled **'Key To The Extraterrestrial Messages'.** The document authored by a **Dr. Campaigne** presented *a series of 29 messages received from outer space in "Extraterrestrial Intelligence".*

The unclassified document describes developing a key to understanding these alien messages confirms the presence of extraterrestrials, but that the US Government has received deep space transmissions from a civilization outside our own solar system!

The following is transcribed from Page 21, Appendix:

"Recently a series of radio messages was heard coming from outer space. The transmission was not continuous but cut by pauses into pieces which could be taken as units, for they were repeated over and over again. The pauses show here as punctuation. The various combinations have been represented by letters of the alphabet so that the messages can be written down. Each message except the first is given here only once. The serial number of the messages has been supplied for each reference." A link to the document is given below for the reader to peruse:
http://www.nsa.gov/public_info/_files/ufo/key_to_et_messages.pdf

This revelation from the NSA disputes the work of **SETI** and **Seth Shostak** that they have not received any more strange alien-like signals from their radio telescopes since **the WOW! Signal** and it confirms **Dr. Greer's** account that Harvard's **BETA Radio Telescope** had also received alien signals on a regular basis that were neither natural nor manmade.

Declassification of UFO documents began by the United Kingdom in 2008. Russia, Ecuador, United States, France, Denmark, Brazil, Sweden, Canada and many countries follow that decision during 2009 making public all the material they had in possession. The government of New Zealand has stated that it will release UFO files in 2010 after personal information is removed. The list below is presented from the most recent to the earliest release of documents:

- United States - Declassified NSA UFO Documents (April 25, 2011)
- United States – FBI UFO Files Released - (April 11, 2011)

- Declassified: the UK's UFO files - A fresh batch of files detailing sightings of unidentified flying objects in the skies of Britain has been declassified by the National Archives. - (March 3, 2011)
- UK National Archives - Newly Released UFO Files - (March 3, 2011)
- New Zealand Releases UFO Files- (December 23, 2010)
- New Zealand UFO Papers to Be Made Public - (January 23, 2010)
- UK UFO Reports - (January 11, 2010)
- Unidentified Aerial Phenomena in the UK, Air Defence Region: Executive Summary (PDF document) - from UK Ministry of Defence website - (December 2000)
- UFO disclosure in Brazil: More top-secret files released by government - (September 30, 2009)
- The British Government and Crop Circles - (August 2009)
- Bill Clinton discusses UFO's, Roswell and Area 51 - Hong Kong Sept. 14, 2005 - Note: Disclosure Project (when it was Project Starlight in the 1990's) provided Bill Clinton with detailed information on this matter. (posted August 22, 2009)
- Russian Navy Reveals Its Secret UFO Encounters - Fox News (July 27, 2009)
- Canada Opens UFO Files (February 15, 2009)
- Secret UFO Archives Opened in Denmark - The Danish Air Force has opened its UFO archives, providing information on over 15,000 reported extraterrestrial sightings to the public - the Copenhagen Post Online - (January 29, 2009)
- Hemmeligt arkiv: 15-årig så rummænd på Fyn [Danish news article about UFO Archives] - Forside (January 28, 2009)
- Above Top Secret Forum discussion with some translations of cases
- Denmark UFO Files Released (in Danish) - documented cases (January 29, 2009)
- NICK POPE AND THE MINISTRY OF DEFENCE ARE TAKEN TO TASK. Crop Circles and UFOs - The Truth we seek - by Colin Andrews (January 2009) -- Response from Nick Pope - (January 24, 2009)
- Tories 'would publish UFO files' - BBC (January 27, 2009)
- RAF 'ordered to shoot down UFOs ' - Daily Telegraph (January 27, 2009)
- Call for Obama to open UFO files - Seacoast Online (January 18, 2009)
- Newly Released UFO files from the UK National Archives (October 20, 2008)
- U.S. pilot was ordered to shoot down UFO - London, Reuters (October 20, 2008)
- A FORMER Top Gun told yesterday how he was ordered to shoot down a massive UFO over NORWICH (October 20, 2008)
- UK releases classified UFO files - Ker Than, NewScientist.com News (May 14, 2008)
- In pictures: The UFO files - BBC News (May 14, 2008)
- UK National Archives UFO Site
- Vatican: It's OK for Catholics to Believe in Aliens - Fox News (May 13, 2008)
- Ovni-USA Article about February 2008 United Nations Meeting: This is a fax of the French article, plus translation into English. Page 1 [JPG file]   Page 2 [JPG file]   Page 3 [JPG file]   Page 4 [text file] - (February 23, 2008)
- MoD to open British UFO X-files (December 24,2007)
- Japan's Top Government Spokesman: UFOs 'Definitely' Exist (December 18, 2007)
- Defense Minister troubled over legal issues if UFO arrives (December 20, 2007) (Note that this could be spun to be xenophobic.)
- Defense Minister Ishiba Considers Japan's Options in UFO Attack (December 21, 2007)

- [Clinton UFO files released by Clinton Library](#) (November 9, 2007)
- [Irish defence forces eyed UFOs for 37 years](#) (September 20, 2007)
- [Ireland's UFO Files Released](#) - A secret 'X-files' style dossier of UFO sightings in Northern Ireland has been made public for the first time. (August 2, 2007)
- [by Nick Pope - Daily Mail, UK - June 30, 2007](#) - includes Rendlesham Forest incident information
- [Daily Telegraph (Australia)](#) - July 1, 2007
- [This is London - July 1, 2007](#)
- [UK MoD opens its files on UFO sightings to public](#) - (The Guardian, May 3, 2007)
- [French space agency to publish UFO archive online.](#) - Reuters (December 29, 2006)

Related Sites:

- [Geipan, the French UAP research and information group](#)
- [EUX TV - France Opens its Official UFO Archives](#) (Thursday, March 22, 2007)
- [New Scientist - France Opens Up its UFO Files](#) (March 22, 2007)
- [physorg.com - France opens secret UFO files covering 50 years](#) (March 22, 2007)
- [Washington Post - French space agency to publish UFO archive online](#) (March 22, 2007)
- [French 'X-Files' to Post Online](#) - Live Science - from Associated Press (March 23, 2007)
- [ABC News - France's Space Agency Puts Its Entire Secret UFO Archive on the Web](#) (March 23, 2007)
- [[UK] Ministry files on UFO sightings open to public - click here to read article Click here to get to Ministry Of Defence page with links to the sightings files](#) - (February 20, 2006)
- [Major UFO Breakthrough in Brazil](#) - (May 22, 2005)
- Mexican Air Force Jets Encounter UFOs - Media Reports: (May 2004)
- [USA Today Report](#)
- [CNN Report](#)
- [Fox News Report](#)
- [CBS News Report](#)
- [COMETA REPORT - Article in the Boston Globe (05-21-2000) COMETA Report - Part 1 - English (PDF file) COMETA Report - Part 2 - English (PDF file)](#)

http://www.disclosureproject.org/countries-releasing-ufo-files.shtml

This second list below contains additional countries that have actually released documents as well as those that are in litigation to disclose all or specific files and countries that have released significant aspects of UFO and alien encounters. There are some countries which are the same as in the above list but, these countries release more than one patch of UFO documents at later times and not all at once, most likely because the files are been compiled and coming from different agencies within the government. The frequency of newly released of files appears to be ramping up in rapid fashion and will continue until we have some undeniable facts brought forth that the UFO and ET phenomenon is real or there will be a UFO event of historic proportions that will be proof positive of ETI existence.

Click on Country for Source:

1. Argentina
2. Australia
3. Brazil
4. Canada
5. Chile
6. China
7. Denmark
8. Finland
9. France
10. Germany
11. India
12. Ireland

13. Japan
14. Mexico
15. New Zealand (Additional Report)
16. Peru
17. Russia
18. Spain
19. Sweden
20. Ukraine (not in English)
21. United Nations
22. United Kingdom
23. Uruguay
24. Vatican City

It seems very clear that humanity is being prepped by 20 plus nations who appear to be cooperatively working together for UFO/ETI disclosure. It is no coincidence that these countries have started partial or full disclosure, even comments and actions from the **Vatican** and the **United Nations** indicate that the "so called powers" want some of this information to become public. http://www.educatinghumanity.com/2011/01/list-of-countries-that-have-disclosed.html

CHAPTER 99

OFFICIAL GOVERNMENT DISCLOSURE
VS. THE PEOPLE'S DISCLOSURE

Many people and UFO researchers believe that at some point, perhaps in the near foreseeable future, the US Government will eventually have an official disclosure announcement or event to reveal publicly what it knows about the UFO phenomenon. However, as the years go by with government changes in political parties and new US Presidents, official disclosure still appears to be a pursuit for the "Holy Grail"!

To date, there is no official US government disclosure on the existence of UFOs and ETI and in all likelihood, there never will be an official public disclosure!!!

Some UFO researchers are starting to realize that no government whether Democratic or Republican or a new President will ever step up to the plate and finally lift the "truth embargo" off the UFO/ETI subject. Any President promising to do so is either threatened and quickly avoids the whole subject completely or when they are interviewed on the **Jimmy Kimmel Show**, they jokingly say that they tried and failed as in the case of the interview with former **President Bill Clinton** or in the case of **President Barrack Obama** who also jokingly said that *"the aliens won't let it happen. We'd reveal all their secrets… that's why they exercise strict control over us!* https://www.youtube.com/watch?v=EYzRY2XpLBk

Author's Rant: Is he kidding me? This is a major public announcement if it is true and Obama jokes about it!

There is a real risk factor to the life to any US President that will disclose the government's knowledge on UFOs. Former **President Jimmy Carter** ran his election campaign to disclose everything the government knew on the UFO matter and then, it was later revealed that when he took office as President that he had been threatened and was told that if he made any official disclosure on UFOs, *"he would end up like Kennedy!"* Thus, Carter never made good on his promise to provide an official government disclosure on UFOs. Since that time every new President voted into office, it was hoped or expected or was petitioned by voters to become the first President of transparency or disclosure on the UFO matter.

To date, there has been no such luck. No President has been brave enough to step up to the plate and become the "**President of Disclosure**"! In fact, many US President were *"cut out of the loop"!* Even **US President Bill Clinton** had a very strong interest in the UFO subject and wanted to know the "secret behind UFOs"!

"Over in the Justice Department Bill Clinton had assigned a close friend to work the UFO disclosure effort for him. Clinton had appointed **Webster Hubbell**, a longtime Arkansas friend, and **Hillary Clinton** associate at the Rose Law Firm, as Assistant Attorney General. Along with the Justice appointment came the UFO assignment.

Hubbell described what Clinton had requested from him in his 1997 book *"Friends in High*

Places". Hubbell recounted Clinton's instructions, *"If I put you over at Justice I want you to find the answers to two questions for me. One, who killed JFK? And two, are there UFOs?"* Hubbell continued, *"Clinton was dead serious. I had looked into both, but wasn't satisfied with the answers I was getting."*

The request for Hubbell not only pointed out Bill Clinton was looking for a UFO answer, it showed clearly that he did not believe the official government position on either UFOs or the Kennedy assassination.

The official government position on UFOs had been spelled out in 1969 when the USAF released its final report on its 21-year investigation of 12618 UFO cases called **Project Blue Book.** The report concluded,

1. No UFO reported, investigated, and evaluated by the Air Force has ever given any indication of threat to our national security.
2. There has been no evidence submitted to or discovered by the Air Force that sightings categorized as "unidentified" represent technological developments or principles beyond the range of present-day scientific knowledge.
3. There has been no evidence indicating the sightings categorized as "unidentified" are extraterrestrial vehicles.

According to **Dr. Steven Greer**, the official Blue Book conclusion was presented to Bill Clinton when he requested a briefing on the subject of UFOs after taking office. President Clinton knew he was getting the runaround and was not happy.
http://www.hillaryclintonufo.net/disclosureefforts.html

The struggle to get official recognition and an unbiased scientific investigation from either the US government or the United Nations has been fraught with a deafening silence and/or sometimes with threats to the US government or UN officials who would participate in any kind of disclosure regarding UFOs or to anyone who would organize such events and campaigns.

The covert and hidden *"powers than be"* supported by the military and various intelligence agencies have done so in the past and will continue to do so in the future, whatever it takes to prevent any public light being shined on the UFO phenomenon. This is no conspiracy theory or irrational paranoia as many people both in officialdom and in the UFO community have experience and will attest such threats are real!

But, every once in awhile, brave and courageous individuals arise knowing the potential circumstances that may befall them and make such disclosures of truth known to the rest of the public! The history of UFO disclosure in the US can be marked by four major events that have involved witnesses from diverse backgrounds coming forward with first-hand testimony and evidence stating the existence of a world-wide phenomenon requiring further research and investigation. These four historic disclosure events include:

I.   The **Symposium on Unidentified Flying Objects** follow the Hearings before the **Committee on Science and Astronautics U.S. House of Representatives**; Ninetieth Congress; Second Session on July 29, 1968, which was chaired by Hon. J. Edward

Roush. It was an opportunity to provided Congressmen and Committee staff with the opportunity to ask questions of the participant scientists to learn about the nature of Unidentified Flying Objects and their implications to mankind.

II.   UN Hearings on UFOs was initiated by the lobbying of **Eric Gairy, the Prime Minister of Grenada** of the United Nations and the efforts of **Lee Speigel** as the primary organizer lead to a hearing on July 14, 1978, by notable scientists to inform **UN Secretary General Kurt Waldheim** about the UFO matter.

III.   **The Disclosure Project** event of May 9, 2001 in Washington, D.C. lead by Dr. Steven M. Greer where over twenty military, intelligence, government, corporate and scientific witnesses came forward at the **National Press Club** in Washington, DC to establish the reality of UFOs or extraterrestrial vehicles, extraterrestrial life forms, and resulting advanced energy and propulsion technologies. The weight of this first-hand testimony, along with supporting government documentation and other evidence, established without any doubt the reality of these phenomena.

IV.   **The Citizens Hearings on UFO Disclosure** arranged by the **Paradigm Research Group**, headed up by lobbyist, **Steven Bassett** was held from April 29 to May 3, 2013 in Washington, DC where researchers, activists, and military/agency/political witnesses representing ten countries gave testimony to six former members of the United States Congress about events and evidence indicating an extraterrestrial presence engaging the human race.

Symposium on Unidentified Flying Objects (1968)

On two previous occasions, the Congress of the United States has conducted open hearings on the subject of Unidentified Flying Objects. On April 5, 1966, the **House Armed Services Committee** held public hearings *(these were not actually open to the public, they were only open to USAF service personnel),* and on July 29th, 1968, the **U. S. House of Representatives' Committee on Science and Astronautics** convened a one-day Symposium on Unidentified Flying Objects, chaired by then-Indiana **Congressman J. Edward Roush**.

However, these two occasions were not the only time that the subject was discussed by legislators. **Project Blue Book** documents, newspaper stories and letters in the **National Investigations Committee on Aerial Phenomena (NICAP)** files show that on a number of occasions UFOs had been privately discussed in an executive session of various committees and subcommittees. However, the July 29, 1968, Symposium on Unidentified Flying Objects was unique in the respect that it provided Congressmen and Committee staff with the opportunity to ask questions of the participants, and the results were made accessible to the public through the government printing office.

Since the late fifties, NICAP had struggled to get Congressional attention focused on the UFO phenomenon and the official handling of UFO investigations. During this period the Project Blue Book files had only been available to a few select individuals. While the Blue Book files contained an extensive collection of UFO reports, they were hardly definitive. In fact, NICAP probably had just as many well-investigated cases in its own files. However, the denial of public access to the Project's files seemed like a cover up, and something on which to focus the request for Congressional action. NICAP developed a number of proposals they hoped Congress would help implement:

(a) The public release of official UFO files from the USAF Project Blue Book and other agencies,

(b) A review and reform of the USAF UFO investigation methods,

(c) An end to the mistreatment of some UFO witnesses, who NICAP felt were unfairly categorized in press statements or ordered into keeping silent about their experiences,

(d) A review of possible threats to US national security, which NICAP thought were being ignored.

Congressman L. C. Wyman requested the type of hearings that NICAP proposed and entered a resolution into the House to authorize the **Committee on Science and Astronautics** to conduct a wide-ranging hearing, complete with witnesses and subpoena powers. **Indiana Congressman J. Edward Roush**, an advocate of serious attention for the UFO problem, thought the action premature, and wanted to wait until the **Condon Committee**, then underway at the University of Colorado, had delivered its final report. In the meantime, he proposed a Symposium and became the driving force behind it. http://www.project1947.com/shg/symposium/shgintro.html

The Symposium that resulted was not what NICAP had hoped for. Rather than examining the USAF's handling of UFO investigations, or the details of the then in-progress University of Colorado study, the discussion was confined to an exchange of views and evidence presented by the participants.

The Symposium consisted of six scientists presenting their views on UFOs to the committee:

- **Dr. J. Allen Hynek**, Chairman, Department of Astronomy, Northwestern University, Evanston, Illinois and at the time a scientific consultant to the USAF on UFOs for almost two decades;
- **Prof. James E. McDonald**, Department of Meteorology, and Senior Physicist at the Institute of Atmospheric Physics, University of Arizona, Tucson, Arizona, who had conducted a multi-year full-time investigation of the UFO problem;
- **Dr. Carl Sagan**, Associate professor of astronomy, Center for Radiophysics and Space Research, Cornell University, Ithaca, New York;
- **Dr. Robert L. Hall**, Head, Department of Sociology, University of Illinois, Chicago, Illinois;
- **Dr. James A. Harder**, Associate professor of civil engineering, University of California; and
- **Dr. Robert M. L. Baker**, Jr. Senior scientist, System Sciences Corp., North Sepulveda Boulevard, El Segundo, California who had also done extensive analysis of UFO films.

    These scientists also participated in discussions with the Congressmen and their staff after the initial presentations and some had written statements read into the record. http://www.project1947.com/shg/symposium/shgintro.html

A number of other scientists who did not appear before the committee but submitted written statements were:

- **Dr. Donald H. Menzel**, Director of the Harvard University Observatory, author of a number of books and articles on UFOs;
- **Dr. R. Leo Sprinkle**, Division of Counseling and Testing, University of Wyoming;
- **Dr. Garry C. Henderson**, Senior Research Scientist, Space Sciences, General Dynamics;
- **Stanton T. Friedman**, Westinghouse Astronuclear Laboratory;
- **Dr. Roger N. Shepard**, Department of Psychology, Stanford University; and
- **Dr. Frank B. Salisbury**, Head, Plant Science Department, Utah State University, NASA consultant and author of UFO articles.

The Symposium represented a variety of opinions on UFOs, from the advocacy of Dr. James McDonald to the skepticism of Dr. Donald H. Menzel, who felt that any consideration of the problem was a complete waste of time.

Representative Roush's defeat in the next Congressional elections was the end of UFO hearings "On the Hill." With the nation's interest consumed by the war in Viet Nam, there could be no further action into public hearings without the support of at least one dedicated Congressman from the committee.

J. Edward Roush maintained his interest in UFOs and later accepted a position on NICAP's Board of Governors.

In his introduction to the *"Scientific Study of Unidentified Flying Objects"*, Dr. Condon pointed out that the Symposium might be viewed as a counterbalance of opposing views to the University of Colorado study. While to a certain extent this may be true, most of the scientists' presentations were introductory in nature and did not go into detailed analyses. Regardless, the Symposium represents diverse opinions on UFOs that were held among a number of scientists who had taken more than a cursory glance at the phenomenon.
http://www.project1947.com/shg/symposium/shgintro.html

The United Nations and UFOs (1977–1979)

In 1977, **Eric Gairy**, the Prime Minister of the Caribbean island nation of Grenada, was struggling to generate U.N. interest in his personal crusade to establish an international committee dedicated to studying and sharing UFO reports.

Lee Speigel, an AOL writer wanted to bring something before the United Nations, to try to get that esteemed group of international leaders to finally come together on a subject that could potentially advance the world in both scientific and sociological ways.

Speigel had been aware of **P.M. Gairy** of the Grenada and his efforts lobbying the U.N. to pay attention to his UFO campaign, offered to help P.M. Gairy produce a presentation with a stacked lineup of UFO experts bound to pique the interest of international leaders. P.M. Gairy quickly agreed and Speigel arranged for several of my UFO "presenters" to meet with Gairy and **U.N. Secretary-General Kurt Waldheim** to discuss our plans.

The experts included astronomer (and former UFO consultant to the Air Force) **J. Allen Hynek**, who coined the term "Close Encounters of the 1st, 2nd, and 3rd Kind." There was also French astronomer **Jacques Vallee**; astrophysicist **Claude Poher**, who headed up the official French UFO research group; psychologist **David Saunders**, a member of the famous U.S. government **Condon Committee**, which discredited UFOs in 1969; and **Gordon Cooper**, one of America's original "right stuff" astronauts, **Stanton Friedman**, nuclear physicist and Army **Lt. Col. Larry Coyne** .

At this highly publicized meeting, Hynek told Waldheim that he and many of his colleagues felt that *"the time has come for official education of the public regarding UFOs. We now recognize that, apart from its potential scientific importance, it has sociological and political significance."*

A large number of pictures, films and documents were collected for the presentation. Among the documents presented was one from Hynek which he'd managed to secure from the Air Force while he was the official scientific consultant. It was a chapter from a 1968 textbook, "Introductory Space Science," created by the USAF Academy department of physics. ***And it was, of course, not for the public eye.*** https://sbeckow.wordpress.com/2010/07/08/lee-spiegel-remembers-his-un-meeting-on-ufos-in-1978/

In the lengthy chapter, titled **"Unidentified Flying Objects,"** Air Force officials laid out for academy cadets clear and concise information about UFOs, including the following, almost shocking, revelations (completely contrary to official explanations):

"What we will do here is to present evidence that UFOs are a global phenomenon which may have persisted for many thousands of years ...

"The most stimulating theory for us is that the UFOs are material objects which are either 'manned' or remote-controlled by beings who are alien to this planet. There is some evidence supporting this viewpoint."

After offering cadets numerous examples of UFO cases that were unexplained, the chapter concludes, *"This leaves us with the unpleasant possibility of alien visitors to our planet, or at least of alien controlled UFOs ... and what questionable data there are suggests the existence of at least three and maybe four different groups of aliens."*

Lee Speigel organized a meeting of military, scientific and psychological experts with U.N. Secretary-General Kurt Waldheim on July 14, 1978, to plan a presentation to the U.N. Special Political Committee.

When the day of the event arrived, Nov. 27, 1978, in addition to speeches given to the special political committee by **Hynek** and **Vallee**, nuclear physicist **Stanton Friedman** spoke about the evidence that favored his theory of extraterrestrial visitation.
https://sbeckow.wordpress.com/2010/07/08/lee-spiegel-remembers-his-un-meeting-on-ufos-in-1978/

And there was also Army **Lt. Col. Larry Coyne**, who electrified the committee with a story about a night in 1973 when he and his Army helicopter crew had a terrifying close encounter with a UFO.

One of the documents **Speigel** obtained for the event was a letter that he asked ex-astronaut Cooper to write, which was distributed to all U.N. delegates. In it, he said, *"I believe these extraterrestrial vehicles and their crews are visiting this planet from other planets, which obviously are a little more technically advanced than we are here on Earth.*

"I would also like to point out that most astronauts are very reluctant to even discuss UFOs due to the great numbers of people who have indiscriminately sold fake stories and forged documents, abusing their names and reputations without hesitation."
https://sbeckow.wordpress.com/2010/07/08/lee-spiegel-remembers-his-un-meeting-on-ufos-in-1978/

Summary: At its 87th plenary meeting, on 18 December 1978, the **UN General Assembly** on the recommendation of the **Special Political Committee** recommended the establishment of an agency or a department of the United Nations for undertaking, co-ordinating and disseminating the results of research into unidentified flying objects and related phenomena.

Specifically: UN General Assembly decision 33/426 was passed at its 87th plenary meeting on 18 December 1978. The General Assembly, on the recommendation of the Special Political Committee, adopted the following text as representing the consensus of the members of the Assembly:

1. The General Assembly has taken note of the statements made, and draft resolutions submitted, by Grenada at the thirty-second and thirty-third sessions of the General Assembly regarding unidentified flying objects and related phenomena.

2. The General Assembly invites interested Member States to take appropriate steps to coordinate on a national level scientific research and investigation into extraterrestrial life, including unidentified flying objects, and to inform the Secretary-General of the observations, research, and evaluation of such activities.

3. The General Assembly requests the Secretary-general to transmit the statements of the delegation of Grenada and the relevant documentation to the Committee on the Peaceful Uses of Outer Space so that it may consider them at its session in 1979.

4. The Committee on the Peaceful Uses of Outer Space will permit Grenada, upon its request, to present its views to the Committee at its session in 1979. the committee's deliberation will be included in its report which will be considered by the General Assembly at its thirty-fourth session."

"The General Assembly invites interested Member States to take appropriate steps to coordinate on a national level scientific research and investigation into extraterrestrial life, including unidentified flying objects, and to inform the Secretary-General of the observations, research, and evaluation of such activities."

While visiting the United Nations in 1979, and before he could present to the General Assembly, Granada Prime Minister Eric Gairy was overthrown and he went into exile. *Under the influence of the United States and other Security Council members, in subsequent years the UN has taken no action regarding alleged extraterrestrial related phenomena.* The matter was for all intent and purpose, dead in the water! http://paradigmresearchgroup.org/dir/chf/un-initiative/

Recommendation to Establish UN Agency for UFO Research
UN General Assembly decision
33/426, 1978

(Nations Unies)

Summary: *At its 87th plenary meeting, on 18 December 1978, the UN General Assembly, on the recommendation of the Special Political Committee recommended the establishment of an agency or a department of the United Nations for undertaking, co-ordinating and disseminating the results of research into unidentified flying objects and related phenomena.*

UN General Assembly decision 33/426, 1978

Establishment of an agency or a department of the
United Nations for undertaking, co-ordinating and disseminating
the results of research into unidentified flying objects and related
phenomena

At its 87th plenary meeting, on 18 December 1978, the General Assembly, on the
recommendation of the Special Political Committee adopted the following text as
representing the consensus of the members of the Assembly:

"1. The General Assembly has taken note of the statements made,
and draft resolutions submitted, by Grenada at the thirty-second
and thirty-third sessions of the General Assembly regarding unidentified
flying objects and related phenomena.

"2. The General Assembly invites interested Member States to take
appropriate steps to coordinate on a national level scientific
research and investigation into extraterrestrial life, including
unidentified flying objects, and to inform the Secretary-General
of the observations, research and evaluation of such activities.

"3. The General Assembly requests the Secretary-general to transmit
the statements of the delegation of Grenada and the relevant documentation
to the Committee on the Peaceful Uses of Outer Space, so that it
may consider them at its session in 1979.

**The UN General Assembly
decision 33/426, 1978 was adopted and passed by the UNGA**
http://paradigmresearchgroup.org/dir/chf/un-initiative/

The Disclosure Project

By now, the majority of American citizens, journalists and the global community at large have probably heard about one of the most important events of the 21st Century to date to have taken place at the National Press Club in Washington, DC on May 9, 2001. In the main ballroom before 100 journalists, 17 camera crews and 100 interested public, the **Disclosure Project**, founded by **Dr. Steven Greer**, presented the testimony of dozens of former military and agency employees which collectively confirms the extraterrestrial presence and the existence of advanced energy technologies which can completely replace any need for fossil fuels or ionizing nuclear power plants.

Dr. Steven M. Greer, International Director of the Disclosure Project
https://www.youtube.com/watch?v=lkswXVmG4xM

Statements were presented in person and via videotape. All witnesses in attendance vowed to repeat their testimony under oath before Congress. This press conference was aired live around the world via the internet and is continuously being downloaded from the Disclosure Project website http://www.disclosureproject.org/ along with the witness testimonies. While the near-term impact of this event was blunted by the **911 terrorist attacks** four months later, it will play a historic role in the disclosure process and will almost certainly be repeated - in or outside of Congress.

The Disclosure Project, a non-profit research organization, is calling for open Congressional hearings on the UFO / Extraterrestrial presence, and for legislation that will ban space-based weapons. Similar, but much smaller Congressional hearings were last held in 1968 by the **House Science and Astronautics Committee** (90th Congress, 2nd Session, Print No. 7. **"Symposium on Unidentified Flying Objects."**)

In 1993, Dr. Steven Greer founded the **Disclosure Project,** which included the documented testimonies of over 400 witnesses with impeccable credentials who had first-hand experience with the UFO/ETI phenomenon. Greer gathered video testimonies of "whistleblowers" within the government, the military, and private sectors as well as from scientists and astronauts through a program known as **Project Starlight.** He then spearheaded the concept of **"disclosure"** using

212

his Starlight witnesses by inviting the mainstream media to the **National Press Club** in Washington, D.C. to document the first public disclosure of the UFO phenomenon. Greer is thus, justifiably regarded the **"Father of Disclosure"** from which many other smaller UFO groups have come forward to disclose additional "insider" information regarding the UFO phenomenon.

Dr. Greer has often been the **"go to person"** on Capitol Hill in Washington. If someone from the military, the intelligence community or government committee wants to know something about UFOs and ETI.

In May 2001, Greer held a press conference at the National Press Club in D.C that featured 22 **Disclosure Witnesses**, among who were top whistleblowers and witnesses to UFO/ET events; some were retired Air Force, Federal Aviation Administration and intelligence officers who demanded that Congress begin hearings on *"secret U.S. involvement with UFOs and extraterrestrials"*.

Among the witnesses attending at the National Press Club are: **John Callahan**, former Division Chief of the Accidents and Investigations Branch, FAA; **Lt. Col. Charles Brown**, US Air Force (Ret.); **Mr. Michael Smith** former Air Traffic Controller US Air Force (Ret.); **Mr. Enrique Kolbach**, Senior Air Traffic Controller; **Commander Graham Bethune**, US Navy (Ret.); **Mr. Dan Willis**, US Navy, Cryptologist; **Mr. Don Phillips**, Lockheed Skunkworks and CIA Contractor; **Captain Robert Sallas,** former SAC Launch Controller, US Air Force and FAA.; **Lt. Col. Dwynne Arnesson**, US Air Force (Ret.); **Harland Bentley**, US Army; **Mr. John Maynard**, Defence Intelligence Agency (Ret.); **Sergeant Karl Wolf**, USAir Force; **Donn Hare**, NASA Employee; **Larry Warren**, US Air Force, Security Officer; **Maj. George A. Filer III**, US Air Force, Intelligence (Ret.); **Sergeant Clifford Stone**, US Army; **Master Sergeant Dan Morris**, former US Air Force and NRO operative; **Mr. Mark McCandlish**, US Air Force; **Dr. Daniel Sheehan**, Attorney; and **Dr. Carol Rosin**, Space Missile Defense Consultant and former spokesperson for **Wernher Von Braun**.

Although there were 22 witnesses at the Club, only the above 19 gave testimonies while three others sat it out. Their testimony was considered so damning and unusual that Dr. Greer at the last moment felt it would be better for the overall success of this first disclosure event to sit them out as silent disclosure witnesses. The news media he felt would not know how to take or make of their testimony in order to give it the serious attention that it deserved. One of those who sat it out disappointedly was **David Adair.** Another was a whistleblower, who cannot be named at this time, worked with as an operative in the intelligence community, who was to reveal insider testimony on an ala' *"Independence Day false flag scenario"* to be perpetrated by MJ-12! At the last moment, he was called away by his handlers and was told not to testify at the disclosure event!

Participants in this phase of the disclosure effort are asking for Congressional, White House, and UN action to allow witnesses to testify under oath in open hearings. The group is requesting a **Presidential Executive Order** to protect witnesses afraid of violating security oaths and to declassify documents and secret projects for the benefit of all world citizens.

The Disclosure Project Witnesses (top to bottom and left to right): John Callahan; Lt. Col. Charles Brown; Mr. Michael Smith; Mr. Enrique Kolbach; Commander Graham Bethune; Mr. Dan Willis; Mr. Don Phillips; Captain Robert Sallas; Lt. Col. Dwynne Arnesson; Harland Bentley; Mr. John Maynard; Sergeant Karl Wolf; Donn Hare; Larry Warren; Maj. George A. Filer III; Sergeant Clifford Stone; Master Sergeant Dan Morris; Mr. Mark McCandlish; Dr. Daniel Sheehan; and Dr. Carol Rosin. Missing was David Adair who did not give his presentation at the time
https://www.youtube.com/watch?v=lkswXVmG4xM

The usual assorted collection of skeptics, naysayers and debunkers came out of the woodwork as well as spokespeople for the U. S. Air Force who through derision and denial maintained that there is no convincing evidence for the speculation that UFOs are alien spacecraft. This is a typical response from professional skeptics and the USAF as these people have been collaborating for decades to cover-up or suppress the truth of the UFO and ETI presence. Their efforts were, however, not effective in preventing the truth from reaching the public.

The Disclosure event was the largest publicly viewed and televised press conference ever held at the **National Press Club**. On a very good day the Press Club would normally have 25,000 people tuned into a press conference announcement, but on the day of **Dr. Greer's Disclosure Project** event, 250,000 people tuned in to watch the proceedings, more than any time in the history of the Press Club! In fact, people who also watched on the internet found half an hour into the press conference that communication signals from the Club were being deliberately jammed! There are only a few organizations that have that capability to do this with any effectiveness and they are either the US military intelligence or one of the US government's intelligence agencies such as the CIA or most probably the NSA!!!

Dr. Greer has stated on numerous podcast and radio station interviews that even some of the news media had reported to him earlier that they were told by intelligent agencies not to give too much press coverage to the event in order to devalue its importance!

These were the basic salient points raised by Dr. Greer at the National Press Club in addition to the Disclosure Witnesses' testimonies:

- The Disclosure Project is a non-profit research effort that has, since 1993 when it began as Project Starlight, been identifying top-secret military, government and other witnesses to UFO and Extraterrestrial events.
- To date, several hundred such witnesses have been identified throughout the world and spanning every branch of the armed services, the NRO, DIA, CIA, NASA, the former USSR, and other agencies and countries. Over 100 have been videotaped, thus far; 70 have been transcribed into edited testimony. A four-hour videotape summary of testimony and an over 500-page briefing document is available that contains excerpts of this historic testimony.
- The weight of this testimony, along with supporting government documents and other evidence, establishes beyond any doubt the reality of extraterrestrial life forms, UFOs, or extraterrestrial vehicles, and advanced energy and propulsion technologies resulting from the study of these vehicles.
- The testimony and evidence prove that these vehicles have been tracked on radar on many occasion, have landed and/or crashed on terra firma, and have been retrieved and studied by specialized and compartmentalized projects. Advanced technologies which have been identified from the study of these vehicles, once disclosed, will replace currently used forms of energy generation and propulsion. These technologies will enable the Earth to attain a sustainable civilization without pollution, energy shortages, or global warming. These technologies are already fully operational. They have been developed within super-secret, unacknowledged special access projects. In short, the definitive solution to the world's energy, pollution, and poverty problems exists within compartmentalized projects that need planned disclosure and relevant legislation.
- The programs controlling this issue are operating outside of legally required Congressional oversight. Even Presidents have been left out of the loop, deliberately deceived, and denied access. Therefore, urgent action is needed on the part of Congress,

the White House, and other institutions to obtain the necessary oversight and control of these operations to ensure that these now-classified technologies are prepared for disclosure and the eventual near-term application for world cooperative energy generation and propulsion.

- A clear and on-going threat to the national security and world peace has arisen because of unauthorized covert actions that have led to the targeting and downing of these extraterrestrial vehicles and to related covert plans to weaponize space. Since it can be proven that we are sharing space with other civilizations, it is critical that a full disclosure of this long-suppressed information take place, and that the National Missile Defense System (NMD/BMD/SDI.) be re-evaluated by policy makers in light of these revelations.

- There is no evidence that these extraterrestrial civilizations are hostile to humanity or the Earth. Rather, the testimony shows that they are very concerned about nuclear and space-based weapons systems, and human warfare. Therefore, a cooperative world policy and law must be immediately established to prohibit the targeting and striking of these vehicles.

- Urgent Congressional, White House, and UN action are needed to allow any and all witnesses to testify under oath so that a full, honest and open disclosure may occur this year, 2001, including witnesses with high-level security clearances.

- A US Presidential Executive Order is needed to protect this military, government, and other witnesses, and to declassify secret projects and their related technologies.

The world community needs to research and develop diplomatic programs and protocols, laws and treaties to address this issue and to interface with these civilizations in a manner that is peaceful, non-violent and mutually beneficial. DISCLOSURE PROJECT BRIEFING DOCUMENT; April 2001 by Dr. Steven M. Greer, Director and Dr. Theodore C. Loder III; The Disclosure Project; PO Box 265, Crozet, VA 22932 and http://www.disclosureproject.org/briefingpoints.shtml

The Citizen Hearing on Disclosure

In 2007, the **Paradigm Research Group**, an independent Ufology group headed by Washington activist **Stephen Bassett**, held a news conference in Washington, D.C. to demand UFO disclosure. Basset as speaker noted the long history of attempts at disclosure and claimed that alien technologies could be beneficial for Earth. This was followed in 2010 with the film release of "The Day Before Disclosure" which was made for free by New Paradigm Films.

From April 29 to May 3, 2013, the Paradigm Research Group held the **"Citizen Hearing on Disclosure"** at the **National Press Club**. The group paid former **U.S. Senator Mike Gravel** and former **Representatives Carolyn Cheeks Kilpatrick**, **Roscoe Bartlett**, **Merrill Cook**, **Darlene Hooley**, and **Lynn Woolsey** $20,000 each. **Joseph Buchman and Daniel Sheehan** were non-paid committee members. The conference, and a documentary (working title: **Truth Embargo**) summarizing its findings were intended to compel Congress, **United Nations** or both to open an investigation. Kilpatrick complained about the lack of transparency at the hearing. http://www.citizenhearing.org/committee.html

216

**Steven Bassett , President of the Citizen Hearings on Disclosure
and founder of the Paradigm Research Group**

http://midnightinthedesert.com/stephen-bassett/

**Committee members of the Citizen Hearing on Disclosure include from top to bottom
and left to right: former Representatives Roscoe Bartlett, Darlene Hooley, former U.S.
Senator Mike Gravel, former Representatives Carolyn Cheeks Kilpatrick, Lynn Woolsey ,
Educator Joseph Buchman and Constitutional Attorney Daniel Sheehan**

Above are the faces of witnesses who gave testimony at the Citizen Hearing on Disclosure
(Google Images)

The Citizen Hearing on Disclosure was an unprecedented event in terms of size, scope and the involvement of former members of the U. S. Congress. With over 30 hours of testimony from 40 witnesses/researchers along with military/agency/political persons of high rank and station came to the National Press Club in Washington, DC to testify before six former members of the United States Congress over five days. The event was the most concentrated body of evidence regarding the Extraterrestrial subject ever presented to the press and the general public at one time. It could be argued that this was a dry-run rehearsal for an eventual official disclosure as envisioned by Basset that would engage current sitting Congressmen and be fully acknowledged by US Congress or the United Nations.

To the extent possible the protocols for congressional hearings were followed. Committee members received written statements from witnesses, heard oral statements and ask whatever questions they wished about the subject matter at hand.

Hearing witnesses testified for 30 hours over five days in five mornings and five afternoon sessions, each composed of two panels of witnesses, each panel lasting approximately 90 minutes. Most panels were centered on specific topics. Some panels covered a range of topics.

The Citizen Hearing on Disclosure was an unprecedented event seeking to reach four targeted audiences - the political media, the United States Congress, the White House and the citizens of all nations. The primary goal of the Hearing is to bring about Disclosure this year by ending the governmental policy of a truth embargo preventing the appropriate institutional engagement of the evidence indicating an extraterrestrial presence.

In service to this goal the organizers of the Citizen Hearing have launched the **Citizen Hearing Foundation (CHF),** a 501(c) 3 non-profit organization. The larger mission of this organization is to educate the public, the media and political leaders regarding controversial issues which do not receive the appropriate degree of consideration by mainstream institutions. In that regard, the CHF will pursue other initiatives which address citizen grievances, abuses of power and secrecy, and media neglect.

The first task before the CHF will be to fund a multi-nation initiative to put a joint resolution before the **United Nations General Assembly** calling for a U. N. backed world conference to assess the evidence pointing toward and extraterrestrial presence engaging the human race. This task was put to the CHF by former members of the U. S. Congress who served on the CHD committee. http://www.citizenhearing.org/aboutchd.html

While not all of the former Congress members at this "mock" Citizen Hearing on Disclosure were convinced that we are being visited by aliens, they all did say they were impressed by the high level of credibility of some of the witnesses they heard from. Because this was a mock Hearing, the former Congress members were paid for their participation time over a five-day period which some critics and debunkers point to the non-credibility of the event.

However, their only requirement was to show up and participate and some supporters of UFO research quickly point out that it would have been much easier for them to collect their checks and simply just agree with reporters that the event was a silly waste of time but, that was not the case. Instead, they expressed their surprise that the credibility of all witnesses were impeccable, than first anticipated, and most committee members came away from the event convinced that ETs were visiting the Earth and that the United Nations needs to investigate it further.

John Podesta Regrets Not Disclosing UFO Files to American Public

Now let's step back a bit and look at a possibility of a US Government disclosure on UFOs in the near future. News of **John Podesta's** appointment as counselor to **US President Barack Obama** has triggered speculation that some of the top secret UFO files may be released soon. The 64 year-old former chief of staff to the **Clinton** administration will be dipping into matters that range from energy to food safety and border enforcement, and who knows, possibly UFO disclosure? According to the Washington Post, Podesta is "one of the Democratic Party's most seasoned political and policy operatives." He is also the founder of the **Center for American Progress (CAP)**, a leading democratic public policy organization. Many people are starting to believe that democrats and republicans are just two wings of the same bird and that the flight path doesn't change, however, it is a known fact that John Podesta is an advocate for UFO disclosure.

In 2002, Podesta spoke at a news conference held at the **National Press Club** in Washington, D.C., where he urged the U.S. government to declassify UFO documents that are more than 25 years old.

John Podesta speaks at the National Press Club urging the U.S. government to declassify UFO documents that are more than 25 years old
http://www.washingtontimes.com/news/2015/feb/15/inside-the-beltway-john-podestas-extraterrestrial-/

"I'm skeptical about many things, including the notion that government always knows best, and that the people can't be trusted with the truth. The time to pull the curtain back on this subject is

long overdue. We have statements from the most credible sources – those in a position to know – about a fascinating phenomenon, the nature of which is yet to be determined." – **John Podesta** (taken from **Leslie Kean's** 2010 New York Times bestseller, "UFOs : Generals, Pilots, And Government Officials Go On The Record," where Podesta wrote the forward).

"I think it's time to open the books, on questions that have remained in the dark on the question of government investigations of UFOs, it's time to find out what the truth really is that's out there. We ought to do it because it's right, we ought to do it because the American people, and people around the world quite frankly can handle the truth, and we ought to do it because it's the law."- John Podesta (Taken from a National Press Club video)

"It is time for the government to declassify records that are more than 25 years old, and to provide scientists with data that will assist in determining the real nature of this phenomenon". – Johan Podesta http://www.collective-evolution.com/2014/01/15/obama-appoints-ufo-disclosure-advocate-into-administration/

Tweets from John Podesta:

Outgoing senior Obama adviser John Podesta reflected on his latest White House stint Friday, listing his favorite moments and biggest regrets from the past year. Chief among them: depriving the American people of the truth about UFOs. Podesta's long time fascination with UFOs is well-documented, as his brief political hiatus following four years as President Bill Clinton's chief of staff freed him up to pursue his otherworldly passion.

At a 2002 press conference organized by the Coalition for Freedom of Information, Podesta spoke on the importance of disclosing government UFO investigations to the public.

"It's time to find out what the truth really is that's out there," he said. "We ought to do it, really, because it's right. We ought to do it, quite frankly, because the American people can handle the truth. And we ought to do it because it's the law." Following Podesta's tweet, Friday, the Washington Post recalled an exchange one of its reporters had with Podesta in 2007. Karen Tumulty had asked Podesta about reports that the Clinton Library in Little Rock, Arkansas, had been bombarded with Freedom of Information Act Requests specifically seeking email correspondence to and from the former chief of staff including terms like "X-Files" and "Area 51."

Podesta's response, through a spokesperson, was "The truth is out there," the tagline for the TV show "The X-Files" of which Podesta was known to be a fan.

Brian Deese ✔ @Deese44
1. Finally, my biggest failure of 2014: Once again not securing the #disclosure of the UFO files. #thetruthisstilloutthere cc: @NYTimesDowd
8:55 AM - 13 Feb 2015
http://news.yahoo.com/outgoing-obama-adviser-john-podesta-s-biggest-regret-of-2014--keeping-america-in-the-dark-about-ufos-234149498.html

@johnpodesta · 10 Dec 2013
The truth is out there, Dave. MT **@daveweigel**: John Podesta returning to the White House to reveal truth about aliens. http://www.slate.com/blogs/weigel/

John Podesta @johnpodesta · 9 Dec 2013
A first: POTUS acknowledges **#Area51** in Kennedy Center remarks; stays mum on UFOs. http://ti.me/18O0mSf

This is about as close as the American public will get in the way of an official government disclosure on UFOs and basically, it's a hand-off from by President Obama because of his failure to be transparent on the most secretive and time-sensitive subject to come across his desk. It is in fact, not an official US government disclosure, but merely an admission from one of his staff, counselor John Podesta who had wished, like the rest of us, that President Obama had done what he had said and promised to do when was voted into office as president.

The conference below was organized by the **Coalition for Freedom of Information** . Last month, TIME magazine listed five topics that Podesta and Obama haven't agreed on. These include drone secrecy, NSA spying, Afghanistan, the American Political System, and UFOs.

Podesta brings up some interesting facts from the quotes above. One of them being that we have statements from the most credible sources, those in a position to *'know'*. He is *referring to generals, pilots, more high ranking political/military personnel, scientists and more.* Not long ago, Former **Canadian Minister of National Defense, Paul Hellyer** said that there are "at least 4 known alien species that have been visiting Earth for thousands of years." This is one statement from hundreds upon hundreds of people that have held positions that might be privy to this type of knowledge.

UFO disclosure in the United States might not be far away, dozens of governments all over the world have already admitted to the UFO phenomenon being real. They've admitted to tracking them on radar and scrambling jets to check them out. For example, here is a batch of files pertaining to the UK government's involvement in UFO investigations. Here is a batch of documents that were declassified by the NSA.

It should be pointed out that if the Obama administration were to disclose this information, it doesn't necessarily mean that they are telling the truth. Manipulation of perception regarding real events has been a common theme throughout history, war is one example. The real story behind the phenomenon is usually something other than what we've been told by corporate media. We don't need the ones "we appointed" to be our source for information, each and every individual is capable of doing it themselves. It doesn't hurt to do a little critical thinking every now and then. At the same time, individuals within the 'elitist,' realm might be going through their own little awakening, throwing judgment at something is never the correct thing to do, and some that play within the game of politics and government might want to see full transparency of information.

There are thousands of interesting, declassified documents out there in the public domain, and you can bet that there are even more that remain classified. http://www.collective-evolution.com/2014/01/15/obama-appoints-ufo-disclosure-advocate-into-administration/

(1) http://www.bloomberg.com/news/2013-12-20/podesta-s-push-for-executive-power-raises-stakes-on-obama-agenda.html
(2) http://www.washingtonpost.com/politics/john-podesta-to-formally-join-obama-inner-circle/2013/12/09/7a4bc430-614f-11e3-bf45-61f69f54fc5f_story.html
(3) http://swampland.time.com/2013/12/10/podesta-and-obama-disagree/?iid=tsmodule
http://www.huffingtonpost.com/2013/12/11/is-ufo-disclosure-on-the-plate-of-john-podesta_n_4428119.html

When we examine the people's recent movement toward disclosure in the UFO/ETI disclosure and the US Government's quasi- disclosure with the release of UFO documents back in the late '60s and early '70s through FOIA requests, the question must be asked: has UFO Disclosure already happened? The answer depends on who you ask as there are those that believe disclosure has already happened and some UFO researchers who say that an official government disclosure as presented by the US President or one of his science advisors has not happened as yet.

The most common perception people have of disclosure is the President comes before the National Press Club or some appropriate venue and makes a special announcement that they have discovered that there is an Extraterrestrial presence on Earth and some UFOs are piloted by Extraterrestrial intelligences. This is the sort of disclosure that the **Citizen Hearing on Disclosure** is seeking in Washington D.C. at the National Press Club and what the general public is demanding from their leaders.

However, this sort of disclosure is wishful thinking, as a lot of these people feel the government is currently lying about the subject. How would we know what we were being told is now the truth? If proponents of this type of disclosure already believe the government cannot be trusted, why would they trust any announcement?

That is a point that is often made by Retired Army Colonel, **Dr. John Alexander** is convinced the UFO phenomena is real and is likely of an extraterrestrial nature but, he points to this very dilemma of trust. However, he feels government disclosure has already happened when the FBI and the Air Force, two government agencies that currently have thousands of files available on their investigations into UFOs which was released back in the late '60s and early '70s and have concluded that they are just as mystified as the rest of us. *Alexander's position on this question makes one wonder, what planet is he coming from, because the FBI and the USAF are two primary agents of disinformation and cover-up on the UFO subject!* (Author's bold italics added for emphasis)

Some people believe that Alexander has an unofficial group of friends inside intelligence circles who have looked into government UFO secrecy and what they related to Alexander is that the topic is a bit of a hot potato. No one wants to deal with it because it is too weird and they fear ridicule. He thinks the real lack of serious investigation by the government is due more to incompetence than anything else. However, he does feel an official investigation is needed and advocates for organized research into the topic.

One UFO investigator, who has his finger on the pulse of UFO/ETI subject, is **Dr. Steven Greer** who has stated repeatedly that *"disclosure"* has already taken place via a people's initiative in which he lead back in May 2001, better known as the **Disclosure Project**! Greer has been the first to bring credible witnesses to the National Press Club in Washington D.C. to address the media in 2001. It was an impressive group that included retired military officers, an FAA official, and credentialed people in the private sector. They testified to their knowledge or involvement with UFO research or sightings. To many, the credentials of these people were sufficient enough to be a sort of disclosure.

Greer has stated that the government is incapable of making an official disclosure announcement because many within the government do not really know what's going on and the few who do are too scared to come forward. They fear personal repercussions to themselves, the government and the country, even the world at large! Such an official disclosure would have a cascading unstoppable domino effect, the implications may actually overturn a lot of the world's most powerful governments so, it is really up to the people to take the leadership role and do what government have failed to do!

Others feel that the secrets are locked up so tight that getting answers is a much more difficult and complicated task. **Richard Dolan**, a UFO historian, thinks that the answers are probably locked up in deep black projects. Even the President may not have the need to know, giving him plausible deniability, and making it true that the White House doesn't know anything about UFOs because they are not in the loop. If we don't know where the secrets are, how do we, or the President or even congress get to them?

The point is that there are many ideas about what disclosure really means, and whether it has already happened, or if it is even necessary, yet there are some US politicians who would like to see the matter of disclosure settled once and for all.

CHAPTER 100

US PRESIDENTIAL QUOTES AND GOVERNMENT COVER-UP

What follows are Presidential quotes and quotes from White House officials, high ranking Military officers, famous and well-known scientists, religious leaders and whistleblowers. These quotes from officialdom indicate the acknowledgements and the seriousness of the UFO/ETI subject taken in many prominent sectors of society. It is known history of how the most powerful men in the world and their **Joint Chiefs of Staff** have dealt with the most highly classified secret of the last century. In short, this is the story of how the U.S. Presidents and the White House have dealt with the mystery of UFOs.

By the quotes made by various Presidents of the United States as well as the military, the scientists and international leaders that they know something elements of the UFO subject matter to be true yet, absolutely NOT ONE of them really knows what is going on the inside the UFO enigma; if they did they would either, disclosure officially its existence or would remain in absolute denial to its reality. ***Therefore, there is reasonable plausibility of denial of their knowing anything about it and thus, how do you disclose that which you really know nothing about?!***

To be sure, there are people within all ranks, sectors and branches of the government, the military, in the science community who do know more than those officials whose statements below are so often quoted in books, magazines, on television and in the movies but, these are the people control the information, provide the spin-doctoring or simply lie, deny, misinform , disinform and obfuscate the truth, so that the general public at large are kept guessing the true nature of the UFO and ETI phenomenon!

Should the reader have read every page within this textbook to the very end, in addition to your own research of this phenomenon then, you will certainly be one of those fortunate few, who will understand and know more of what's going on than many of those government officials! Read the official quoted positions of those leaders who represent you and decide for yourself if they know what's really going on or they are as in the dark about the whole subject matter as the general public.

Quotes from United States Presidents About UFOS

Dwight Eisenhower
"The last time I heard this talked to me, a man whom I trust from the Air Forces said that it was, as far as he knew, completely inaccurate to believe that they came from any outside a planet or otherwise". - **December 1954 White House News Conference**

"In the counsels of Government, we must guard against the acquisition of unwarranted influence, whether sought or unsought, by the Military Industrial Complex. The potential for the disastrous rise of misplaced power exists and will persist. We must never let the weight of this combination endanger our liberties or democratic processes. We should take nothing for granted. Only an alert and knowledgeable citizenry can compel the proper meshing of the huge industrial and

military machinery of defense with our peaceful methods and goals so that security and liberty may prosper together." - **January 1961.** http://www.presidentialufo.com/ufo-quotes

Harry S. Truman
"*I can assure you the flying saucers, given that they exist, are not constructed by any power on earth.*" - **White House Press Conference, Washington DC, April 4, 1950.**

John F. Kennedy
"*We seek a free flow of information... we are not afraid to entrust the American people with unpleasant facts, foreign ideas, alien philosophies, and competitive values. For a nation that is afraid to let its people judge the truth and falsehood in an open market is a nation that is afraid of its people.*" - **Nov. 21, 1963.**

Lyndon Johnson
"*While we would not be in a position to send an expert in response to each such request, we are always willing to investigate all responsible reports or claims of evidence.*" - **Writing as Vice-President to a letter about a UFO photo written by Ronald Anstee - September 21, 1961**

Gerald Ford
"*No doubt, you have noted the recent flurry of newspaper stories about unidentified flying objects. I have taken special interest in these accounts because many of the latest reported sightings have been in my home state of Michigan... Because I think there may be substance to some of these reports and because I believe the American people are entitled to a more thorough explanation than has been given them by the Air Force to date, I am proposing that either the Science and Astronautics Committee or the Armed Services Committee of the House schedule hearings on the subject of UFOs and invite testimony from both the executive branch of the Government and some of the persons who claim to have seen UFOs... In the firm belief that the American public deserves a better explanation than that thus far given by the Air Force, I strongly recommend that there be a committee investigation of the UFO phenomena. I think we owe it to the people to establish credibility regarding UFOs and to produce the greatest possible enlightenment on this subject.*" - **From a letter, he sent as a Congressman to L. Mendel Rivers, Chairman of the Armed Services Committee, on March 28, 1966.**

"*During my public career in Congress, as Vice President and President, I made various requests for information on UFOs. The official authorities always denied the UFO allegations.*" - **March 17, 1998, letter to George Filer** http://www.presidentialufo.com/ufo-quotes

Jimmy Carter
"*I don't laugh at people anymore when they say they've seen UFOs. It was the darndest thing I've ever seen. It was big, it was very bright, it changed colors and it was about the size of the moon. We watched it for ten minutes, but none of us could figure out what it was. One thing's for sure I'll never make fun of people who say they've seen unidentified objects in the sky. If I become President, I'll make every piece of information this country has about UFO sightings available to the public and the scientists.*" - At a Southern Governors Conference describing an alleged UFO sighting, he had in October of 1969 to reporters while campaigning in 1976.

"I am convinced that UFOs exist because I have seen one." - **Former U.S. President, five-time nominee for the Nobel Peace Prize.** http://www.presidentialufo.com/ufo-quotes

"I don't see any reason to keep information like that secret, but there may be some aspects of the UFO information which I am not familiar that might be related to some secret experiments that we were doing that involve national security or new weapons systems. I certainly wouldn't release that". - **1976 presidential campaign**

"I think it is impossible in my opinion – some people disagree, to have space people from other from other planets or other stars to come here. I do not believe that is possible". - **November 2007 speaking to CNN following the Kucinich UFO question in the Democratic debate on October 30.**

Ronald Reagan

"... when you stop to think that we're all God's children, wherever we may live in the world, I couldn't help but say to him, just think how easy his task and mine might be in these meetings that we held if suddenly there was a threat to this world from some other species from another planet outside in the universe. We'd forget all the little local differences that we have between our countries and we would find out once and for all that we really are all human beings here on this earth together." - **White House transcript of "Remarks of the President to Fallston High School Students and Faculty," December 4, 1985.**

"In our obsession with antagonisms of the moment, we often forget how much unites all the members of humanity. Perhaps we need some outside, universal threat to make us recognize this common bond. I occasionally think how quickly our differences worldwide would vanish if we were facing an alien threat from outside this world." - **Speech to the United Nations General Assembly, Forty-second session, "Provisional Verbatim Record of the Fourth Meeting", September 21, 1987.)** http://www.presidentialufo.com/ufo-quotes

"I occasionally think how quickly our differences, worldwide, would vanish if we were facing an alien threat from outside this world." - **From a speech with President Mikhail Gorbachev, in 1988. He made almost the same comment on many occasions.**

"I was in a plane last week when I looked out the window and saw this white light. It was zigzagging around. I went up to the pilot and said, "Have you ever seen anything like that?" He was shocked and he said, "Nope." And I said to him: "Let's follow it!" We followed it for several minutes. It was a bright white light. We followed it to Bakersfield and all of a sudden to our utter amazement it went straight up into the heavens. When I got off the plane I told Nancy all about it. But we didn't file a report on the object because for a long time they considered you a nut if you saw a UFO..." - **1974, Reagan was often quoted referring to the possibility of an alien threat.**

US Presidents who have known or dealt with the UFO issue
Google Images

Google Images

George H. W. Bush Sr.
"Yes. If I can find out what it is... it really is." - **1988 in reply to releasing the UFO files if elected as President**

"I know some. I know a fair amount." - **George Bush replying in 1988 to a question about UFOs while campaigning to become President.** *"*

Bill Clinton
"I want you to find the answers to two questions for me. One, who killed JFK. And Two, are there UFOs?" - **From "Friends in High Places", by Webster Hubbell, Clinton's associate attorney general.**

"As far as I know, an alien spacecraft did not crash in Roswell, New Mexico, in 1947......If the United States Air Force did recover alien bodies, they didn't tell me about it either and I want to know." - **In reply to a letter from a child asking about the Roswell Incident**

"If we were being attacked by space aliens, we wouldn't be playing these kinds of games."
- Bill Clinton told educators visiting Washington D.C. http://www.presidentialufo.com/ufo-quotes

"Well I don't know if you all heard this, but, there was actually, when I was president in my second term, there was an anniversary observance of Roswell. Remember that? People came to Roswell, New Mexico from all over the world. And there was also a site in Nevada where people were convinced that the government had buried a UFO and perhaps an alien deep underground because we wouldn't allow anybody to go there. And uhm... I can say now, 'cause it's now been released into the public domain... This place in Nevada was really serious, that there was an alien artifact there. So I actually sent somebody there to figure it out."

"I did attempt to find out if there were any secret government documents that revealed things. If there were, they were concealed from me too. And if there were, well I wouldn't be the first American president that underlings have lied to, or that career bureaucrats have waited out. But there may be some career person sitting around somewhere, hiding these dark secrets, even from elected presidents. But if so, they successfully eluded me...and I'm almost embarrassed to tell you I did (chuckling) try to find out." - **September 2005 to CLSA group in Hong Kong** http://www.presidentialufo.com/ufo-quotes

"I got a letter from 13-year-old Ryan from Belfast. Now, Ryan, if you're out in the crowd tonight, here's the answer to your question. No, as far as I know, an alien spacecraft did not crash in Roswell, New Mexico, in 1947. And, Ryan, if the United States Air Force did recover alien bodies, they didn't tell me about it, either, and I want to know." - **Belfast Northern Ireland November 1995**

"You know, I've always been really interested in this stuff, and I'm going to read this." - **August 2007 to Hollywood producer Paul Davids when he was given the books "Witness to Roswell" and "The Roswell Legacy", and a new affidavit that had been released following the death of Roswell witness Walter Haut.**

"Sarah, there's a government inside the government, and I don't control it." - **as quoted by senior White House reporter Sarah McClendon in reply to why he wasn't doing** http://www.presidentialufo.com/ufo-quotes

George W. Bush
"This man (Dick Cheney) knows. He was Secretary of Defense...And was a great one. It will be the first thing he (pointing to Cheney) will do. He'll get right on it." - **in reply to if he will release the truth on UFOs.**

"I heard you had reports this morning of an unidentified aircraft. Don't worry it was just me".
- **speaking to the Military Academy in Roswell, New Mexico January 22, 2004**

"My alien encounter is further proof of my commitment to expanding the Republican Party's appeal. New faces new voices. It goes to show that I am willing to reach across certain demographic lines." - **to reporters on his campaign plane in response to a Weekly World News from page story showing George Bush shaking hands with an alien who was predicting his victory in the 2000 election.**

"Sure I will," - **In a news conference on July 29, 2000, televised on CNN, George W. Bush reply when a citizen asked if he would tell the public "what the hell is going on" with UFOs. Bush indicated that vice-presidential candidate Dick Chaney would use his experience as Secretary of Defense to address the issue.**
http://www.presidentialufo.com/ufo-quotes and https://www.youtube.com/watch?v=Q-w9-Y2JMZw

Quotes from United States Senators and Congressman about UFOs

U.S. Representative Thomas L. Ashley
"I share your concern over the secrecy that continues to shroud our intelligence activities on this subject." - **In a letter on the subject of UFOs to NICAP (National Investigations Committee on Aerial Phenomena.**

U.S. Representative William H. Ayres
"Congressional investigations ... are still being held on the problem of unidentified flying objects and the problem is one in which there is quite a bit of interest... Since most of the material presented to the Committees is classified, the hearings are never printed." - **1958**

U.S. Senator Robert Byrd
"Look, this subject - with a top-secret clearance and subpoena, I have not been able to penetrate it. You're dealing with the varsity team of all super-secret projects. Good luck." [laughs]...- **West Virginia's chief counsel and attorney who had a top-secret clearance and subpoena power to Dr. Steven Greer**

U.S. Vice-President Cheney
U.S. Vice-President Cheney was taken aback by the very first question posed to him" by a caller to Diane Rehm's national radio show. The caller asked Cheney whether the administration "has

230

developed a policy on 'UFOs' and the little creatures flying them? Raising an eyebrow, Cheney no doubt made the caller's day by replying that if he did attend such a meeting on UFOs, it most certainly would be 'classified,' and therefore he'd be unable to discuss it. **Snips from Washington Times, 4-13-01** http://www.ufoevidence.org/documents/doc1358.htm

Mike Huckabee
"I believe in G-O-D, not U-F-O". - **Former Governor of Arkansas and 2008 Republican candidate for President**

U.S. Senator Robert Kennedy
"I am keeping myself abreast of information developed on this subject." - **Letter to UFO researcher Robert Barrow, October 2, 1965**

"If it were desirable to hold congressional hearings on the question of unidentified flying objects, I would certainly agree that they should be open to the public." - **Letter to Ralph Rankow, August 15, 1965**

U.S. Senator Robert F Kennedy
"As you may know, I am a card-carrying member of the Amalgamated Flying Saucers Association. Therefore, like many other people in our country I am interested in the phenomenon of flying saucers. It is a fascinating subject that has initiated both scientific fiction fantasies and serious scientific research. I watch with great interest all reports of unidentified flying objects, and I hope that someday we will know more about this intriguing subject. Dr. Harlow Shapley, the prominent astronomer, has stated that there is a probability that there is other life in the universe. I favor more research regarding this matter, and I hope that once and for all, we can determine the true facts about flying saucers. Your magazine can stimulate much of the investigation and inquiry into this phenomenon through the publication of news and discussion of material. This can be of great help in paving the way to acknowledge of one of the fascinating subjects of our contemporary world." - **In a letter to Gray Barker Publisher, Saucer News May 9, 1968".** http://www.presidentialufo.com/ufo-quotes

Henry Kissinger
"The subject is the biggest hot potato of all time". - **speaking of UFOs**
U.S. Representative Gerald R. Ford

"I think there may be substance in some of these reports ... I believe the American people are entitled to a more thorough explanation than has been given them by the Air Force to date. I think we owe it to the people to establish credibility regarding UFOs and to produce the greatest possible enlightenment of the subject." - **March 1966**

"In the firm belief that the American public deserves a better explanation than that thus far given by the Air Force, I strongly recommend that there be a committee investigation of the UFO phenomena. I think we owe it to the people to establish credibility regarding UFOs and to produce the greatest possible enlightenment of the subject." - **Gerald Ford (during his years as a US Congressman).**

U.S. Senator Barry Goldwater,
"It is true that I was denied access to a facility at Wright-Patterson Air Force Base in Dayton, Ohio because I never got in. I can't tell you what was inside. We both know about the rumors (concerning a captured UFO and crew members). I have never seen what I would call a UFO, but I have intelligent friends who have." - **From a letter April 1979**
"I think some highly secret government UFO investigations are going on that we don't know about--and probably never will unless the Air Force discloses them."

"I remember the case in Georgia in the 1950's of a National Guard plane going after a UFO and never returning. And I recall the case in Franklin, Kentucky when four military planes investigated a UFO. One of them exploded in midair and no one knows why."

"Hell no, you can't go. I can't go, and don't ask me again." - **Senator Goldwater quoting General Curtis LaMey's response to the senator's request to visit the "Blue Room" at Wright-Patterson Air Force Base where Goldwater claims he was told physical evidence exists confirming the existence of alien spacecraft.** http://www.presidentialufo.com/ufo-quotes

"Yes." - **Senator Goldwater's response to Larry King's question:** *"Do you think our government knows UFOs are real and are keeping this fact from the American public?"*

"I certainly believe in aliens in space. They may not look like us, but I have very strong feelings that they have advanced beyond our mental capabilities. " and added, *"I think some highly secret government UFO investigations are going on that we don't know about and probably never will unless the Air Force discloses them."* - **He said he was refused permission to check the Air Force files on UFOs.**

"While flying with several other USAF pilots over Germany in 1957, we sighted numerous radiant flying discs above us. We couldn't tell how high they were. We couldn't get anywhere near their altitude."

"While working with a camera crew supervising flight testing of advanced aircraft at Edward's Air Force Base, California, the camera crew filmed the landing of a strange disc object that flew in over their heads and landed on a dry lake nearby. A camera crewman approached the saucer; it rose up above the area and flew off at a speed faster than any known aircraft."

U.S. Senator Daniel K. Inouye
"There exists a shadowy Government with its own Air Force, its own Navy, its own fundraising mechanism, and the ability to pursue its own ideas of national interest, free from all checks and balances, and free from the law itself."

232

Dr. Scott Jones
"From what I have been told, a Pentagon spokesman was less than civil in a verbal response to a GAO investigator when asked about Roswell. The alleged response was "Go shit in your hat." If that is accurate, it is my opinion an unfortunate reply. It could reflect that he was having a bad day, or it could be a recent example that unwillingness to discuss the subject". - **Head Human Potential Foundation and Aide to Senator Claiborne Pell who was part of the Rockefeller UFO Initiative to the Clinton White House – writing to President Clinton's science Advisor February 17, 1994,** http://www.presidentialufo.com/ufo-quotes

Melvin Laird
"The whole question of declassification of any government projects which might have been associated with unidentified flying objects seems to be on the right track... I am sure that should classification be lifted some individuals will be disappointed as certain of these phenomena will be pretty well explained. Any review will certainly disappoint some individuals who have built up some rather extreme antidotal and uncollaborated accounts which the removal of classification might discredit to a large extent." - **(Former Secretary of Defense under Richard Nixon) May 9, 1994, Letter to Laurance Rockefeller**

U.S. Representative John W. McCormack, Former Speaker of the House.
"Some three years ago, (1957), as chairman of the House Select Committee on Outer Space out of which came the recently established NASA, my Select Committee held executive sessions on the matter of 'Unidentified Flying Objects.' We could not get much information at that time, although it was pretty well established by some in our minds that there were some objects flying around in space that were unexplainable." - **In a November 4, 1960, letter to Major Donald Keyhoe**

"I feel that the Air Force has not been giving out all the available information on the Unidentified Flying Objects. You cannot disregard so many unimpeachable sources." - *January 1965* http://www.presidentialufo.com/ufo-quotes

U.S. Representative Jerry L. Pettis
"Having spent a great deal of my life in the air, as a pilot... I know that many pilots... have seen phenomena that they could not explain. These men, most of whom have talked to me, have been very reticent to talk about this publicly, because of the ridicule that they were afraid would be heaped upon them... However, there is a phenomena here that isn't explained." - **July 29, 1968, during the House Committee on Science and Astronautics hearing on UFOs.**

U.S. Senator Richard Russell, Head of Senate Armed Services Committee.
"I have discussed this matter with the affected agencies of the government, and they are of the opinion that it is not wise to publicize this matter at this time." - **Regarding his sighting of a UFO during a 1955 trip to the Soviet Union.**

U.S. Representative Steven H. Schiff
"I wrote to the Dept. of Defense, laying out these allegations and asking them if someone could come over with the file and brief me on it. My intent was to simply release this back to whoever inquired, which is very routine in Congress." - **In response to inquiries from his constituents**

in 1993 concerning a possible cover-up of the crash of an alleged UFO outside Roswell, NM in 1947, requested information from the Department of Defense.

U.S. Representative Steven H. Schiff

"The response I got was not routine. The response I got was a very brief letter from the Air Force saying that my request had been referred to the National Archives, without any further comment... and without any offer of any kind of assistance in retrieving it... So I went to the National Archives and the National Archives wrote a letter back to me saying they didn't have anything in their files on the Roswell incident... I just have to say this much: the way the Dept. of Defense has responded has not been routine."

"Having been given a "runaround" in his search, he instigated an inquiry by the GAO (General Accounting Office) in 1994 into the handling of Air Force files relating to this matter."

"I did not ask the General Accounting Office to try once and for all to resolve this matter... What I asked the GAO to do was to assist me in locating whatever Air Force and Defense Department files would have existed on the subject or an accounting of what happened to them. "

"To me, the issue is government accountability. I think that people who want to see government records are entitled to see government records or to get an explanation of what happened to them, regardless of their reason, regardless of the subject matter. It was my intention simply to make that information public if I could... unless there is a present security reason why not - and I have to add real fast if the matter is classified 'military secret,' we members of Congress can't just go monkeying around in there anytime we want. There are procedures for us too and that's fine with me."

"I was not told that we have a file that's classified. I was simply referred to an agency which I have to believe - now that I know the prominence of the Roswell incident - I have to believe the Dept. of Defense knew very well that I wasn't going to find anything in the National Archives when they sent me there twice."

"It's difficult for me to understand even if there was a legitimate security concern in 1947, that it would be a present security concern these many years later. Frankly, I am baffled by the lack of responsiveness on the part of the Defense Dept. on this one issue, I simply can't explain it."

- Excerpts of Congressman Schiff's remarks on CBS radio's The Gil Gross Show, February 1994. http://www.presidentialufo.com/ufo-quotes

Congressman William Stanton (Pennsylvania)

"The Air Force failed in its responsibility in thoroughly investigating this incident (April 17, 1966, sighting in Pennsylvania) ...Once people entrusted with the public welfare no longer think people can handle the truth, then the people, in turn, will no longer trust the government."
Ravena Record Courier **- April 1966 "Presidential Briefing Document" by Dr. Steven Greer**

CHAPTER 101

QUOTES FROM U. S. MILITARY WITNESSES ABOUT UFOS

Edwin Bauhan
"It came directly overhead and was no more than five hundred feet high, so we got an excellent view of it. It had no motors, no rigging, it was noiseless. . . .a rose or sort of flame color.......I could observe no windows ...... We all experienced the weirdest feeling of our lives, and sat in our tent puzzling over it for some time." - **Edwin Bauhan, one of several soldiers at Rich Field in Waco Texas, 1918, who observed a 100-150 foot long cigar-shaped object after leaving the mess hall.**

Thomas E. Bearden, US Army, Ph.D. in Nuclear Engineering, Colonel
"Probably 50 inventors have invented [virtually cost-free energy systems]. If we use these systems, we can clean up this biosphere. But what we have is a situation where the entire structure of science, industry, and the patent office are against you. And behind this, we have a few people who are quite wealthy. The more powerful the agency, the more they will resort not only to legal but to extra-legal means to suppress their competition. Lethal force is used".
-From Disclosure, Steven M. Greer

Major Jeremiah Boggs
"We were naturally anxious to get hold of one of the [UFOs]. We told pilots to do practically anything in reason, even if they had to grab one by the tail." - **Major Jeremiah Boggs was in the U.S. Air Force.**

US Air Force, Colonel Charles Brown
"I was getting 20 to 30% improvement in efficiency on an internal combustion engine. I sponsored the US Army race team on a racing car, [and] we won a race. [Then] the Federal Trade Commission performed an illegal act. I lost my vehicle, about $100,000 worth of equipment, and a test vehicle was stolen. ...So in three weeks, psychologically I was wiped out."
- From Disclosure, Steven M. Greer

General George S. Brown
"I don't know whether this story has ever been told or not. They weren't called UFOs. They were called enemy helicopters. And they were only seen at night and they were only seen in certain places. They were seen up around the DMZ [demilitarized zone] in the early summer of '68. And this resulted in quite a little battle. And in the course of this, an Australian destroyer took a hit and we never found any enemy, we only found ourselves when this had all been sorted out. And this caused some shooting there, and there was no enemy at all involved but we always reacted. Always after dark. The same thing happened up at Pleiku at the Highlands in '69." - **Brown, as U.S.A.F. Chief of Staff, addressing the appearance of UFOs during the Vietnam War at a press conference in Illinois, October 16, 1973.**

Colonel Joseph J. Bryan III
"These UFOs are interplanetary devices systematically observing the earth, either manned or under remote control, or both."

"Information on UFOs, including sighting reports, has been and is still being officially withheld." - **Bryan was founder of the CIA's psychological warfare staff, special assistant to the secretary of the Air Force, advisor to NATO, and board member of the National Investigations Committee on Aerial Phenomenon (NICAP).**

General Wesley Clark

"We need to look at the realms of applied and higher mathematics. I still believe in E = mc squared. But I can't believe that in all of human history, we'll never ever be able to go beyond the speed of light to reach where we want to go. I happen to believe that mankind can do it. I've argued with physicists about it. I've argued with best friends about it. I just have to believe it. It's my only faith-based initiative."- **Sept. 27, 2003, in New Castle, New Hampshire**

"I heard a bit. In fact, I'm going to be in Roswell, New Mexico tonight. There are things going on. But we will have to work out our own mathematics". **- Reno Nevada, October 30, 2004, in reply to whether or not he was briefed on UFO**
http://www.aliensthetruth.com/Aliens_quotes.php?view=1&category=Military#.VS7soZP7OVo

Dr. Brain T. Clifford

"...that contact between U.S. citizens and extra-terrestrials or their vehicles is strictly illegal"
- The Star, New York, Oct. 5, 1982, Dr. Brain T. Clifford of the Pentagon announced at a press conference

Lieutenant Colonel Lou Corbin

"The UFOs are no figment of the imagination." **- Corbin was Army Intelligence.**

Colonel Philip J. Corso

"If you suppress the truth it becomes your enemy...if you expose the truth it becomes your weapon." "I had the evidence that a crash did happen. I ask you this, were you there with me? Did you have the clearances? They can't answer these questions, they simply criticize with no evidence."

"I had the evidence that a crash did happen here....Give this information to the young people of the world and this country..They want it. Give it to them. Don't hide it and tell lies and make stories. They're not stupid. . It's their information. It doesn't belong to the Army or the Department of Defense. If it's classified, take the classification off and give it to them!"
- Colonel Corso made this impassioned plea not long before his death in the summer of 1998. He was a member of President Eisenhower's National Security Council and later went on to become head of the U.S. Army Research and Development department's Foreign Technology Desk where he claimed to have worked with General Arthur Trudeau in "seeding" American military and industrial institutions with technology under his stewardship which came from a crashed alien craft in Roswell New Mexico.

"I was most interested in the file descriptions accompanying a two-piece set of dark elliptical eyepieces as thin as skin. The Walter Reed pathologists said they adhered to the lenses of the extraterrestrial creatures' eyes and seemed to reflect existing light, even in what looked like

236

complete darkness, so as to illuminate and intensity images in the darkness to allow their wearer to pick out shapes. The reports had said that the pathologists at Walter Reed hospital who autopsied one of these creatures tried to peer through them in the darkness to watch the one or two army sentries and medical orderlies walking down a corridor adjacent to the pathology lab. These figures were illuminated in a greenish orange, depending upon how they moved, but the pathologists could see only their outer shape." - Col. Philip J. Corso Sr. - 'The Day After Roswell"

Lieutenant Colonel Lawrence J. Coyne

"With the aircraft under my control, I observed the red-lighted object closing upon the helicopter at the same altitude at a high rate of speed. It became apparent a mid-air collision was about to happen unless evasive action was taken."
http://www.aliensthetruth.com/Aliens_quotes.php?view=1&category=Military#.VS7soZP7OVo

"I looked out ahead of the helicopter and observed an aircraft I have never seen before. This craft positioned itself directly in front of the moving helicopter. This craft was 50 to 60 feet long with a grey metallic structure. On the front of this craft was a large steady bright red light. I could delineate where the red stopped on the structure of this craft because red was reflecting off the grey structure. The design of this craft was symmetrical in shape with a prominent aft indentation on the undercarriage. From this portion of the undercarriage, a green light, pyramid-shaped, emerged with the light initially in the trail position. This green light then swung 90 degrees, coming directly into the front windshield and lighting up the entire cockpit of the aircraft. All colors inside the cabin of the helicopter were absorbed by this green light. That includes the instrument panel lights on the aircraft."

"As a result of my experience, I am convinced this object was real and that these types of incidents should require a thorough investigation. It is my own personal opinion that worldwide procedures need to be established to effectively study this phenomena through an international cooperative effort. The establishment of a Transponder Code for aircraft flying worldwide is needed, to identify to ground controllers that a pilot is indeed experiencing a UFO phenomena and that pilot anxiety can be reduced to provide safe effective flying, knowing he is under radar control." **- Lt. Col. Coyne was a U.S. Army Reserve helicopter pilot with 3,000 hours of flying time. He and other three airmen had a close encounter with a UFO on the night of October 18, 1973, while flying in a U.S. Army Bell Huey utility helicopter in the vicinity of Mansfield, Ohio. Lt. Coyne described his experience at a United Nations UFO hearing in 1978.** http://www.aliensthetruth.com/Aliens_quotes.php?view=1&category=Military#.VS7soZP7OVo

Sergeant Major Robert O. Dean

"Wright-Patterson Air Force Base... was the headquarters for the foreign technology division of the U.S. Air Force. It later became the headquarters for the alien technology division of the U.S. Air Force. Wright-Patterson, for many years, was the central repository of not only the hardware but some of the little bodies and even some of the living crew members who had been retrieved. But it became very clear after a time that there wasn't enough room at Wright-Patterson Air Force Base. We literally filled up hangar after hangar with hardware. We're

storing it now in at least three different Air Force bases and much of it is being kept underground at a place not too far from Las Vegas just beyond Nellis Air Force Base which is repeatedly referred to as Dreamland, Groom Lake, or Site 51. That today is one of the biggest repositories of hardware."

"...They made...a recommendation to General Limnitzer, the American four-star general that I worked for. And they suggested this is so sensitive ... The conclusions that we have reached, we believe could be...substantially earthshaking to the people unless they're prepared for it. We believe at this point that this should be given the highest classification NATO has' [which] at that time was and still is Cosmic Top Secret." - **From a videotaped interview in which Dean discussed SHAPE's alleged report entitled "An Assessment" based on a 3-year investigation of UFOs being tracked on radar over central Europe. Dean was a former NATO intelligence analyst for SHAPE (Supreme Headquarters Allied Powers Europe)**

Colonel Thomas Jefferson Dubose
"There were orders to ship the material from Roswell directly to Wright Field by special plane."
- Dubose was adjutant to Brig. General Roger Ramey at the time of the Roswell Incident. This statement contradicts the official story that the material was first flown to Fort Worth, Texas, where Ramey posed with Major Marcel and pieces of a weather balloon for the media. http://www.ufoevidence.org/documents/doc1743.htm

Admiral Delmar Fahrney
"Reliable reports indicate there are objects coming into our atmosphere at very high speeds and controlled by thinking intelligences." - **A public statement, 1957.**
"No agency in this country or Russia is able to duplicate at this time the speeds and accelerations which radars and observers indicate these flying objects are able to achieve."
- Printed in New York Times. Admiral Fahrney was former head of the Navy's guided-missile program.
Lieutenant Frederick Fox, US Navy Pilot

"There is a [military] publication called JANAP 146E that has a section that says you will not reveal any information regarding the UFO phenomenon under penalty of $10,000 fine and ten years in jail. So, the secret has been kept." - **From Disclosure, Steven M. Greer,**

Lieutenant George Gorman *"I am convinced there was thought behind the thing's manoeuvres."* - **Gorman was in the North Dakota Air National Guard. In October 1948 Gorman (flying an F51) chased a "ball of light some eight inches wide" for about thirty minutes, through a series of twists, turns, and circles, nearly colliding with it on at least one occasion. The object was also witnessed from the control tower at the Fargo Airport through high-powered binoculars.**

Lieutenant Walter Haut
"The story I put out was very simple, to the effect that we had, in our possession a flying saucer. It was found on a ranch up north of Roswell. It was being flown to General Ramey's office. The information was given to me, almost verbatim, by Colonel Blanchard. He said, "I want you to give it to the local newspapers and radio stations, and do it post haste" ... The cover-up was

pretty well orchestrated. I think the thought of handling it that way came down from Washington, through channels. We were told that we were all wrong, that it was just a weather balloon."
- **Lieutenant Walter Haut, Roswell Army Air Base Public Information Officer**

Lt. Colonel Richard Headrick
"Saucers exist, I saw two. They were intelligently flown or operated (evasive tactics, formation flight, hovering). They were mechanisms, not United States weapons, nor Russian. I presume they are extraterrestrial." - **Headrick was a radar bombing expert, 1959.**
http://www.aliensthetruth.com/Aliens_quotes.php?view=1&category=Military#.VS7soZP7OVo

Vice-Admiral R.H. Hillenkoetter
"Behind the scenes, high-ranking Air Force officers are soberly concerned about the UFOs. But through official secrecy and ridicule, many citizens are led to believe the unknown flying objects are nonsense. To hide the facts, the Air Force has silenced its personnel." - **Vice-Admiral R.H. Hillenkoetter, Pacific Commander of Intelligence in W.W. II and later Director of the CIA, NICAP news release, February 27, 1960, The New York Times, Sunday, February 28, 1960: "Air Force Order on 'Saucers'**

US Air Force Lieutenant / Professor Robert Jacobs
"So this thing [UFO] fires a beam of light at the warhead, hits it and then it moves to the other side and fires another beam of light. And the warhead tumbles out of space. What message would I interpret from that? [The UFOs were telling us] don't mess with nuclear warheads. Major Mannsman said, "You are never to speak of this again." After an article [about the incident years later], people would call and start screaming at me. One night somebody blew up my mailbox." - **From Disclosure, Steven M. Greer**

Major Donald E. Keyhoe
"The Air Force had put out a secret order for its pilots to capture UFOs."

"If, in fact, we are able to find life or to answer the question 'Are we alone?' then that certainly is grand enough and noble enough to be the enduring legacy of our civilization." - **NASA, October 1999.**

"With control of the universe at stake, a crash program is imperative. We produced the A-bomb, under the huge Manhattan Project, in an amazingly short time. The needs, the urgency today are even greater. The Air Force should end UFO secrecy, give the facts to scientists, the public, to Congress." "Once the people realize the truth, they would back, even demand a crash program... for this is one race we dare not lose..." - **Statement in 1953.**

"Russia and the U. S. have announced they are definitely planning several space machines. So it's quite possible that the first space ships or satellites may encounter other interplanetary machines, manned or otherwise. Our space devices may even be closely approached by such alien machines."

"For the last six months, we have been working with a congressional committee investigating official secrecy concerning proof that UFOs are real machines under intelligent control." **- From a live national broadcast, on CBS in 1958. Keyhoe had an approved script to follow, but when deviated unexpectedly from it with this astonishing statement, the audio was cut-off in the middle of his sentence,** *"...for reasons of national security."* **Keyhoe was in the United States Marine Corp.**

Major General Joe W. Kelly
"Air Force interceptors still pursue Unidentified Flying Objects as a matter of national security to this country and to determine technical aspects involved." **- Kelly made this statement in 1957.**

General Robert B. Landry
"I was called one afternoon [in 1948] to come to the Oval Office – the President wanted to see me... I was directed to report quarterly to the President after consulting with Central Intelligence people, as to whether or not any UFO incidents received by them could be considered as having any strategic threatening implications .." **- Landry was an aide to President Harry S. Truman.**
http://www.aliensthetruth.com/Aliens_quotes.php?view=1&category=Military#.VS7soZP7OVo

Major General E.B. LeBaily
"Many of the reports that cannot be explained have come from intelligent and technically well-qualified individuals whose integrity cannot be doubted." **- As USAF Director of Information, in a September 28, 1965, letter to USAF Scientific Advisory Board.**

General Curtis LeMay
"We had a number of reports from reputable individuals (well-educated serious-minded folks, scientists, and fliers) who surely saw something." **- As Air Force Chief of Staff, in his 1965 autobiography, 'Mission With LeMay,' stated that although the bulk of UFO reports could be explained as conventional or natural phenomena, some could not.**

"Many of the mysteries might be explained away as weather balloons, stars, reflected lights, all sorts of odds and ends. I don't mean to say that, in the unclosed and unexplained or unexplainable instances, those were actually flying objects. All I can say is that no natural phenomena could be found to account for them... Repeat again: There were some cases we could not explain. Never could." **- Statement from 1965 autobiography Mission With LeMay, with MacKinlay Kantor, New York: Doubleday, 1965.**

General Stephen Lovekin
"Colonel Holomon brought out a piece of what appeared to be metallic debris. He went on to explain that this was material that had come from a New Mexico crash in 1947 of an extraterrestrial craft, and that was discussed at length...I got an opportunity to travel with the President [Eisenhower]. He was very, very interested in what made [the UFOs] go. But what happened was that Eisenhower got sold out. He realized that he was losing control of the UFO subject. He realized that the [study of these technologies] was not going to be in the best hands.

240

That was a real concern." - **From Disclosure, Steven M. Greer, General Lovekin was in the US Army.**

General Douglas MacArthur
"The nations of the world will have to unite, for the next war will be an interplanetary war. The nations of the earth must someday make a common front against attack by people from other planets." - **The New York Times, October 8, 1955**

"You now face a new world - a world of change. The thrust into outer space of the satellite, spheres, and missiles marked the beginning of another epoch in the long story of mankind , the chapter of the space age... We speak in strange terms: of harnessing the cosmic energy... of the primary target in war, no longer limited to the armed forces of an enemy, but instead to include his civil populations; of ultimate conflict between a united human race and the sinister forces of some other planetary galaxy." - **An address by General Douglas MacArthur to the United States Military Academy at West Point, May 12, 1962.**

Mark McCandlish, US Air Force, Aerospace Illustrator.
"This [US made] antigravity propulsion system this flying saucer was one of three that were in this hangar at Norton Air Force Base. They called [it] the Alien Reproduction Vehicle [ARV], also nicknamed the Flux Liner." - **From Disclosure, Steven M. Greer. McCandlish was a US Air Force aerospace illustrator.**
http://www.aliensthetruth.com/Aliens_quotes.php?view=1&category=Military#.VS7z9ZP7OVo

Captain Thomas Mantell
"It appears to be a metallic object.......tremendous in size....directly ahead and slightly above ... I am trying to close for a better look." - **Mantell, a USAF pilot, reporting to the tower at Goodman Air Force base, a UFO they had picked up on radar and requested him to investigate. They were his last known words. Mantell's plane was later found strewn across a stretch of ground just southwest of Franklyn, Kentucky.**

Major Jesse Marcel
"I was amazed at what I saw. The amount of debris that was scattered over such an area... The more I saw of the fragments, the more I realized it wasn't anything I was acquainted with. In fact, as it turned out, nobody else was acquainted with it....There was a cover-up someplace about this whole matter." - **Major Marcel, a U.S. Army Intelligence Officer, in a videotaped interview. Among the first to arrive at the crash site in Roswell, Marcel was well acquainted with all the weather balloons launched by the 509th Bomb Group, presumably including the Mogul balloons, one of which the U.S. Government now claims accounts for the Roswell wreckage.**

Lt. Colonel James McAshan
"In concealing the evidence of UFO operations, the Air Force is making a serious mistake."
- **McAshan was USAF.**

Sergeant Dan Morris

"UFOs are both extraterrestrial and manmade...It's not that our government doesn't want us to know that there are other people on other planets. What the people in power don't want us to know is that this free energy [from energy generators developed with UFO technology] is available to everybody. So secrecy about the UFOs is because of the energy issue. When this knowledge is found out by the people, they will demand that our government release this technology, and it will change the world." - **From Disclosure, Steven M. Greer, Sergeant Morris was a US Air Force NRO operative.**

Captain Eddie Rickenbacker

"Flying saucers are real. Too many good men have seen them, that don't have hallucinations."
- **Captain Rickenbacker was known as, "American Ace of Aces," medal of honor-winning commander of the 94th Aero Pursuit Squadron in WWI, with 26 "kills".**

Captain Edward J. Ruppelt

"Every time I get skeptical, I think of the other reports made by experienced pilots and radar operators, scientists, and other people who know what they are looking at. These reports were thoroughly investigated and they are still unknowns."

"We have no aircraft on this earth that can at will so handily outdistance our latest jets... The pilots, radar specialists, generals, industrialists, scientists, and the man on the street who have told me, 'I wouldn't have believed it either if I hadn't seen it myself,' knew what they were talking about. Maybe the Earth is being visited by interplanetary spaceships."

"When four college professors, a geologist, a chemist, a physicist, and a petroleum engineer report seeing the same UFOs on fourteen different occasions, the event can be classified as, at least, unusual. Add the fact that hundreds of other people saw these UFOs and that they were photographed, and the story gets even better. Add a few more facts, that these UFOs were picked up on radar and that a few people got a close look at one of them, and the story begins to convince even the most ardent skeptic." - **Ruppelt, Chief of Project Blue Book, from his book, The Report on Unidentified Flying Objects, 1956**
http://www.aliensthetruth.com/Aliens_quotes.php?view=1&category=Military#.VS7z9ZP7OVo

Captain Robert Salas

"[The security guard called and] said, "Sir, there's a glowing red object hovering right outside the front gate. I've got all the men out here with their weapons drawn." We lost between 16 to 18 ICBMs [nuclear-tipped InterContinental Ballistic Missiles] at the same time UFOs were in the area...[A high-ranking Air Force Officer] said, "Stop the investigation; do no more on this and do not write a final report." I heard that many of the guards that reported this incident were sent off to Viet Nam." - **From Disclosure, Steven M. Greer, Captain Robert Salas was a US Air Force officer.**

Colonel Carl Sanderson

"From their manoeuvres and their terrific speed, I am certain their flight performance was greater than any aircraft known today." - **Sanderson, USAF, commenting on his sighting of**

two circular silver UFOs in close proximity to his plane over Hermanas, New Mexico. The UFOs were said to make a series of seemingly impossible manoeuvres before disappearing at an astonishing speed and showing up again over El Paso, Texas.

Lt. Frank H. Schofield

"Three objects appeared beneath the clouds, their color a rather bright red. As they approached the ship they appeared to soar, passing above the broken clouds. After rising above the clouds they appeared to be moving directly away from the earth. The largest had an apparent area of about six suns. It was egg-shaped, the larger end forward. The second was about twice the size of the sun, and the third, about the size of the sun. Their near approach to the surface appeared to be most remarkable. That they did come below the clouds and soar instead of continuing their southeasterly course is also curious. The lights were in sight for over two minutes and were carefully observed by three people whose accounts agree as to the details." - **Lt. Frank H. Schofield, later to become Commander-in-Chief of the Pacific Fleet, aboard the U.S.S. Supply off of the eastern coast of Korea, February 28, 1904**

Secretary of Air Force

"The Air Force maintains a continuous surveillance of the atmosphere near Earth for Unidentified Flying Objects." - **Secretary of Air Force to Base Commanders, August 15, 1960.**

Colonel Charles Senn

"I sincerely hope that you are successful in preventing a reopening of UFO investigations." - **In a letter from, to Lieutenant General Duward Crow of NASA, dated 1 September 1977. Colonel Senn was Chief of the Air Force, Community Relations Division.**

Brig. General George Shulgen

"It is the considered opinion of some elements that the object may, in fact, represent an interplanetary craft of some kind." - **From a Draft Intelligence Collections Memorandum by Brig. General George Shulgen, Oct. 28, 1947,**
http://www.aliensthetruth.com/Aliens_quotes.php?view=1&category=Military#.VS7z9ZP7OVo

Major Gerald Smith

"[There was] something definite in the sky...If it had proved to be hostile we would have destroyed it." - **Smith, USAF, was one of the F-106 pilots scrambled under orders from NORAD (North American Air Defense Command) to investigate a UFO over West Palm Beach, Florida on September 14, 1972. The UFO was viewed through binoculars by the FAA supervisor, George Morales, sighted by an Eastern Airlines captain, police, and several civilians, as well as being tracked on radar by Miami International Airport and Homestead AFB.**

Lieutenant D.A. Swimley

"And don't tell me they were reflections, I know they were solid objects." - **Swimley, USAF, commenting on a sighting of eight disc shaped objects he and several fellow officers watched circling over Hamilton AFB, California, on August 3, 1953. The objects were also**

picked up on radar and spotted by many civilian pilots. F-86 Sabres were scrambled to intercept the objects, but the jets were apparently too slow.

General Nathan Twining

"The reported operating characteristics such as extreme rates of climb, maneuverability (particularly in roll), and action which must be considered evasive when sighted ... lend belief to the possibility that some of the objects are controlled." - **Twining as Head of Air Material Command (AMC), 1947.**

"The phenomenon is something real and not visionary or fictitious. There are objects approximating the shape of a disc, some of which appear flat on bottom and domed on top. These objects are as large as man-made aircraft and have a metallic or light-reflecting surface. Further, they exhibit extreme rates of climb and maneuverability with no associated sound and take action which must be considered evasive when contacted by aircraft and radar."
- **Twining, in a declassified letter to the Pentagon. General Twining was the Chairman of the Joint Chiefs of Staff 1957-1960.**

Captain Bill Uhouse

"The [flight] simulator was for the extraterrestrial craft they had - a 30 meter one that crashed in Kingman, Arizona, back in 1953. I was inside the actual alien craft for a start-up...There are probably two or three dozen [ARVs] that we built." - **From Disclosure, Steven M. Greer. Captain Bill Uhouse was in the US Marine Corps.**

Corporal Jonathan Weygandt

"[The UFO] was buried in the side of a cliff. When I first saw it, I was scared. I think the creatures calmed me...[Later] I was arrested [by an Air Force officer]. He was saying, "Do you like the Constitution?" I'm like, "Yeah." He said, "We don't obey. We just do what we want. And if you tell anybody [about us or the UFO], you will just come up missing." - **From Disclosure, Steven M. Greer. Corporal Weygandt was in the U.S. Marine Corps.**

Major Robert White

"There are things out there! There absolutely is!" - **White exclaiming over the radio about a UFO encounter taking place on a 58 mile high X-15 flight on July 17, 1962.** http://www.aliensthetruth.com/Aliens_quotes.php?view=1&category=Military#.VS7z9ZP7OVo

"I have no idea what it could be. It was greyish in color and about thirty to forty feet away."
- **He later reported**

Colonel Robert Willingham

"Headquarters wouldn't let us go after it and we played around a little bit. We got to watching how it made 90 degree turns at this high speed and everything. We knew it wasn't a missile of any type. So then, we confirmed it with the radar control station, and they kept following it, and they claimed that it crashed somewhere off between Texas and the Mexico border."

- Willingham, USAF, from an affidavit filed in the 1970s. Willingham and his navigator were test flying an F-94 on Sept.6, 1950 out of San Angelo, Texas when they were alerted by radar control operators of a UFO in their area.
Colonel Steve Wilson

"I have no feelings, truthfully. My association with MJ-12 has left me dead inside. I feel myself still cold and calculating. I never let anyone get close to me. I feel like a human robot. I have killed mercilessly and lied for the good of the country, or so I believed at the time."

"The things I have seen are beyond human understanding and totally unbelievable. I only have a desire to help humanity somehow through what is bound to come soon." - **Col. Wilson, USAF, revealed that he was in charge of Project Pounce, the unit tasked to retrieve downed UFOs and prevent civilian access to them. He also revealed the designations and manufacturers of U.S. antigravity craft.**
hhttp://www.aliensthetruth.com/Aliens_quotes.php?view=1&category=Military#.VS7z9ZP7OVo

CHAPTER 102

QUOTES FROM INTERNATIONAL MILITARY WITNESSES ABOUT UFOS

Major General Vasily Alexeyev

"As a rule, [places where UFOs appear] are objects of strategic significance... [The Air Force] came up with a table with pictures of all the shapes of UFOs that had ever been recorded about fifty-ranging from ellipses and spheres to something resembling spaceships...The study of UFOs may reveal some new forms of energy to us, or at least bring us closer to a solution." - ***Major General Alexeyev was attached to the Russian Space Communications Center***

Commander Juan Barrera

"That it could be an aircraft constructed on this earth, I do not believe possible." - **Statement while in command of Aquirre Cerda Airbase, commenting on a UFO he allegedly witnessed.**

General Carlos Castro Cavero

"Everything is in a process of investigation both in the United States and in Spain, as well as in the rest of the world... Look, as a General, as a military man, I have the same position as the one officially held by the Ministry [of Defense]. Now, from a personal position, as Carlos Castro Cavero, I believe that UFOs are spaceships or extraterrestrial craft... The nations of the world are currently working together in the investigation of the UFO phenomenon. There is an international exchange of data. Maybe when this group of nations acquire more precise and definite information, it will be possible to release the news to the world."

"I myself have observed one [UFO] for more than an hour... It was an extremely bright object, which remained stationary there for that length of time and then shot off towards Egea de los Caballeros, covering the distance of twenty kilometers in less than two seconds. No human device is capable of such a speed." - **He added that the Spanish Air Ministry investigated UFO cases, including instances in which pilots had flown alongside UFOs, but when they tried to get closer, the UFOs moved away at fantastic speeds. The investigations were kept confidential at the time, but in a 1976 interview with journalist J.J. Benítez, he acknowledged that UFOs were taken quite seriously by the Spanish military. Cavero, General in the Spanish Air Force. In 1992 the Spanish Air Force finally began to declassify its UFO files systematically.**
http://www.bibliotecapleyades.net/ciencia/ufo_briefingdocument/quogov.htm

General Lionel M. Chassin

"The number of thoughtful, intelligent, educated people in full possession of their faculties who have 'seen something' and described it, grows every day... We can... say categorically that mysterious objects have indeed appeared and continue to appear in the sky that surrounds us." He observed that some UFO sightings by different persons over a 24-hour period when plotted on a map, revealed that the UFO appeared to travel in either a straight line or a large circle. Concerning these patterns, he concluded:

"Webs and networks... unmistakably suggest a systematic aerial exploration and cannot be the result of chance. It indicates purposive and intelligent action." - **General Chassin was Commanding General of the French Air Forces, and General Air Defense Coordinator, Allied Air Forces, Central Europe (NATO). From 1964 until his death in 1970, he was president of the French private UFO research group GEPAN.**

Air Marshal Azim Daudpota

"This was no ordinary UFO. Scores of people saw it. It was no illusion, no deception, no imagination." (The Times, London, August 3).
On July 22, 1985, in western Zimbabwe, a UFO was witnessed by dozens of persons on the ground and in the control tower at Bulawayo Airport, as well as by the pilots of two Hawk jets that were scrambled to pursue it. The UFO was also tracked on radar. The UFO was very bright and rounded, with a short cone above it, and evaded the Hawk jets.

Major General Wilfred De Brouwer

"In any case, the Air Force has arrived to the conclusion that a certain number of anomalous phenomena has been produced within Belgian airspace. The numerous testimonies of ground observations were reinforced by the reports of the night of March 30-31 [1990], have led us to face the hypothesis that a certain number of unauthorized aerial activities have taken place. Until now, not a single trace of aggressiveness has been signaled; military or civilian air traffic has not been perturbed or threatened. We can, therefore, advance that the presumed activities do not constitute a concrete menace."

"The day will come undoubtedly when the phenomenon will be observed with technological means of detection and collection that won't leave a single doubt about its origin. This should lift a part of the veil that has covered the [UFO] mystery for a long time. A mystery that continues to the present. But it exists, it is real, and that in itself is an important conclusion." - **De Brouwer was Deputy Chief, Royal Belgian Air Force.'**

Air Chief Marshall Lord Hugh Dowding

"Of course, UFOs are real--and they are interplanetary.... The cumulative evidence for the existence of UFOs is quite overwhelming and I accept the fact of their existence." - **Dowding, commanding officer of the Royal Air Force during WWII, and during the Battle of Britain made this statement in August of 1954.**
http://www.bibliotecapleyades.net/ciencia/ufo_briefingdocument/quogov.htm

"More than 10,000 sightings have been reported, the majority of which cannot be accounted for by any "scientific" explanation... I am convinced that these objects do exist and that they are not manufactured by any nation on Earth." "I can, therefore, see no alternative to accepting the theory that they come from some extraterrestrial source." - **Printed in Sunday Dispatch, London, July 11, 1954.**

Colonel João Glaser

"From 1969 to 1972, the Ufological activities of this organization, were most varied, including the elaboration of information bulletins, a draft of SIOANI regulations, contacts with interested parties, panels, catalogs of contacts and others, always attempting to contribute in this field of

research that was already well known in Brazil." - **Colonel Glasera was with a specialized UFO bureau called System of Investigation of Unidentified Aerial Objects**

Lt. Col. Peter Grunnet
"We had many adventures flying under primitive conditions in the frozen north, but none compared with this." "I looked back and saw something that didn't make sense," "It was nothing like flying machines of that period," "It was hexagonal, flat, and seemingly made of aluminum or some other metal, with no breaks in the surface and no rivets." "At the time, I had a spooky feeling. I can't explain it. It was as if I 'felt' the presence of whoever was inside that craft--and the feeling was hostile. In the years since I've realized that the craft was 'saucer' shaped, and I believe it really was a flying saucer." - **Lt. Col. Peter Grunnet, Royal Danish Air Force in 1932 while flying an H. E. 8 seaplane dispatched to the east coast of Greenland conducting a photogrammetric survey. The Saga UFO Report for October 1977 "UFO Crisis over Greenland"** http://www.bibliotecapleyades.net/ciencia/ufo_briefingdocument/quogov.htm

Colonel Fuijo Hayashi
"UFOs are impossible to deny...It is very strange that we have never been able to find out the source for over two decades." - **Colonel Hayashi, Commander of the Air Transport Wing of Japan's Air Self-Defense Force, Made this statement was sometime in the 1960s.**

Paul Hellyer, Former Canadian Minister of Defence
"UFOs are as real as the airplanes that fly over your head....... I'm so concerned about what the consequences might be of starting an intergalactic war, that I just think I had to say something. . . . The secrecy involved in all matters pertaining to the Roswell incident was unparalleled. The classification was, from the outset, above top secret, so the vast majority of U.S. officials and politicians, let alone a mere allied minister of defence, were never in-the-loop." - **Paul Hellyer, Canada's Defence Minister from 1963-67, from a speech at the University of Toronto, September 25, 2005.**

Lieutenant General Akira Hirano
"We frequently see unidentified objects in the sky. We are quietly investigating them." **Lieutenant General Hirano was Chief of Staff of Japan's Air Self-Defense Force, September 1977.**
http://www.bibliotecapleyades.net/ciencia/ufo_briefingdocument/quogov.htm

General Kanshi Ishikawa
"Much evidence tells us UFOs have been tracked by radar; so, UFOs are real and they may come from outer space... photographs and various materials show scientifically that there are more advanced people piloting the saucers and motherships." - **Ishikawa, Chief of Air Staff of Japan's Air Self-Defense Force and Commander of the 2nd Air Wing, Chitose Air Base, made this statement in 1967.**

George Keleti
"I believe that we are not alone in the universe and other galaxies are also carrying life on the planets. I never saw any alien green men here on the Earth. Yes, I was a columnist [in Budapest's Ufomagazin] and I published UFO cases that were observed and registered within

248

the Hungarian armed forces. I never stated that we are preparing any kind of action against UFO forces, I only pointed out to the public that, as a civilization, we would be unable to defend ourselves here on the Earth... Around Szolnok many UFO reports have been received from the Ministry of Defense, which obviously and logically means that they [UFOs] know very well where they have to land and what they have to do. It is remarkable indeed that the Hungarian newspapers, in general newspapers everywhere, reject the reports of the authorities." (Lenart, Attila, "Ask a Question to the Minister of Defense, George Keleti, are you afraid of a UFO invasion?" - **Nepszava, Budapest, August 18, 1994. George Keleti was the Hungarian Minister of Defense**

Major Shiro Kubuta

"It was made and flown by intelligent beings." - **Major Shiro Kubuta, of Japan's Air Self-Defense Force (I think). Kubuta and his pilot, Lt. Colonel Toshio Nakamura, were scrambled in an F-4EJ to intercept what they were told was a Soviet Bomber. Once Airborne they were informed that their target was actually a UFO which had been sighted by ground and was being tracked on radar. When they closed upon the red, disk-like UFO, it began to maneuver around the plane, causing Nakamura to take evasive action. A dogfight of twists, turns and dives ensued, which after several minutes resulted in a collision, which caused the jet to crash. Both men ejected. Tragically, Nakamura's parachute caught fire and he fell to his death.**

Captain Sánchez Moreno

"Between 1950 and 1965, personnel of Argentina's Navy alone made 22 sightings of unidentified flying objects that were not airplanes, satellites, weather balloons or any type of known (aerial) vehicles. These 22 cases served as precedents for intensifying that investigation of the subject by the Navy. In the past two years, nine incidents have been recorded that are being studied by Captain Pagani and a team of military and civilian scientists and collaborators. Likewise, a meticulous questionnaire was drafted, printed and distributed to different bases. In a short time, the Service of Naval Intelligence was in possession of a stack of highly significant reports of testimonies. On the basis of this important documentation, it was possible to obtain a coherent overview of the problem." - **From a document titled "Official UFO Report" prepared by Captain Sánchez Moreno Naval Air Station Comandante and witnessed by Comandante Espora in Bahía Blanca.**
http://www.bibliotecapleyades.net/ciencia/ufo_briefingdocument/quogov.htm

Air Marshall Roesmin Nurjadin

"UFOs sighted in Indonesia are identical with those sighted in other countries. Sometimes they pose a problem for our air defense and once we were obliged to open fire on them." - **Nurjadina as Commander-in-Chief of the Indonesian Air Force.**

Brigadier General João Adil Oliveira

"I wish to give you a summary of what is known in the world about 'flying discs,' of what is known about the opinion of qualified experts who have dealt with this matter. The problem of 'flying discs' has polarized the attention of the whole world, but it's serious and it deserves to be treated seriously. Almost all the governments of the great powers are interested in it, dealing with it in a serious and confidential manner, due to its military interest." - **In a briefing to the**

Army War College in Rio de Janeiro on November 2, 1954, Col. Oliveira's briefing included short summaries of several UFO incidents in the USA and Brazil. Later promoted to the rank of Brigadier General, he was interviewed by the Brazilian press on February 28, 1958.

"It is impossible to deny anymore the existence of flying saucers at the present time... The flying saucer is not a ghost from another dimension or a mysterious dragon. It is a fact confirmed by material evidence. There are thousands of documents, photos, and sighting reports demonstrating its existence. For instance, when I went to the Air Force High command to discuss the flying saucers I called for ten witnesses - military (AF officers) and civilians - to report their evidence about the presence of flying saucers in the skies of Rio Grande do Sul, and over Gravataí AFB [Air Force Base]; some of them had seen UFOs with the naked eye, others with high-powered optical instruments. For more than two hours the phenomenon was present in the sky, impressing the selected audience: officers, engineers, technicians, etc." - **Brazilian press interview on February 28, 1958, Brigadier General João Adil Oliveira, Chief of the Air Force General Staff Information Service (with the rank of Colonel), led the first official military UFO inquiry in Brazil in the mid-50s.**

Captain Engineer Omar R. Pagani
"The unidentified flying objects do exist. Their presence and intelligent displacement in the Argentine airspace has been proven. Their nature and origin is unknown and no judgment is made about them." - **Pagani was Director of the Argentine Navy UFO investigation team in the 1960s He disclosed this at a press conference, as a result of a series of observations at Argentine and Chilean meteorological stations on Deception Island, Antarctica, in June and July 1965.**

"The case of UFO interference with our naval transport, the Punta Mendota, was but one of fifteen such cases which the Argentine Navy has reported since 1963." - **Lt. Commander O.R. Pagini, special assistant to Sec. of Argentine Navy, letter to NICAP, September 1965.**

Captain D.A.Perissé
"From the Navy post at the South Orkney Islands comes a message of extreme importance: during the passage of the strange object over the base [earlier the same day], two magnetometers in perfect working condition registered sudden and strong disturbances of the magnetic field (at 17:03 hrs.), which were recorded on their tapes." - **From the Argentine Navy Bulletin #172 of July 7, 1965.**
http://www.bibliotecapleyades.net/ciencia/ufo_briefingdocument/quogov.htm

Air Commander J. Salutun
"I am convinced that we must study the UFO problem seriously for reasons of sociology, technology, and security... " - **Letter published in UFO News, Vol. 6, No. 1, 1974, CBA International, Yokohama, Japan.**

Air Commodore J. Salutun
"The most spectacular UFO incident in Indonesia occurred when during the height of President Sukarno's confrontation against Malaysia, UFOs penetrated a well-defended area in Java for

250

two weeks at a stretch, and each time were welcomed with perhaps the heaviest anti-aircraft barrage in history."

"The study of UFOs is a necessity for the sake of world security in the event we have to prepare for the worst in the space age, irrespective of whether we become Columbus or the Indians." Salutun was on the National Aerospace Council of Indonesia, and Indonesian Parliament Member." - **Salutun as Member of Parliament and Secretary of the National Aerospace Council of the Republic of Indonesia.**

Colonel Sergio Candiota da Silva
"His Excellency recognizes the importance of the [UFO] matter, to the extent that within the Ministry of Aeronautics there exists a Bureau in charge of studying the matter, receiving, analyzing and archiving chronologically the phenomena observed in Brazilian airspace that comes to the attention of this Ministry." - **From a letter to the Minister of Aeronautics to Brazilian UFO researcher Irene Granchi, dated December 19, 1988**

Air Commodore David Thorne
"Although not speaking officially], as far as my Air Staff is concerned, we believe implicitly that the unexplained UFOs are from some civilization beyond our planet." - **Thorne who was Director General of Operations in an October 1985 letter to Timothy Good.**
http://www.bibliotecapleyades.net/ciencia/ufo_briefingdocument/quogov.htm

CHAPTER 103

QUOTES FROM U. S. GOVERNMENT WITNESSES ABOUT UFOs

Harry G. Barnes
"For six hours ... there were at least ten unidentifiable objects moving above Washington. They were not ordinary aircraft." - **Barnes was Senior Air Traffic Controller for the C.A.A. in 1952.**

Harland Bentley
"I was in a facility in California doing classified work. Our astronauts were doing a loop around the moon. I heard them say they had a bogey coming in at 11:00. It was another type of ship. There were portals there that they could see in. They could see beings of some sort. They just took photographs. After a few thousand miles, they took off from the capsule and went away. This happened before the lunar landing. This event was unedited because of where I was."
- **Harland Bentley, NASA, Department of Energy**
http://www.ufoevidence.org/documents/doc1737.htm

John Callahan
"The UFO was bouncing around the 747. [It] was a huge ball with lights running around it...Well, I've been involved in a lot of cover-ups with the FAA. When we gave the presentation to the Reagan staff, they had all those people swear that this never happened. But they never had me swear it never happened. I can tell you what I've seen with my own eyes. I've got a videotape. I've got the voice tape. I've got the reports that were filed that will confirm what I've been telling you." - **John Callahan was FAA Division Chief of Accidents and Investigations. From a video interview with The UFO Disclosure Project.**

Lambros Callimahos
"If 'they' discover you, it is an old but hardly invalid rule of thumb, 'they' are your technological superiors." - **Callimahos was a National Security Analyst, 1968.**

Dr. H. Marshall Chadwell
"Sightings of unexplained objects at great altitude and traveling at high speeds in the vicinity of major US defense installations are of such nature that they are not attributable to natural phenomena or known types of aerial vehicles." - **Chadwell, former assistant director of the CIA's Office of Scientific Intelligence, in a December 1952 memo to then-director of the CIA, General Walter B. Smith.**

Albert M. Chop
"I've been convinced for a long time that the flying saucers are real and interplanetary. In other words, we are being watched by beings from outer space." - **Chop was deputy public relations director at NASA and former US Air Force spokesman for Project Blue Book.**

Dr. Brain T. Clifford
"Contact between U.S. citizens and extra-terrestrials or their vehicles is strictly illegal"
- **Dr. Brain T. Clifford, a Pentagon official, from a press conference ("The Star", New**

York, Oct. 5, 1982)

Allen Dulles
"Maximum security exists concerning the subject of UFOs." - **Dulles was CIA Director, 1955.**

James V. Forrestal
"The truth was known by only a very few persons. Of the original group that were the first to learn, several committed suicide, the most prominent of which was Forrestal, who jumped to his death from a 16th story hospital window. Secretary Forrestal's records are sealed to this day."
- Forrestal was Secretary of Defense

Roscoe Hillenkoetter
"Behind the scenes, high-ranking Air Force officers are soberly concerned about UFOs. But through official secrecy and ridicule, many citizens are led to believe the unknown flying objects are nonsense." - **From a public statement, 1960.**
http://www.ufoevidence.org/documents/doc1737.htm

"Unknown objects are operating under intelligent control... It is imperative that we learn where UFOs come from and what their purpose is... " - **From, "What The Admiral Knew: UFO, MJ-12 and R. Hillenkoetter," International UFO Reporter, Nov./Dec., 1986.**

"It is time for the truth to be brought out in open Congressional hearings. Behind the scenes, high-ranking Air Force officers are soberly concerned about the UFOs. But through official secrecy and ridicule, many citizens are led to believe the unknown flying objects are nonsense."
- Statement in a NICAP news release, February 27, 1960. Admiral Hillenkoetter was the first Director of the CIA, 1947-50. In 1957, he joined the Board of Governors of the National Investigations

J. Edgar Hoover
"I would do it [aid the Army Air Force in its investigations] but before agreeing to it we must insist upon full access to the discs recovered. For instance, in the LA* case the Army grabbed it and would not let us have it for cursory examination." - **From a handwritten notation at the bottom of a now-declassified memo. (asterisk added, see below) * -- "LA" in this case may refer to Los Alamos, where the first atomic bombs were developed. There is some corroboration for this in the newly released, but unofficial MJ-12 documents.**

"The Federal Bureau of Investigation has been requested to assist in the investigation of reported sightings of flying disks..."

"[UFOs are] considered top secret by intelligence officers of both the Army and the Air Forces."
- From a declassified 1949 FBI document from the San Antonio FBI office, to J. Edgar Hoover.

"An investigator for the Air Force stated that three so-called flying saucers had been recovered in New Mexico. They were described as being circular in shape with raised centers. Approximately 50 feet in diameter. Each one was occupied by three bodies of human shape but only 3 feet tall. Dressed in metallic cloth of a very fine texture. Each body was bandaged in a manner similar to the blackout suits used by speed flyers and test pilots." - **From a March 22, 1950, memo to J. Edgar Hoover from the Washington FBI Office, released in 1976 under the freedom of information act. Hoover was Director of the FBI**
Victor Marchetti

"We have, indeed, been contacted - perhaps even visited - by extraterrestrial beings, and the U.S. government, in collusion with the other national powers of the earth, is determined to keep this information from the general public." http://www.ufoevidence.org/documents/doc1737.htm

"The purpose of the international conspiracy is to maintain a workable stability among the nations of the world and for them, in turn, to retain institutional control over their respective populations. Thus, for these governments to admit that there are beings from outer space... with mentalities and technological capabilities obviously far superior to ours, could, once fully perceived by the average person, erode the foundations of the earth's traditional power structure. Political and legal systems, religions, economic and social institutions could all soon become meaningless in the mind of the public. The national oligarchical establishments, even civilization as we now know it, could collapse into anarchy."

"Such extreme conclusions are not necessarily valid, but they probably accurately reflect the fears of the 'ruling classes' of the major nations, whose leaders (particularly those in the intelligence business) have always advocated excessive governmental secrecy as being necessary to preserve 'national security."- **How the CIA Views the UFO Phenomenon", Victor Marchetti , May 1979. Marchetti was a former CIA official.**
John Maynard

"I said, "Now this is supposed to be a system that tracks radar anomalies on Earth, right?" He says, "Yep, that's what it does." So I ask, "Then why are half of them pointed toward outer space, towards the moon, towards areas that are just blank space? What are they looking for?" He says, "Well, you've got to have a need to know to know about that"...We go back to a comment made by one of the astronauts when he stepped on the moon. It was the day after they got there. It was the original flight, and he says, "You're right, they're already here." It got out on the airwaves. I know several people who recorded it. It was quickly taken out of all tapes that were public broadcast." - **John Maynard, Defense Intelligence Agency, Military Intelligence Analyst**

Walter Bedell Smith
"The Central Intelligence Agency has reviewed the current situation concerning unidentified flying objects which have created extensive speculation in the press and have been the subject of concern to Government organizations... Since 1947, approximately 2,000 official reports of sightings have been received and of these, about 20% are as yet unexplained."

254

"It is my view that this situation has possible implications for our national security which transcend the interests of a single service. A broader, coordinated effort should be initiated to develop a firm scientific understanding of the several phenomena which apparently are involved in these reports..." - **From a 1952 memorandum to the National Security Council. Smith was Director of the CIA from 1950 to 1953.**

http://www.ufoevidence.org/documents/doc1737.htm

CHAPTER 104

INTERNATIONAL QUOTES FROM WORLD LEADERS AND OFFICIALS ABOUT UFOS

Roger Bacon
Byname ***Doctor Mirabilis*** **(Latin: "Wonderful Teacher")**, (born *c.* 1220, Ilchester, Somerset, or Bisley, Gloucester?, England—died 1292, Oxford?), English Franciscan philosopher and educational reformer who was a major medieval proponent of experimental science. Bacon studied mathematics, astronomy, optics, alchemy, and languages. He was the first European to describe in detail the process of making gunpowder, and he proposed flying machines and motorized ships and carriages. https://www.britannica.com/biography/Roger-Bacon

"Flying machines as these were of old, made even in our days" - **Friar, 13th century.**

https://en.wikipedia.org/wiki/Roger_Bacon

Monsignor Corrado Balducci
"We can no longer think... is it true? Is it not true? Are they truths or are they lies--if we believe or if we don't believe--no! There are already numerous considerations which make the existence of these beings into a certainty we cannot doubt." - **Monsignor Corrado Balducci is a Vatican theologian "close to the pope", who has been charged with studying reports of UFOs sent in from Vatican embassies around the world. This quote comes from one of Father Balducci's many appearances on Italian television.**

Sir Francis Chichester

"An object 'like an oblong pearl' drew steadily closer until perhaps a mile away when, right under my gaze as it were, it suddenly vanished. ..... But it reappeared close to where it had vanished. .....It drew closer. I could see the dull gleam of light on nose and back. It came on, but instead of increasing in size, it diminished as it approached! When quite near, it suddenly became its own ghost. For one second I could see clear through it and the next ..it had vanished." **- June 10, 1931, Tasman Sea. Sir Francis Chichester, a famous aviator, sailor, and author, reporting on a strange sighting he had while flying his Gypsy Moth.**

Sir Winston Churchill

"What does all this stuff about flying saucers amount to? What can it mean? What is the truth?"

L. Clerebaut

"Scientifically we eliminate the simple hypotheses: It's not a plane. It's not a helicopter. It's not a natural phenomenon because the descriptions don't match. Therefore this global phenomenon resists any other explanation. The only remaining hypothesis is the hypothesis of extraterrestrial origin." - **Clerebaut was Secretary General, Belgian Government, in charge of investigating UFO sightings.**

Lord Davies of Leek

"If one human being out of tens of thousands who allege to have seen these phenomena is telling the truth, then there is a dire need for us to look into the matter." - **Lord Davies was a Member of the House of Lords.**

Sir Eric Gairy

"I was privileged to see what I consider an Unidentified Flying Object very far out in the sky and it was going at tremendous speed. I could not identify any movement within the object. It was just a tremendous light. ..We would be naive to accept any theory that earth is the total estate of God's domain. There must be other planets. There must be other creatures. There must be other living things. We are not the most intelligent of God's creatures".- **Prime Minister of Grenada 1974-1979 repeatedly called for a special group within the UN to investigate UFO's**

258

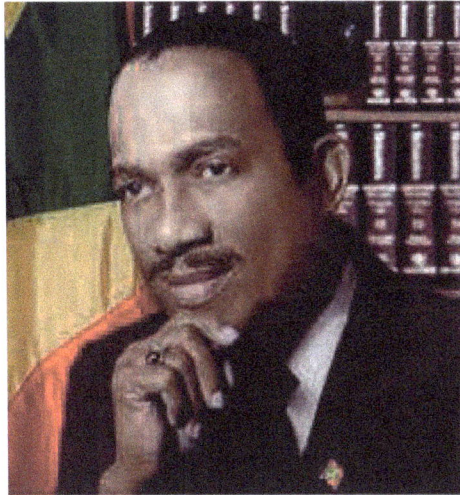

M. Robert Galley

"I must say that if listeners could see for themselves the mass of reports coming in from the airborne gendarmerie, from the mobile gendarmerie, and from the gendarmerie charged with the job of conducting investigations, all of which reports are forwarded by us to the National Center for Space Studies, then they would see that it is all pretty disturbing."

"I believe that the attitude of spirit that we must adopt vis-à-vis this phenomena is an open one, that is to say, that it doesn't consist in denying apriori, as our ancestors of previous centuries did deny many things that seem nowadays perfectly elementary." - **From an interview by Jean-Claude Bourret, on February 21, 1974. Galley was French Minister of Defense**

Mikhail Gorbachev

"In spite of all the differences between us, we must all learn to preserve our one big family of humanity. At our meeting in Geneva, the U.S. President said that if the earth faced an invasion by extraterrestrials, the United States and the Soviet Union would join forces to repel such an invasion. I shall not dispute the hypothesis, though I think it's early yet to worry about such an intrusion." **- From a speech to the International Forum, "For a Nuclear-Free World and the Survival of Humanity," at the Grand Kremlin Palace in Moscow on February 16, 1987.**

"I know that there are scientific organizations which study this problem."
In reply to the question, "Does the USSR government study UFOs? - Asked while visiting the Uralmash plant in Sverdlovsk on April 26th, 1990
"The phenomenon of UFOs does exist, and it must be treated seriously." **- From Soviet Youth, May 4, 1990.**

Paul Hellyer, former Canadian Minister of Defence (the only G8 senior defence minister to come out publicly announce the reality of the UFO phenomenon)
"UFOs, are as real as the airplanes that fly over your head,"

"I'm so concerned about what the consequences might be of starting an intergalactic war, that I just think I had to say something."

"The United States military are preparing weapons which could be used against the aliens, and they could get us into an intergalactic war without us ever having any warning."

"The time has come to lift the veil of secrecy, and let the truth emerge, so there can be a real and

informed debate, about one of the most important problems facing our planet today." - **Paul Hellyer former Canadian Minister of Defence at the Toronto Exopolitics Symposium, September 25, 2005**

"To turn us in the direction of re-unification with the rest of creation Alfred Lambremont Webre is proposing a 'Decade of Contact' - an 'era of openness, public hearings, public funded research, and education about extraterrestrial reality.' That could just be the antidote the world needs to end its greed-driven, power-centered madness." - **Paul Hellyer at University of Toronto**

Lord Hill-Norton
"The evidence that there are objects which have been seen in our atmosphere, and even on terra firma, that cannot be accounted for either as man-made objects or as any physical force or effect known to our scientists seems to me to be overwhelming... A very large number of sightings have been vouched for by persons whose credentials seem to me unimpeachable. It is striking that so many have been trained observers, such as police officers and airline or military pilots. Their observations have in many instances... been supported either by technical means such as radar or, even more convincingly, by... interference with electrical apparatus of one sort or another..."
- **From the foreword to a book written by British UFO researcher Timothy Good, Above Top Secret, in 1987. Lord Hill-Norton (GCB), Chief of Defense Staff, Ministry of Defense, Britain; Chairman, Military Committee of NATO; Admiral of the Fleet; Member of House of Lords.**

"I have frequently been asked why a person of my background a former Chief of the Defense Staff, a former Chairman of the NATO Military Committee - why I think there is a cover-up [of] the facts about UFOs. I believe governments fear that if they did disclose those facts, people would panic. I don't believe that at all. There is a serious possibility that we are being visited by people from outer space. It behooves us to find out who they are, where they come from, and what they want." - **British Royal Navy, Video, and Disclosure, pp. 305 - 307**

Prime Minister Toshiki Kaifu

"Japan does not have such organizations at a government level... If young people display a serious interest in similar phenomena, we should perhaps think of forming a UFO-data collecting group under the auspices of the Ministry of Education." - **From an interview with students of Waseda University in Tokyo in November 1989 on the question of whether Japan had an official UFO organization**

"First of all, I told a magazine this past January that, as an underdeveloped country with regards to the UFO problem, Japan had to take into account what should be done about the UFO question, and that we had to spend more time on these matters. In addition, I said that someone had to solve the UFO problem with far-reaching vision at the same time. Secondly, I believe it is a reasonable time to take the UFO problem seriously as a reality... I hope that this Symposium will contribute to peace on earth from the point of view of outer space, and take the first step toward the international cooperation in the field of UFOs. From the point of view of 'people' in outer space, all human beings on earth are the same people, regardless of whether they are American, Russian, Japanese, or whoever." - **Kaifu, who was Prime Minister, concerning an upcoming Symposium on Space and UFOs.**

262

Earl of Kimberly
"UFOs defy worldly logic... The human mind cannot begin to comprehend UFO characteristics: their propulsion, their sudden appearance, their disappearance, their great speeds, their silence, their maneuver, their apparent anti-gravity, their changing shapes." - **Kimberly was former Liberal Party spokesman on aerospace and member of the House of Lords**.
Prince Phillip, His Royal Highness, Duke of Edinburgh
"There are many reasons to believe that they (UFOs) do exist: there is so much evidence from reliable witnesses."

Nick Pope
"I concentrate on the science. I'm interested in the UFOs seen by the police and military witnesses. I'm interested in the near misses that pilots report, where their aircraft nearly collide with these things. I'm interested in the visual sightings backed up by radar. I'm interested in the military bases that are overflown by these things. I'm interested in the cases where you have radiation readings on the ground. These are no lights in the sky. These are not misidentifications of fantasy prone individuals. This is a cutting-edge technology being reported by reliable, trained observers, and it is something that goes beyond what we can do. That to me suggests that if it is not ours, it belongs to someone else. If that technology is better than ours, then the extraterrestrial hypothesis seems to me the best explanation."

"Certainly, when I socialized with my RAF colleagues, I would find that they were a little bit more receptive to the idea of UFOs--and by that I mean perhaps even an extraterrestrial explanation for this -- than you might have supposed. One of the reasons for that was that so many RAF pilots had actually seen things themselves. Many of them have never made an official report. I had one chap tell me that he had seen something over the North Sea. I asked him why he

hadn't reported it, and he said, 'I don't want to be known as Flying Saucer Fred for the rest of my career.'"

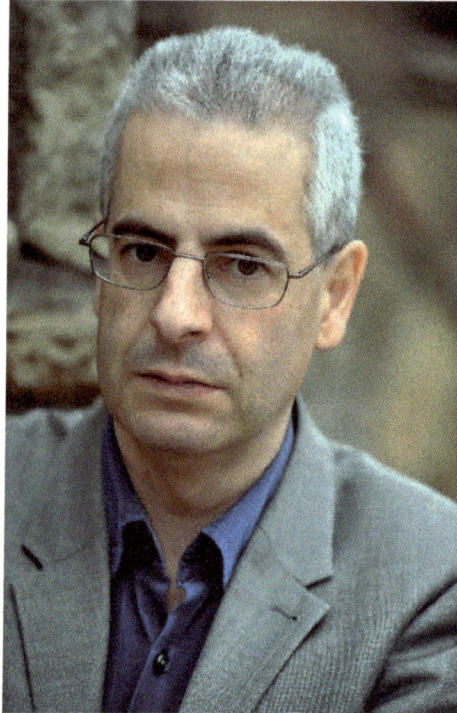

"We were asking the Americans, 'Are you operating a prototype aircraft in our airspace?' That, of course, was nonsense. You simply would not do that from a diplomatic and political point of view. It would undermine the entire structure of NATO if you were putting things through someone else's airspace, particularly a close ally, without seeking the proper diplomatic clearance. But we had to ask. And the Americans, having had similar reports, I guess, since the Hudson Valley wave [New York state, the mid-1980s], had been quietly asking us if we had some large, triangular shaped object that could go from 0 to Mach 5 in a second. Our response was that we wished we did. This was the bizarre situation: that we were chasing the Americans, and the Americans were chasing us."

"The official line from the Ministry of Defense is, 'Yes, this happened. No, we don't know what it is, but we say that it is of no defense significance.' How can it possibly be of no defense significance when your best jet is left for standing by a UFO? And, again, how can it be of no defense significance when your air defense region is routinely penetrated by structured craft?"
Excerpts British Ministry of Defense from 1991-1994, and has served in other departments of the Ministry of Defense since 1985.

Lord Rankeillour
"Many men have seen them [UFOs] and have not been mistaken. Who are we to doubt their word?... Only a few weeks ago a Palermo policeman photographed one, and four Italian Navy officers saw a 300-foot long fiery craft rising from the sea and disappearing into the sky... Why

264

should these men of law enforcement and defense lie?" - **Lord Rankeillour was a Member of the House of Lords.**

Air Marshall Nurjadin Roesmin

"UFOs sighted in Indonesia are identical with those sighted in other countries. Sometimes they pose a problem for our Air Defence and once we were obliged to open fire on them." - **Air Marshall Roesmin was Commander-in-Chief of the Indonesian Air Force, 1967.**

Wilbert Smith

"The matter is the most highly classified subject in the United States Government, rating higher even than the H-bomb. Flying saucers exist. Their modus operandi is unknown but a concentrated effort is being made by a small group headed by Doctor Vannevar Bush. The entire matter is considered by the United States authorities to be of tremendous significance." - **Smith was a senior radio engineer with the Department of Transport, headed Project Magnet, the first Canadian government UFO investigation in the 1950s.**

http://luforu.org/wilbert-b-smith/

Earl Alexander of Tunis

"There are of course many phenomena in this world which are not explained and it is possible to say that the orthodox scientist is the last person to accept that something new (or old) may exist which cannot be explained in accordance with his understanding of natural laws." - **Earl Alexander was British Minister of Defense.**

Jean-Jacques Velasco

"There are cases which remain unexplained... Let's say simply that the events which were registered and measured, particularly at Trans-en-Provence, but also in the case of l'Amarante [a CE-II on Oct. 21, 1982] and two others, allow us to suppose that there are phenomena which escape our understanding completely. I must say that this permits us to suppose that there is an intelligence behind the phenomena. But I believe it would be largely speculation to go beyond this point." **- Velasco, the last head of GEPAN and director of SEPRA at CNES Headquarters in Toulouse, in an interview with the French magazine Phénomèna. Velasco also stated that SEPRA's primary task was tracking "satellite re-entries, which are more and more numerous, and secondly, to continue the activities of GEPAN, stopped in 1988.**

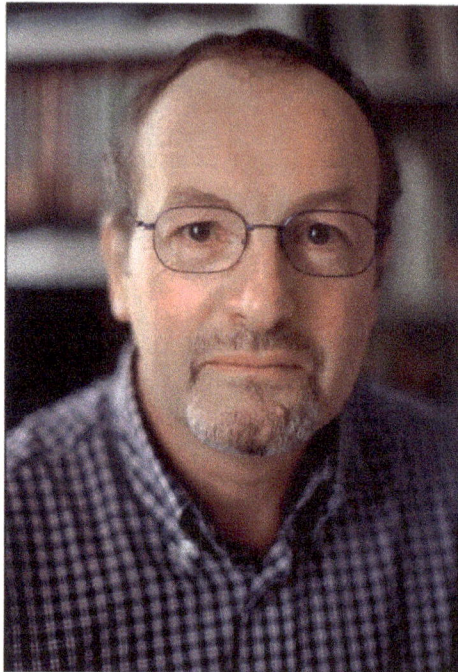

https://www.ouillade.eu/agenda/le-soler-conference-roswell-la-contre-enquete-par-jean-jacques-velasco/148204

CHAPTER 105

QUOTES FROM ASTRONAUTS AND COSMONAUTS ABOUT UFOS

NASA and the CIA instructed the American astronauts to *"never use the word UFO upon sighting the UFO (s) while broadcasting or transmitting from outer space. Instead, the code and secret name 'Santa Claus' is to be exclusively used upon encountering flying saucers."* This was confirmed by the US astronauts in person and in writing.

In the field of UFO, investigation eyewitness testimony that comes from astronauts and cosmonauts are ranked highest because of the intensive training these exceptional men and women must undergo in order to go into space. Part of that training includes their observational skills which must be honed to the highest degree so as to be aware of anything and everything in their unique environment. These keen powers of observation could mean the difference between life and death or possible collision mishaps of the lost of expensive and irreplaceable equipment.

With every space mission since the early '60s, these brave heroic pioneers of space have returned to Earth with unusual accounts of alien craft in orbit with them or when they walked upon the Moon, even in defiance of NASA's and other government agencies efforts to keep these reports from the public. Contrary to what skeptics and debunker would have you believe, we present the best irrefutable testimonies from the best-trained observers in the world for the existence of UFOs and the ET presence in orbit around the Earth.

Cosmonaut Victor Afanasyev

"It followed us during half of our orbit. We observed it on the light side, and when we entered the shadow side, it disappeared completely. It was an engineered structure, made from some type of metal, approximately 40 meters long with inner hulls. The object was narrow here and wider here, and inside there were openings. Some places had projections like small wings. The object stayed very close to us. We photographed it, and our photos showed it to be 23 to 28 meters away." - **Afanasyev commenting on a UFO sighting that occurred while en route to the Solyut 6 space station in April of 1979.**

Buzz Aldrin

NASA: *What's there?*
Apollo11: *These "Babies" are huge, Sir! Enormous! OH MY GOD! You wouldn't believe it! I'm telling you there are other spacecraft out there, Lined up on the far side of the crater edge! They're on the Moon watching us!* - **Transmission from Apollo 11 on the Moon on July 21, 1969**

"Now, obviously, the three of us were not going to blurt out, 'Hey Houston we got something moving along side of us and we don't know what it is,' observed Aldrin. "We weren't about to do that, cause we know that those transmissions would be heard by all sorts of people and who knows what somebody would have demanded that we turn back because of Aliens or whatever the reason is, so we didn't do that but we did decide we'd just cautiously ask Houston where, how far away was the S-IVB? And a few moments we decided that after a while of watching it (UFO), it was time to go to sleep and not to talk about it anymore until we came back and (went through)

debriefing."- **Aldrin, commenting after the Apollo 11 flight.**
http://www.gravitywarpdrive.com/UFO_Testimonies.htm

Neil Armstrong
"We have no proof, But if we extrapolate, based on the best information we have available to us, we have to come to the conclusion that ... other life probably exists out there and perhaps in many places..." - **From a statement in October 1999.**

"It was incredible, of course, we had always known there was a possibility, the fact is, we were warned off! (by the Aliens). There was never any question then of a space station or a moon city. I can't go into details, except to say that their ships were far superior to ours both in size and technology - Boy, were they big!...and menacing! No, there is no question of a space station? **-Neil Armstrong on encounters with extraterrestrial life during the Apollo 11 Moon Landing.**

Scott Carpenter
"At no time, when the astronauts were in space were they alone: there was a constant surveillance by UFOs." - **Carpenter photographed a UFO while in orbit on May 24, 1962. NASA still has not released the photograph.**

Eugene Cernan
"...I've been asked about UFOs and I've said publicly I thought they were somebody else, some other civilization." - **Cernan commanded the Apollo 17 Mission-The quote is from a 1973 article in the Los Angeles Times**. http://www.gravitywarpdrive.com/UFO_Testimonies.htm

Colonel L. Gordon Cooper
"I wanted to convey to you my views on our extra-terrestrial visitors popularly referred to as ' UFOs,' and suggest what might be done to properly deal with them."

"I believe that these extraterrestrial vehicles and their crews are visiting this planet from other planets which obviously are a little more technically advanced than we are here on Earth. I feel that we need to have a top-level, coordinated program to scientifically collect and analyze data from all over the earth concerning any type of encounter, and to determine how best to interface with these visitors in a friendly fashion. We may first have to show them that we have learned to resolve our problems by peaceful means, rather than warfare, before we are accepted as fully qualified universal team members. This acceptance would have tremendous possibilities of advancing our world in all areas. Certainly, then it would seem that the UN has a vested interest in handling this subject properly and expeditiously."

"If the UN agrees to pursue this project, and to lend their credibility to it, perhaps many more well-qualified people will agree to step forth and provide help and information." - **Cooper, addressing a UN panel discussion on UFOs and ETs in New York in 1985.**

"For many years I have lived with a secret, in a secrecy imposed on all specialists and astronauts. I can now reveal that every day, in the USA, our radar instruments capture objects of form and composition unknown to us. And there are thousands of witness reports and a quantity

268

of documents to prove this, but nobody wants to make them public." - **Cooper testifying to a U.N. committee.**

"As far as I am concerned, there have been too many unexplained examples of UFO sightings around this Earth for us to rule out the possibilities that some form of life exists out there beyond our own world."

"I know other astronauts share my feelings, and we know the government is sitting on hard evidence of UFOs!"

"We thought they could have been Russian we regularly had MiG-15s overflying our base. We scrambled our Sabre jets to intercept and got to our ceiling of 45,000 feet . . . and they were still way above us traveling faster than we were. These vehicles were in formation like a fighter group, but they were metallic silver and saucer shaped. Believe me, they weren't like any MiGs I'd seen before! They had to be UFOs." - **Cooper said he first encountered UFOs as a military pilot in Germany in the early 1950s when unidentified craft were spotted over an air base.**
http://www.gravitywarpdrive.com/UFO_Testimonies.htm

"I had a camera crew filming the installation when they spotted a saucer. They filmed it as it flew overhead, then hovered, extended three legs as landing gear, and slowly came down to land on a dry lake bed! These guys were all pro cameramen, so the picture quality was very good. The camera crew managed to get within 20 or 30 yards of it, filming all the time. It was a classic saucer, shiny silver and smooth, about 30 feet across. It was pretty clear it was an alien craft. As they approached closer it took off." When his camera crew handed over the film, Cooper followed standard procedure and contacted Washington to report the UFO and *"all heck broke loose,"* he said. *"After a while, a high-ranking officer said when the film was developed, I was to put it in a pouch and send it to Washington. He didn't say anything about me not looking at the film. That's what I did when it came back from the lab and it was all there just like the camera crew reported."* When the Air Force later started Operation Blue Book to collate UFO evidence and reports, Cooper says he mentioned the film evidence. *"But the film was never found supposedly. Blue Book was strictly a cover-up anyway."* - **In 1957, Cooper was one of an elite band of test pilots at Edwards Air Force Base in California, in charge of several advanced projects, including the installation of a precision landing system.**

"I had a good friend at Roswell, a fellow officer. He had to be careful about what he said. But it sure wasn't a weather balloon, like the Air Force cover story. He made it clear to me what crashed was a craft of alien origin, and members of the crew were recovered." - **Cooper revealed he's convinced an alien craft crashed at Roswell, N. Mex., in 1947 and aliens were discovered in the wreckage.**

"It started in World War 2 when the government didn't want people to know about UFO reports in case they panicked," said Cooper. *"They would have been fearful it was superior enemy technology that we had no defense against. Then it got worse in the Cold War for the same reason. So they told one untruth, they had to tell another to cover that one, then another, then another . . . it just snowballed. And right now I'm convinced a lot of very embarrassed government officials are sitting there in Washington trying to figure a way to bring the truth out.*

They know it's got to come out one day, and I'm sure it will. America has a right to know!"
- A statement when asked why has the government kept its UFO secrets for so many years? Cooper was a Mercury 9 and Gemini-5 astronaut.

American astronauts who saw UFOs in space
Google Images

"I should point out that I am not an experienced UFO professional researcher. I have not yet had the privilege of flying a UFO, nor of meeting the crew of one. I do feel that I am somewhat qualified to discuss them since I have been into the fringes of the vast areas in which they travel. Also, I did have occasion in 1951 to have two days of observation of many flights of them, of different sizes, flying in fighter formation, generally from east to west over Europe. They were at a higher altitude than we could reach with our jet fighters of that time."

"I would also like to point out that most astronauts are very reluctant to even discuss UFOs due to the great numbers of people who have indiscriminately sold fake stories and forged documents abusing their names and reputations without hesitation. Those few astronauts who have continued to have participation in the UFO field have had to do so very cautiously. There are several of us who do believe in UFOs and who have had occasion to see a UFO on the ground, or from an airplane. There was only one occasion from space which may have been a UFO."

"Several days in a row we sighted groups of metallic, saucer-shaped vehicles at great altitudes over the base, and we tried to get close to them, but they were able to change direction faster than our fighters. I do believe UFOs exist and that the truly unexplained ones are from some other technologically advanced civilization. From my association with aircraft and spacecraft, I think I have a pretty good idea of what everyone on this planet has and their performance capabilities, and I'm sure some of the UFOs at least are not from anywhere on Earth." - **Omni, Vol. 2, No. 6, March 1980.**

John Glen
"I believe certain reports of flying saucers to be legitimate."
http://www.gravitywarpdrive.com/UFO_Testimonies.htm

James Irwin
"Look, I have a pension to worry about. I have a family to take care of, and they told me to just back away from this entirely or else." - **Apollo 15 astronaut to Frank Stranges after backing out of speaking at a 1976 UFO convention where he was going to "inform us of the strange things he saw on the surface of the moon.**

"He told me the story here in the hanger…as soon as we were by ourselves. He turned to me and he looked at me and said, "I'm going to say something that is Top Secret. If you repeat it I will deny ever having said it." He said they weren't there (on the moon surface) an hour and a saucer landed a mile away from them and we asked Houston if we could motor over and say howdy. They said 'no ignore them and pretend they're not there, and carry on about your business.' He said the all the time we were there they saw no sign of movement, and they were still there when we left". - **(on tape) to a famous Canadian pilot**
http://www.gravitywarpdrive.com/UFO_Testimonies.htm

Yevegni Khrunov
"Is the presence of extraterrestrial civilizations conceivable? Of course. Before the uniqueness of the earth is demonstrated, this assumption should be taken as quite legitimate. As regards UFOs, their presence cannot be denied: thousands of people have seen them. It may be that their source is optical effects, but some of their properties, for instance, their ability to change course by 90

degrees at great speed, simply stagger the imagination." - **Sputnik, "UFOs Through the Eyes of Cosmonauts," December 1980. Yevegni Khrunov was the Soyuz-5 spacecraft pilot in 1969.**

Russian Cosmonauts Victor Afanasyev, Yevegni Khrunov,Major General Vladimir Kovalyonok, and Major General Pavel Popovich who saw UUFOs in space.
Google Images

Major General Vladimir Kovalyonok
"On May 5, 1981, we were in orbit [in the Salyut-6 space station]. I saw an object that didn't resemble any cosmic objects I'm familiar with. It was a round object which resembled a melon, round and a little bit elongated. In front of this object was something that resembled a gyrating depressed cone. I can draw it, it's difficult to describe. The object resembles a barbell. I saw it becoming transparent and like with a ' body' inside. At the other end, I saw something like gas

272

discharging, like a reactive object. Then something happened that is very difficult for me to describe from the point of view of physics. Last year in the magazine Nature I read about a physicist... we tried together to explain this phenomenon and we decided it was a ' plasmaform.' I have to recognize that it did not have an artificial origin. It was not artificial because an artificial object couldn't attain this form. I don't know of anything that can make this movement... tightening, then expanding, pulsating. Then as I was observing, something happened, two explosions. One explosion, and then 0.5 seconds later, the second part exploded. I called my colleague Viktor [Savinykh], but he didn't arrive in time to see anything."

"What are the particulars? First conclusion: the object moved in a suborbital path, otherwise I wouldn't have been able to see it. There were two clouds, like smoke, that formed a barbell. It came near me and I watched it. Then we entered into the shade for two or three minutes after this happened. When we came out of the shade, we didn't see anything. But during a certain time, we and the craft were moving together." - **Videotaped interview with Giorgio Bongiovanni in the village of Kosnikov, near Moscow, 1993.**
http://www.gravitywarpdrive.com/UFO_Testimonies.htm

Dr. Jerry Linenger
"In five months in space, I have seen unidentified flying objects for sure. Sometimes I looked out of the window and I could see a metallic thing like a spoon flying methodically." - **Dr. Linenger was a NASA astronaut and during five months in he logged 50 million miles - the equivalent of over 110 round trips to the moon, travelling at an average speed of 18,000 miles per hour. Dr. Linenger was in Dubai to speak at the BurJuman Retail Conference at the Emirates Towers Hotel.**

James Lovell
Lovell: *Bogey at 10 o'clock high.*
Capcom: *This is Houston. Say again 7.*
Lovell: *Said we have a bogey at 10 o'clock high.*
Capcom: *Gemini 7, is that the booster or is that an actual sighting?*
Lovell: *We have several...actual sightings.*
Capcom: *...Estimated distance or size?*
Lovell: *We also have the booster in sight...* - **James Lovell in conversation with mission control during his flight on Gemini 7.**
"Mission Control, please be informed, there is a Santa Claus." - **James Lovell, who was commander of the ill-fated Apollo-13 mission, made this transmission after coming around the far side of the moon on the Apollo-8 mission on or around Christmas in 1968 with Frank Borman and William Anders. Although it was Christmas time, this statement has caused considerable controversy as "Santa Claus" was apparently a codeword used to indicate a UFO or other unusual sighting.**

James McDivitt
"At one stage we even thought it might be necessary to take evasive action to avoid a collision." - **James McDivitt commenting on an orbital encounter he and Ed White had with a "weird object" with arm-like extensions which approached their capsule. Later in the flight, they saw two similar objects over the Caribbean.**

Edgar D. Mitchell

"We all know that UFOs are real. All we need to ask is where do they come from?" - **From a statement in 1971**

"I've talked with people of stature of military and government credentials and position and heard their stories, and their desire to tell their stories openly to the public. And that got my attention very, very rapidly... The first-hand experiences of these credible witnesses that, now in advanced years are anxious to tell their story, we can't deny that, and the evidence points to the fact that Roswell was a real incident, and that indeed an alien craft did crash, and that material was recovered from that crash site." http://www.gravitywarpdrive.com/UFO_Testimonies.htm

"The U.S. Government hasn't maintained secrecy regarding UFOs It's been leaking out all over the place. But the way it's been handled is by denial, by denying the truth of the documents that have leaked. By attempting to show them as fraudulent, as bogus of some sort. There has been a very large disinformation and misinformation effort around this whole area. And one must wonder, how better to hide something out in the open than just to say, 'It isn't there. You're deceiving yourself if you think this is true.' And yet, there it is right in front of you. So it's a disinformation effort that's concerning here, not the fact that they have kept the secret. They haven't kept it. It's been getting out into the public for fifty years or more."

"I have been over the years very skeptical like many others. But in the last ten years or so, I have known the late Dr. Alan Hynek, who I highly admire. I know and currently, work with Dr. Jacques Vallee. I've come to realize that the evidence is building up to make this a valid and researchable question. Further, because my personal motivation has always been to understand our universe better, and my own theoretical work has convinced me that life is everywhere in the universe that has been permitted to evolve, I consider this a very timely question... By becoming more involved with the serious research field, I've seen the evidence mount towards the truth of these matters. I rely upon the testimony of contacts that I have had - old timers - who were involved in official positions in government and intelligence and military over the last 50 years. We cannot say that today's government is really covering it up - I think that most of them don't know what is going on anymore than the public..." - **From an interview with MSN 1998.**

"The evidence points to the fact that Roswell was a real incident and that indeed an alien craft did crash and that material was recovered from that crash site," says Mitchell, who became the sixth man on the moon in the Apollo 14 mission. Mitchell doesn't say he's seen a UFO. But he says he's met with high-ranking military officers who admitted involvement with alien technology and hardware. - **Captain Mitchell was an Apollo 14 Astronaut.**

Storey Musgrove

"Statistically, it's a certainty there are hugely advanced civilizations, intelligences, life forms out there. I believe they're so advanced they're even doing interstellar travel. I believe it's possible they even came here."

"I try to communicate with the life that's out there. I'm serious. It is not that far out. When I'm circling around out there, I try in whatever ways I can to get them to come down here and get

me." - **The Houston Post, December 1, 1993. Story Musgrave was the Space Shuttle astronaut who flew on the repair mission of the Hubble Space Telescope.**

Dr. Brian O'Leary
"We have contact with alien cultures."

Major General Pavel Popovich
"Today it can be stated with a high degree of confidence that observed manifestations of UFOs are no longer confined to the modern picture of the world... The historical evidence of the phenomenon... allows us to hypothesize that ever since mankind has been co-existing with this extraordinary substance, it has manifested a high level of intelligence and technology. The UFO sightings have become the constant component of human activity and require a serious global study... The scientific study of the UFO phenomenon should take place in the midst of other sciences dealing with man and the world." - **Popovich, P., MUFON 1992 International Symposium Proceedings.** http://www.gravitywarpdrive.com/UFO_Testimonies.htm

"The influence the UFO has on people, as well as the effects it produces, should become the items of special research. The UFO's interaction with the environment, the behavior that it motivates, and its genesis, also present interesting areas for concentrated study. Today, many specialists have come to the opinion that [UFO] phenomenon research should be taken up along with understanding and comprehension of other unexplained phenomena... The development of new approaches for the identification and study of energy and information processes will allow for an enthusiastic move toward the comprehension of the phenomenon. The results of these studies should aid the survival of the people on earth..."

"It's necessary to carry out the popular Ufological enlightenment, since the probability for a meeting of a person with a UFO exists, and this person should be ready for this event. Precautionary measures are especially important. It's necessary, to tell the truth, which has been distorted previously by the politically engaged sciences and most recently by ufological dilettantes. The main purpose of the primary local groups that of controlling the ufological situation mustn't be forgotten. The Ufologists should know all the UFO's landing places and contacts in their regions. They should have relations with the local authorities, and in particular, with the police, the civil defense bodies, as well as information, scientific, and medical organizations." - **Popovich, P., "Ufology in the Commonwealth of Independent States: Organization Problems," in the MUFON 1992 International UFO Symposium Proceedings. Major General Popovich was a pioneer Cosmonaut, "Hero of the Soviet Union", and President of All-Union Ufology Association of the Commonwealth of Independent States.**

Donald Slayton
"I was testing a P-51 fighter in Minneapolis when I spotted this object. I was at about 10,000 feet on a nice, bright, sunny afternoon. I thought the object was a kite, then I realized that no kite is gonna fly that high. As I got closer it looked like a weather balloon, gray and about three feet in diameter. But as soon as I got behind the darn thing it didn't look like a balloon anymore. It looked like a saucer, a disk. About the same time, I realized that it was suddenly going away from me and there I was, running at about 300 miles per hour. I tracked it for a little way, and

then all of a sudden the damn thing just took off. It pulled about a 45-degree climbing turn and accelerated and just flat disappeared." - **Donald Slayton, Mercury astronaut, in a 1951 interview.**

Joseph Albert Walker

"I don't feel like speculating about them. All I know is what appeared on the film which was developed after the flight." - **Joseph Walker from May 11, 1962 commenting during a lecture at the Second National Conference on the Peaceful Uses of Space Research in Seattle, Washington after he had filmed five or six UFOs , (he stated one his tasks was to detect UFOs), during his record breaking fifty-mile-high flight in April, 1962 while piloting an X-15.**

Major Robert White

"There ARE things out there! There absolutely is!" - **White exclaiming over the radio about a UFO encounter taking place on a 58 mile high X-15 flight on July 17, 1962.**
http://www.gravitywarpdrive.com/UFO_Testimonies.htm

"I have no idea what it could be. It was greyish in color and about thirty to forty feet away." - **Major Robert White, on July 17, 1962, during his fifty-eight-mile high flight of an X-15.**

Al Worden

"And a literal translation describes very clearly a spacecraft with the ability to land vertically and take off vertically, and it was an object that looked very much like the Lunar Module that we used on the Moon; and if it's going to land vertically and take off vertically, it had to come from some place and go back some place." - **Worden discussed his views that Earth was probably visited in the past by extraterrestrial explorers. He began by commenting on the well known "UFO interpretation" of the vision of the prophet Ezekiel in the Bible.**

"In my mind, the universe has to be cyclic, so that in one galaxy if there is a planet maybe that has arrived at the point of becoming unlivable, you will find in another part of a different galaxy a planet that has just formed which is perfect for habitation. I see some kind of intelligent being, like us, skipping around the universe from planet to planet as, let's say, the South Pacific Indians do on the islands, where they skip from island to island. When the first island blows up due to a volcano, they will have their progeny on all these other islands and they will be able to continue the species. I think that's what the [alien] space program is all about."

"I think we may be a combination of creatures that were living here on Earth sometime in the past, and having a visitation, if you will, by creatures from somewhere else in the universe, and those two species getting together and having progeny. I am not at all convinced that we are not the result of that particular union some many thousands of years ago. If that is the case, in fact, a very small group of explorers could land on a planet and create successors to themselves that would eventually take up the pursuit of, let's say, inhabiting the rest of the universe." - **Excerpts from his interview in the documentary "The Other Side of the Moon," produced by Michael G. Lemle, and broadcast by PBS in July 1989. Al Worden was an Apollo 15 astronaut who later became a poet.**

276

Unidentified Russian Cosmonauts

Female Cosmonaut: *"I'll take it and hold it with my right hand. Look out the peephole! I have it!"*

Male Cosmonaut: *"There is something! If we do not get out the world will never know about this!"* - **From the final transmission of a pair of Cosmonauts whose scheduled seven-day mission was interrupted by a malfunction of unknown origins. This piece of conversation was recorded on February 24, 1961, while they were trying to repair the damage. The two Cosmonauts were never heard from again. UFO comments by the world's top military leaders.** http://www.gravitywarpdrive.com/UFO_Testimonies.htm

CHAPTER 106

QUOTES FROM SCIENTISTS AND SCIENCE ORGANIZATIONS ABOUT UFOS

American Institute of Aeronautics and Astronautics UFO Subcommittee
The AIAA established a subcommittee in 1967 to look into the UFO question. The UFO Subcommittee issued several reports and statements, including in-depth studies of two UFO incidents. The UFO Subcommittee stated that its "most important conclusion" was that government agencies consider funding UFO research:

"From a scientific and engineering standpoint, it is unacceptable to simply ignore substantial numbers of unexplained observations... the only promising approach is a continuing moderate-level effort with emphasis on improved data collection by objective means... involving available remote sensing capabilities and certain software changes."

The Encyclopedia of UFOs, Ronald D. Story, New York: Doubleday, 1980. The Subcommittee of the American Institute of Aeronautics and Astronautics criticized the conclusion of The Condon Report as the personal views of Dr. Condon, and added:
"The opposite conclusion could have been drawn from The Condon Report's content, namely, that a phenomenon with such a high ratio of unexplained cases (about 30 percent) should arouse sufficient scientific curiosity to continue its study."

Brookings Institution
"If the intelligence of these creatures were sufficiently superior to ours, they might choose to have little, if any, contact with us." - **Brookings Institution report on extraterrestrial life. Quoted in New York Times, December 15, 1960.**

Professor Gabriel Alvial
"There is scientific evidence that strange objects are circling our planet. It is lamentable that governments have drawn a veil of secrecy around this matter." - **Professor Gabriel Alvial, Cerro Calan Observatory, quoted by Reuters on August 26, 1965. http://www.bibliotecapleyades.net/ciencia/ufo_briefingdocument/quosci.htm**

Prof. Claudio Anguila
"We are not alone in the universe!" - **Prof. Claudio Anguila, director of Cerro Calan Observatory, quoted by Reuters, August 26, 1965.**

Dr. Robert M. L. Baker, Jr.
"The system is partially classified and, hence, I cannot go into great detail... Since this particular sensor system has been in operation, there have been a number of anomalistic alarms. Alarms that, as of this date, have not been explained on the basis of natural phenomena interference, equipment malfunction or inadequacy, or man-made space objects." - **In 1968, he made this statement concerning the one U.S. radar system in operation at that time that, to his knowledge, exhibited sufficient continuous coverage to reveal UFOs operating above the earth's atmosphere during 1968 Congressional Hearings. He has specialized in the study of motion pictures of UFOs and anomalistic radar images and has concluded that two of the most famous UFO motion pictures, taken in the 1950s, cannot be explained in terms of**

conventional phenomena. Dr. Robert Baker was President of West Coast University; author of two astrodynamics textbooks; head of Lockheed's Astrodynamics Research Center (1961-64) and member of the faculty of Astronomy and Engineering at UCLA (1959-71).

Merccelin Berthelot
"From now on the universe is without further mystery" - **1887**

Dr. Maurice Biot
"The least improbable explanation is that these things UFO's are artificial and controlled. My opinion for some time has been that they have an extraterrestrial origin." - **Biot was one of the world's leading aerodynamicists and mathematical physicists. Life, April 7, 1952.**

Louis Breguet
"The discs use a means of propulsion different from ours. There is no other possible explanation. Flying saucers come from another world." - **Breguet was a French aircraft designer and manufacturer.**

Chinese Academy of Social Sciences
"One of the branches of the Chinese Academy of Social Sciences is the China UFO Research Organization (CURO). As of 1985, CURO had 20,000 members, and two publications, the Journal of UFO Research and Space Exploration. The Journal's first issue in 1981 included an article by Comrade Bang Wen-Gwang of the Chinese Academy of Sciences' Beijing Astronomical Research Society. The article stated in part: "In this field [Ufology], prejudice will take you farther from the truth than ignorance... But with a topic such as UFOs, where does the scientific method begin? And where does it end? This grand endeavor would consist of the serious recording of the enormous available data and the use of all scientific procedures for the purpose of analysis... China is so vast, and UFOs are certainly being witnessed again and again all throughout China, and China most definitely will evolve her own indigenous school of UFO researchers. This is our sincerest and deepest hope." - **Wen-Gwang, B., "The Aspirations & Hopes of the Chinese UFO Investigator," The Journal of UFO Research, No. 1, People's Republic of China, 1981.**
http://www.bibliotecapleyades.net/ciencia/ufo_briefingdocument/quosci.htm

China Daily
"UFO Scientific Conference in Darlian. In 1985, the government newspaper, China Daily, reported that a UFO Scientific Conference was held in Darlian, with some forty papers presented on various aspects of UFO research. Professor Liang Renglin of Guangzhou Jinan University, Chairman of CURO, stated in the Darlian Conference that more than 600 UFO reports had been made in China during the past five years. The article concluded: "UFOs are an unresolved mystery with profound influence in the world." - **"UFO Conference Held in Darlian," China Daily, August 27, 1985; quoted in Good, T., ibid.) Camille Flammarion, The systematic denial of unexplained facts has never advanced science by one single step -, French Astronomer**

Professor Gabriel Alvial

Dr. Robert M. L. Baker, Jr.

Merccelin Berthelot

Dr. Maurice Biot

Louis Breguet

Dr. Paul Czysz

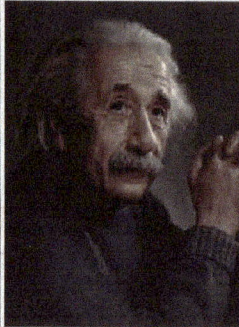
Albert Einstein

Richard Feynman

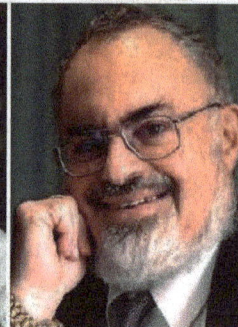
Stanton T. Friedman

Dr. Pierre Guérin

Dr. Richard F. Haines

Stephen Hawking

Dr. J. Allen Hynek

Dr. Carl Gustav Jung

Robert J. Low

Dr. John E. Mack

Dr. J. C. MacKenzie

Dr. Eugene Mallove

Clark McClelland

Dr. James E. McDonald

Scientists who believed that Unidentified Flying Objects
were Extraterrestrial in origin
Google and Yahoo Images

280

Dr. Paul Czysz

"When I was at Wright-Patterson Air Force Base, we had flying saucers that covered the distance from Columbus to Detroit in the equivalent of about 20,000 miles per hour ... Zero-point energy represents about 40-50 megawatts of power per cubic inch of space. That's a lot of power. If you could tap it at will, then no one would have to sell gasoline or oil anymore ... Depending on the secrecy level, you have to go through a significant background check. When you do that, if you're in a very tight compartment, you sign a statement that you will not divulge the existence of the project or even answer a question that could acknowledge the existence of the project. I know people today that worked on one of the things I worked on, and if you asked them about it, they would say, "No, I have no idea what you're talking about." They're in their seventies now, but they still absolutely would never admit that they even know what you're talking about. If there were non-earthbound sources of information, the people who were doing the design or analysis work would never have any idea of where it came from." - **Dr. Paul Czysz, McDonnell-Douglas, Professor of Aeronautical Engineering**

Albert Einstein

"Mr. President, anyone who can cross millions of miles of space will be able to take care of themselves when they get there. Don't start something you can't finish".- **Talking to President Truman about the "shoot-them-down" order of UFOs flying over Washington DC on July 19, 1952 - As quoted by national radio host Frank Edwards**

Richard Feynman

"I think that it is much more likely that the reports of flying saucers are the results of the known irrational characteristics of terrestrial intelligence than of the unknown rational efforts of extra-terrestrial intelligence." - **Richard Feynman an influential American physicist and a key player the development of the atomic bomb.**
http://www.bibliotecapleyades.net/ciencia/ufo_briefingdocument/quosci.htm

Stanton T. Friedman

"1. To what conclusions have you come with regard to UFOs? I have concluded that the earth is being visited by intelligently controlled vehicles whose origin is extraterrestrial. This doesn't mean I know where they come from, why they are here, or how they operate.

"2. What basis do you have for these conclusions? Eyewitness and photographic and radar reports from all over the earth by competent witnesses of definite objects whose characteristics such as maneuverability, high speed, and hovering, along with definite shape, texture, and surface features, rule out terrestrial explanations.

"3. Were there any differences between the unknowns and the knowns? A 'chi square' statistical analysis was performed comparing the unknowns in this study to all the knowns. It was shown that the probability that the unknowns came from the same population of sighting reports as the knowns, was less than 1%. This was based on apparent color, velocity, etc... Maneuverability, one of the most distinguished characteristics of UFOs, was not included in this statistical analysis." - **Stanton T. Friedman was a nuclear physicist and well-known UFO researcher responsible for the original investigation of the Roswell, New Mexico incident. From a prepared statement submitted to the House Science and Astronautics Committee UFO Hearings in 1968, he posed and answered a series of key questions about the UFO phenomenon.**

Dr. Pierre Guérin

"At the very least, it is already possible to show scientifically the evidence for physicochemical modifications affecting sometimes the ground of alleged landing sites, as well as the effects produced on the vegetation. Such research has already begun and doesn't necessarily require large sums."

"The UFO problem in its totality, nevertheless, cannot be really understood unless our science someday is able to propose physical models that take into account the observed phenomena. We are not able to know if this will ever occur, and in any event, we are still very far from that stage." - **Dr. Pierre Guérin, senior researcher at the French National Council for Scientific Research (CNRS), has written extensively about the need for scientific research in the UFO field. He was concluding a summary of the UFO evidence published in Sciences & Avenir in 1972. Guérin, P., "Le Dossier des Objets Volants Non Identifiés," Sciences & Avenir, No. 307, Paris, September 1972.**
http://www.bibliotecapleyades.net/ciencia/ufo_briefingdocument/quosci.htm

Dr. Richard F. Haines

"What I found [in doing research for the book Project Delta] was compelling evidence to claim that most of these aerial objects far exceeded the terrestrial technology of the era in which they were seen. I was forced to conclude that there is a great likelihood that Earth is being visited by highly advanced aerospace vehicles under highly 'intelligent' control indeed."

"We're not dealing with mental projections or hallucinations on the part of the witness but with a real physical phenomenon."

"Reports of anomalous aerial objects (AAO) appearing in the atmosphere continue to be made by pilots of almost every airline and air force of the world in addition to private and experimental test pilots. This paper presents a review of 56 reports of AAO in which electromagnetic effects (E-M) take place on-board the aircraft when the phenomenon is located nearby but not before it appeared or after it had departed.

"Reported E-M effects included radio interference or total failure, radar contact with and without simultaneous visual contact, magnetic and/or gyro-compass deviations, automatic direction finder failure or interference, engine stopping or interruption, dimming cabin lights, transponder failure, and military aircraft weapon system failure." - **Observing UFO's, Haines, Dr. Richard, Chicago: Nelson-Hall, 1980. Haines, a retired NASA senior research scientist at Ames Research Center and the Research Institute for Advanced Computer Science where he worked on the International Space Station , from the preface of his book, CE-5, 1998. Dr. Haines was also a psychologist specializing in pilot and astronaut "human factors" research for the Ames NASA Research Center in California, from where he retired in 1988 as Chief of the Space Human Factors Office.**
http://www.bibliotecapleyades.net/ciencia/ufo_briefingdocument/quosci.htm

Dr. Frank Halstead

"Many professional astronomers are convinced that saucers are interplanetary machines."
- **Halstead was with the Darling Observatory, Minnesota in 1957.**

282

Stephen Hawking
"Of course, it is possible that UFO's really do contain aliens as many people believe, and the government is hushing it up" - **Comment by Hawking on C-Span Television. Stephen Hawking was the guest lecturer at the second Millennium Evening at the White House on March 6, 1998**.

Dr. J. Allen Hynek
"When I first got involved in this field, I was particularly skeptical of people who said they had seen UFOs on several occasions and totally incredulous about those who claimed to have been taken aboard one. But I've had to change my mind."

It reminds me of the days of Galileo when he was trying to get people to look at the sunspots. " They would say that the sun is a symbol of God; God is perfect; therefore the sun is perfect; therefore spots cannot exist: therefore there is no point in looking." - **Hynek in Newsweek, Nov. 21, 1977.**

"I was there at [Project] Bluebook and I know the job they had. They were told not to excite the public, not to rock the boat... Whenever a case happened that they could explain--which was quite a few--they made a point of that and let that out to the media. Cases that were very difficult to explain, they would jump handsprings to keep the media away from them. They had a job to do, rightfully or wrongfully, to keep the public from getting excited."
"When one gets reports from scientists, engineers and technicians whose credibility by all common standards is high and whose moral caliber seems to preclude a hoax, one can do no less than hear them out, in all seriousness." - **From, "The UFO Gap", Hynek, J. Allen, Playboy, Vol. 14, No. 12, December 1967.**

"There exists a phenomenon... that is worthy of systematic rigorous study... The body of data point to an aspect or domain of the natural world not yet explored by science... When the long awaited solution to the UFO problem comes, I believe that it will prove to be not merely the next small step in the march of science but a mighty and totally unexpected quantum jump." -**From Hynek, J. Allen, The UFO Experience: A Scientific Inquiry, Chicago: Regnery Co., 1972. Hynek was former Chairman of the Dept. of Astronomy at North Western University and scientific advisor to Project Bluebook from 1952-1969.**
http://www.bibliotecapleyades.net/ciencia/ufo_briefingdocument/quosci.htm

"Despite the seeming inanity of the subject, I felt that I would be derelict in my scientific responsibility to the Air Force if I did not point out that the whole UFO phenomenon might have aspects to it worthy of scientific attention." - **From Hearings on Unidentified Flying Objects, Committee on Armed Services, House of Representatives, Eighty-ninth Congress, Second Session, 1966.Dr. Hynek was Chairman of the Department of Astronomy at Northwestern University and scientific consultant for Air Force investigations of UFOs from 1948 until 1969 (Projects Sign Grudge and Blue Book). Over his long career, he made numerous comments about the scientific implications of the UFO phenomenon.**

"I have begun to feel that there is a tendency in 20th Century science to forget that there will be a 21st Century science, and indeed a 30th Century science, from which vantage points our

knowledge of the universe may appear quite different than it does to us. We suffer, perhaps, from temporal provincialism, a form of arrogance that has always irritated posterity." - **Hynek, J. Allen, letter to Science magazine, August 1, 1966.**

Dr. Carl Gustav Jung

"A purely psychological explanation is ruled out... the discs show signs of intelligent guidance, by quasi-human pilots... the authorities in possession of important information should not hesitate to enlighten the public as soon and as completely as possible." - **Dr. Carl Jung on Unidentified Flying Objects," Flying Saucer Review, Vol. 1, No. 2, 1955.**

"It remains an established fact, supported by numerous observations, that UFOs have not only been seen visually but have also been picked up on the radar screen and have left traces on the photographic plate."

 "Unfortunately, however, there are good reasons why the UFOs cannot be disposed of in this simple manner. It remains an established fact, supported by numerous observations, that UFOs have not only been seen visually but have also been picked up on the radar screen and have left traces on the photographic plate. It boils down to nothing less than this: that either psychic projections throw back a radar echo, or else the appearance of real objects affords an opportunity for mythological projections." - **"A Fresh Look at Flying Saucers," Time, August 4, 1967.**

Lee Katchen

"UFO sightings are now so common, the military doesn't have time to worry about them ... when a UFO appears, they simply ignore it. .Unconventional targets are ignored because apparently, we are only interested in Russian targets, possibly enemy targets. Something that hovers in the air then shoots off at 5,000 miles per hour, doesn't interest us because it can't be the enemy. UFOs are picked up by ground and air radar, and they have been photographed by gun camera all along. There are so many UFOs in the sky that the Air Force has had to employ special radar networks to screen them out." - **Katchen, NASA atmospheric physicist, in an announcement on June 7, 1968, in which he stated that he believed, based on his examination of 7,000 reports, that UFOs have an extraterrestrial origin.**
http://www.bibliotecapleyades.net/ciencia/ufo_briefingdocument/quosci.htm

Robert J. Low

"The trick would be, I think, to describe the project so that, to the public, it would appear a totally objective study but, to the scientific community, would present the image of a group of non-believers trying their best to be objective, but having an almost zero expectation of finding a saucer." - **Low, project coordinator of the Colorado University UFO Project (a.k.a. The Condon Committee), in a memorandum of instruction from August 9, 1966. This telling quote gives an impression as to what may have been the goal of the Project: to either get the thing out of the way without hurting any of the scientists' credibility or to comply with a rumored Air Force directive to produce a report showing UFOs to be unworthy of scientific consideration.**

Dr. John E. Mack

"I will stress once again that we do not know the source from which the UFOs or the alien

284

beings come (whether or not, for example, they originate in the physical universe as modern astrophysics has described it). But they manifest in the physical world and bring about definable consequences in that domain." - **Abduction - Human Encounters With Aliens, Mack, J., New York: Scribners, 1994. Dr. John E. Mack, Professor of psychiatry at The Cambridge Hospital, Harvard Medical School, and founding director of the Center for Psychology and Social Change. A 1977 Pulitzer Prize winner for his biography of Lawrence of Arabia.**

Dr. J. C. MacKenzie

"It seemed fantastic that there could be any such thing. At first, the temptation was to say it was all nonsense, a series of optical illusions. But there have been so many reports from responsible observers that they cannot be ignored. It seems hardly possible that all these reports could be due to optical illusions." - **MacKenzie was Chairman of the Canadian Atomic Energy Control Board and former president of the National Research Council.**

Dr. Eugene Mallove

"I was the Chief Science Writer at the MIT news office when the cold fusion story out of Utah broke on March 23, 1989. It turns out the cold fusion effect was real. In fact, what Pons and Fleischmann found was only the tip of an iceberg. There are huge quantities of technical literature published by proponents and a much smaller amount by the people who found so-called negative results...One day while at MIT, I inadvertently was looking through some piles of paper by physicists doing their repeat of the Pons-Fleischmann experiment. To my utter astonishment, I can remember sitting at my desk and actually seeing two sheets of paper, one dated July 10 and another July 13. The July 10 control experiment showed in the raw data excess heat. But then, on July 13, it was shifted completely. It was altered. Clear fraud...no question. I asked for a review at MIT. I got nowhere. Yet today, MIT data is held up. There has been an extraordinary abrogation of legal responsibility at the Patent Office and the Department of Energy on the matter of cold fusion. There is serious criminal activity going on that ultimately must be rooted out if the cold fusion and new energy revolution are to go forward."
- **Dr. Eugene Mallove, MIT Chief Science Writer. See Dr. Mallove's website** www.infinite-energy.com **http://www.bibliotecapleyades.net/ciencia/ufo_briefingdocument/quosci.htm**

Clark McClelland

"As the Gemini Capsule entered orbit, the RCA world tracking team began to realize that 'our' capsule was not alone as viewed through their incoming telemetry, visual theodolite, and other high-powered optical data. Our capsule had four 'visitors'. The RCA team was ordered to run a recheck of the situation to be certain ghost images were not the cause. The Titan II stages were also excluded as causing the images After much huddling and discussion, the intelligent determination was that we had other physical objects up there with our Gemini capsule   The official NASA determination was that the objects were the torn particles or remains of the Titan upper stage that apparently entered orbit with the Gemini capsule. I was at the news conference and I nearly began to laugh. How could a broken stage overtake the capsule and stop slightly ahead of the capsule to accompany it an entire orbit around the earth? But I held my laugh to save my job." - **Commenting on the April 9, 1964, unmanned launch of the Gemini-Titan 2.**

"The day will arrive when the governments of earth will finally admit we are not alone, that humans have come face to face with other lifeforms from the cosmos." - **These quotes come**

from Clark's website, The Stargate Chronicles. McClelland was an Aerospace Engineer and Technical Assistant to the Apollo Program Manager during the Apollo moon landings, also assisted in almost six hundred launches at Cape Canaveral, and in addition to working in the Mercury and Gemini programs, Space Lab and the Space Station was heavily involved in the Space Shuttle program

Dr. James E. McDonald

"The type of UFO reports that are most intriguing are close-range sightings of machine-like objects of unconventional nature and unconventional performance characteristics, seen at low altitudes, and sometimes even on the ground. The general public is entirely unaware of the large number of such reports that are coming from credible witnesses... When one starts searching for such cases, their number are quite astonishing. Also, such sightings appear to be occurring all over the globe." - **Symposium on Unidentified Flying Objects," Hearings before the Committee on Science and Astronautics, U.S. House of Representatives, July 29, 1968.**
"I have absolutely no idea where the UFO's come from or how they are operated, but after ten years of research, I know they are something from outside our atmosphere." - **Dr. McDonald was Senior Physicist at the Institute of Atmospheric Physics at the University of Arizona.**

Dr. Margaret Mead

"There are unidentified flying objects. That is, there are a hard core of cases - perhaps 20 to 30 percent in different studies - for which there is no explanation... We can only imagine what purpose lies behind the activities of these quiet, harmlessly cruising objects that time and again approach the earth. The most likely explanation, it seems to me, is that they are simply watching what we are up to."

"There are unidentified flying objects. That is, there are a hard core of cases - perhaps 20 to 30 percent in different studies - for which there is no explanation... We can only imagine what purpose lies behind the activities of these quiet, harmlessly cruising objects that time and again approach the earth. The most likely explanation, it seems to me, is that they are simply watching what we are up to." - **"UFOs - Visitors from Outer Space?", Mead, Margaret, Redbook, vol. 143, September 1974. Dr. Mead is a world-renowned anthropologist**
http://www.bibliotecapleyades.net/ciencia/ufo_briefingdocument/quosci.htm

Dr. Harry Messel

"The facts about saucers were long tracked down and results have long been known in top secret defense circles of more countries than one." - **Dr. Harry Messel, Professor of Physics at Sydney University, Australia, in a 1965 statement.**

Dr. Auguste Meessen

"There are too many independent eyewitness reports to ignore. Too many of the reports describe coherent physical effects, and there is an agreement among the accounts concerning what was observed... But of course, there are also physical effects. The Air Force report [of the F-16 jet scramble incident on the night of March 30-31, 1990] allows us to approach the problem in a rational and scientific way. The simplest hypothesis is that the reports are caused by extraterrestrial visitors, but that hypothesis carries with it other problems. We are not in a rush to form a conclusion, but continue to study the mystery." - **From an interview with French**

journalist, Marie-Therese de Brosses. Dr. Auguste Meessen was Professor of physics at the Catholic University in Louvain and one of the scientific consultants for the Belgian Society for the Study of Space Phenomena (SOBEPS)

C. B. Moore
"Based on the descriptions, I can definitely rule this out. There wasn't a balloon in 1947 or today that could account for this incident." - **Moore, General Mills Meteorologist, and expert on weather balloons, when asked whether he believed the Roswell Incident could be explained by a Mogul balloon.**

Dr. Herman Oberth
"UFOs are conceived and directed by intelligent beings of a very high order, and they are propelled by distorting the gravitational field, converting gravity into useable energy. There is no doubt in my mind that these objects are interplanetary craft of some sort. I and my colleagues are confident that they do not originate in our solar system, but we feel that they may use Mars or some other body as sort of a way station. They probably do not originate in our solar system, perhaps not even in our galaxy." - **1954 American Weekly of Oct. 24, 1954.**

"It is my thesis that flying saucers are real and that they are spaceships from another solar system."

*"We cannot take the credit for our record advancement in certain scientific fields alone. **We have been helped!** I think that they possibly are manned by intelligent observers who are members of a race that may have been investigating our earth for centuries. I think that they possibly have been sent out to conduct systematic, long-range investigations, first of men, animals, vegetation, and more recently of atomic centers, armaments, and centers of armament production."* - **Flying Saucers Come From A Distant World", Oberth H., The American Weekly, October 24, 1954.**
http://www.bibliotecapleyades.net/ciencia/ufo_briefingdocument/quosci.htm

"They are flying by means of artificial fields of gravity... They produce high-tension electric charges in order to push the air out of their paths, so it does not start glowing, and strong magnetic fields to influence the ionized air at higher altitudes. First, this would explain their luminosity... Secondly, it would explain the noiselessness of UFO flight... Finally, this assumption also explains the strong electrical and magnetic effects sometimes, though not always, observed in the vicinity of UFOs ." - **Dr. Hermann Oberth discusses UFOs," Fate, May 1962.**

"It is my conclusion that UFOs do exist, are very real, and are spaceships from another or more than one solar system. They are possibly manned by intelligent observers who are members of a race carrying out long-range scientific investigations of our earth for centuries." – **UFO News, 1974**

Dr. Claude Poher
"Taking into account the facts that we have gathered from the observers and from the location of their observations, we concluded that there generally can be said to be a material phenomenon

behind the observations. In 60% of the cases reported here, the description of this phenomenon is apparently one of a flying machine whose origin, modes of lifting and/or propulsion are totally outside our knowledge. " - **From a report on UFOs for French officials.**

"The phenomenon seems to be real... The general coherence of sighting reports worldwide should not leave researchers indifferent. One does not conceive objective arguments to justify an attitude that would avoid at all cost these observations... The risk is, at worst, to confirm the existence of unknown vehicles appearing erratically into our atmosphere - a hypothesis that seems to explain nearly all reported aspects of the phenomenon and could be linked to the current (1970) exobiology branch of space research."

"Given the volume of the objects described in the observations... I can affirm that our futuristic space generators are far from being able to produce the amount of energy seen by the UFO witnesses. The light power seen is probably the tip of the iceberg, because no thermodynamic system can produce energy without dissipating a part of it. The megawatts of observed light are most likely the energy 'leak' from the energy conversion system used by the flying object, which means that the useful energy produced is much greater than what is seen." - **1971 Statistical Study prepared for the CNES and French officials.**

"The knowledge of such an energy production method is crucial for the future of mankind. The UFO observation reports tells us that ambitious, entirely new, solutions are possible [underlined in the original]. This is very important." - **From a letter to Marie Galbraith, November 26, 1995. Dr. Poher earned a Ph.D. in astronomy, was an expert in aeronautics, and astronautics, an engineer at the French Space Agency (CNES) for thirty years, specializing in rocket propulsion and nuclear space energy; former Chairman of many working groups in the International Astronautical Federation; founder of GEPAN in May 1977 and its first Director until 1979. Before creating GEPAN, he had studied the UFO phenomenon for many years and had access to French military and police UFO files, including classified reports.** http://www.bibliotecapleyades.net/ciencia/ufo_briefingdocument/quosci.htm

Dr. Walther Riedel

"I am completely convinced that [UFOs] have an out-of-world basis."

"First, the skin temperatures of structures operating under the observed conditions would make it impossible for any terrestrial structure to survive. The skin friction of the missile at those speeds at those altitudes would melt any metals or nonmetals available."

"Second, consider the high acceleration at which they fly and maneuver... In some descriptions the beast spirals straight up. If you think of the fact that the centrifugal force in a few minutes of such a maneuver would press the crew against the outside, and do likewise to the blood, you see what I mean. Third.... There are many occurrences where they have done things that only a pilot could perform but that no human pilot could stand. Fourth, in most of the reports, there is a lack of visible jet. Most observers report units without visible flame.......and no trail. If it would be any known type of jet, rocket, piston engine, or chain-reaction motor, there would be a very clear trail at high altitude. It is from no power unit we know of....."

288

"The least improbable explanation is that these things are artificial and controlled ... My opinion for some time has been that they have an extraterrestrial origin." - **From LIFE Magazine, April 7, 1952. Riedel was research director and chief designer at Germany's rocket center in Peenemunde and also worked on classified projects for the U.S. after WW2.**

Dr. Carol Rosin

"Von Braun [founder of modern rocket science] told me [in 1974] that the reasons for space-based weaponry were all based on a lie. He said that the strategy was to use scare tactics - that first the Russians, then the terrorists are going to be considered the enemy. The next enemy was asteroids. "The last card is the alien card. We are going to have to build space-based weapons against aliens, and all of it is a lie."...I was at a meeting in Fairchild Industries in the War Room. The conversation [was] about how they were going to antagonize these enemies and at some point, there was going to be a Gulf War. Now this is 1977!" - **From a UFO Disclosure Project video. Dr. Carol Rosin was Corporate Manager of Fairchild Industries and spokesperson for Wernher von Braun.**
http://www.bibliotecapleyades.net/ciencia/ufo_briefingdocument/quosci.htm

Dr. Carl Sagan

"It now seems quite clear that Earth is not the only inhabited planet. There is evidence that the bulk of the stars in the sky have planetary systems. Recent research concerning the origin of life on Earth suggests that the physical and chemical processes leading to the origin of life occur rapidly in the early history of the majority of planets. The selective value of intelligence and technical civilization is obvious, and it seems likely that a large number of planets within our Milky Way galaxy - perhaps as many as a million - are inhabited by technical civilizations in advance of our own. Interstellar space flight is far beyond our present technical capabilities, but there seems to be no fundamental physical objections to preclude, from our own vantage point, the possibility of its development by other civilizations." - **Unidentified Flying Objects", Sagan, Carl, The Encyclopedia Americana, 1963. Dr. Sagan was Professor of Astronomy and Space Sciences at Cornell University:**

"After I give lectures - on almost any subject - I am often asked, "Do you believe in UFOs?". I'm always struck by how the question is phrased, the suggestion that this is a matter of belief and not evidence. I'm almost never asked, "How good is the evidence that UFOs are alien spaceships?" - **Carl Sagan , 'The Demon Haunted World,' 1996**

Dr. Frank B. Salisbury

"I must admit that any favorable mention of the flying saucers by a scientist amounts to extreme heresy and places the one making the statement in danger of excommunication by the scientific theocracy. Nevertheless, in recent years I have investigated the story of the unidentified flying object (UFO), and I am no longer able to dismiss the idea lightly." - **Paper on "Exobiology" presented at the First Annual Rocky Mountain Bioengineering Symposium, held at the United States Air Force Academy, in May 1964. Quoted in Fuller, John G., Incident at Exeter, Putnam, 1966. Dr. Salisbury was Professor of Plant Physiology at Utah State University.** http://www.bibliotecapleyades.net/ciencia/ufo_briefingdocument/quosci.htm

Dr. Margaret Mead   Dr. Harry Messel   Dr. Auguste Meessen   Dr. Herman Oberth   Dr. Claude Poher

Dr. Walther Riedel   Dr. Carol Rosin   Dr. Carl Sagan   Dr. Frank B. Salisbury   Wilbert Smith

Dr. Leo Sprinkle   Dr. Peter A. Sturrock   Clyde Tombaugh   Dr. Jacques Vallee   Werner Von Braun

Dr. Alfred Webre   Zhang Zhousheng   Dr. Felix Y. Zigel   Dr. Mitrofan Zverev

**More Scientists who believed that UFOs are extraterrestrial
in origin from other star systems**
Google and Yahoo Images

Dr. Paul Santorini

"We soon established that they were not missiles. But, before we could do anymore, the Army, after conferring with foreign officials, ordered the investigation stopped. Foreign scientists flew to Greece for secret talks with me... A world blanket of secrecy surrounded the UFO question because the authorities were unwilling to admit the existence of a force against which we had no possibility of defense." - **"UFOs: Interplanetary Visitors", Fowler, R., New York: Bantam Books, 1974. Dr. Santorini was a Greek physicist and engineer credited with developing the proximity fuse for the Hiroshima atomic bomb, two patents for the guidance system used in the U.S. Nike missiles, and a centimetric radar system. He has also stated that he believes UFOs are under intelligent control. In 1947, he investigated a series of UFO reports over Greece that were initially thought to be Soviet missiles.**

Dr. John Sathco

"There are in excess of 200 reports of the type that we had from down in Louisiana, from people claiming that they have had direct contact with a spacecraft full of aliens. I mean 200 reports from witnesses who are as reliable or more so than these people. I'm not counting the reports from the obvious crackpots that have an axe to grind...If you accept them at face value then you're forced to accept that we have been visited.- **Sathco was an Astronomer at the University of Southern California in 1973.**

Wilbert Smith

"The matter is the most highly classified subject in the United States Government, rating higher even than the H-bomb. Flying saucers exist. Their modus operandi is unknown but concentrated effort is being made by a small group headed by Dr. Vannevar Bush." - **From a declassified Canadian government memorandum dated Nov. 21, 1950.**

"...it soon became apparent that there was a very real and quite large gap between this alien science and the science in which I had been trained. Certain crucial experiments were suggested and carried out, and in each case, the results confirmed the validity of the alien science. Beyond this point, the alien science just seemed to be incomprehensible." - **In a speech concerning experiments allegedly suggested by EBEs (Extraterrestrial Biological Entities); March 31, 1958.**

"If, as appears evident, the Flying Saucers are emissaries from some other civilization, and actually do operate on magnetic principles, we have before us the Fact that we have missed something in magnetic theory but have a good indication of the direction in which to look for the missing quantities. It is therefore strongly recommended that work on Project Magnet be continued and expanded to include experts in each of the various fields involved in these studies" - **From an interim report dated 25th June 1952 Wilbert Smith was the electrical engineer who convinced the Canadian government to establish Project Magnet to study the UFO phenomenon and later served as engineer-in-charge of the project.**
http://www.bibliotecapleyades.net/ciencia/ufo_briefingdocument/quosci.htm

Dr. Leo Sprinkle

"We watched it for quite a few minutes. We could see it was larger than the headlights of the cars below. And we could see it was not attached to anything. And there was no sound. I became

frightened actually, because it wasn't anything I could understand... from a personal viewpoint, I am pretty well convinced that we are being surveyed." - **Flying Saucers," Special Issue of Look magazine, 1967. Dr. Sprinkle, Professor of psychology at the University of Wyoming had his first UFO sighting in 1951 when he and a friend saw "something in the sky, round and metallic looking." In 1956, he had a second sighting while driving with his wife near Boulder, Colorado.**

Dr. Peter A. Sturrock

"The definitive resolution of the UFO enigma will not come about unless and until the problem is subjected to open and extensive scientific study by the normal procedures of established science. This requires a change in attitude primarily on the part of scientists and administrators in universities." - **Sturrock, Peter A., Report on a Survey of the American Astronomical Society concerning the UFO Phenomenon, Stanford University Report SUIPR 68IR, 1977.**

"Although... the scientific community has tended to minimize the significance of the UFO phenomenon, certain individual scientists have argued that the phenomenon is both real and significant. Such views have been presented in the Hearings of the House Committee on Science and Astronautics [and elsewhere]. It is also notable that one major national scientific society, the American Institute of Aeronautics and Astronautics, set up a subcommittee in 1967 to 'gain a fresh and objective perspective on the UFO phenomenon.' In their public statements (but not necessarily in their private statements), scientists express a generally negative attitude towards the UFO problem, and it is interesting to try to understand this attitude. Most scientists have never had the occasion to confront evidence concerning the UFO phenomenon. To a scientist, the main source of hard information (other than his own experiments' observations) is provided by the scientific journals. With rare exceptions, scientific journals do not publish reports of UFO observations. The decision not to publish is made by the editor acting on the advice of reviewers. This process is self-reinforcing: the apparent lack of data confirms the view that there is nothing to the UFO phenomenon, and this view works against the presentation of relevant data."
- **"An Analysis of the Condon Report on the Colorado UFO Project", Sturrock, Peter A., Journal of Scientific Exploration, Vol. 1, No. 1, 1987. Dr. Sturrock was Professor of Space Science and Astrophysics Deputy Director of the Center for Space Sciences and Astrophysics at Stanford University, and Director of the Skylab Workshop on Solar Flares in 1977.** http://www.bibliotecapleyades.net/ciencia/ufo_briefingdocument/quosci.htm

Clyde Tombaugh

"The illuminated rectangles I saw did maintain an exact fixed position with respect to each other, which would tend to support the impression of solidity. I doubt that the phenomenon was any terrestrial ... I do a great deal of observing (both telescopic and unaided eye) in the backyard and nothing of the kind has ever appeared before or since." -**Tombaugh, the astronomer who discovered Pluto in a letter dated September 10, 1957. The phenomenon was also witnessed by his wife.**
U.S.S.R.

Institute of Space Research of the Soviet Academy of Sciences published in 1979, a 74-page statistical analysis of over 250 UFO cases reported in the Soviet Union. After stating that hallucinations, errors, and conventional explanations (aircraft, satellites, etc.) could not account for many of the reports, the study concluded:

"Obviously, the question of the nature of the anomalous phenomena still should be considered open. To obtain more definite conclusions, more reliable data must be available. Reports on observations of anomalous phenomena have to be well documented. The production of such reports must be organized through the existing network of meteorological, geophysical, and astronomical observation stations, as well as through other official channels... In our opinion, the Soviet and foreign data accumulated so far justifies setting such studies." - **Gindilis, L.M., Men'kov, D.A. & Petrovskaya, I.G., "Observations of Anomalous Atmospheric Phenomena in the USSR: Statistical Analysis," USSR Academy of Sciences Institute of Space Research, Report PR 473, Moscow, 1979.**

U.S.S.R. Scientific Commissions

"Of special value are the archives set up by the Commission. They contain over 13 thousand reports connected with PEs [Paranormal Events] and with UFOs in particular... UFOs have been seen to hover over ground objects, to chase or fly side by side with airplanes and cars, to follow geometrically regular trajectories, and to send out ordered flashes of light. In other words, such 'paranormals' behave, from the viewpoint of human beings, quite often showing capabilities yet beyond the reach of the machines built on the Earth." - **Faminskaya, T. & Petukhov, A., "At 4.10 Hours and After," Almanac Phenomenon 1989, Moscow Mir, 1989. The Soviet press was informed in the mid-80s that the All-Union Council of Scientific and Technical Societies (now the Council of Scientific and Engineering Societies) had set up a non-governmental Commission on Paranormal Events, headed by V.S. Troitsky, a Corresponding Member of the USSR Academy of Sciences.**
http://www.bibliotecapleyades.net/ciencia/ufo_briefingdocument/quosci.htm

Dr. Jacques Vallee

"Skeptics, *who flatly deny the existence of any unexplained phenomenon in the name of 'rationalism,'* ***are among the primary contributors to the rejection of science by the public. People are not stupid and they know very well when they have seen something out of the ordinary.*** *When a so-called expert tells them, the object must have been the moon or a mirage, he is really teaching the public that* ***science is impotent or unwilling to pursue the study of the unknown."*** - **Vallee, J., Confrontations, New York: Ballantine Books, 1990.**

"It is unusual for scientists to keep diaries and even more unusual for them to make them public... I have followed this rule of silence for the last thirty years, but I have finally decided that I had no right to keep them private anymore... They provide a primary source about a crucial fact in the recent historical record: the appearance of new classes of phenomena that highlighted the reality of the paranormal. These phenomena were deliberately denied or distorted by those in authority within the government and the military. Science never had fair and complete access to the most important files. The thirteen years covered here, from 1957 to 1969, saw some of the most exciting events in technological history... Behind the grand parade of the visible breakthroughs in science, however, more private mysteries were also taking place:... all over the world, people had begun to observe what they described as controlled devices in the sky. They were shaped like saucers or spheres. They seemed to violate every known principle in our physics."

"Governments took notice, organizing task forces, encouraging secret briefings and study groups, funding classified research and all the time denying before the public that any of the phenomena might be real... The major revelation of these Diaries may be the demonstration of how the scientific community was misled by the government, how the best data were kept hidden, and how the public record was shamelessly manipulated." - **Vallee, J., Forbidden Science, Berkeley: North Atlantic Books, 1992. Dr. Jacques Vallee, astrophysicist, computer scientist and world-renowned researcher and author on UFOs and paranormal phenomena. He worked closely with Dr. J. Allen Hynek.**

Werner Von Braun

"We find ourselves faced by powers which are far stronger than we had hitherto assumed, and whose base is at present unknown to us. More I cannot say at present. We are now engaged in entering into closer contact with those powers, and in six or nine months time it may be possible to speak with some precision on the matter." - **This comment comes from "News Europa" Jan. 1959 and refers to mysterious events during the re-entry phase of the Juno 2 rocket during a test flight.**

"...it is as impossible to confirm them (UFOs) in the present as it will be to deny them in the future." - **In a comment to NASA scientist, Clark McClelland. Von Braun was a rocket scientist who was instrumental in the development of Nazi Germany's V2 rocket and later, the American space program.**

Dr. Alfred Webre

"I worked on the 1977 Carter White House Extraterrestrial Communication Project. It called for creation of central and regional databases under independent control on UFOs and EBEs—that is Extraterrestrial Biological Entities. The full management staff and the research institute had signed off knowingly on the proposal ...I flew back from my meeting with the White House, at which this final approval had been given. And when I arrived back at my offices at SRI (Stanford Research Institute), I was called back into the office of the senior SRI official. The project was to be terminated. They had received direct communication from the Pentagon that if the study went forward, SRI's contracts would be terminated. These contracts were a substantial part of SRI's business at the time. The senior Pentagon liaison stated that the project was terminated because "There are no UFOs." Here we have a President of the United States who came to office under a pledge to open up the UFO issue, and an open study in the White House, and that was squelched." - **Dr. Alfred Webre, Stanford Research Institute, Senior Policy Analyst**
http://www.bibliotecapleyades.net/ciencia/ufo_briefingdocument/quosci.htm

Dr. Weisberg

"Like a turtle's back, with a cabin space some fifteen feet in diameter. The bodies of six occupants were seared and the interior of the disc had been badly damaged by intense heat."
- **Dr. Weisber, from a memo by the director of the Borderland Science Research Foundation, Layne Meade, in 1949 concerning a description given by Dr. Weisberg, a Canadian physics professor who apparently examined some retrieved discs for the U. S. Air Force at Edwards AFB.**

Zhang Zhousheng

"What was especially important was that, at a distance of 180 kilometers apart, the records about the direction of movement of the strange aerial body in space, made independently by at least two different observers was basically the same ......To the present time, this strange phenomenon has not been satisfactorily explained, yet there were thousands of good observers who had seen it." - **Zhang Zhousheng an astronomer at the Yunnan Observatory in Chengdu City, China. Zhousheng and others nearby watched a strange glowing, spiral object moving steadily across the sky for about five minutes on the evening of July 26, 1977.**
http://www.bibliotecapleyades.net/ciencia/ufo_briefingdocument/quosci.htm

Dr. Felix Y. Zigel

"The important thing now is for us to discard any preconceived notions about UFOs and to organize on a global scale a calm, sensation-free and strictly scientific study of this strange phenomenon. The subject and aims of the investigation are so serious that they justify all efforts. It goes without saying that international cooperation is vital." **(Zigel, F., "Unidentified Flying Objects," Soviet Life, No. 2 (137), February 1968.) - Dr. Felix Y. Zigel, Professor of mathematics and astronomy at the Moscow Aviation Institute, father of Russian Ufology: "Unidentified Flying Objects," Soviet Life, February 1968**

"Unidentified flying objects are a very serious subject which we must study fully. We appeal to all viewers to send us details of strange flying craft seen over the territories of the Soviet Union. This is a serious challenge to science and we need the help of all Soviet citizens."
"Observations show that UFOs behave 'sensibly.' In a group formation flight, they maintain a pattern. They are most often spotted over airfields, atomic stations and other very new engineering installations. On encountering aircraft, they always maneuver so as to avoid direct contact. A considerable list of these seemingly intelligent actions gives the impression that UFOs are investigating, perhaps even reconnoitering... The important thing now is for us to discard any preconceived notions about UFOs and to organize on a global scale a calm, sensation-free and strictly scientific study of this strange phenomenon. The subject and aims of the investigation are so serious that they justify all efforts. It goes without saying that international cooperation is vital." - **"Unidentified Flying Objects", Zigel, F., Soviet Life, No. 2 (137), February 1968.**

"We have seen these UFOs over the USSR; craft of every possible shape: small, big, flattened, spherical. They are able to remain stationary in the atmosphere or to shoot along at 100,000 kilometers per hour... They are also able to affect our power resources, halting our electricity generating plants, our radio stations, and our engines, without however leaving any permanent damage. So refined a technology can only be the fruit of an intelligence that is indeed far superior to man." - **From an interview with Henri Gris in 1981, Gente, July 31, 1981, and August 7, 1981. Dr. Zigel was Professor of mathematics and astronomy at the Moscow Aviation Institute, known as the father of Russian Ufology. In a November 10, 1967, broadcast on Moscow Central Television, with Soviet Air Force General Porfiri Stolyarov.**

Dr. Mitrofan Zverev

"Something unknown to our understanding is visiting this Earth." - **Dr. Mitrofan Zverev (USSR), quoted by Reuters, August 26, 1965.**
http://www.bibliotecapleyades.net/ciencia/ufo_briefingdocument/quosci.htm

CHAPTER 107

WHAT IS EXOPOLITICS AND THE INSTITUTE
FOR COOPERATION IN SPACE (ICIS)?

In western societies, if you want to keep your families intact and your friends always close to you, then you never talk about either politics or religion, particularly in the same breath! In this book, we cross both boundaries bravely disregarding the fall-out from both sectors of society in order to understand the true nature of the UFO/ETI question hoping it will provide enlightenment for the masses!

Politics in the traditional historic sense has its origins that go back to the times of the ancient Greeks (**Plato's** *Republic* and **Aristotle's** *Politics)* as well as the Chinese (**Confucius**). Politics is the practice and theory of ***influencing*** other people on a civic or individual level. Rightfully or wrongly in the more narrow sense, it refers to achieving and exercising positions of governance — organized control over a human community, particularly a state.

A variety of methods are employed in politics, which include promoting one's own political views among people, negotiation with other political subjects, making laws, and exercising force, including warfare against adversaries. Politics is exercised on a wide range of social levels, from clans and tribes of traditional societies, through modern local governments, companies, and institutions up to sovereign states, to the international level.

Modern political discourse focuses on democracy and the relationship between people and politics. A political system , therefore, is a framework which defines acceptable political methods within a given society. http://en.wikipedia.org/wiki/Politics

Now, within the larger framework of the UFO/ETI phenomenon, how does politics relate to this matter, more specifically, **exopolitics**?

If one accepts the existence of other sentient beings in the universe besides ourselves whether within our own solar neighbourhood or intergalactically, then the need for exopolitics arises naturally. Any intelligent, sentient beings, wherever they exist will naturally have their own needs, desires, interests and agendas. Galvanizing a global social order via unity of a common purpose requires politics and it is likely what interstellar civilizations that are space faring have achieved. So, naturally, it becomes necessary to look at our neighbors politically, although, it may still be premature at this stage in our evolutionary development on Earth. We have not as yet, achieved a global society or civilization on this planet that has equal footing with other interstellar civilizations on the intergalactic stage of politics!

Remember, we are not even a **Type I planetary civilization;** thus any review of this subject is an exercise of theories and hypotheses which nonetheless, hasn't stopped people from trying to project political values upon Extraterrestrial Intelligences for which we have yet to meet in a one on one, face to face ambassadorial meeting. A meeting between humans and ETI will have profound implications and no doubt will be a wake-up call to our perceptions of the life, the universe, our moral and societal values and everything we cherish about life in general.

Alright, what does exopolitics mean to the average human being and should we even care at this point?

Michael Sokolov defines exopolitics as the name suggests, as extraterrestrial politics. (The use of the "exo-" prefix to refer to extraterrestrial matters is already established, for example, by the term "exobiology" meaning the study of extraterrestrial life.)
http://www.bibliotecapleyades.net/exopolitica/esp_exopolitics_v.htm

The term exopolitics, in the meaning of political relations within the scope of the universe, was discussed as early as in 1977 by **Timothy Leary**. Exopolitics is a direct logical extension of conventional politics to the interplanetary theatre. **Dr. Alfred Webre**, who formally introduced exopolitics as a discipline of study, defined it as the study of law, governance and politics in the Universe. http://en.wikipedia.org/wiki/Alfred_Webre

According to **Dr. Michael Salla** on his website exopolitics.org, the term Exopolitics is the study of the key individuals, political institutions, and processes associated with Extraterrestrial life. http://exopolitics.org/about/welcome/

Dr. Alfred Webre believes that there is intelligent Extraterrestrial life in our universe, a concept that is widely acknowledged by the general public. He is the author of the online e-book, *Towards a Decade of Contact* (2000) (which contains his first discussion of exopolitics) and the book *Exopolitics: Politics, Government, and Law in the Universe* (2005) and most recently (2020) his latest book, *"Emergence of the Omniverse"*. The exopolitics model functionally maps the operation of politics, government, and law in an intelligent universe, and provides an operational bridge between models of terrestrial politics, government and law, and the larger models of politics, government, and law in the proposed society of the greater universe. http://en.wikipedia.org/wiki/Alfred_Webre

By analogy with political parties on Earth, which are aggregations of individuals who share a certain common political agenda, i.e., a certain set of policies they mutually seek to bring about, we can introduce the notion of an exopolitical party, which we shall define as any grouping of members of the interplanetary community with a specific exopolitical agenda, i.e., a specific set of policies toward other members of the interplanetary community.

But, there is a problem with this new concept of exopolitics which Salla rightfully brings to our attention which first and foremost is the very acknowledgement of an Extraterrestrial presence and the associated alien technology that is controlled by **MJ-12** or the **Military Industrial Complex.** On the America political stage and no doubt supported by other governments, there is officially is a denial of any exopolitical relations between Earth's most powerful governments and military forces and some ET groups.

According to the testimony of Disclosure Project Witnesses who are government and military whistle-blowers that broken ranks and come forward with classified information, there is a strong case that supports that some type of relationship exists with ETs. As to the nature of that relationship, it is speculative and may be a part of a disinformation campaign to discredit UFO researchers. Such hearsay accounts at present time focus on abduction scenarios with no real emanate from clandestine government/military like Area 51 and Dulce, New Mexico. Information concerning Extraterrestrial life and technology is kept secret from the general public, elected political representatives and even senior military officials. The supporting evidence is overwhelming in scope and shows that decision making is restricted on a strict *"need to know"* basis. http://exopolitics.org/about/welcome/

The main job of exopolitics research then is first to declassify and disclose the presence of the Extraterrestrial presence suppressed by M.I.C. then, identifying the existing members of our immediate neighborhood and classifying them by their exopolitical agendas, thus establishing a picture of the existing exopolitical parties in our immediate neighborhood. However, all of this may be a moot point if some sort of diplomatic relationship has already be established with ETs within the covert **"black world government"** of the M.I.C.

It is conceivable that the **"Ancient Alien Hypothesis" (AAH)** as postulated by **Erich von Daniken, Giorgio A. Tsoukalos, Zecharia Sitchin**, and **Maurice Chatelain**, indicates that ETs have visited the Earth and have intervened in Earth affairs since ancient times, it follows, therefore, that humanity may already have been engaged in exopolitical relations with one or more ET groups, i.e. exopolitical parties may have existed in ancient times. This hypothesis has been confirmed by scores of researchers of alternative archaeology, alternative history, and ancient mysteries.

Both of the above assertions imply that humankind has already been engaged in some exopolitical relations, and therefore exopolitics as a discipline of study is not merely hypothetical, but actually relevant to past, present, and future world events. http://www.bibliotecapleyades.net/exopolitica/esp_exopolitics_v.htm

Again, everything is moot at this time as we can only theorize about hypothetical situations or scenarios but, let us play along and perhaps, at some future time, we will have all the bases covered for any and all eventualities that may unfold with ET contact and diplomatic relations.

The most commonly asked question about any visiting ET civilization to Earth or using exopolitically correct terminology… *exopolitical party* is whether they are friendly or hostile. It depends on which side of the fence you sit, in other words, your personal religious perceptions, your state of mind, your experience in life and whether you like or dislike anything different than you!

Those exopolitical parties that may have best interests of humanity at heart could be classified as an exopolitical party that is friendly to humanity, exopolitical parties that do not have best interests of humanity may be classified as an exopolitical party as unfriendly or hostile. The word "hostile" should be defined further as no diplomacy whatsoever; the aliens just open fire and obliterate humanity and the Earth! It is conceivable that some or many exopolitical parties may consider us backwards in social development and may determine the affairs of humanity in a parental fashion much like a child needs to be educated or be disciplined in order to grow.

This analysis only considers those exopolitical parties that are in some way concerned with our planet and/or species. It would be extremely naive and egocentric to assume that every party in the Universe has an agenda or policy concerning us. Most probably don't care about us one way or another, we are not that important.
http://www.bibliotecapleyades.net/exopolitica/esp_exopolitics_v.htm

We must expect the full range of exopolitical agendas when first contact arrives on our doorstep if we accept ancient evidence, present day contactee testimony and the information gathered from remote viewing and channeling, some of which lie in line with the best interests of humanity while others don't.

The spectrum of diversity in exopolitical differences marries with natural human expectations just as there is diversity in human experiences, temperament, perceptions, and learning, etc. An environment as diverse as the interplanetary community would be naturally expected to host many widely different individuals and groups. Why would we expect an interplanetary community to be different?

Now, we are getting ahead of ourselves of our treatment of contact and communications with ETI which follows this section but, in exopolitics, we must and separate out any and all forms of racism and xenophobia from the task at hand which is communications and diplomacy! The common bonds between interstellar civilizations are sentience consciousness and the recognition of a supreme force or power or divine creator in the universe, these are the bonds that tie!

Every individual regardless of race or planet of origin has the power and responsibility of free choice and must be held to the standards of moral rights and responsibilities independently of any others. Just like not every German forced to live under Hitler's regime during World War II was a Nazi, even if a certain planet were determined to be ruled by a tyrant with a hostile exopolitical agenda that would not make everyone from that planet an "evil alien".

Conversely, there can be bad apples in good bunches, and it is possible for advanced and positive societies to have rogue members.
http://www.bibliotecapleyades.net/exopolitica/esp_exopolitics_v.htm

Exopolitics is a very new and emerging discipline of study, and a lot of facts remain to be uncovered and ascertained. The work lying ahead before exopolitics researchers is to identify which exopolitical parties have involved themselves in Earth affairs in our history and which ones are involved with us and/or our planet now.

Their exopolitical agendas need to be examined and a determination made as to whether they act in the best interests of humanity or not. The politically engaged citizenry must force full disclosure out of its leaders and rulers regarding all contacts and deals with any exopolitical parties.

Our immediate neighborhood should be searched for any potential allies or enemies of humanity.
http://www.bibliotecapleyades.net/exopolitica/esp_exopolitics_v.htm

As Dr. Greer has often stated at his lectures and in his "**Ambassadors to the Universe**" training seminars, *"If there are hostile ETs out there that do not have our best interests at heart, they would be the first ones I want to meet. Why? Because like any diplomatic relation between the US and with nations like Russia, China, Korea or Iran, it always requires more work than it does with a relationship with nations like Canada or Britain!"*

Alfred Lambremont Webre (born May 24, 1942) is an American author, lawyer, futurist, peace activist, environmental activist, and a space activist who promotes the ban of space weapons. He was a co-architect of the **Space Preservation Treaty** and the **Space Preservation Act** that was introduced to the U.S. Congress by **Congressman Dennis Kucinich** and is endorsed by more than 270 NGO's worldwide.

Dr. Michael E. Salla is a pioneer in the development of '**Exopolitics**', the political study of the key actors, institutions, and processes associated with extraterrestrial life. His interest in exopolitics evolved out of his investigation of the sources of international conflict and its relationship to an extraterrestrial presence that is not acknowledged to the general public, elected officials or even senior military officials. His groundbreaking *Exopolitics: Political Implications of the Extraterrestrial Presence* (2004) was the first published book on exopolitics and explained the political implications of extraterrestrial life. In *Exposing U.S. Government Policies on Extraterrestrial Life* (2009) he revealed how the world's most powerful nation secretly manages information concerning extraterrestrial life and technology. In his most recent book, *Galactic Diplomacy: Getting to Yes with ET* (2013) he shows how humanity can negotiate with extraterrestrial civilizations in a way that protects our vital interests. Dr. Salla founded the **Exopolitics Institute** (2005) and the **Exopolitics Journal** (2006). He has co-organized four international conferences on extraterrestrial life and Earth Transformation on the Big Island of Hawaii. http://exopolitics.org/about/founder/

Alfred Lambremont Webre and Michael Salla are primary promoters of Exopolitics
http://exopolitics.blogs.com/exopolitics/2011/07/ and http://drjradiolive.com/09-08-2016-dr-michael-salla-exopolitics-org-secret-space-program-update/

He helped draft the **Citizen Hearing** in 2000 with **Stephen Bassett** and serves as a member of the Board of Advisors. Webre is also the congressional coordinator for **The Disclosure Project** and is a judge on the **Kuala Lumpur War Crimes Commission**.

In 1977, he joined **SRI International** in Menlo Park, California, as a futurist for the **Center for the Study of Social Policy**. His responsibilities were the studies in alternative futures, innovation diffusion, and social policy applications for clients including the **Carter White House Extraterrestrial Communications Study**, the **National Science Foundation, U.S. Congress (Office of Technology Assessment),** the **U.S. Department of Energy,** and the **State of California (Energy Plan).** http://en.wikipedia.org/wiki/Alfred_Webre

Alfred Webre and **Dr. Carol Rosin** founded the **Institute for Cooperation in Space (ICIS)** in 2001, as an outgrowth of the former **(ISCOS) Institute for Cooperation and Security in Space**. The ICIS mission is to educate decision-makers and the grassroots about why it is important to ban space weapons. Through the help of former **Congressman Dennis Kucinich,** the **Space Preservation Act** was originally introduced into the 107th Congress on October 2, 2001 (HR 2977) and included provisions banning *"extraterrestrial"* weapons, as well as *"chemtrails"* and *"exotic weapons systems"* such as **HAARP**. A revised Space Preservation Act (HR 3657) eliminating the prohibitions on space-based extraterrestrial, chemtrails, and exotic weapons systems was introduced to the 108th Congress on January 23, 2002. ICIS continues to lobby for a **Space Preservation Treaty** conference where leaders of the world would gather to ban space weapons. Former **Canadian Defence Minister Paul Hellyer** believes that this treaty would help put a cap on the war industry and open the door for international cooperation in outer space exploration. The end result would then be to transform the *"war "based"* economy into a *"peace based"* economy.

The ICIS board was made up of various prominent individuals such as former astronauts **Edgar Mitchell** and **Dr. Brian O'Leary**, and formerly the late **Arthur C. Clarke, General Counsel Daniel Sheehan,** and **John McConnell** who is the founder of **International Earth Day**. Alfred Webre resigned from the board of directors of ICIS on January 1, 2011, to focus on a treaty to ban HAARP, which he alleges to be a weapon.

Webre believes that as exopolitics posits, the truest conception of our human circumstance may be that we are on an isolated planet in the midst of a populated, evolving, highly organized interplanetary, intergalactic, multi-dimensional universe society. He believes that we live on a planet that has been *quarantined* (the **Zoo Hypothesis**) and that we are now being given an opportunity to join the rest of the spiritually evolved universe society in peace, thus an opportunity to avoid environmental global self-destruction or global self-destruction through war.

In the first decade of **Exopolitics** (2000–2010) about 30 nations have released their secret extraterrestrial and UFO files and to date, exopolitical organizations are active in approximately 40 nations. http://en.wikipedia.org/wiki/Alfred_Webre

Dr. Carol Rosin, Congressman Dennis Kucinich, Canadian Defence Minister Paul Hellyer Edgar Mitchell, Dr. Brian O'Leary, Arthur C. Clarke, General Counsel Daniel Sheehan, and **John McConnell**
Google Images

Explanation Detour: The **Zoo Hypothesis** is one of many theoretical explanations for the **Fermi Paradox**. The hypothesis speculates as to the assumed behavior and existence of technically advanced extraterrestrial life and the reasons they refrain from contacting Earth. One interpretation of the hypothesis argues that *intelligent alien life ignores Earth to allow for natural evolution and sociocultural development*. The hypothesis seeks to explain the apparent absence of extraterrestrial life despite its generally accepted plausibility and hence the reasonable expectation of its existence.

Aliens might, for example, *choose to allow contact once the human race has passed certain technological, political, or ethical standards*. They might *withhold contact until humans force contact upon them*, possibly by sending a spacecraft to planets they inhabit. Alternatively, a *reluctance to initiate contact could reflect a sensible desire to minimize risk*. An alien society with advanced remote-sensing technologies may conclude that direct contact with neighbors confers added risks to oneself without an added benefit.
http://en.wikipedia.org/wiki/Zoo_hypothesis

In the context of exopolitics as used by Dr. Webre, its meaning is used in the sense *"once the human race has passed certain technological, political, or ethical standards"* and given our predilection for war, the quarantine imperative becomes a greater concern for more advanced Extraterrestrial Intelligences imposing such restrictions on humanity!

There are some basic assumptions here that need to be explained. The zoo hypothesis assumes that aliens have great reverence for independent, natural evolution and development. Assuming that intelligence is a physical process that acts to maximize the diversity of a system's accessible futures, a fundamental motivation for the zoo hypothesis would be that premature contact would "unintelligently" reduce the overall diversity of paths the universe itself could take.

These ideas are perhaps most plausible if there is a relatively universal cultural or legal policy among a plurality of extraterrestrial civilizations necessitating isolation with respect to civilizations at Earth-like stages of development. In a Universe without a hegemonic power, random single civilizations with independent principles would make contact. This makes a crowded Universe with clearly defined rules seem more plausible.

If there is a plurality of alien cultures, however, this theory may break down under the uniformity of motive concept because it would take just a single extraterrestrial civilization to decide to act contrary to the imperative within our range of detection for it to be abrogated, and the probability of such a violation increases with the number of civilizations. This idea, however, becomes more plausible if all civilizations tend to evolve similar cultural standards and values with regard to contact much like *convergent evolution* on Earth has independently evolved eyes on numerous occasions, or all civilizations follow the lead of some particularly distinguished civilization . . . the first civilization. http://en.wikipedia.org/wiki/Zoo_hypothesis

The UFO/ETI phenomenon has been evolving over the decades moving from one stage to the next following the socio-geopolitical and technical development of humanity. The phenomenon has evolved from being the strange reports of nocturnal lights and daylight discs seen by every level and sector of society to being tracked by military and commercial ground and air radar.

These mysterious alien craft which no country has admitted to having the technical capability of building, at least so we are told by officialdom were being seen globally by millions of people.

These unusual craft have landed and left ground traces of their presence; they have caused odd interferences to electrical and electronic systems upon cars, trucks, buses, trains and planes, even shutting down major electrical grid systems and power stations which in turn have caused massive blackouts in various areas of America. People have filed reports of seeing various EBEs in, out and around landed saucer shape craft, even on occasion of interacting with them and receiving peaceful messages of love or warnings of possible nuclear war!

ETVs have been routinely monitoring every military air, naval base, missile site on the planet; every war and social conflict anywhere in the world seems to attract their attention, every major engineering, scientific development, and construction site is under surveillance, even when humanity is at play during the winter and summer Olympic games or in celebration at festive firework events, ET spacecraft have been there watching, monitoring and assessing everything we do.

Extraterrestrial Intelligences have engaged people randomly and sometimes not so randomly but with purposeful intent as reported by "contactees" and "experiencers". Many people have also reported being the victims of so-called alien abduction events which appear to be alien in nature but nevertheless, have all the earmarks of a military intelligence covert operation. These ETs are frequently described as short, gray in complexion with large heads and large black almond eyes. Compounding matters are the cattle mutilations that also attributed to these same malevolent "Grey" ETs, but a military presence is always nearby!

When crop circles began appearing in Britain in the late '70s and which continue to this day, always increasing in number and in geometric complexity there many theories as to their origin but the final conclusion reached by the experts settled on the ET hypothesis as the most tenable explanation of their creation. It would seem that ETI were reaching out to humanity in various modalities of communication right here on Earth in almost every country, even when scientist like SETI have failed in their efforts to receive intelligent signals from the vast reaches of space with their billion dollar array of radio telescopes, perhaps they need to look closer to home!

When mankind left the planet and flew into space orbiting the Earth and eventually walking upon the Moon, the Extraterrestrial presence was there in the flying saucer craft; when we sent space probes to other planets the ET presence followed or were there in orbit or in nearby space about those planets ever monitoring our space explorations!

Finally, when a few clever and courageous souls began to figure out it was possible to initiate contact and communicate directly with ETI, that's when the Military Intelligence Complex stepped in to prevent such citizen efforts. Back in the '50s, the MIC discredited contactee claims of ET contact but, when the **"Rosetta Stone"** in ET communications was decoded by a few citizens in the early '90s, the military increased their efforts of intervention with more determination. The next logical step in the evolvement of human - ET engagement had been discovered, it is the **(CE-5) Close Encounters of the Fifth Kind, ... human initiated contact and communications with Extraterrestrial Intelligences!!!**

304

CHAPTER 108

CHOOSING THE BEST METHODS TO COMMUNICATE
WITH EXTRATERRESTRIAL INTELLIGENCES

This section of the book can best be summed up by some insightful observations made by my wife, who has frequently stated on numerous occasions, whenever I was going out into some remote area of the countryside leading a team of like-minded individuals on a field expedition to establish contact and hopefully communications with visiting **ETI**:

*"If there is life in the universe, for which I have no doubt there is and if they are coming to this planet then, they are searching for signs of **intelligent life**, not some yahoos out in the middle of God knows where signalling with lights and calling 'come friendly aliens!' **No, they are looking for real signs of intelligent life on Earth and there isn't any!!!"***

This may be the real crux of any effort by humans to reach out to make contact with another intelligent species coming to this planet. If they are here, as most Ufologists and some scientists believe then, will we humans be recognized by an advanced interstellar civilization as being sufficiently intelligent enough for them to want to contact and communicate with us? From everything we have discussed so far in this book, it not only appears from all the evidence accumulated over the decades of research and investigations into the **UFO/ETI phenomenon** that Extraterrestrials have contacted us but, are trying to actually communicate with us on a reasonably intelligent level. In fact, these communications from ETI with human beings appears to be escalating particularly among those humans who are not only open-minded to the concept but, actually wish to engage in this inter-species communications and have pro-actively sought it out!

The need to communicate with each other is an evolutionary imperative to our survival as rational, intelligent, spiritual, sentient beings. We communicate with each other to impart ideas, hopes, fears, cares, concerns, and love. We communicate to enlist the help and cooperation of others to aid in constructing our environment or in educating others or in organizing our families and societies. We will even communicate with our pets as if they are able to understand the complexity of our languages. We will communicate with the animals in the wild, to the birds, the trees, the plants and insects around us and even, when we are alone, we will communicate to the walls or other inanimate objects or with ourselves when we look into the mirror. It is our need to communicate whether verbally or in silent mind or with a physical expression of loving affection to those closest to us. In moments of aloneness, we may also find ourselves communicating through prayer and meditation to the forces and powers of the universe asking for guidance, affirmation, or giving back gratitude and praise for our existence.

The fact is that as intelligent rational beings, we crave communication in some form or another to assure ourselves that this existence is not one that is experienced alone but, in the company of another or in a community of others similar to ourselves. Communities are assemblages of like-minded people who communicate, ergo, communities communicate! Communities will seek out other communities of like-minded individuals or even, not so like-minded in order to expand their awareness and understanding of the world in which they live in. Communities that communicate with other communities grow and evolve to higher levels of civilization by sharing ideas, hopes, fears, concerns and love for each other.

It is only natural that in this day and age those communities have grown to become countries that communicate with other countries and we can now see with absolute certainty that we are evolving into a global community or world commonwealth. Therefore, the natural process of evolution mandates that as a global civilization, we communicate not only among ourselves but that we reach out to communicate with other civilizations, one that is from the stars!
The question is then, by what method do we use to communicate with other intelligent species or civilization? Do we use geometrical forms carved into the ground or planted using cereal crops or trees? Do we build mammoth structures miles in length, width, and height that can easily be seen from space? Do we use smoke or fire signals or city lights or do we transmit by means of radio/microwave signals from large radio telescope arrays such as used by **SETI**?

What should be the content of the message, should it be simple or complex? Do we even want to send any transmissions into space or is it too late, fearing as **Stephen Hawking** suggested that we may attract a hostile ET species to our planet that may not have our best intentions at heart? What are we to make of all the mountains of UFO sightings and accounts reported by witnesses from around the world? As we have already demonstrated in previous sections of this textbook UFOs are spacecraft piloted by intelligent beings from other star systems, perhaps these UFOs by their very presence in our atmosphere and in near orbit about the Earth are already here to engage and communicate with mankind or they are biding their time, until there appears to be true *"signs of intelligence"*.

It may actually come as a surprise to many people including governments, military departments, scientists, and religious officials that contact and communications with ETI has already randomly taken place among small groups of people globally and the method of communications are not in the anthropocentric manner in which we may have anticipated!

In this section we will look at some of the ways humans have tried to reach out to other intelligences in the universe using a variety of methods, some successful, most not so successful. Some of the methods have been touched upon briefly in other sections while proving the existence of UFOs and ETI and so they will be covered in more detail here. Some methods were seriously considered and explored with an open scientific approach to the possibility that other sentient life besides our own does exist elsewhere in the universe. It will be shown that some methods that are trumpeted as being scientific in their approach and fully supported by the government and the **DOD (Dept. of Defense)** were seriously flawed from the start, as mere fronts designed to distract people's interest away from the real exploration going on behind closed doors in covert secret programs.

Later, we will examine the more controversial methods of communication that fly in the face of accepted scientific standards and methods yet, they will be proven to be not only highly effective in contact and communication with **ETI (Extraterrestrial Intelligences),** but will be shown to be scientific in their methodology and reproducible by anyone!!!

We will also look at the possible language modalities of Extraterrestrial Intelligences who are visiting the Earth at this time and once again, it appears that ETs do not always communicate in the traditional earthly manner of verbal syntax.

CHAPTER 109

SOME CONTROVERSIAL METHODS OF EXTRATERRESTRIALS CONTACT AND COMMUNICATIONS

We've looked at some of the more obvious methods that humans have tried to communicate with Extraterrestrial Intelligences that may be in other star systems or within our Solar system or that are orbiting our planet and landing on terra firma with more frequency. Some of these methods are outlandish and useless attempts that will never produce results, no matter how altruistic our intentions to reach out and make contact. Some are a great waste of financial resources, materials, and methodology. They are often trumpeted as the most common sense approach to ET communications because they are government funded and military sanctioned which in the minds of these officials gives such projects legitimacy and thus, they become the go-to agency for "official" information.

The fact that none of these methods has worked so far, *unless ETs have been within our solar neighbourhood when these signals were transmitted and the replies kept secret from the public,* is proof positive that some alternative method must be considered that can transcend the technological boundaries and limits of the speed of light that are currently been employed by SETI in the form of microwaves and radio waves.

The alternative methods when factoring in these limits of light speed based upon current scientific principles must be able to obliterate this drag coefficient that limits the effectiveness of communications over the immense vastness of intergalactic space. The methods have to be instantaneous or nearly instantaneous and they have to be understood by the recipient, regardless of where in the universe they may be located. Time and space must be obliterated to effectively communicate with another ET civilization!

Are there effective alternative methods of communications that fit this description? Apparently so and the military of the various superpowers take these alternative methods very seriously. When it comes to anything that is paranormal or in the realm of psychic ability, there are universities and military departments both in Russia and in the US where money, time and research are spent developing programs to militarize such abilities. The most famous psychic development program in the US is the **Stargate program** where soldiers and officers develop and practice the ability of **Remote Viewing. Stanford Research Institute** in California was also heavily involved with the military in developing remote viewing techniques with certain people who already had demonstrated psychic tendencies. **The Monroe institute (TMI)** in Virginia is known not only throughout the US but worldwide as another research and educational organization dedicated to enhancing the uses and understanding of human consciousness without any military affiliations.

Let's look at some of the more questionable or controversial methods that some people claim to receive ET communications by, to see if there is any merit to any of them.

Primary Psychic Abilities

Here is list of the primary psychic abilities of which the many various "mancies" and "scopies" are a subset of these primary functions and are therefore not included in this short list:

Astral Projection - The ability to leave one's body and travel in spirit to another location.

Aura Reading – is the ability to see the energy fields that emanate from living beings. Psychic ability can often reveal itself through the seeing of auras.

Automatic Writing - Writing through the subconscious mind without conscious thought, or through the guidance of an outside intelligence.

Bilocation - is the ability of being in two or more **(multilocation)** different locations simultaneously which is usually triggered through a meditative state of higher consciousness.

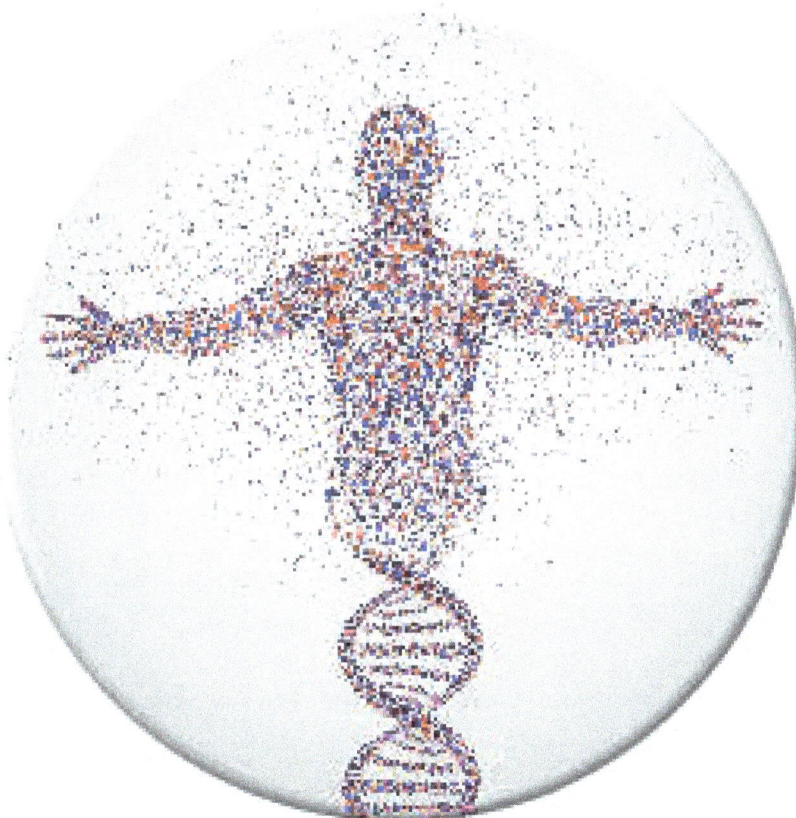

The psychic abilities are inherent within each person which may lay dormant or become active, triggered by some event, object or person
http://uk.iacworld.org/full-list-of-different-types-of-psychic-abilities-here/

Channeling - Associated with mediums, this is the ability to act as a channel or vessel for an outside intelligence.

Clairaudience - Put simply, this type of ability is used to hear what is "inaudible". For example, someone with this ability could be a thousand miles way and "hear" a loved one's cry of distress.

Clairvoyance - Usually confused with Precognition, this ability actually has much more in common with "Remote Viewing", True clairvoyance is not the ability to see into the future, but the psychic ability to see visions of that which is hidden or far away.

Clairsentience - In this instance, the psychic has an insight or "knowing" of and a hidden or forgotten fact.

Divination - A broad term that includes fortune telling, precognition, prophesy, and other methods used in an effort to predict the future.

Dowsing - Also known as "water witching", dowsing involves the use of a rod, sticks, or pendulum to locate water or lost objects.

Empathy - The talent to sense the needs, drives, and emotions of another. As with Aura Reading, psychic ability can often reveal itself through the development of empathy.

E.S.P. - Extra Sensory Perception(s) is the awareness of information about events external to the psychic that are not gained through the senses and not deducible from previous experience. Often used to describe clairvoyance, precognition, telepathy, etc...

Intuition - Similar to clairsentience, this is the power or faculty of attaining direct knowledge or cognition without rational thought or inference.

Levitation - The ability to cause one's body to hover off the ground. One of the more well-documented cases of levitation involved St. Theresa of Avila during the 16th century.

Mind Over Body - Suppressing or mentally satisfying the need for water, food, or sleep. There is some debate over whether this is actually a psychic ability since many of those who are associated with this trait (monks, yogis, mystics, etc...) are not generally called "psychics".

Precognition - Quite simply, "knowing the future". However, since time is a dynamic construct, no one psychic can ever know every detail about the future. Usually, this ability refers to knowing general outcomes of specific courses of action, with occasional flashes of detailed insight.

Psychometry - Also known as "object reading", psychometry enables a psychic to pick up on psychic impressions (vibrations) left on an object by someone connected with it. Someone with this ability could use an unfamiliar object to reveal much about its owner.

Pyrokinesis - The ability to start fires with one's mind.

Remote Viewing - is the ability to see things at a distance with the mind while in a meditative state of higher consciousness. In military or intelligence concepts allows a perceiver (a "viewer") to describe or give details about a target that is inaccessible to normal senses due to distance, time, or shielding.

Telekinesis - Also known as psychokinesis, the ability to move objects with one's mind.

Telepathy - The ability to communicate mind-to-mind with another.
http://www.thelostfound.com/psychic-abilities.html

Channelling – Do Spirits or Aliens Actually Speak Through Anyone?

Perhaps, of all the physics abilities that people claim to possess, "**channelling**" where one person permits an entity to take control of their body or their conscious mind or vocal cords for a brief period of time, is by far the most difficult ability to accept as a genuine method of communications.

The channeller often goes into an altered state (trance or higher vibration) through which an entity on another plane or outside of the Earth's sphere can communicate with the Earthlings. Some channellers state that they raise their own frequency and that the ET being lowers his own frequency so that they meet in the middle. Others say they receive telepathic pulses and emotion-packages which they translate into their language and concepts. There are many problems inherent in this type of communication method which is highly suspicious in its authenticity.

It matters not whether a person claims to channel **spirits** of the long departed or the recently departed or if the entity is a saint, an **ascended master,** an **avatar, a minor or major prophet of God** or even God himself or in this day and age they are receiving communications from **Extraterrestrials beings** on another planet, a spaceship or an **inter-dimensional reality**, the experience from the observers perspective can be strange and somewhat uncomfortable. The client or the audience may not know what to make of the experience or the information being expressed unless one is already leaning in the direction of alternative pseudo-guidance, other than what can be found in everyday religion or from a good councillor or psychiatrist. The entities that are channelled by spiritual mediums usually do not include historical persons whom mediums claimed to channel after they died unless this is the main reason why this person became famous. http://en.wikipedia.org/wiki/Category:Channelled_entities

The major problem is that the information that is being channelled is that the information is usually not anything new or relevant that one could not just as easily look up in a book from the library or from talking to someone else beforehand or from information gleaned from TV media, etc. Even for an individual to fabricate an original running dialog off the top of their head is difficult at best but, channellers claim that it is original information however, they don't necessarily claim that it would be current or totally new information that has never been heard before. At best they may state that they did not know or remember what was being relayed by the entity possessing their body at that moment.

Channellers are often viewed as cynical where their messages often seem to be vague lacking any real substance of useful information ("We come in peace, make love not war! Don't pollute the Earth!") or pretentious ("We've come to help save you from yourselves") and humorless or

unaccepting of criticism (cultish). **Channellers** can, on the other hand, be self-doubting or self-deprecating about the information they provide. Some messages on occasion have the appearance of being genuine, sometimes even useful, practical, accurate in some predictions, and generally humorous and intelligent. (See statements from early contactees below)
http://www.abovetopsecret.com/forum/thread353245/pg1

Another problem is that most **channelled entities** of the departed spirit kind usually all sound the same with either, a pronounced deepen male voice inflection or high feminine inflection which is harder to do if the channeller is a male. If the channelled entity happens to be Extraterrestrial then, that ET entity somehow knows how to speak English or whatever the native language of the channeller happens to be. Even, the name of the entity is strange in its pronunciation like **Ramtha, Quasqar, Ashtar, Bashar, Zandor, Aiwass**, etc.

If one happened to be in the company of a someone who is a channeller when they give a reading or channelling session, the entity will often speak in a manner which sounds almost condescending and like a bloody know-it-all with a definite air of otherworldliness.

Here is a partial list of some well-known entities and their mediums or channellers:

- *Abraham* (group of entities), channeled by **Esther Hicks**
- *Aiwass* channeled by **Aleister Crowley**
- *Alexander* channeled by **Ramon Stevens**
- **Arten** and *Pursah* appeared to **Gary Renard**
- *Ashtar (extraterrestrial being)* communicated telepathically to **George Van Tassel**
- **Bashar** channeled by **Darryl Anka**
- *The Cassiopaeans* channeled by **Laura Knight-Jadczyk**
- *The Galactic Federation of Light* (alien beings), channeled by various mediums at LightWorkers.org (The most prolific is *SaLuSa* channeled by **Mike Quinsey**)
- *God* channeled by **Barbara Rose**
- *God* channeled by **Eileen Caddy**
- *God and Jesus Christ* channeled by **Joseph Smith, Oliver Cowdery**, and other early Mormon leaders
- **God** channeled by Neale **Donald Walsch**
- *The Guide* channeled by **Eva Pierrakos**
- *Hathor* (from an ascended civilization) channeled by **Tom Kenyon**
- *Jehovih* through automatic writing by **John Ballou Newbrough**
- *Jesus* channeled by **Helen Schucman**
- *Kirael* (7th Dimension Master Guide) channeled by **Fred Sterling**
- **Kryon** entity, channeled by **Lee Carroll**
- *Melchezidek* entity, channeled by **Kathryn E Cole**
- *Maitreya* channeled by **Margaret McElroy**
- *Michael* (The Michael Teachings), written by **Chelsea Quinn Yarbro**
- *Oth* channeled by **Ellen Rauh**
- *Pleiadeans* entity, channeled by **Barbara Marciniak**
- *Pleiadian Collective* (9th Dimensional), channeled by **Wendy Kennedy**
- *Ra* channeled by **Carla Rueckert**

312

- *Ramtha*: Ascended Master, channeled by **JZ Knight**
- **Seth** entity, channeled by **Jane Roberts**
- *Speakers of the Sirian High Council*, 6th dimension beings, channeled by **Patricia Cori**
- *Thoth* (Ascended Master from Atlantis, Egypt, and Greece) channeled by **Drunvalo Melchizedek**
- *Tobias* (of the Crimson Council) channeled by **Geoffrey Hoppe**. (He also channels ascended masters *Adamus Saint-Germain and Kuthumi*)
- **Various psychic surgeons**, such as *Zé Arigó,* claim to work as channels for deceased surgeons.
- *YHWH* channeled by **Arthur Fanning** (He also channels the being *Jehovah*)
- *Zoosh* channeled by **Robert Shapiro**
  http://en.wikipedia.org/wiki/List_of_modern_channelled_texts

Of all the channelled literature that has come into print, **The Urantia Book** stands out as the best known. The exact circumstances of the origin of *The Urantia Book* are unknown. The book and its publishers do not name a human author. Instead, it is written as if directly presented by numerous celestial beings appointed to the task of providing an "epochal" religious revelation. For each paper, either a named celestial being, an order of being, or a group of beings is credited as its author.

The Urantia Book is a spiritual and philosophical book that discusses God, Jesus, science, cosmology, religion, history, and destiny. It possibly surfaced sometime between the period of 1924 and 1955 in Chicago, Illinois. It is 2,097 pages long and consists of an introductory foreword followed by 196 "papers" divided into four parts.
http://en.wikipedia.org/wiki/List_of_modern_channelled_texts

Is there any creditability that can be given to channelling as a legitimate method to contact and communicate with Extraterrestrial Intelligences? The answer to that question rests with the type of information that is being channelled, it genuineness and relevance and whether any of that communiqué can be substantiated by some practical means or by other persons or by one or more correlated events. In which case, it is this author's opinion that the medium is not channelling but is communicating via some telepathic means with the entity being channelled.

In the early 1950's many contactees often channelled information from the ET entities that curiously spoke about the *dangers that were facing our civilization, like the build up of armaments, the potential for thermal nuclear war, the rapacious consumption of Earth's natural resources, pollution and the destruction of the environment, and over-population, etc.* All of these admonitions by channelled ET entities, when considered at this current time, seem to be not only apropos but, almost prophetic in their warnings. Could this be genuine communication via the method of channelling or could some other technique of communication been employed?

There is a flip side to this channelling conundrum: how can you really know that a thought has been sent telepathically to you or that another thought is really your own? What are the types of telepathy humans have thus far encountered? Here are several types of telepathy:

1. **Channeling is a form of telepathy.** Some channels go into deep trances like old-fashioned spiritualists (**trance mediums)** in order to receive messages from aliens of the higher realms (or possibly, the lower realms). Do these channels, some of whom are well-known personalities, hear the alien's voice in their heads or do "concepts from outside" simply land in their heads as thoughts? Or are they faking? For the purpose of this article, we'll assume the phenomenon of channeling is sometimes not faked. Whether they hear a voice or simply receive thoughts, the channel is receiving a form of telepathy.

2. **Another type of telepathy is experienced by people who were contacted at some point, then feel an entanglement of consciousness with a UFO occupant.** This is an on-going, probably lifelong form of telepathy which perhaps is not pure telepathy but rather is the blending of two consciousnesses through quantum manipulation.

Quantum entanglement occurs when two or more objects share an unseen link bridging the space between them. A hypothetical pair of dice, for instance, would always land on matching numbers when rolled simultaneously, even if one was rolled on Earth and the other was rolled on Mars. "Entanglement" is a neutral term used by scientists to describe an astounding phenomenon of quantum physics where a photon experiences an interaction from some outside force and another photon somewhere else, even at a great distance is through entanglement affected by the same force and reacts in the same manner as the first photon.

Albert Einstein called Quantum entanglement as *"spooky action at a distance."* Einstein muttered that it violated his famous speed limit which states, nothing goes faster than the speed of light in his **Theory of Relativity**.

Physicists have recently managed to entangle two diamonds at the quantum level, so it has been proven that this finicky phenomenon is not limited to tiny, ultra-cold objects. Quantum entanglement has now been achieved for macroscopic diamonds on a quantum level. Everyday objects like diamonds can be placed into a quantum state.

In this experiment, the two diamonds sub-atomic molecules became entangled and therefore a **phonon** (the quantum of acoustic or vibrational energy e.g. sound which is considered a discrete particle) was not confined to either diamond; instead, the two diamonds entered an entangled state in which they shared one phonon. If we crushed one diamond on Earth, would the diamond on Mars also crush in the same second? https://en.wikipedia.org/wiki/Phonon

If UFO occupants use quantum entanglement to entangle the consciousness of an alien mind and a human mind (which amounts to a **shared consciousness** and feeling of mutual identity), then *this is an on-going telepathy of sorts* involving the thought processes and the actual *sentience* of both alien (UFO occupant) and human. In a sense, this is a long-term, comprehensive telepathy.

"I feel that UFO occupants use this scientific procedure fairly often, thus some humans feel a connection to one particular UFO occupant and even a spiritual, emotional pull toward his or her identity as if it were their own – a beckoning. Humans with this telepathic connection often "channel" also, but perhaps, it is not actually channeling but simply reaching for the Other Being within their Own Being – reaching for the knowledge he or she already knows – *a shared consciousness*." http://ufodigest.com/article/telepathy-and-how-it-fits-ufo-puzzle

314

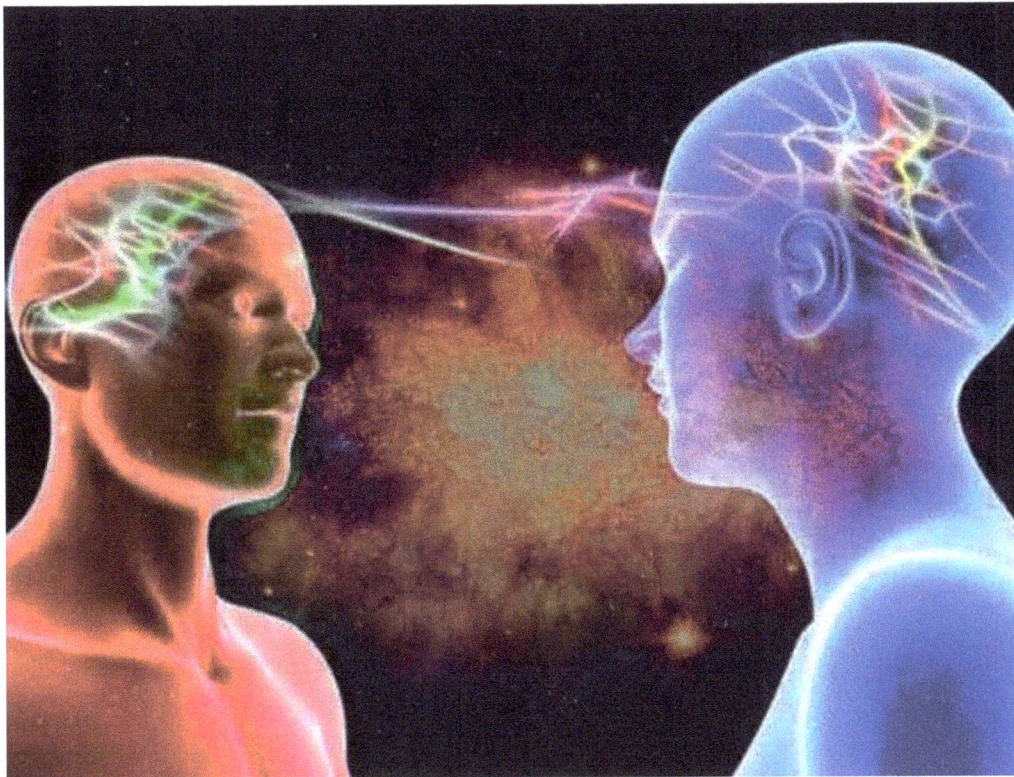

**Is channelling a form of telepathy or a quantum entanglement of
two separate consciousnesses or something else entirely?**
https://www.themystica.com/mystica/articles/t/ten_steps_to_master_telepathy%20.html

More investigation is needed to determine if channelling is telepathy or if it is a quantum entanglement of consciousnesses or something entirely. It is still too close to the charlatan practices of mediums back in the late 1800s and early 1900s which were often exposed as fraudulent in nature.

Astral Travel/Projection – Out of Body Soul or Mind Travelling

Astral projection (or **astral travel**) is an interpretation of **out-of-body experience (OBE)** that assumes the existence of an "**astral body**" separate from the physical body and capable of traveling outside it. Astral projection or travel denotes the astral body leaving the physical body to travel in the astral plane.

The idea of astral travel is rooted in common worldwide religious accounts of the afterlife in which the consciousness or the soul's journey or "ascent" is described in such terms as "an out-of-body experience, wherein the spiritual traveler leaves the physical body and travels in his/her subtle body (or **dream body** or **astral body**) into 'higher' realms." It is therefore associated with near death experiences and is also frequently reported as spontaneously experienced in association with sleep and dreams, illness, surgical operations, drug experiences, sleep-paralysis and forms of meditation.

Astral travel or out-of- body (Soul) travel often occurs during sleep or the dream state
http://www.poklat.com/astral-projection-spiritual-travel-time-space/

It is sometimes attempted out of curiosity, or may be believed to be necessary to, or the result of, some forms of spiritual practice. It may involve "travel to higher realms" called astral planes but is commonly used to describe any sensation of being "out of the body" in the everyday world, even seeing one's body from outside or above. It may be reported in the form of an apparitional experience, a supposed encounter with a **doppelgänger**, some living person also seen somewhere else at the same time.

Through the 1960s and 1970s, surveys reported percentages ranging from 8 percent to as many as 50 percent (in certain groups) of respondents who state they had such an experience. The subjective nature of the experience permits explanations that do not rely on the existence of an "astral" body and plane.

The theme is treated in anthropological or ethnographic literature on witchcraft and shamanism, in classical philosophy and in various myths and religious scriptures. One such religion is the new age cult religion of **Eckankar**, founded in Chanhassen, Minnesota, USA which uses astral travel as one of its main tenants to teach adherents of Eckankar how to leave the body to discover self-realization and to co-partner with the source of their being, God.
http://en.wikipedia.org/wiki/Astral_projection and http://en.wikipedia.org/wiki/Eckankar

316

Many people who practice OBE or astral travel have reported the ability to explore other realms of existence, to be able to communicate with other astral travelling souls or with the departed. Some claim to communicate with ascended masters, saints, and avatars, while others state that they have communicated with Extraterrestrials beings either in this physical realm of existence or from another dimension. Once again, this sound similar to the communication practices derived from channeling. The only real difference is that one must astral travel to communicate with other beings or the God-head or experience alternative realities.

Astral projection is more common than most people think as it is an experience that millions of people worldwide have practiced with intention because of their involvement with a particular religion or organization or from some life and death situation or from a highly charged emotional event which induces a spontaneous out-of- body experience. In fact, one religion, the **Baha'i Faith** says that everyone who dreams, which is most of us, experiences it to some degree each night we go to bed.

In one of the writings of **Baha'u'llah**, the **Manifestation of God** for this day and age and founder of the Baha'i Faith states the nature of astral travel while explaining the power and comprehension of the human spirit:

"Know that the power and the comprehension of the human spirit are of two kinds: that is top say, they perceive and act in two different modes. One way is through instruments and organs: thus with this eye, it sees, with this ear, it hears, with this tongue or talks. Such is the action of the spirit, and the perception of the reality of man, by means of organs. That is to say, that the spirit is the seer, through the eyes; the spirit is the hearer, through the ear; the spirit is the speaker, through the tongue.

The other manifestation of the powers and actions of the spirit is without instruments and organs. For example, in the state of sleep without eyes it sees, without ears, it hears, without a tongue, it speaks, without feet it runs. Briefly, these actions are beyond the means of instruments and organs. How often it happens that it sees a dream in the world of sleep, and its signification becomes apparent two years afterwards in corresponding events. In the same way, how many times it happens that a question which one cannot solve in the world of wakefulness, is solved in the world of dreams. In wakefulness, the eyes sees only for a short distance, but in dreams, he who is in the East sees the West: awake he sees the present, in sleep he sees the future. In wakefulness, by means of rapid transit, at the most he can travel only twenty farsakha an hour; in sleep, in the twinkling of an eye, he traverses the East and West. For the spirit travels in two different ways: without means, which is spiritual traveling; and with means, which is material traveling: as birds which fly, and those which are carried.

In the time of sleep this body is as though dead; it does not seen or hear, it does not feel, it has no consciousness, no perception: that is to say, the powers of man have become inactive, but the spirit lives and subsists. Nay, its perception is increased, its flight is higher and its intelligence is greater". **Baha'i World Faith, Selected Writings of Baha'u'llah and 'Abdu'l-Baha; by National Spiritual Assembly of the United States, 1943, 1956; Baha'i Publishing Trust; ISBN 0-87743-043-8; pg. 326**

And in Baha'u'llah's book "Seven Valleys and Four Valleys", from the "Valley of wonderment":

"Indeed, O Brother, if we ponder each created thing, we shall witness a myriad perfect wisdoms and learn a myriad new and wondrous truths. One of the created phenomena is the dream. Behold how many secrets are deposited therein, how many wisdoms treasured up, how many worlds concealed. Observe, how thou art asleep in a dwelling and its doors are barred; on a sudden thou findest thyself in a far-off city, which thou enterest without moving thy feet or wearying thy body; without using thine eyes, thou seest; without taxing thine ears, thou hearest; without a tongue, thou speakest. And ten years are gone, thou wilt witness in the outer world the very things thou hast dreamed tonight.

Now there are many wisdoms to ponder in the dream, which none but the people of this Valley can comprehend in their true elements." **The Seven Valleys and the four Valleys, by Baha'u'llah; NSA of the USA; 1945 and 1952; Baha'i Publishing Trust; pg. 32-33**

The fact is that it is hard to argue against the existence of astral travel when one major religion supports its existence and when tens of millions of people worldwide will tell you that they have experienced it in some form or fashion.

Are they all delusional, simply misinterpreting some other everyday common event? Some scientists don't think so, as there are case studies going on at various universities in most countries to understand this phenomenon, along with many other paranormal studies. In fact, the military takes this subject very seriously, why do you think one of the names given to **Area 51** is…***"Dreamland"***?!

Dr. Steven Greer recounts how one of his **Disclosure Witnesses** who worked in Area 51, the S-4, S-5 and "Dreamland" areas of Nevada who has a military background tells how he was taught to astral travel or have an out of body experience and was having some difficulty with it but, his commanding officer felt he was very close to having an experience. That night this officer went home and when he fell asleep, he immediately had an out of body experience. He reported that he shot out of his body and went flying up into space where he collided with an ET flying saucer causing it to rock back and forth!

Now, that's very interesting on a number of levels!!!

He passed through the hull of the ship and landed on the floor in his astral body state, where the Extraterrestrial beings on board, sensing him casually turned around, looked at him and told him to *"Watch where you are going!"*

It would seem by this officer's sworn testimony that communications with ETI in the astral state is possible and proof that this can be an effective method of communications not only between humans but with ETs as well.

Author's Rant: I have personally experienced astral projection and/or travel on several occasions and can vouch for its authenticity. The first such time occurred with a young lady

who became my first girlfriend. We were at a beach one evening on Vancouver Island, British Columbia passionately kissing each other. When quite by accident I felt myself emerge out through by back and was floating a couple of feet above the young lady and myself observing what I was doing. At that moment, I was gazing at some animalistic form operating only on carnal desires and instinct. It seemed that it wasn't me but, someone else and I turned. I turned in my astral body and glanced out to the waves that gently broke upon the nearby shoreline and became entranced by the sound of their crashing. I turned once more and gazed upon myself and my girlfriend and felt myself merged back into my body, at that moment I physically came up for air! A few days later, I told her what had happened while we were necking on the beach.

The second time I experienced astral travel was after reading a book given to me by a dearly departed friend in which I practiced for a few weeks until one evening in bed I felt myself levitate out of my body above my bed and then passed out through a closed window into the open air. I didn't recall much of the experience as I fell into a deep unconscious sleep.

On the third occasion, I heard a buzzing sound like bees as I reached a semi wake, semi-sleep state in my bed and almost right away I started to leave my body but feet first! It was an odd sensation as the rest of my body could not emerge or levitate, just my feet. I recalled at that moment from the book on "Leaving the Body" that if this should happen. One merely has to use their astral hands and arms and touch the floor or a bedside night table and simply roll out of their body to an upright position. This seemed to work and I decided to go downstairs to look at my two cats to prove that I was indeed astral travelling. I went over to the male cat to pet it with my astral hand and I remember that it hissed at me! Very unusual! It must have been aware of my presence in some way. I then went back upstairs almost instantly and went out through the bedroom window again but, this time, I had a destination in mind. I was going to go to the top of a nearby mountain where I had seen a UFO hovering in that area many months earlier. However, as I moved toward the mountain, I once again lost consciousness and fell into a deep sleep. Such were my experiences in astral travel.

Telepathy – Mind to Mind Communication

Telepathy (from the ancient Greek *tele* meaning "distant" and *pathe* or *patheia* meaning "feeling, perception, passion, affliction, experience") is the transmission of information from one person to another without using any of our known sensory channels or physical interaction. The term was coined in 1882 by the classical scholar **Frederic W. H. Myers**, a founder of the **Society for Psychical Research**, and has remained more popular than the earlier expression *thought-transference.*

Scientific consensus does not view telepathy as a real phenomenon. Many studies seeking to detect, understand, and utilize telepathy have been done, but according to the prevailing view among scientists, telepathy lacks replicable results from well-controlled experiments.
http://en.wikipedia.org/wiki/Telepathy

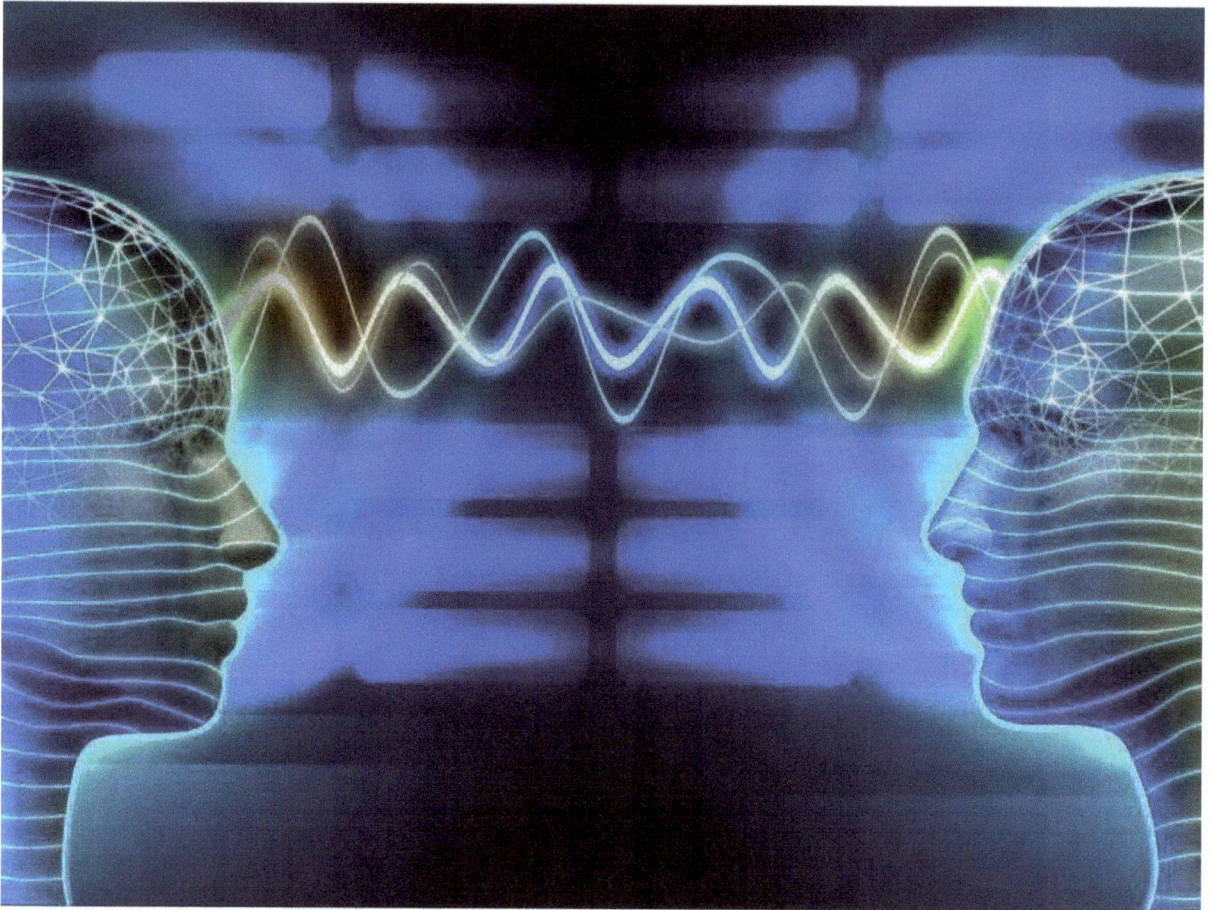

Telepathy is a synchronization of brain wave frequencies between two people

Within the field of parapsychology, telepathy is considered to be a form of **ESP (extra-sensory perception)** or anomalous cognition in which information is transferred through **Psi (Psychic Factor)**. It is often categorized similarly to precognition and clairvoyance. Experiments have been used to test for telepathic abilities. Among the most well-known are the use of **Zener cards** and the Ganzfeld experiment.

Zener cards are marked with five distinctive symbols. When using them, one individual is designated the "sender" and the other, the "receiver". The sender selects a random card and visualizes the symbol on it, while the receiver attempts to determine that symbol using Psi. A repeated successful score rate significantly higher than 20% is a demonstration of telepathic ability.

320

Zener cards are typically used in parapsychology experiments to determine telepathy
(Google Images)

When using the **Ganzfeld experiment** to test for telepathy, one individual is designated the receiver and is placed inside a controlled environment where they are deprived of sensory input, and another is designated the sender and is placed in a separate location. The receiver is then required to receive information from the sender. The nature of the information may vary between experiments.

A physical model of telepathy, whether described as radiational or in other terms, assumes that transference is effected by means of a vibratory current linking one brain to another. **William Crookes** proposed a "brain wave" theory in which he claimed telepathy might occur due to high frequency vibrations of the ether. Crookes had stated that there may be parts of the human brain that may be capable of sending and receiving electrical rays of wavelengths. **William Fletcher Barrett** and **Frederic William Henry Myers**, however, pointed out problems in a physical theory for telepathy and instead advocated psychical theories.

In the early 20th century, there were two other prominent concepts of telepathy: the spiritualist position which claimed telepathy was the result of external spirits and a view claiming interactions between two or more subconscious minds. The subconscious mind view was advocated by psychical researcher Thomson Jay Hudson who wrote that the mind is a duality that consists of two minds: the *objective (conscious)* and the *subjective (subconscious)*.
http://en.wikipedia.org/wiki/Telepathy

The psychical researcher **John Arthur Hill** wrote regarding telepathy *"No physical theory of telepathy has been worked out — there are no 'brain-waves' known, and no receiving stations yet discovered inside our skulls."* **George N. M. Tyrrell** also claimed that a physical basis for telepathy was untenable as ideas cannot be transmitted from one mind to another by any physical means without being first translated into a code. **H. H. Price** suggested that telepathy was incompatible with any material explanation, as a physical theory of telepathy would reveal radiations detectable on physical instruments, but none have ever been detected.

Some parapsychologists proposed that telepathy may have a physical explanation. The Italian neurologist **Ferdinando Cazzamali** in the 1920s had claimed that telepathic communication

321

occurred due to a type of electromagnetic radiation. However, the neurophysiologist **William Grey Walter** in his book *The Living Brain* (1953) wrote that electrical 'brain-waves' are too weak to explain telepathy. **Hans Berger** also held this view but extended the theory by proposing that telepathy occurs when ***"electrical energy in the agent's brain is transformed into 'psychic energy' which can be diffused to any distance, passing through obstacles without attenuation"***.

In 1974 **Michael Persinger** proposed that **extremely low-frequency (ELF)** electromagnetic waves may be able to carry telepathic and clairvoyant information. **Johnjoe McFadden** has written, *"the EM field outside the head is far too weak and it is highly unlikely that any other brain could detect it, and still more unlikely that the other brain could decode the EM field information that was encoded by your brain"*.

Gerald Feinberg suggested that telepathy may exist due to as yet undiscovered elementary particles which he called **'psychons'** or **'mindons'**.

In recent years the parapsychologist **Charles Tart** has accepted the existence of telepathy but claims that it is ***nonphysical in nature and cannot be fitted into any physical theory***.

Parapsychology describes several forms of telepathy:

- **Latent telepathy**, formerly known as "deferred telepathy", is described as the transfer of information, through Psi, with an observable time-lag between transmission and reception.
- **Retrocognitive, precognitive, and intuitive telepathy** is described as being the transfer of information, through Psi, about the past, future or present state of an individual's mind to another individual.
- **Emotive telepathy**, also known as remote influence or emotional transfer, is the process of transferring kinesthetic sensations through altered states.
- **Superconscious telepathy** involves tapping into the superconscious to access the collective wisdom of the human species for knowledge.
  http://en.wikipedia.org/wiki/Telepathy

Although not a recognized scientific discipline, people who study certain types of paranormal phenomena such as telepathy refer to the field as parapsychology. Parapsychologists claim that ***some instances of telepathy are real.***
Outside of parapsychology, telepathy is generally explained as the result of fraud, self-delusion and/or self-deception and not as a paranormal power.

A variety of tests have been performed to demonstrate telepathy, but there exists no scientific evidence that the power exists. The scientific community considers parapsychology a pseudoscience. The field of parapsychology can be filled with landmines for scientists to navigate through. At best, they may be viewed as a fringe group of science; at worst, they can be lumped together with astrologers and fortune tellers.

We find that in this light of scientific inquiry or the lack of it, scientists find themselves divided on the question of the reality of telepathy or any extra sensory perception (ESP) with the majority favouring the negative naysayers due to the lack of measurable quantitative proof to its existence.

Officially, scientists position themselves in the manner that if there is no rational proof of something, meaning physical proof, whether it's in the field of the paranormal, cryptozoology or the study of UFOs and ETI, etc., it simply does not exist and is not worth any real serious research effort. But, get those same scientists aside having coffee or a few drinks together and if the conversation arises on one of these topics then, those scientists will tell you they believe that there is evidence to support research on the subject matter.

The problem arises for these scientists when they are threatened with the loss of their professional tenure; their reputation becomes the subject of ridicule by fellow scientists, the news media, and the general public. The end result is they become people discredited, now on the outside looking in at what could have been a positive future if they only kept their mouths shut.

But, if there is really nothing defining about telepathy and it is considered pseudoscience by the scientific community then, why does the CIA and the military train people to become psychic spies? Why is the military spending millions of dollars on R&D programs to develop psychotronic weaponry and warfare? Why is the Army interested in mind to mind communications among soldiers in the battlefield?

Recently, online news media websites are reporting that the US Army through DARPA's R&D are planning to have the infantry equipped in the very near future with a specialized helmet that will allow mind to mind communications among soldiers through what is termed synthetic telepathy.

"At least, that's the hope of researchers at the Pentagon's mad-science division **DARPA (Defense Advanced Research Projects Agency)**. The agency's budget for the next fiscal year includes $4 million to start up a program called "**Silent Talk**". The goal is to "allow user-to-user communication on the battlefield without the use of vocalized speech through analysis of neural signals." That's on top of the $4 million the Army handed out last year to the University of California to investigate the potential for computer-mediated telepathy.

Sounds familiar, does it? It should be if the reader recalls Dr. Greer's usage of the terms: CAT – Conscious Assisted Technology and TAC – Technology Assisted Consciousness!

Before being vocalized, speech exists as word-specific neural signals in the mind. DARPA wants to develop technology that would detect these signals of "pre-speech," analyze them, and then transmit the statement to an intended interlocutor. DARPA plans to use EEG to read the brain waves. It's a technique they're also testing in a project to devise mind-reading binoculars that alert soldiers to threats faster the conscious mind can process them.

The project has three major goals, according to DARPA. First, try to map a person's EEG patterns to his or her individual words. Then, see if those patterns are generalizable — if everyone has similar patterns. Last, "construct a fieldable pre-prototype that would decode the signal and transmit over a limited range."

The military has been funding a handful of mind-tapping technology recently, and already have monkeys capable of telepathic limb control. Telepathy may also have advantages beyond covert battlefield chatter. Last year, the **National Research Council** and the **Defense Intelligence Agency** released a report suggesting that neuroscience might also be useful to "make the enemy obey our commands." The first step, though, may be getting a grunt to obey his officer's remotely-transmitted thoughts." http://www.wired.com/dangerroom/2009/05/pentagon-preps-soldier-telepathy-push/

Remote Viewing, Psychotronic Warfare (Psi Wars) and Psychotronic Weaponry

Now, there have been rumours around for decades that army intelligence was determined to use mind control technology to their advantage but, the cloud of secrecy shrouding it was a mixture of truth and fiction.

Take, for example, the tongue-in- cheek movie "The Men Who Stare At Goats" from the book of the same name, starring **George Clooney** which depicts concepts like mind control used during interrogation exercises in the last three decades. This movie did nothing to educate the public about the nature of the subject or the methodology of psychic ability other than staring at a goat to kill it with malevolent thoughts. Now it can be rightfully argued that this is a movie for entertainment purposes and was not created to educate but, we have already seen the hidden agendas and motivations behind Hollywood's movie moguls and mafia where most alien movies depict **Grey ETs** as sinister beings hell-bent on malevolence toward humans. This stereotype casting of Grey ETs as the "bad guys" is deliberate to instill a subconscious fear into the general public much like looking at spiders and crawling insects or ants brings out an irrational fear in people.

If they had made the movie instead, about the work of two physicists, **Russell Targ** and **Dr. Hal Puthoff**, who helped the CIA start a **remote-viewing** research program in 1972 at the **Stanford Research Institute** in California, with a leading clairvoyant, **Ingo Swann,** this would have been a real eye-opener for public audiences into the mind and the operations of the CIA. But, for the CIA or any other intelligence agency that would be hitting too close to home!

The movie could also have focused on the work of the US Army setting up its own small remote-viewing program. A program called the **Stargate Program**... *a name that would sure to be Hollywood blockbuster movie title*... operating under orders from the Army's Assistant Chief of Intelligence in the **Pentagon,** code-named **project Grill Flame.**

The **Stargate program**, which was originally set up in 1977 to assess what intelligence information an enemy could tap into by psychic means using a small group of army personnel. These were six gifted psychics: **Mel Riley, Joe McMoneagle, Ken Bell, Fern Gauvin, Hartleigh Trent** and **Nancy Stern** who were tested by the **Stanford Research Institute** for the **Grill Flame project**.
http://www.bibliotecapleyades.net/vision_remota/esp_visionremota_9c.htm

324

But, what did movie audiences get instead, a movie about killing goats with evil malice of aforethought that did little to explain telepathy, remote viewing or **telekinesis,** other than to make it an entertaining, quasi-military illusionist' trick. But, then again, what did we really come to expect? Did we really think we were going to get real insider information into the workings of the CIA and Military mindset? Fat chance! After all, the CIA and the US Military must be seen as the good guys in the movies, not the bad guys. It would be, according to the CIA and Military way of thinking, too upsetting for the general public to know that the CIA and the Military were really no better than the bullies living down the street from them.

Parapsychology and mind control began, so we are told, with scientists in pre-Revolutionary Russia who were studying the area of parapsychology. Later in 1922, a commission composed of psychologists, medical hypnotists, physiologists, and physicists like **V.M. Bekhterev, A.G. Ivanov-Smolensky,** and **B.B. Kazhinsky** worked on parapsychology problems at the Institute for Brain Research in Petrograd (Leningrad). Work flourished throughout the thirties with research being reported in the literature in 1934, 1936, and 1937. After 1937 further experiments in the field of parapsychology were forbidden. During Stalin's time, any attempt to study paranormal phenomena might have been interpreted as a deliberate attempt to undermine the doctrines of materialism. So stated the 1972 DIA report '**Controlled Offensive Behavior - USSR**'

The US Army, through the **Defense Intelligence Agency** carried out intelligence work for the Pentagon and according to an official **CIA** paper written by **Gerald K. Haines**, the historian of the **National Reconnaissance Office** (**NRO**):

'There is a **DIA Psychic Center** and the **NSA (National Security Agency)** studies parapsychology, that branch of psychology that deals with the investigation of such psychic phenomena as clairvoyance, extrasensory perception, and telepathy.

In 1960 the Stalinist taboo that prohibited research into the paranormal was lifted and the **KGB** and **GRU (Soviet military intelligence)** began a scientific exploration of the weapons potential of psychic energy.

Soviet interest in **psi** was reawakened in February 1960 by a story which appeared in French magazine *Science et Vie* (Science and Life).
The story was entitled '*The Secrets of the Nautilus*' and it claimed that the US government had secretly used telepaths to communicate with the **first nuclear submarine** ever constructed, **the Nautilus**, while it was under the Arctic ice pack. This telepathy project involved, according to the article, **President Eisenhower**, the **US Navy**, the **US Air Force**, **Westinghouse**, **General Electric**, **Bell Laboratories** and the **Rand Corporation**. Communicating with submarines is difficult as radio waves do not penetrate to the depths of the ocean.

Extremely low frequency (ELF) waves are used to signal the submarine to come to the surface to receive a message - these super-long waves penetrate almost anything including water but carry little information - so if telepathy could work it would be a perfect method of communicating with submarines while still submerged. The story was almost certainly a *hoax* but the Soviets were spurred into action, according to the **DIA**:

Ship-to-shore telepathy, according to the French, blipped along nicely even when the Nautilus was far under water.

'Is telepathy a new secret weapon? Will ESP be the deciding factor in future warfare? Has the American military learned the secrets of mind power?'

In Leningrad, the Nautilus reports went off like a depth charge in the mind of **Dr. L.L. Vasilev.** In April of 1960, Doctor Vasilev, while addressing a group of top Soviet scientists stated:

"We carried out extensive and until now completely unreported investigations under the Stalin regime. Today the American Navy is testing telepathy on their atomic submarines. Soviet scientists conducted a great many successful telepathy tests over a quarter of a century ago. It's urgent that we throw off our prejudices. We must again plunge into the exploration of this vital field."

Psychotronics is a term coined in 1967 by **Zdeněk Rejdák** for the study of parapsychology. Extensive research programs and numerous conferences into the field during the 1970s and 80s sparked Cold War fears of mind control and other psychotronic weaponry being developed by Eastern Bloc countries which led to the popularization of the term in the West.

Since the mid-1990s, rumors of secret research into psychological warfare have led to a number of conspiracy theories. Campaign groups in Russia and the US have alleged that their governments are using psychotronic weapons against them to torture them, track their movements or control their minds. These campaigns are typically dismissed by psychologists as being a delusional response to auditory hallucinations similar to accounts of alien abductions. http://en.wikipedia.org/wiki/Psychotronics

This response by psychologists is a typical disinformation and spin doctoring campaign to minimized the information and marginalize the programs from the inquiring public who may want to investigate further.

In fact anytime sensitive information from a secret black program is leaked into the public sector by an enthusiastic news reporter or by an insider, there is sometimes a free fall of information before damage control is implemented to contain the veracity of the leak. There are threats, intimidation, and punishment to those who leak information to the public.

As part of a surge in research into parapsychology during the 1970s, regular conferences were held in Eastern Europe and the former Soviet Union although the word parapsychology was discarded in favour of the term psychotronics. The stated objectives of psychotronics were to verify telepathy, clairvoyance, and psychokinesis in order to discover new principles of nature. One significant promoter of psychotronics was Czech scientist **Zdeněk Rejdák**, who promoted psychotronics as a physical science on the world-wide scale for many years, organizing conferences and presiding over the **International Association for Psychotronic Research.**

Against the background of the Cold War, this research sparked concerns in the US that Eastern Bloc countries were successfully developing mind control technology and other psychotronic

weaponry, with one report studying so-called "psychotronic generators" developed by the Czech researcher **Robert Pavlita**. Pavlita created devices which were "allegedly able to amass human mental energy and release it mechanically or electromagnetically". A report from 1975 the United States' **Defense Intelligence Agency** took the device seriously as a potential weapon, reporting that "when flies were placed in the gap of a circular generator, they died instantly" and that Pavlita's daughter had become dizzy when the device was pointed at her from a distance of "several yards". These fears diminished as it proved impossible to replicate Pavlita's machines and he died in 1991 without revealing how they had worked.
http://en.wikipedia.org/wiki/Psychotronics

In Russia, a group called "**Victims of Psychotronic Experimentation**" attempted to recover damages from the Federal Security Service during the mid-1990s for alleged infringement of their civil liberties including "beaming rays" at them, putting chemicals in the water, and using magnets to alter their minds. These fears may have been inspired by revelations of secret research into psychological warfare during the early 1990s, with Lopatkin, a State Duma committee member in 1995, surmising "Something that was secret for so many years is the perfect breeding ground for conspiracy theories."

In the US, there are a growing number of people who hear voices in their heads that claim the government is using *"psychotronic torture"* against them, and who campaign to stop the use of alleged psychotronic and other mind control weapons. These campaigns have received some support from *government representatives* including **Dennis Kucinich** and **Jim Guest**. Yale psychiatry professor **Ralph Hoffman** notes that people often ascribe voices in their heads to external sources such as government harassment, God, and dead relatives, and it can be difficult to persuade them that their belief in an external influence is delusional.
http://en.wikipedia.org/wiki/Psychotronics

Again, the spin-meisters belittle the outcries and concerns of the public as just delusional but, reality is often framed by such outlandish absurdities that should not exit but do!

In the late 1990s, the **American Psychological Association (APA)** had to review its list of disorders in which people reported voices in their head, they had to include a new entry to this list based upon **psychotronic warfare**. When a person reported that they heard voices in their heads, they may actually be hearing voices in their head because they were being targeted by some government agency with psychotronic weaponry. They were not being delusional but relating a real event.

Recall **Martin Cannon's** paper *"The Controllers"* discussed earlier in this book which described CIA's secret mind control program **MK Ultra** which used brain implant transmitters and microwaves RF transmitters upon unsuspecting victims as a means of manipulating their minds and controlling their actions.

Cannon recounts the **Pascagoula Incident** involving **Charles Hickson** and **Kevin Parker,** two fishermen who became the unsuspecting targeted victims of a CIA psychotronic weapon incident which essentially resulted in one of the best news media stories of an alleged UFO and ETI encounter or "abduction". It was later determined through Martin Cannon's investigations that the whole UFO account that was an intelligence agency fabricated story induced through psychotropic drugs and mind altering RF microwaves using a portable, handheld psychotronic emitter. In other words, the whole story was a deliberate misleading account about UFOs and ETI to discredit the whole UFO/ETI phenomenon.

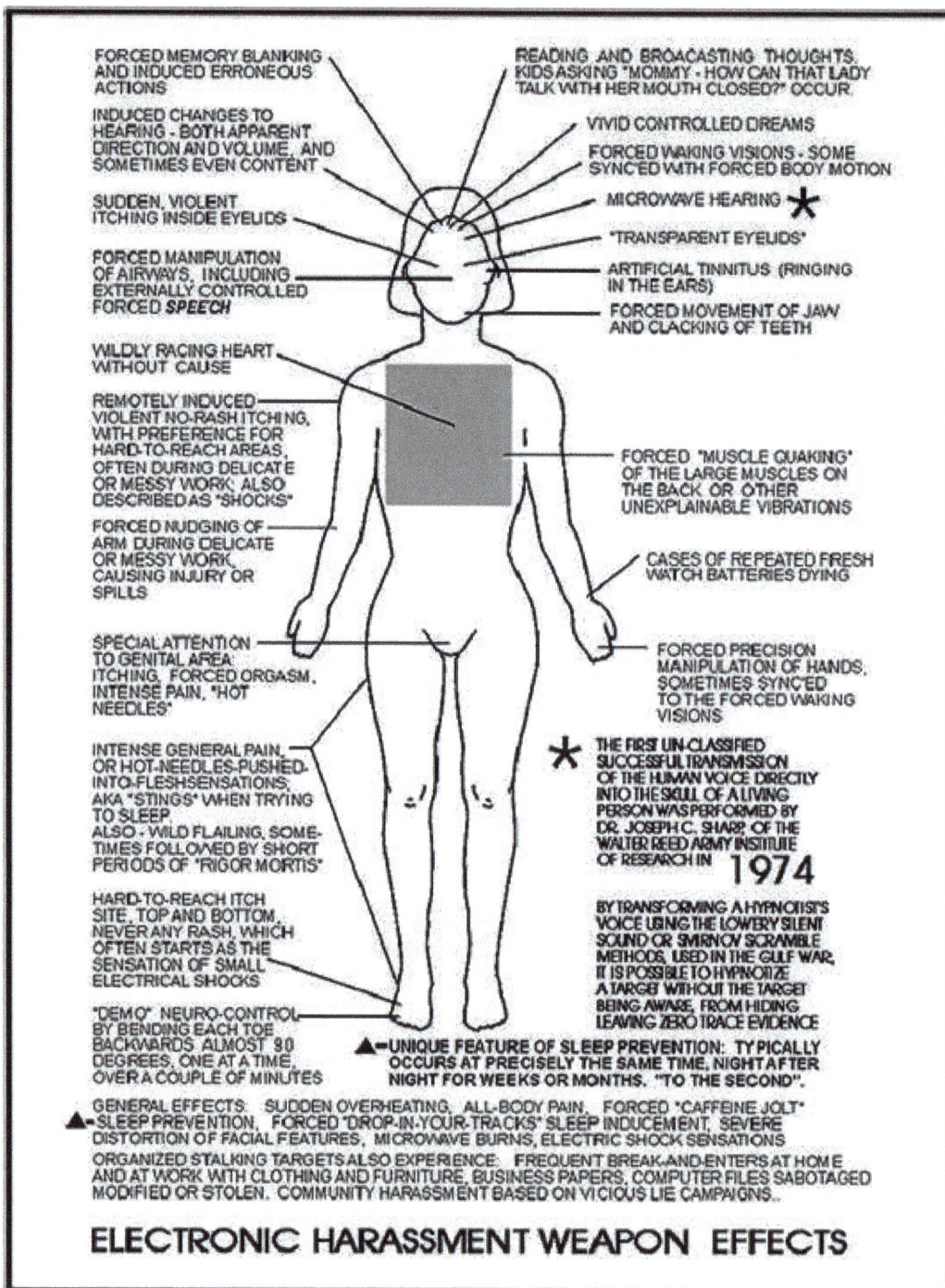

**Targeted areas of the human body affected by an
electronic harasment psychotronic weapon**
http://www.wakeupkiwi.com/images/electronic-harassment-body.jpg

These telepathic inducing devices exist, no question about it, they have been around since the '60s and perhaps, even earlier!

Author's Rant: Back in the summer of 2001, I once worked with one of Dr. Steven Greer's Disclosure Witnesses, Alfred Webre on planning a Disclosure event in Vancouver, BC for Dr. Greer and some of his other Disclosure Witnesses. He confided in me that he had been targeted with one of these psychotronic weapons on more than one occasion while working with the President Carter administration in developing the Space Preservation Treaty and the Space Preservation Act. The effect was that he heard voices in his head and felt light-headed or dizzy from the experience.

Conspiracy groups use news stories, military journals and declassified national security documents to support their allegations that governments are developing weapons intended to send voices into people's heads. Psychotronic weapons were reportedly being studied by the Russian Federation during the 1990s with military analyst **Lieutenant Colonel Timothy L. Thomas** saying in 1998 that there was a strong belief in Russia that weapons for attacking the mind of a soldier were a possibility, although no working devices were reported.
http://en.wikipedia.org/wiki/Psychotronics

Such reports like these are a typical ploy that American intelligence agencies used to deflect interest away from their own psychic research programs claiming that they need to start their own catch-up programs to be on par with the Russians. In truth, the US has had such programs in operation for a long time and this tactic is a way to find out by confirmation or denial from the other side, the degree of their involvement in such programs. It is a sort of testing the waters of experimental research programs; finding out what each side knows is a game all super powers play.

In 1987, a **U.S. National Academy of Sciences** report commissioned by the **Army Research Institute** noted psychotronics as one of the "colorful examples" of claims of psychic warfare that first surfaced in anecdotal descriptions, newspapers, and books during the 1980s. The report cited alleged psychotronic weapons such as a "hyperspatial nuclear howitzer" and beliefs that Russian psychotronic weapons were responsible for Legionnaire's disease and the sinking of the USS Thresher among claims that "range from incredible to the outrageously incredible". The committee observed that although reports and stories, as well as imagined potential uses for such weapons by military decision makers, exist, "Nothing approaching scientific literature supports the claims of **psychotronic weaponry**".

This last statement is not true as **Colonel Tom Bearden,** a nuclear engineer was asked by the US Army to investigate this area of Russian psychical research. By then, the **DIA** were discussing Soviet psychokinesis at length:

All the Soviet and Czech research on PK is significant, especially that associated with the spectacular **Soviet psychics Kulagina**, **Vinogradova,** and **Ermolayev.**

- Kulagina's highly publicized ability to affect living tissues might be applied against human targets

- in like manner, Vinogradova's power to move objects
- Ermolayev's levitational ability could possibly be used to activate or deactivate power supplies or to steal military documents or hardware

Col. Bearden talks extensively in many of his books and lectures about the use of **Scalar and Psychotronic Weaponry** by the Russians in their war against Afghanistan, before the Americans undertook their war, much later with the same country. The US and the Russians, as well as a few other countries, have these types of weapons; they have been around probably since the '80s and perhaps earlier. When you throw enough money at a project or program, anything is possible and technological breakthroughs occur all the time, it's just that the public never hears about, until 20 or 30 years later and sometimes, never!

HAARP as previously discussed much earlier in this book is a prime example of the employment scalar physics to alter weather, as a means of communications with submarines and in the use of mind control over large populations of society. The statement, "Nothing approaching scientific literature supports the claims of psychotronic weaponry" is nothing more than a bold face lie created as a cover for their own psychotronic weaponry programs.

In 2012, **Russian Defense Minister Anatoly Serdyukov** and **Prime Minister Vladimir Putin** commented on plans to draft proposals for the development of psychotronic weapons. **NBC News Science Editor Alan Boyle** dismissed notions that such weapons actually existed, saying, "there's nothing in the comments from Putin and Serdyukov to suggest that the Russians are anywhere close to having psychotronic weapons." **http://en.wikipedia.org/wiki/Psychotronics**

If telepathy and other paranormal phenomenon are just pseudo-sciences as posited by the science community then, why are the militaries of the most powerful nations spending millions or billions of dollars on research and development programs into this phenomenon?

By now, the reader should be well aware of the methods used by the military and intelligence departments that discredit or lie, deny, misinform or spin doctor disinformation in order to confuse the public as to what is really going on in covert programs. Particularly with projects and programs that employ potentially offensive weapons against enemy nations or has the potential to be used against its own people.

The point here is that everyone needs to independently investigate for themselves to determine the truth of a matter and not simply rely upon so-called "trusted sources" to get the "honest truth" from the information they seek.

The question that still needs to be answered, is telepathy an effective way to communicate with Extraterrestrial Intelligences? It is obvious that various forms of telepathy and its associated psychic perceptions are being utilized by the intelligence communities and the military departments, so they must be efficacious but, have they been used as a means to communicate with ETI? What evidence is there to support telepathy with Extraterrestrials?

The evidence for telepathy with Extraterrestrials comes from accounts of individuals who have had a close encounter with an unidentified flying object which has either landed, hovering or is in very close proximity to the individual and/or there is the appearance an extraterrestrial biological entity also in close proximity to the individual where contact with the EBE is imminent. This is the preferred method for contact and communication with Extraterrestrial Intelligence, however, it is a seldom the way such events unfold.

People have reported receiving telepathic messages in many cases from non-visible entities that according to the "contactee" or experiencer, the entity identifies itself as an extraterrestrial being or angel or demonic being. In these circumstances, the proof of evidence is missing and cannot be verified without some physical proof, however, if there are more than one witness, preferably three or more who all claim a communicative contact with the same being then, this subjective, circumstantial evidence may be tested with a polygraph test of the individuals involved in the event to determine its validity.

Information gleaned from the recall of the individual and the other participants can be analyzed for subject content to see if the educational, mathematic and scientific information was imparted or if there were admonitions or threats, or devotional expressions of brotherly love and joy or possibly invitational greetings to participate in some major event.

Because no ETIs showed their presence to the contactee does not mean there was no experience or that no telepathic message transmitted. Frequently, telepathy falls into the category of "high strangeness" often associated with ET craft or their presence or some other related event. A typical event that has been reported states that the witness or experiencer saw a craft land or already on the ground, saw one or more ET beings outside, around the craft and was either approached by the ET beings or the witness was given an invitation to come aboard the spacecraft for a possible off planet experience. All this was conveyed without any auditory-vocal exchange between ETs and the human but, transmitted by telepathy in which the ET beings sent a mental message to the human and the human clearly "hears" the mental thoughts of the ET. In other words, the informational exchange was through telepathy, mind to mind, whereby the sender transmits a mental message and the recipient clearly receives and understands it!

Extraterrestrial Intelligence as reported historically in the UFO literature when encountered either physically or in some out-of-phase, holographic form (inter-dimensional or altered form) appear to be very great transmitters of telepathic information, whereas humans appear to be good receivers of this telepathic information. Human, on the other hand, are not considered very good as yet, in sending messages telepathically, although this is open for debate. There is ample reported evidence to support that when there have been UFOs in the vicinity of humans when they were flying rapidly by, frequently when the observer was extremely curious and wanted to get a second look and wished the craft would re-appear or come closer, the UFO would respond to the individuals mental thought or wish and re-appear then, it would fly off into the distance.

This is a strong intention by humans sent mentally where ETs aboard their craft have responded to those intentions whether spoken or in many cases unspoken. It, therefore, appears that humans can both send and receive mental thoughts via telepathic communications, sometimes naturally without much mental exertion and at other times it is a necessary learned and practiced ability

These are skills that can be learned by anyone given sufficient knowledge and practice. It is like a muscle that has not been sufficiently used which becomes atrophied but, with a little exercise it can become healthy and rejuvenated, reaching its full potential.

There are two classic examples known in the UFO literature and among Ufologists of this telepathic ability been deliberately used to communicate with Extraterrestrials. Both examples use telepathy either singularly by an individual or more often in group settings. Two well-known people have emerged over the years from the global UFO community with this telepathic ability in very dynamic and profound ways. They are **Sixto Paz Wells** and **Dr. Steven M. Greer**!

Most people know or have heard about Dr. Steven Greer and CSETI, where he has successfully lead teams of people out into remote desert areas to establish contact and communications with Extraterrestrial Intelligences visiting the Earth.

Remote Viewing (RV) and the Stargate Project

Remote viewing (RV) is a mental faculty that allows a perceiver (a "viewer") to describe or give details about a target that is inaccessible to normal senses due to distance, time, or shielding. For example, a viewer might be asked to describe a location on the other side of the world, which he or she has never visited; or a viewer might describe an event that happened long ago or may happen in the future; or describe an object sealed in a container or locked in a room; or perhaps even describe a person or an activity; all without being told anything about the target -- not even its name or designation.

From this explanation, it is obvious that remote viewing is related to so-called **psi** (also known as "psychic" or "parapsychological") phenomena such as clairvoyance or telepathy. Whatever it is that seems to make it possible for human beings to do remote viewing is probably the same underlying ability that makes such things as clairvoyance work. http://www.irva.org/remote-viewing/definition.html

332

**Remote viewing is the ability to see things at distance with the mind
while in a meditative state of higher consciousness**
(Google Images)

It is a matter of historical record that **remote viewing (RV)** has been used operationally in the past with considerable success by the U.S. Army and the CIA for espionage purposes, There has been a lot written in science papers, journals and in the UFO literature about such remote viewing programs. Apart from the negative reports and the debunking that has come out from some government agencies as well as from professional skeptics like **James Randi**, there has, however, been an official and largely positive evaluation of a significant part of the government's early **remote-viewing program**. Current levels of governmental support for remote-viewing research and operations in the United States are not publicly known, although interest in its potential military and intelligence application has never really subsided thus, utilizing remote viewing for espionage purposes continues today, both in the U.S. and elsewhere.

Currently, there is a great deal of research being done globally to better understand remote-viewing processes. Understanding the underlying mechanism of psi functioning has kept researchers busy trying to develop theories that explain various known and repeatable phenomena associated with remote viewing.

There are many types of phenomena that we still consider to be false precepts but, have since been proven to be a true reality long ago. This has been the basis of the great divide between the covert black world of science and the current world of white science. These realities have far-reaching implications, and shedding light on them further confirms new concepts of reality that are yet to receive the attention they deserve in our scientific community. If we come across information of a metaphysical or paranormal nature that intrudes into the realm of physical sensory perceptions, does it cease to become scientific? Absolutely not! Humanity is steadily acknowledging the non-physical nature of reality which quantum physics has confirmed the in past years.

"The day science begins to study non-physical phenomena, it will make more progress in a decade than in all the previous centuries of its existence" – **Nikola Tesla**

Science has studied non-physical phenomena before, unfortunately, a majority of it remains classified with many discoveries associated with non-physical phenomena continuing to remain unknown by the general public but known in the classified world. Multiple studies and research confirm the validity of remote viewing and other paranormal phenomena, concepts that have been demonstrated and proven in multiple laboratories globally.

Some background history is needed to understand what remote viewing is and how it is used. In 1995, the CIA declassified and approved the release of documents related to the **Stargate Project,** a $20 million research program sponsored by the US government in 1975 in order to determine any potential military application of psychic phenomena, like the remote viewing program that was conducted at the **Stanford Research Institute** in Menlo Park, California. The program was designed to determine if agencies like the CIA could use such phenomena for "intelligence collection." The research conducted by Stanford University and the CIA lasted for decades and confirmed the fact that the intelligence community has a high interest and involvement in parapsychological phenomena. http://www.princeton.edu/~pear/pdfs/1979-precognitive-remote-viewing -stanford.pdf and http://www.lfr.org/lfr/csl/library/AirReport.pdf

The term remote viewing was originally coined by physicists **Russell Targ** and **Harold Puthoff**, parapsychology researchers at **Stanford Research Institute**, to distinguish it from clairvoyance which became a synergy created between *telepathy* and *clairvoyance.* It is like a psychic version of *"I spy with my little eye something beginning with the map co-ordinates..."* The monitor in this psychic-spying game travels mentally to that specific location, and the guesser attempts to obtain a mental image of that location and then sketches what he sees. The Stargate program, however, was terminated in 1995 after it *allegedly* failed to produce any useful intelligence information. http://en.wikipedia.org/wiki/Remote_viewing

Russell Targ and Harold Puthoff , parapsychology researchers at Stanford Research Institute who worked for the CIA Stargate Program

https://redice.tv/red-ice-radio/inception-of-remote-viewing-and-the-reality-of-esp
and https://www.youtube.com/watch?v=EoNOBLDeI2o

The study was compromised of department of defence personnel and psychic-gifted individuals like **Ingo Swann**, **Mel Riley, Joe McMoneagle, Ken Bell, Fern Gauvin, Skip Atwater, Hartleigh Trent** and **Nancy Stern** and **Uri Geller,** just to name a few. Research was conducted by Russell Targ and Harold E. Puthoff, the physicists who founded the 23-year long study at Stanford University alongside Defence Intelligence Agencies and Army Intelligence.
http://www.princeton.edu/~pear/pdfs/1979-precognitive-remote-viewing -stanford.pdf and
http://www.lfr.org/lfr/csl/library/AirReport.pdf and
http://www.scientificexploration.org/journal/jse_10_1_puthoff.pd

Ingo Swann was the **Stargate Program's** star remote viewer who helped developed a new form of remote viewing employing geographical coordinators **(CRV - Coordinate Remote Viewing).** It helps increase Ingo Swann's efficiency to meaningful levels, and the CIA became interested enough to increase their initial funding of the project. When Puthoff gave Swann the co-ordinates of a place just east of California's Mount Shasta, the psychic's response was, *'Definitely see mountain to south-west, not far, also east.'* The co-ordinates of a point 20 miles east of Mount Hekla volcano in southern Iceland produced: *'Volcano to south-west, I think I'm over ocean.'* When Puthoff gave the co-ordinates of the middle of Lake Victoria in Africa, Swann described: *'Sense of speeding over water, landing on land. Lake to west, high elevation.'* Puthoff thought Swann had described the target inaccurately until he consulted the Times Atlas of the World and found his co-ordinates were those of the Tanzanian village of Ushashi, some 30 miles inland from Lake Victoria's south-eastern shore.

Ingo Swann – Remote Viewer Extraordinaire
http://www.dailygrail.com/Mind-Mysteries/2013/2/Vale-Ingo-Swann-Remote-Viewing-Pioneer

Ingo Swann talks of an incident that occurred between 1975 and 1976 when he was asked to remotely view Soviet submarines:*

'This was one of those "big test" things that went on, with witnesses, and the room was filled with top brass. Oh my God! Hal, I don't know what to do. I think that this submarine has shot down a **UFO** *or the UFO fired on her. What shall I do? And Puthoff was as pale as anything you know, and he looked at me and whispered, "Oh Christ! It's your show. You do what you think you should do."*

So I sketched out this picture of this UFO and this brass (two or three star general) sitting on my right grabbed it and said, "What's that, Mr. Swann?" I said, "Sir, I think it's rather obvious what that is." And he took the paper and stood up, and when he stood up, everybody else stood up except me and Puthoff and he walked out of the room, and so did the others. So Puthoff and I went back to the hotel and I said, "Oh Christ, we've blown the program." So we went out and got drunk on margaritas and things like that. Three days later Puthoff got a call. The call said, "OK, how much money do you want?"'

* **Ingo Swann interview on 'Dreamland' transcribed organization, University of Wisconsin, 12 December 1996. Quoted from "Remote Viewing and the US Intelligence Community" Armen Victorian (Lobster magazine June 1996 No. 31)**

Results such as these enabled **Puthoff** to get funding from the CIA *Technical Services Division in the Directorate of Operations* and from the CIA *Office of Research and Development. Successful replication of this type of remote viewing in independent laboratories has yielded considerable scientific evidence for the reality of the (remote viewing) phenomenon. Adding to*

336

the strength of these results was the discovery that a growing number of individuals could be found to demonstrate high-quality remote viewing, often to their own surprise. The CIA even participated as remote viewers themselves in order to critique the protocols. CIA personnel generated successful target descriptions of sufficiently high quality to permit blind matching of descriptions to targets by independent judges. http://www.lfr.org/lfr/csl/library/AirReport.pdf

After confirming the reality using objects hidden in envelopes and identifying physical characteristics at locations a few hundred kilometres away, **Ingo Swann** suggested carrying out an experiment to remote view the planet Jupiter.

It just so happens that the NASA pioneer 10 was about to make a flyby of the planet. Before the flyby, Ingo was able to view a specific ring around Jupiter before NASA was able to take pictures of it. Mr. Swann was correct. As he had claimed, he successfully remote viewed the ring around Jupiter. This result was published by Stanford University in advance of the rings discovery . Many other anomalies of Jupiter were described before being given scientific substantiation.

The Jupiter experiment was not to be an official one. It was wrapped up in very stringent protocols, but the remote viewing raw data had to be recorded somehow so that it could be established that it existed prior to the NASA vehicles getting to the planet. The raw data were circulated far and wide, offered to and accepted by many respected scientists in the Silicon Valley area, including two at Jet Propulsion Laboratories. - Ingo Swann
http://ia600605.us.archive.org/30/items/PenetrationTheQuestionOfExtraterrestrialAndHumanTelepathy/Penetration_Ingo_Swann.pdf

The intelligence services are heavy players, they required an active picture of Psi potentials somewhat larger than standard parapsychology could provide. Because of these unusual circumstances, I got dragged into realms of often idiotic secrecy, into endless security checks conducive of paranoia, into all kinds of science fiction dreamworks, into intelligence intrigues whose various formats were sometimes like toilet drains, and into quite nervous military and political ramifications - **Ingo Swann**
http://ia600605.us.archive.org/30/items/PenetrationTheQuestionOfExtraterrestrialAndHumanTelepathy/Penetration_Ingo_Swann.pdf

Ingo was taken to mysterious underground places, transported and blindfolded. He encountered some very shady intelligence officers and was subject to mistreatment and intimidation like tactics. He was examined, tested and also used in an experiment to remote view the moon. After all of his work within the intelligence community, Ingo was dumbfounded. He successfully remote viewed objects, structures, and bases on the dark side of the moon. Presumably, the intelligence community was already aware of this phenomenon. Ingo gave reference to the fact that the intelligence community did not need his input.
http://ia600605.us.archive.org/30/items/PenetrationTheQuestionOfExtraterrestrialAndHumanTelepathy/Penetration_Ingo_Swann.pdf

It's one thing to read about UFOs and stuff in the papers or in books. It is another to hear rumours about the military or government having an interest in such matters, rumours which say

they have captured extraterrestrials and downed alien spacecraft. But it's quite another matter to find oneself in a situation which confirms everything. I found towers, machinery, lights buildings; humanoids busy at work on something I couldn't figure out.
http://ia600605.us.archive.org/30/items/PenetrationTheQuestionOfExtraterrestrialAndHumanTelepathy/Penetration_Ingo_Swann.pdf

Stanford University and the Intelligence community are not the only entities to confirm the remote viewing reality. An evaluation of remote viewing and its research applications was prepared by the **American Institute for Research (AIR).** They outline how remote viewing could serve as a tremendous potential utility for the intelligence community. A panel was assembled of multiple professors from multiple universities. They examined approximately 80 separate publications and determined that they provided very strong evidence for the remote viewing phenomenon.

A study published by the Princeton Engineering Anomalies Research department also confirmed the reality of remote viewing

"As the program expanded, in only a very few cases could the clients' identities and program tasking be revealed." http://www.princeton.edu/~pear/pdfs/1979-precognitive-remote-viewing-stanford.pdf and http://www.lfr.org/lfr/csl/library/AirReport.pdf

Remote viewing programs have been validated on numerous occasions, yet there is a tremendous amount of secrecy surrounding its findings. The department of defence loves to classify information. Fortunately, the intelligence gathered from the practice of remote viewing has the potential to lift the lid of secrecy that continues to plague planet Earth
http://www.themindunleashed.org/2013/10/shocking-discoveries-made-studies.html

In the late 1970s and early 1980s, another psychic spy, Joe McMoneagle was known as Remote Viewer 001 "Remote Viewer No. 1" in the US Army's psychic intelligence unit at Fort Meade, Maryland which used soldiers with psychic talents to peer across borders and spy on enemies.

McMoneagle has been featured in *Newsweek, Time, Reader's Digest*, and on ABC's *Nightline* and CBS's *48 Hours*, and on prime-time British and Japanese television. He's the author of a number of books on remote viewing. He now teaches remote viewing at **The Monroe Institute** in Virginia, which was started by his father-in-law, author and consciousness researcher **Robert Monroe**.

In October 1965, **McMoneagle** witnessed a UFO while stationed in the Bahamas. Later, in the 1980s, he remote viewed the face on Mars and claims to have contacted **Martians** who were trying to survive an environmental cataclysm of their planet.
http://www.bibliotecapleyades.net/sociopolitica/hambone_info/People1.html#Joseph_McMoneagle

When remote viewers weren't engaged in psychic spying on the Russian s or on other countries, they would hone their skills by remote viewing other things, places and events such as the Moon, Mars and even, UFOs around the Earth.

**Some of the Remote Viewers of the Stargate Project (left to right and top to bottom),
Mel Riley, Joe McMoneagle, Ingo Swann, Lyn Buchanan,
F. Homes "Skip" Atwater and Pat Price**

http://www.oocities.org/remoteviewerorg/remoteviewing/riley.htm and http://redwheelweiser.com/detail.html?id=9781571741592
and http://www.crviewer.com/lyn.php and http://bestebookreviews.blogspot.ca/2012/07/skip-atwaters-captain-of-my-ship-master.html

With the help of **F. Holmes "Skip" Atwater,** another psychic spy who initiated the remote viewing intelligence program now known to the world by the code name **Stargate,** McMoneagle decided to hone his remote viewing skills on a non-intelligence RV exercise. Atwater serviced as his guide:

Holmes: Using the information in the envelope, focus on 40.89 degrees North, 9.55 degrees West.

Joe: I see what looks like a, no it sort a looks. I've got an oblique view of a, ah pyramid, a pyramidal form. Um, it's very high. It's kind a sitting in a large depressed area.

Holmes: All right.

Joe: It's yellowish, ah ocher-colored.

Holmes: All right. Move in time to the time indicated in the envelope I've provided you and describe what's happening.

Joe: Got an impression of severe, severe clouds, more like dust storm, a geologic problem. Um, seems to be like a, uh... Just a minute, I've got to iron this out. It's a little weird.

Holmes: Just report your raw perceptions at this time. You're still early in the session.

Joe: I'm looking at an after-effect from a major geologic problem.

Holmes: Okay. Go back to the time before the geologic problem.

Joe: Um, total difference. It's a, before there's no a, a no... Oh hell. It's like mounds of dirt appear and they disappear when you go before. I see a, large flat surfaces — very a, smooth — angles, walls. They're really large though. I mean they're megalithic...

Holmes: All right, all right. At this period in time now, before the geologic activity, look around — in and around, this area. See if you can find any activity.

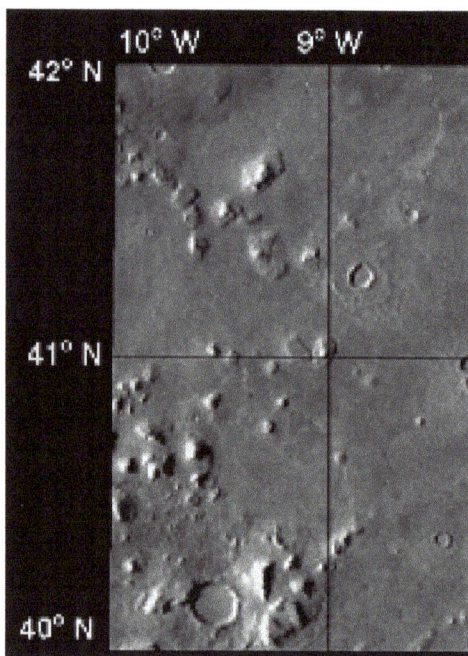

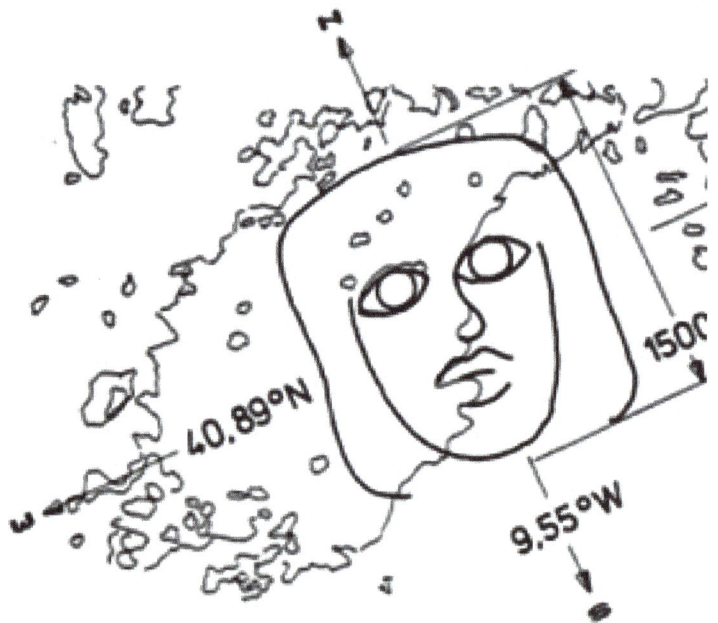

Martian reconnaissance photo from USGS (left) and "Mars Face" structure located in Cydonia on 40.89 deg. North and 9.55 deg. West on Mars (right)

**Actual 1976 NASA photo image of the Cydonia Region
and the Face on Mars, in upper right corner**
http://humansarefree.com/2016/12/3-ex-nasa-scientists-claim-giant-face.html

Joe: I'm seeing a... it's like *a perception of a shadow of people. Very tall, thin but it's only a shadow*. It's as if they were there and they're not, not there anymore.

Holmes: Go back to a period of time where they are there.

Joe: Impression of... It's like I get a lot of static on the line and everything. It's a, breaking up all the time. Very fragmentary pieces.

Holmes: Just report the raw data. Don't try to put things together. Just report the raw data.

Joe: *I keep seeing very large people. They appear very thin and tall but they're very large. Ah, wearing some kind of strange clothes*.

Joe went on to describe *eight different locations on the planet Mars* and was actually able to communicate "telepathically" with one of the beings he encountered.
http://www.bibliotecapleyades.net/vision_remota/esp_visionremota_35.htm

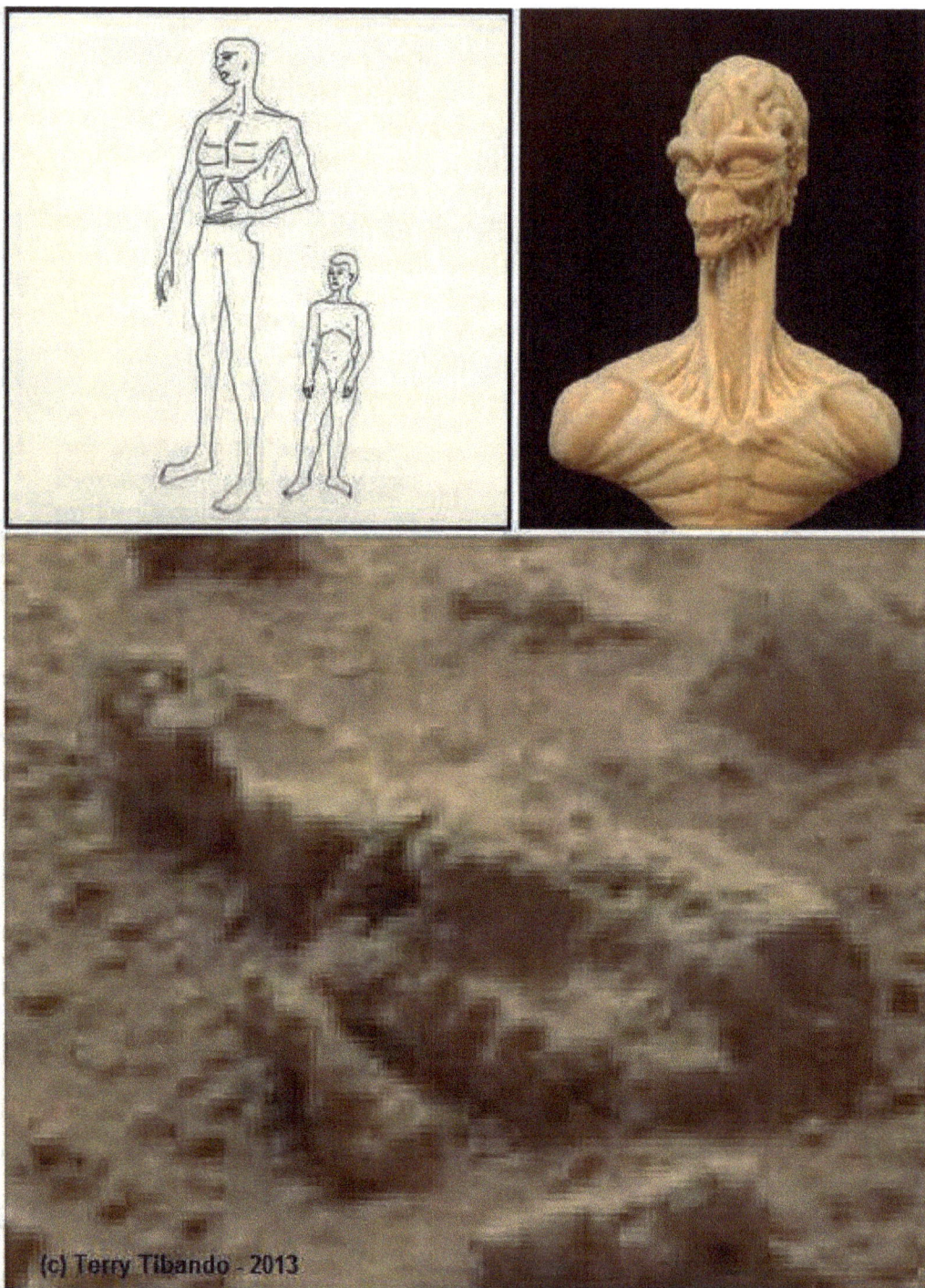

Joe McMoneagle's drawing (left) of a tall Martian being, an artist conception in bust form (right) of a Martian and this author's discovery of a tall Martian corpse (bottom) found on February 2013 from raw NASA photographic images
(c) Terry Tibando

McMoneagle gives a more detail account from a 1984 transcript of the above session which was obtained by **Rob and Trish McGregor** in an interview for their book *The Synchronicity Highway*: *Exploring Coincidence, the Paranormal, & Alien Contact.*

Here are McMoneagle's comments:

"On this one occasion, I was taking a nap during lunch hour inside the controlled isolation chamber in the lab, when Bob woke me up by announcing that he had a target for me. Lieutenant Atwater had brought him a card with seven sets of coordinates on it, and an envelope which was sealed. Bob told me he had the target envelope in his shirt pocket and that he would read off the coordinates to me one at a time, and I was to describe what I saw at each set. I agreed.

What I remember is that the first coordinate was a huge pyramid, like none I'd ever seen before. I asked him if this was a new discovery because it seemed this was larger than the one at Giza, Egypt. He said he didn't know, all he had was the sealed envelope and the coordinates. So, I described it to him. He gave me another coordinate and this one appeared to be some kind of a ruin. And on it went.

I remember at one point looking up at the location and getting a very strange impression of the sun. I told Bob, "The sun, it looks very weird."

He said; "I'm not interested in the sun, I'm interested in what's at the coordinate."

So, onward we went. At the end of the session neither he nor I could figure out what this target was – it was mostly ruins, a few pyramid shapes, and feelings like the whole thing had to do with the preservation of life, the need to pass along a great deal of information.

I began seeing a race of people who were very much like us, but much larger – like, huge larger – over ten feet in height. And these people were fighting to stay alive, were building hibernation chambers inside pyramids, and trying to put aside information for those who might come later, informing them of what went wrong.

In any event, when we finished the remote viewing effort, **Skip Atwater** asked Bob to open the envelope and tell us what was inside. The card within the envelope said; "MARS ONE MILLION BC." The coordinates were for specific locations on a certain area of Mars, which included what appeared to be ruins, lots of pyramids of different shapes and designs. I asked Skip where the coordinates had come from. He said they originated with the **Jet Propulsion Laboratory** (NASA).

When I was doing the viewing, I kept getting a really sad feeling – these people were losing their home, and a handful had volunteered to stay behind to try and set up messages for those who might come after them. I got the distinct feeling that the pyramids were being set up to be used as hibernation chambers, and some point at some time in the near future they had some expectation that someone would eventually find them and understand what they did to save their people. It was very moving. I don't think I expected such a powerful response to the remote viewing.

In any event, when we were finished with the viewing and the discussion of the results, Bob wasasked to open the envelope. Inside we discovered the "Mars, one million BC" targeting instruction. It really surprised us both."
http://www.synchrosecrets.com/synchrosecrets/?p=17114

According to **McMoneagle**, remote viewing is possible and accurate outside the boundaries of time. He believes he has remote viewed into the past, present, and future and has predicted future events. Among the subjects he claims to have remote-viewed are a Chinese nuclear facility, the Iranian hostage crisis, the **Red Brigades**, and **Muammar Qadhafi**. He writes that he predicted the location and existence of the Soviet "Typhoon"-class submarine in 1979 and that in mid-January 1980, satellite photos confirmed those predictions. McMoneagle says the military remote viewing program was ended partly due to stigma: ***"Everybody wanted to use it, but nobody wanted to be caught dead standing next to it. There's an automatic ridicule factor. 'Oh, yeah, psychics.' Anybody associated with it could kiss their career goodbye.*** http://en.wikipedia.org/wiki/Joseph_McMoneagle and https://www.youtube.com/watch?v=IBcQ8RDIe9w

From the above-recorded session, the reader can get a basic *"feel"* of what is happening when one remote views a place or event. Each individual's session is different and unique to that person's perceptions and abilities.

What we can take away from this paranormal ability is that it can produce some remarkable insights and perceptions that are physically not possible to see from one's location particularly when the place, object or event is halfway around the world or even on another planet. Confirmations from photographs or other raw data collected through intelligence or government agencies or even from other remote viewers operating independently of each other is the key to knowing whether the remote viewer's perceptions are accurate or not. Confirmation may not come right away, it could come weeks, months or years later.

The fact is that remote viewing works as a form of communications and it has been tested and proven to work based upon many university studies that are still going on globally. Theories abound but, university studies still do not understand the mechanism of how it works, just that it works!

There does not appear to be any limits to what can be remote viewed including off planet perceptions whether in this Solar System or beyond and even, time travel perceptions of both past and future events. Remote viewing it can be seen from the above examples is a proven and tested method in seeing planetary surface geometry and in communicating with Extraterrestrials although, such type of accounts are probably classified by government intelligence agencies.

From the release of intelligence documents like the **Stargate Program** and **Project Grill Flame** into the public domain, the distillation of parapsychology with all its associated paranormal psi abilities has finally reached the UFO community

Bi-Location - Being in Two Places at the Same Time

Bilocation is the term given to the paranormal effect of being in two different locations at once which is usually triggered, either unintentionally or deliberately, often through a meditative state of higher consciousness. The physical manifestation of an ethereal body appears during an episode of bilocation, it is all those things that take on a physical form apart from the material plane of existence. Early accounts about this parapsychological subject dates back to the writings of Paracelsus, who taught the concept of the existence of astral planes and astral bodies.

344

According to Paracelsus, the astral body cannot die, as it is composed of minuscule indestructible particles which ascend to the heavens after death. Current beliefs uphold the traditional beliefs of Paracelsus' time that the astral body is also a place of extraordinary energy. Astral projection or astral travel is the ability to separate oneself from their physical body and therefore, material existence, in an astral body onto the astral plane. Some parapsychologists also view the near-death experiences as similar to astral projection and bilocation.

There is a long history of bilocation acknowledged by parapsychologists which seem to occur frequently in **Christendom**, particularly among many Catholic saints. One such individual is **Padre Pio,** who was said to have been seen in various places around the world at the same time. It is said that **Padre Pio's** bilocation ability was utilized to save people from life-threatening situations ranging from automobile accidents to stepping in to prevent suicides and bring the unfortunate back into the fold of the **Catholic Church**. What makes this all the more amazing is the fact that Padre Pio was said to have never stepped outside the environs of the monastery in which he lived. http://voices.yahoo.com/what-astral-projection-bilocation-796569.html

Bilocation seems to be the ability then, of being in two locations simultaneously either physically or astrally and interacting or communing with other people, places or events.

Author's Rant: I have personally experienced remote viewing and possibly bilocation as a side effect of the remote viewing session which was induced by achieving higher consciousness through meditation. I began my meditation and remote viewing session alone, right after breakfast in my hotel room and within minutes I had remote viewed the inside of an ET spacecraft. I seem to be looking out of a "window" or "viewing wall" at the Earth below. I could see a desert area far below straddled between two mountain ranges that somehow, with an inner knowing, this area looked vaguely familiar, though I have never seen it before from this perspective, nor have I flown over this area at any time in the past.

I suddenly realized that the desert area below was the National Sand Dunes Park situated between the Sangre de Cristo Mountains and Mount Blanca. It is a desert park area approximately ten square miles in size and using this known estimate, I judged from satellite maps of that area, that the ET spacecraft I was in was at an altitude between 150 to 200 + miles up in low Earth orbit! I continued looking out the spacecraft's "window" and I could see the ground coming up quickly as we descended toward the Sand Dunes.

The large ET spaceship landed and immediately by consciousness shifted to the outside of it, a few miles away and I noticed that it had landed on a slight angle to the desert floor as if pointing toward the Sangre de Cristo Mountains. My consciousness shifted once more, but this time, much further away perhaps, 15 to 20 miles from the craft and at this point I observe the craft lift off the desert slightly and move toward me. It moved toward me at least a couple of times and then, I suddenly came out of the remote viewing session with the feeling that this event experience was going to happen that day.

I arose and went down to the debriefing conference room in the hotel and shared my insights with Dr. Greer and the other CSETI team members. It turned out that a lady in

our team also had the exact same remote viewing experience and Dr. Greer realized at that moment that we needed to act upon this information. Long story short, we discover that the event did not unfold that day or that night, even though, we did see some ET spacecraft being pursued by USAF jet fighters. However, the following night, the event did unfold exactly as I and the lady had remote viewed!!

It turns out that we had experienced not only an accurate remote viewing event but, that we had glimpsed into the future to see this event unfold as it did!!!

Did I actually bilocate aboard the spacecraft? I don't know for sure but, it sure felt like I was there, even if it was with just my mind!

Bilocation has been claimed by many people historically and even in laboratory test conditions, **Technical Remote Viewing (TVR)** (as bilocation is sometimes known), runs the full gambit of experiences. It stretches the perception and the creditability of the RV experience and even the acceptance of its reality by scientists, yet these experiences have persisted down through the ages.

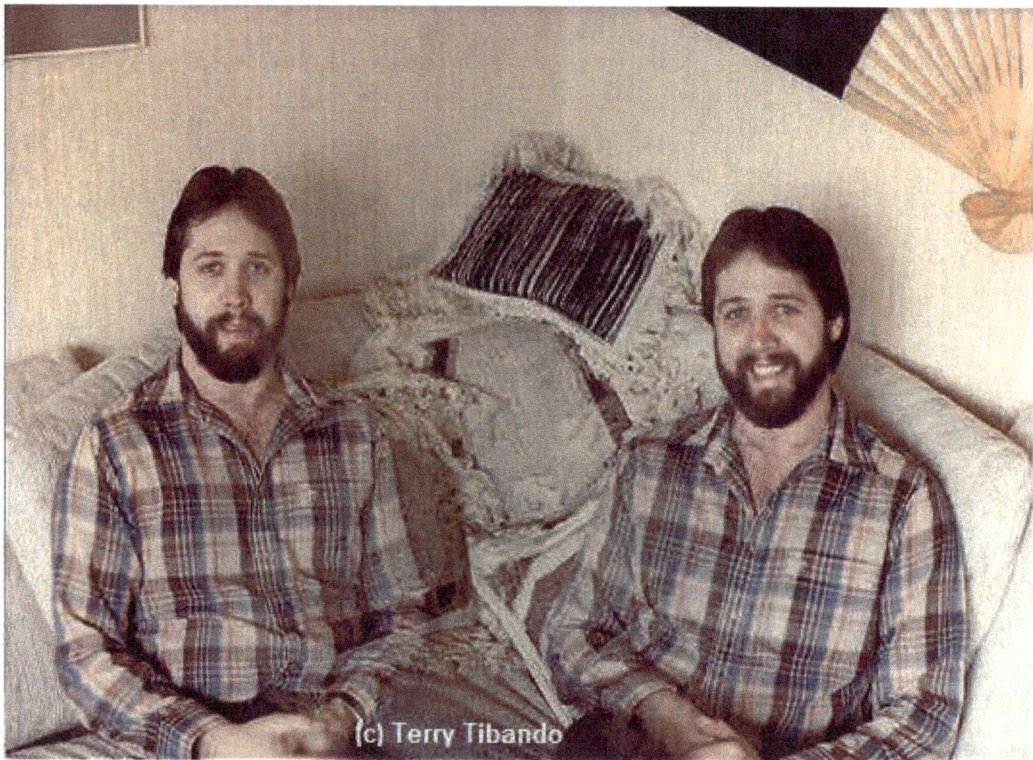

The author in earlier years, bilocating beside himself or is it trick photography?
(c) Terry Tibando

Bilocation is often associated with remote viewing and mistakenly people often refer to bilocation as remote viewing, which it is not. Bilocation historically can be a separate experience from remote viewing, although it can be frequently associated with RV.

For our purposes in exploring viable methods of communication with Extraterrestrials, bilocation may be a candidate as a communications protocol, but further research is required to understand its true nature before making its acceptance a scientific reality.

Remote Vectoring - Inviting ETI to Your Location
Via Coherent Thought Sequencing (CTS)

Another possible technique in communications with ETI is the ability to reach with through the mind into deep space via remote viewing and locate an ET spacecraft or a civilization on a planet and to invite the Extraterrestrial civilization to Earth, to your specific location through what is termed **remote vectoring.** Again, this is done in association with remote viewing and it should be understood that many of the psi abilities spring board off, one or more other psi abilities in a progressive fashion in order to achieve the desired result or effect.

Remote Vectoring is simply the ability showing and guiding a target or subject (in our case ETI) to where you are situated. The guidance system of remote vectoring is known as **Coherent Thought Sequencing** which is the logical traversing of distant places until the intended subject can see your precise present location.

This can compared to the ancient Islamic oral tradition that says, *"If Muhammad won't go to the mountain then, bring the mountain to Muhammad."* This analogy is a proactive position that is based upon the premise that intelligent life exists elsewhere in the universe, that it is interstellar and that it has entered our solar system and is currently in orbit around the Earth landing frequently on terra firma to make random contact with some of its inhabitants. Therefore, ETI have come to us, we do not need to at this time, go to them but, we can be diplomatic and invite them to communicate with us through one or more of the possible psychic methods of communication.

Sixto La Paz Wells

Sixto La Paz Wells is perhaps not as well known in North America having not received the same media exposure as has **Dr. Greer**, but in Central and South America as well as Europe, he is known as the man who calls down Extraterrestrials to communicate with on a regular basis. Sixto Paz Wells is a very well-known Ufologists and "contactee" who communicates with ETs through meditation and telepathy and thus, is able to bring them into a pre-determined location. The term "contactee" originally coined in the US and given to individuals, who claimed contact experience with ET beings, may be an unfair moniker to label Sixto Paz Wells with, as he employs a similar technique of close encounters protocols, the **human initiated CE-5** developed by Dr. Steven Greer. However, even Greer is sometimes labeled a contactee which he says he is not, but prefers the term an **"ambassador to the universe!"**

Sixto La Paz Wells is a Peruvian citizen, born in Lima, Perú on December the 12th, 1955 and is apparently a relative of **Orson Wells**, the late Hollywood actor. In 1974 he had an encounter and communication with putative extraterrestrial beings thus, Paz is considered a "putative contactee". He currently heads and directs a contactee group known as **Mission Rahma.**

Initially, during meditation on the night of January 22, 1974, Sixto with his sister and mother, received a psychographic message, which is a form of telepathic channeling through automatic, but conscious writing, from **Extraterrestrial Intelligences** telling him to take a young group of adolescent friends to travel out to an isolated area of the Chilca desert, south of Lima, Peru to keep an appointment of an unusual kind. On a February night around 9 pm in 1974, in the Chilca desert, there suddenly emerged an object full of lights from behind the dark hills, moves slowly toward the now panic-stricken group, descending to several metres above them. The friends are now terrified, but they are experiencing simultaneous communication as if spoken directly into their ears, saying they don't know how to control their emotions, and needed to prepare for the next contact. Eight months later, they try again. This time, the Manta Ray-like craft, lands, and from the interior, a being descends, some 2.5 metres in height, of Scandinavian appearance, with blue eyes, which communicated with them on a mental level. This initial experience was to set the tone of Sixto's life for years to come. http://ufodigest.com/news/0708/sixto.html

His background may have prepared him a little. His father is the founder of the **'Peruvian Institute for Interplanetary Relations'**, so Sixto grew up in an atmosphere where talk of UFOs and meetings with contactees was something of a norm. However, he had no idea that in time, he too would become as passionately involved in things extraterrestrial, and got on with life, getting married, working for a local bank, and raising a family.

According to Joseph Burkes MD, a member of the **American Public Health Association (APHA)** and a former senior team investigator for **CSETI (Center for the Study of Extraterrestrial Intelligence),** Rahma is an international organization of contact workers primarily a Spanish language based group that was established in Lima, Peru in 1974. Rahma's initial activities involved the ET contacts of a then, young man by the name of Sixto Paz Wells. Burkes was told that their group had facilitated over 25,000 individual human encounters with non-human intelligence of a presumed extraterrestrial nature.

Rahma's continuing success relied on the efforts of scores of talented Latin American contactees from many countries. Most of them were recruited into the Rahma contact network as young people in their 20s and early 30s.Their leaders were not only well versed in advanced meditation techniques but also possessed impressive psychic abilities. http://www.contactunderground.com/ and https://www.facebook.com/ContactUnderground

Sixto's main contact was **Oxalc**, who came from inside **Morlen** (aka. **Ganymede**, one of the moons of Jupiter). Sixto was subsequently to visit Ganymede, which he considered one of the pinnacle events of his life, a journey, in the physical ship, with Oxalc. Sixto maintains that his experience is not unique, but he chooses to be an ambassador and share his experiences. His wife and family likewise have had their fair share of connections with beings from other worlds.

The object of the contact, as Sixto explains is to create a bridge: to engender awareness and understanding, in an effort to bring peace and harmony to our planet. It seems these beings have, for want of a better term, a karmic link with humanity, and are here to help, to share, as much as is possible, at the current level of human evolution, information for our progression, on all levels. http://ufodigest.com/news/0708/sixto.html

Sixto La Paz Wells

Over the years Sixto has been involved in many groups, giving interviews around the world, and arranging for the world's media to be present at pre-arranged UFO meetings. People from many countries have joined him at these events, and some have been able to go through a **Xendra** or *"dimensional doorway"* as it manifests during some gatherings, appearing as a vivid blue or gold light. This *"doorway"* allows interdimensional experience, a sort of **ISP (Interdimensional Service provider (?))** to the cosmos. Those who go into it, are often seen to vanish, and return 15 or 20 minutes later, ***often feeling or experiencing as if they have been away for a couple of weeks****!*

Ganymede

There are too many elements to write of here, about Sixto's journey to Ganymede, but he noted that it's a monogamous community, where partners choose each other through aura compatibility, though he said it was difficult to know who was with whom, as everyone expressed care and love for each other. Sixto has written several books about his experiences, though mostly in Spanish, there are some English translations. The main book in English is called "The Invitation" and is of his experiences from the beginning. He also features in the excellent 2005 DVD, "The Cosmic Plan." **http://ufodigest.com/news/0708/sixto2.html**

What's interesting about Paz's case, and what makes a difference with other cases of so-called "contactees" is that Paz has invited famous journalists and writers to be present in ***pre-programmed*** encounters with UFOs. The aliens supposedly communicate to him a pre-arranged future UFO sighting event with specific date, time and specific location, whereby Sixto may invite independent witnesses to be present on that date and location). This pre-arranged or pre-programmed event permits the independent testimonies of such people to be used as objective evidence and validation for Paz's putative communication with the UFO's occupants.

For example, in 2009, Paz invited famous lawyer and TV presenter **Ana Maria Polo** (from the Telemundo's TV program called Closed Case or Caso Cerrado) and journalists to witness a pre-programmed encounter with UFOs in Chical, Peru.

350

The journalists were convinced that they saw UFOs, and even published the news in local and national newspapers, for example the journalists even filmed on video the UFOs in order to record and document the evidence. There is another interesting video where Sixto Paz, immediately after arriving at the Pascua Island (La Isla de Pascua), shows a UFO in the sky in the presence of the airport's civil and military authorities (they couldn't identify the object, which is the crucial point of the video. You can watch the airport's authorities conceding that they can see the UFO but don't know what object it is. The fraud hypothesis, in this case, is pretty unlikely since it's hard to think that Sixto or anyone else could fool the civil and military authorities in their own territory and in front of their faces).
http://subversivethinking.blogspot.ca/2010/11/sixto-paz-wells-contemporary-contactee.html
See the video here: http://www.youtube.com/watch?v=TkC7ShSdQ0I

So, Paz's case of putative contact and communication with aliens is interesting like Dr. Greer's experiences where the UFOs will either show up ahead of time waiting for the humans to arrive or when the leader of the groups like Paz Wells or Greer show up then, the UFOs or ETI suddenly make their appearance. **(This author has seen this unusual aspect occur repeatedly in CSETI lead groups, whenever he was in the presence of Dr. Steven Greer).**

A definitive opinion about the authenticity or falsehood of Sixto Paz's claims is left to those who were present with him when the ET craft showed up as pre – arranged. As *they* say, **"Seeing is believing!"**

Telepathy, therefore, is a viable method of communications with Extraterrestrial Intelligences, particularly when actual physical events unfold based upon information received mentally as in the above example of Sixto Paz Wells. Although, the reader may find it hard to accept the claims of Sixto Paz, particularly is meeting with the humanoid, Oxalc and his travels to the base inside of the Jovian moon, Ganymede, nevertheless, one must give him the benefit of the doubt with regard to his claims, when he can call upon ET spacecraft to come down to his location.
In the next section of this book, we will look at the work of Dr. Steven Greer and CSETI in greater detail. It should be stated here, that the UFO/ETI work and methods of both men are somewhat similar with the exception that Dr. Greer's work is far more ambitious and target specific. The fact that their methods and claims appear to be similar to the early claims of contactees of the '50s and '60s does not justify them being labeled contactees. The early contactees were passive in their communications with ETs, whereas Sixto Paz and Dr. Greer use human-initiated contact protocols (CE-5) to establish communications with Extraterrestrial Intelligences and are, therefore, pro-active in their approach

The Chilean newspaper Contra Corriente with front page headlines
of Sixto Paz's UFO contact and communications
(Google Images)

CHAPTER 110

THE SCIENCE OF CONSCIOUSNESS OR COSMIC CONSCIOUSNESS

What is consciousness, other than awareness, wakefulness, and sentience?!

Consciousness is the quality or state of being aware of an external object or something within oneself. It has been defined as sentience, awareness, subjectivity, the ability to experience or to feel, wakefulness, having a sense of selfhood, and the executive control system of the mind. http://en.wikipedia.org/wiki/Consciousness

It is the faculty through which the external world is apprehended usually through the sensory faculties. It is sentience usually associated with intelligence in which we are aware of awareness, whether the reality of the experience is tangible or intangible, ephemeral or eternal (the timely and the timeless), finite or infinite; whether physical or spiritual, dimensional less or multi-dimensional. It is the ability to perceive and comprehend the Nothingness.

Higher consciousness is that state of mind that is aware of a higher state of awareness. It is a higher cognitive process that presupposes the availability of knowledge and the ability to put it to use.

"In cosmic consciousness, the psyche or individual consciousness expands to a cosmic or universal level. The small personality with its identification with the body, the mind and relationships makes a radical shift so that the self now identifies with the non-local, timeless existence of the cosmos. So it is called cosmic consciousness." – **Deepak Chopra** https://www.deepakchopra.com/blog/view/1288/cosmic_consciousness

In this context, **Higher consciousness** also called *Siddhis* (a Sanskrit noun which can be translated as "perfection", "accomplishment", "attainment", or "success") which are spiritual, magical, supranormal, paranormal, or supernatural or psychic powers acquired through a **sadhana** (spiritual practices), such as Meditation and Yoga. People who have attained siddhis are formally known as **siddhas**.

It is also known as **Super Consciousness** (Yoga), **Objective Consciousness** (Gurdjieff), **Buddhic Consciousness** (Theosophy), **Cosmic Consciousness**, **God-Consciousness** (Islam, Hinduism), **Christ Consciousness** (Christian Mysticism) and **Super- Human Consciousness** are expressions used in various spiritual and intellectual traditions to denote the consciousness of a human being who has reached a higher level of development and who has come to know reality as it is (**Sanskrit**: *Yatha bhuta*). It also refers to the awareness or knowledge of an 'ultimate reality' which traditional theistic religion has named God and which **Gautama Buddha** referred to as the unconditioned element. http://en.wikipedia.org/wiki/Siddhi

Evolution in this sense is not that which occurs by natural selection over generations of human reproduction but evolution brought about by the application of spiritual knowledge to the conduct of human life, and of the refinement of the mind brought about by spiritual practices. Through the application of such knowledge (traditionally the preserve of the world's great religions) to practical self-management, the awakening and development of faculties dormant in the ordinary human being are achieved. These faculties are aroused by and developed in conjunction with certain virtues such as lucidity, patience, kindness, truthfulness, humility, and forgiveness towards one's fellow man – qualities without which, according to the traditional teachings, higher consciousness is not possible. As an inter-connected group, it is called **Collective Consciousness** in Philosophy. http://en.wikipedia.org/wiki/Higher_consciousness

Cosmic Consciousness
http://bohemian-bitchess.tumblr.com/

There is an emergence of a new science taken placing in the hallowed halls of academia of Western societies, a science of consciousness that seems to be related to cognitive psychology yet; it is also related to philosophy, neuroscience, phenomenology, and physics. While recent discoveries in neuroscience are providing us with new insights into the workings of the brain, a comprehensive science of the mind is only just beginning to emerge. However, the science of consciousness in the East as it is understood and practiced in the Vedic traditions of East Indian

354

beliefs and in many of the world's religions is nothing new, where consciousness is associated with spirit, sentience, rational mind, soul, ego, intelligence and knowledge.

Western understanding of the world segregates knowledge into sciences, disciplines, and orders of knowledge, whereas Eastern cultures have traditionally seen knowledge as also interrelated with spiritual, philosophical and metaphysical realities. If we look at any of the continental indigenous cultures of Europe, Africa, Australia, Indonesia, Asia or the Americas, whether at the communal level or at a shamanistic level, there have been the keepers of sacred knowledge, tribal elders or an elite religious order, who preserve and practice a close spiritual reverence for the Earth, the environment, and the universe.

Some people may see a divergence at this point between science and religion or mysticism; between Western thinking which see most things in a reductionist perspective, as yet to be discovered, rationalized and compartmentalized and the Eastern thinking which holds to old traditional values, philosophies, mysticism and faith in the unexplainable preserving and handing these values down from generation to generation.

That fact is that both the East and the West are in essence practicing the same values and disciplines, whether they are couched in science or spiritualism; it is in reality, they are multiple forms of higher consciousness, expressions of investigative discoveries and inspirational revelations of the divine universal knowledge.

If higher consciousness is attained through investigation, theorizing, and discovery from deduced or observable facts or in the quiet solitude of prayer and meditative reflection imploring divine enlightenment and guidance, then both methods are practicable and laudable. They have endured and have become the wellsprings of knowledge for the advancement of civilization.

Cosmic Consciousness is the idea that the universe exists as an interconnected network of consciousness, with each conscious being linked to every other. Sometimes this is conceived as forming a collective consciousness which spans the cosmos, other times it is conceived of as an Absolute or Godhead from which all conscious beings emanate.
http://en.wikipedia.org/wiki/Cosmic_consciousness

Universal Mind is the universal higher consciousness or source of being in some forms of esoteric or New Thought and spiritual philosophy. It may be considered synonymous with the subjective mind or it may be referred to in the context of creative visualization, usually with religious or spiritual themes

Universal mind may be defined as the nonlocal and temporal hive mind of all aggregates, components, knowledges, constituents, relationships, personalities, entities, technologies, processes and cycles of the Universe.

The nature of the Universal mind is omniscient, omnipotent, omnificent and omnipresent.

It's also the human nature. It's believed that one has access to all knowledge, known and unknown. Through the Universal Mind, people have access to an infinite power; one then is able to tap into the limitless creativity of the One. All these attributes are present within one at all times in their potential form. http://en.wikipedia.org/wiki/Universal_mind

"There is a single, intelligent **Consciousness** that pervades the entire Universe - the **Universal Mind**. It is all knowing, all powerful, all creative and always present. As it is present everywhere at the same time, it follows that it must also be present in you - that it is you. Your mind is part of the one Universal Mind. This is not simply a philosophical ideal passed down to us through the ages. It is an exact scientific truth. Know it, believe it, apply it and you will see your life transform in miraculous ways.

Albert Einstein told us that "everything is energy"; that "a human being is a part of the whole called by us [the] Universe". His words echoed the most ancient of spiritual and philosophical teachings and still underpin today's cutting-edge scientific discoveries. The Universal Mind goes by many names. In the scientific world, we know of the **Unified Field**, in spiritual philosophy we refer to **The All** or **Universal Consciousness** and in religion we call upon **God** who Himself goes by many names - **Jehovah, Allah** and **Brahman** to mention but a few. The name is relevant only in so far as it resonates with you.

Whichever way you cut it, you come to this one unavoidable conclusion: there is but **One Consciousness** of which your consciousness must be apart and "a part", as **Charles Haanel** said, "must be the same in kind and quality as the whole, the only difference being one of degree".

The nature of the Universal Mind is **Omniscience** (all-knowing), **Omnipotence** (all-powerful), **Omnificence** (all-creative) and **Omnipresence** (all-present). Know that this too is your nature. You have access to all knowledge, known and unknown; you have access to an infinite power for which nothing is impossible; you have access to the limitless creativity of the **One Creator**. All these attributes are present within you at all times in their potential form. http://www.mind-your-reality.com/universal_mind.html#Part_2

It is up to you to know and act upon your own nature. The inscription on the ancient Greek **Temple of Apollo at Delphi** left no room for misunderstanding: ***"Know thyself and thou shalt know all the mysteries of the gods and the universe"***. It is through the power of your subconscious mind and your higher self that you can learn to align yourself with the Omnipotence, Omniscience and Omnificence of the Universal Mind at all times.

Each and every one of us is a manifestation of this single Universal Consciousness. There is profound truth in the ancient teaching that we are all One. We are all connected - not only to each other but to all of Nature and to everything in the Universe. This is the Law of One. What you do to others, you do to yourself. The way you treat Nature, you, in fact, treat yourself. The separateness you "see" is an illusion of the personality ego. The true nature of reality is non-dualistic, meaning that while things may appear distinct, they are not separate.

You are able to create your ideal reality because you are already connected to everything you want. Nothing and no one is separate from you. You can experience happiness, true love, perfect health, abundance, wealth and anything else you intend. All you have to do is bring yourself into

356

vibrational harmony with the nature of that which you want to experience through the creative power of your thoughts. To become the master of your destiny, you must master your thoughts.

In a nutshell, there is a single Consciousness, the Universal Mind, which pervades the entire Universe. It is all knowing, all powerful, all creative and always present everywhere at the same time. Your consciousness is part of it - it *is* It. All is One. You are connected to everything and everyone. You are already connected what you want. To the degree that you truly comprehend and internalize this Truth, you will be able to become the master of your mind and the director of your life." http://www.mind-your-reality.com/universal_mind.html#Part_2

It should not come as a surprise to the reader that throughout this textbook, this author has been leading up to a crescendo of thought and evidence that has focused a spotlight on the next evolutionary step in UFO and ETI research. It is the imminent contact and communications with Extraterrestrial Intelligences currently visiting this planet!

Its primary leader and current pioneer in this emerging ETI contact and communication field is Dr. Steven M. Greer! In the next section we will see how Dr. Greer uses **the "Rosetta Stone" of ETI contact and communications** utilizing these psychic modalities to lead other people to go out to remote locations globally and establish through CSETI's **Close Encounters of the Fifth kind (CE5) Initiative and Protocols**, contact and communications with Extraterrestrial Intelligences. https://www.youtube.com/watch?v=tEH6sQJd7CE

CHAPTER 111

IN WHAT LANGUAGE DO EXTRATERRESTRIAL INTELLIGENCES COMMUNICATE?

What do you call a language that is extraterrestrial in origin? It will probably depend on to who you are asking the question. A professor of linguistics would probably respond with the term **Exolinguistics,** a biologist might refer to an ET language as **xenolinguistics** and an astronomer may reply with **astrolinguistics.** The common man on the street, if asked the same question would probably respond with that **"Alien" language** or the response may be Martian, Venusian or Klingon based upon our limited understanding of planetary science or science fiction. The study of an alien language is for most scientists and the general public, a hypothetical exercise that at this time has no apparent answer unless one has had a close encounter with an Extraterrestrial Intelligence.

The question of what form an alien language might take, and whether humans would recognize it as a language if they encountered it, has been approached from several perspectives. Consideration of such questions forms part of the linguistics and language studies programs at some universities.

Life on Earth employs a variety of non-verbal methods of communication, and these might provide clues to hypothetical alien language. Amongst humans alone, these include many visual signals such as *sign language, body language, facial expression and writing (including pictures*), and it is possible that some extraterrestrial species may have no spoken language. Amongst other creatures, there are some which use other forms of communication, such as *cuttlefish and chameleons,* which can alter their body color in complex ways as a method of communication, and *ants* and *honeybees,* which use *pheromones* to communicate complex messages to other members of their *hives*. http://en.wikipedia.org/wiki/Alien_language

Dutch mathematician **Hans Freudenthal** in a 1960 book described **Lincos**, a constructed language which includes a dictionary that uses basic mathematics as "common ground" to develop a working vocabulary. A point of consideration is that cereal crop formations or crop circles use mathematics and geometry as a communication modality!

Language develops among a species when it acquires meaning through a community of speakers using it as part of their way of life and would only be meaningful to them and not to us. Hence beings with a radically different way of life would not be able to make sense of the others' utterances.

However, if aliens evolved under pressure of natural selection, we would expect them to have the same drive to survive and reproduce that we have and thus, there would be a common ground of understanding between each other. http://en.wikipedia.org/wiki/Alien_language

The nature and form of such languages according to current thinking remains purely speculative because it is premised upon the false notion that so far, no government sponsored program has detected signs of intelligent life beyond Earth. It is a false notion because most scientists refuse to become involved with any investigative program that considers seriously the global phenomenon of UFOs and ETIs.

There are myriads of accounts that have been and continue to be reported by eyewitnesses who have had interaction with non-earthly intelligences and their spacecraft. Such accounts often tell of communications between humans and ETI. These eyewitness accounts are opportunities for scientists to understand not only the types of ET intelligences visiting this planet but, a chance to understand their languages and to come to know something about their civilizations. However, their lack of interest in the possibility of future contact with intelligent extraterrestrial life, because it doesn't fit neatly in with the scientific concepts of reality, has made the question of the structure and form of a potential alien language, a topic of scientific and philosophical discussion. A discussion that is usually considered pseudoscience.

We obviously cannot rule out the advancements made in the covert black world of science which have sought vigorously to understand the UFO /ETI phenomenon but, have used that knowledge gained by those scientists who work within the **Military Industrial Complex** for practical application by developing military weapons assets and aerospace hardware.

Understanding the language of another sentient intelligent species visiting the Earth permits one to understand the thinking processes of another culture or in this case another civilization and communicating in a shared language allows for the exchange of ideas and perhaps, of materials goods; it is a catalyst for the advancement of one's own civilization.

Extraterrestrial Presence and Interaction as a Language of Communications

The reports of ET encounters with people fill the pages of books in UFO literature; they recount the attempts at interspecies communication in one form or another whether desired or not. Extraterrestrial spacecraft in our atmosphere by their very presence communicates *"We are real!"* and *"Here we are!"* This is a very basic but, straight forward approach to communicating. ETI spacecraft design and configuration as well as their movement through our skies or out in space also communicates that they have an advanced technology and are from some other star system. The communiqué from ETs is *"We are technically advanced!"*

In the First World War, many British air pilots of the **Royal Flying Corps** and the **Royal Naval Air Service (RNAS)** patrolled the skies over London and other major British cities in search of German airships or **Zeppelins**, which were intent on bombing London. Radar was not invented until the Second World War and patrolling the skies was the only way to ensure the enemy limits its attack capability. Beginning in 1916, when pilots went on air patrol missions, they would report lights in the sky thinking they were German airships and would give chase only to find that these mysterious *"airship"* lights had disappeared. These pilots would be debriefed upon landing and would include in their reports these strange encounters with lights. These reports became tagged as "phantom airships" or "ghost planes" and were never really taken seriously by the British Military Authorities.

"One year later, GHQ issued a secret Intelligence Circular which concluded there was "no evidence on which to base a suspicion that this class of enemy activity ever existed." It said an investigation by Intelligence officers had satisfactorily explained 89 percent of the reports received and the authors attacked "the groundless rumours regarding the presence of hostile airships over Great Britain which of late have become very frequent." In addition, the Military Authorities decided to impose severe penalties upon what it called "irresponsible persons" who were originating and circulating such stories. They would be dealt with, it threatened, "under the Defence of the Realm regulations" which included imprisonment. One year later, GHQ issued a secret Intelligence Circular which concluded there was "no evidence on which to base a suspicion that this class of enemy activity ever existed." It said an investigation by Intelligence officers had satisfactorily explained 89 percent of the reports received and the authors attacked "the groundless rumours regarding the presence of hostile airships over Great Britain which of late have become very frequent." In addition, the Military Authorities decided to impose severe penalties upon what it called "irresponsible persons" who were originating and circulating such stories. They would be dealt with, it threatened, "under the Defence of the Realm regulations" which included imprisonment." http://www.uk-ufo.org/condign/hist1916.htm

What can be concluded here and what was to be repeated again and again in the major theatres of war during WWII by countless air pilots on both sides of the war was that these mysterious objects appeared to be monitoring human predilection for war and on a global scale at that. This monitoring of wartime activities can also be considered as a form of communications that clearly states, **"We are watching you!"**

When the atom bomb was exploded for the first time in the American Southwest and then, later over the cities of Nagasaki and Hiroshima of Japan, Extraterrestrials Intelligences sat up and paid immediate attention to these locations, particularly the **White Sand Proving Grounds** (test site for most of the American atomic bombs).

Inadvertently, we were communicating back to these visiting ETI. We were saying, *"Hey, we found the matches!"* and *"Boy! That was a big explosion!"* and *Oh! By the way, we're down here!"*

The old adage: *"Actions speak louder than words"* has never been truer in this increasingly dynamic relationship between ETI and humans. Extraterrestrials ramped up the surveillance on humanity as it emerged out of its second world war with itself, overflying all military bases and weapons sites and in many cases shutting down missile sites. They were growing increasingly concerned with humanities testing of atomic and nuclear weapons as well as the build-up in most nations of armaments. The message that ETI were sending to mankind with their over-flight actions, *"We are alarm by your aggressive actions and we don't want you to harm yourselves or your planet with possible nuclear annihilation!"* Also, *"We can, if necessary intervene and shut down your weapons systems should you threaten each other with nuclear war!!"* *"We are still watching you!!!"*

UFO sightings were increasing at an almost exponential level and it wasn't long before there were reported accounts of UFO landings along with random encounters with ETI. There were frequent reports that ETs were taking samples of vegetation and various life forms with a chance encounter with an intrusive but, curious human or two. The message being communicated was

"If you are going to threaten each other with nuclear Armageddon then, we will see to it that some of the life on this beautiful planet survives!"

As more and more people started to see small humanoid beings of various kinds not just the typically described short Grey being but there were frequent descriptions that included human-like beings that essentially could be our **"cosmic cousins"**. These chance encounters with humans took on less randomness and more of an opportunistic interaction with a specific message that needed to be delivered to humanity through one or more individuals.

Extraterrestrial Linguistics - Did You Say Something?

These messages began to be verbalized in the language of the individual's mother tongue or country usually taking place in remote areas of the country away from prying eyes like desert areas, long stretches of open highways or deserted winding countryside roads. These contact and communication encounters usually followed a basic protocol of friendship, peace, and harmony with the same repeated message given to the humans that were encountered: *"Don't be afraid of us. We come in peace!"* But, more importantly, the message would have a series of global concerns and admonitions along these lines: *"We see that you are exhausting your resource and polluting your planet, that you threaten each other with war and conflict. Your actions have a profound effect on the universe. We are concerned for you. You need to correct human behaviour and to move away from violence toward a peaceful, unified civilization"*.

These people who reported their unusual encounters with people from another planet were marginalized as **"contactees"** and their outlandish stories as the delusions of a raving madman or crackpot who sought public attention and notoriety. At some point in the late '50s or early '60s these types of human ET contacts with human beings of Earth ceased being reported and by the late '60s and early'70s the prominence of the little Grey aliens became the common ET being that was being sighted. There are, of course, other xeno-type intelligences besides the Grey ETs that were being seen that ranged from insect-like, reptilian, robotic, exotic and animal-like in appearance. All beings encountered had a message for mankind whether it was understood or not by the sometimes, hapless humans who were the intended subject of ETI interest.

With such a diverse federation of Extraterrestrial intelligences coming to earth, it goes without saying that each species would have its own language, therefore, how do they communicate with each other? How do they understand each other in a coordinated mission? Do they use some kind of *"universal translator"* as used in the **Star Trek** television serious? Do they learn each other's home-planet common language or are they so intelligently gifted as to know all languages becoming multi-linguistic like some people have become on this planet? Is there a **Universal Language** of the universe? Humanity has some emerging languages that may become universal to this planet like English which is becoming universal in business and trade and many nations already speak it even as a secondary language to their own mother tongue. There are of course other languages like Chinese with all it numerous dialects or Hindi of India but, these are chiefly spoken within their own countries and not a part of any other nations language, even as a secondary language. https://www.youtube.com/watch?v=KTc3PsW5ghQ

Is it possible that ETIs have far superior intellects that they are accomplished in all human languages as well as the multitude of other alien languages or is there another way to communicate effectively with humans? Perhaps, the solution to make things easier when visiting planets is a specialization of purpose in which each interstellar race, partners in space exploration make contact with similar species to their own thus, making the transition in contact and communication smoother. This might explain why human (contactees) reported encounters with human-like unless that whole time period of ETI contact was nothing more than a covert black ops deception program hoaxed upon the general public! This is a very real possibility.

It would seem, that if we accept the historical accounts of eye witnesses and contactees that many people experienced interaction with human-like ETs initially and then, over a short period of time, with different humanoid ETs beings. There are reports that ET beings used guttural howls, growls, high pitch noises much like humans do when in some emotional state like having sex or suffering injury or when expressing excitement and joy. Sometimes the method of communication may be by gestures or with a firm, but non-aggressive touch or grip to guide a human to a specific area or to look at something.

We must consider the possibility that humans are not the brightest light bulbs on the galactic block as we are a young species and therefore, it is necessary that ETI must communicate to humanity on multiple levels using various modalities to ensure we understand whatever message is being communicated to us.

Crop Circle Formations as an Extraterrestrial Language

Almost at the same time, another silent but, potent source of ET communication began showing up in the unlikely medium of plants or more specifically, mysterious geometric formations known as **crop circles** were being etched into the farm fields of cereal crops throughout the British countryside. These agriglyphs took many farmers, by complete surprise, with some scratching their heads wondering how it all happened, while other less amused and annoyed farmers thinking it was teenage pranksters, immediately destroyed these beautiful geometric designs by ploughing their fields and these formations over.

The curious public and the British military suddenly got wind of these incredible formations and started to come out into the farmers' field uninvited causing more damage to their fields. Military helicopters also began flying over these farm fields, their low altitudes were a source of loud noise and caused stress to milking cows and other livestock. Public concern was raised by farmers to the government that something had to be done to stop this chaos and destruction of their livelihood.

Initially, it was thought that theses crop circles were the work of two drunken Brits with more time on their hands than good common sense. Indeed many crop circles were the creative shenanigans of **"Doug Bower and Dave Chorley"** and other like-minded individuals looking for public attention and as a source of laughter and discussion over a few pints of beer. However, cereal crop investigators took a closer look and discovered that some crop circles had no human intrusion into the wheat, rye, barley and rapeseed fields, whatsoever. Crops that were bent over and laid down in various spirals and intricate weaved patterns displayed no damage to the individual plant stalks at all but, remained living viable plants!

Do ETI communicate with humans using Crop Circles? https://i.pinimg.com/564x/c8/bc/46/c8bc46a46097ff23a71b9c66261fc194.jpg

These agriglyphs were thought to be due to whirlwinds, localized thunderstorms, upwellings of magnetic disturbances, earth spirits, insect damage, plant viruses and mutations, as well as the obvious hoaxes, as already discussed. The UFO/ETI factor was also thought to be the likely cause of these formations as sightings had been reported on more than one occasion in these particular farm areas usually situated near ancient monuments and ruins like Stonehenge. There are videos of small white objects flying around these crop fields in England and Germany and then within seconds a crop circle is formed. In one video a small UFO is seen flying over a field with a British military helicopter in pursuit of the strange object, the farm area has been a known crop circle location.

There was even a report by a police constable who was driving by a farm area on patrol when he glanced over toward a field and could see a crop circle formation with what he took to be two very tall humans dress in white uniforms. He stopped his patrol car and got out to approach these strange people who immediately ran very quickly toward the centre of the formation and then simply disappeared as if they stepped through a doorway. When the constable got to where the strange men disappeared, there was nothing to see or find in the area. He left mystified and reported the incident to his superiors at police headquarters.

Crop circle formations have been reported globally in twenty-eight countries and in areas far removed from roads and human habitation and in fields with no irrigation tramlines or any entry or exits from the fields.

Science in its typical official position maintains the status quo by stating that the theories posited by Ufologists and anomalistic investigators are nothing more than pseudoscience. This official, but false assumption by scientists prevents any real scientific inquiry or research but, nonetheless, UK military intelligence is not discouraged by the science community's position on the subject. What do they know that keeps them aggressively interested in this phenomenon when science shows no interest whatsoever? http://en.wikipedia.org/wiki/Crop_circle

Could this be another method in which ETs are communicating with humans through crop circle formations as it is now becoming the standard hypothesis behind this enigma? ET saucer craft and amber glowing lights continue to be seen and crop circles continue to be made in farm fields.

The designs created and left in the cereal fields have evolved from simple circles, crosses, squares with various types of connecting lines to become highly complex formations that utilize mathematics, physics, astronomy, meteorology, biology, chemistry, linguistics, history and music diatonic and chromatic scales, etc. Such knowledge is beyond the capacity of Doug and Dave and they certainly don't travel the world creating these crop circles in other countries.

The crop circles, if not the practical jokes of two British pub buddies then they must be hoaxes which are being perpetrated by a highly positioned intelligence group to discredit those borderline sciences like Ufology by ridiculing the subject matter and getting the public to turn its attention to other things.

The message contained within crop circles formations as determined by many Ufologists and cereal crop investigators is that it was created by extraterrestrials and is knowledge-based that covers many of Earth sciences from ancient to current times as well as possible knowledge about Extraterrestrial life forms and their technology.

Telepathy - The Language of Extraterrestrial Intelligences

These various ET xeno-types seldom, if ever talk but when they do, it will be either in the native language of the individual in whose country the encounter has taken place or it will unintelligible to the human ear, in other words in their alien tongue. When human contact does occur, ETI prefer, it would seem, to communicate through telepathy. This preferred method of communications is their "language" of choice as it by-passes the confusion and complexity of each civilization's language. It can be understood by almost every reasonably evolved intelligent sentient being, because it operates mind to mind on the level of non-locality, outside of the three-dimensional space-time reality. Images and words, including emotions and vast informational downloads could be conveyed from "sender" to "receiver" almost instantaneously.

Telepathy is like a laser in that it is a coherent energy form that is amplified by the intention or will of the sentient being based upon their evolutionary development and their society's practice of this method of communications. A well-trained mind can not only transmit his thoughts over vast distances but also, into other dimensions or realities and it can be target-specific to a single individual or to a whole planet. The well-trained telepath is also a good receiver of transmitted thought and can operate like a radio or television set by discriminating incoming information selectively or when necessary, completely silencing the mind to all incoming thought. Such are the ways of the telepath regardless of intelligent species.

The fact that humans don't engage in telepathy as a modality of communications is that we surround ourselves with mental clutter, material distraction, and loud noises on a daily basis. We fill ourselves with chaos, stress and negativity then, in turn, reflect this back into the environment and toward others. How often has an individual entered into a home or work environment where there is an overwhelming sense of negativity and is quickly overcome by it in those surroundings. Again, in a home where there is positive expression of emotions, attitude, and behaviour the energy output from such a place can permeate into the environment that it can literally be felt by an attuned individual or society, tens or hundreds of miles away! This is the type of energy that can be generated by telepathy and intention.

Author's Rant: I have experienced this telepathic energy form on several occasions, to the point that even inanimate objects can become imbued with this energy, that they seem to be alive!!

Frequent eyewitness reports tell how the individual, whether in close proximity to an ET being outside near the individual's home or in the countryside or on board their spacecraft or even when they cannot be seen, have stated that the ETs never seem to move their lips or mouths but, that they could hear them clearly within their heads. There are cases where human – ETI telepathic communications have been across distances of hundreds or thousands of miles and even much further, interstellar in fact. There particular incidences were reported by various **CSETI (Center for the Study of Extraterrestrial Intelligence)** teams when pro-actively engaged with contact and communications with ETI via deep meditation and higher consciousness sessions.

As more and more UFO researchers begin to "connect the dots" on the UFO/ETI phenomenon, it is interesting to recall that **Ben Rich**, director **of Lockheed Skunk Works** had admitted in his **Deathbed Confession** that Extraterrestrial UFO visitors are real and the U.S. Military *has the capability to* travel among stars. *(Author's bold italics added for emphasis).*

The military does not travel among the stars but merely has the capability. In reality, Extraterrestrial Intelligences have quarantined Earth's ability for interstellar travel among the stars! This is an important fact to remember as it will be discussed in the next section of this book.

According to an article published in May 2010 issue of the **MUFON UFO Journal - Ben Rich**, the "Father of the Stealth Fighter-Bomber" and former head of **Lockheed Skunk Works**, had once let out information about *Extraterrestrial UFO Visitors Are Real And U.S. Military Travel To Stars.*

"We already have the means to travel among the stars, but these technologies are locked up in black projects, and it would take an act of God to ever get them out to benefit humanity. Anything you can imagine, we already know how to do."

When Rich was asked how UFO propulsion worked, he said:

"Let me ask you. How does ESP work?"

The questioner responded with, *"All points in time and space are connected?"*

Rich then said, *"That's how it works!"*

This is a very interesting statement by someone who definitely knows the inside story. Nevertheless, Ben Rich did not spill the beans as to precisely how it may be accomplished, other than to say that Lockheed *understood* and had the *capability* to take ET home!

CHAPTER 112

WHY ARE ETS VISITING THE EARTH –
IS THERE AN ALIEN AGENDA?

An alien agenda would by its very nature be premised upon the fact that Extraterrestrial vehicles or spacecraft commonly but, incorrectly referred to as UFOs are piloted by Extraterrestrial Intelligences visiting the Earth. The evidence for their reality is overwhelming and unless, you have been living in a cave for the last hundred years or have never bothered to look up occasionally into the night sky or have never read a newspaper headline or simply mused to yourself, could life really exist elsewhere in the universe. In which case, you may have assessed that contrary to all the evidence of their existence, such notions and perceptions must be the delusions of a few troubled minds in our midst or that the world is in a far more chaotic and dysfunctional state than we realized.

In all likelihood your perceptions and assessments of those things and events of chaos that are happening around you are probably not too far wrong and you can take some small comfort in that knowledge but, it doesn't alter the nearly eight decades of accumulated facts and evidence as presented in this book, that **UFOs and ETI are REAL! We are not alone in the universe and never have been!!**

With this startling realization that has hit us like a massive electric surge to our neurons, it has shaken us to our very being but, it is intended to wake us out of our complacency to respond with action and to start questioning those officials in authority who probably know what's really going on. There are those who will prefer to bury their heads in disbelief, not wishing to have their comfortable worldview altered by some new alien paradigm of reality but, it is too late for that anthropocentric, self-centeredness existence.

Questions must be answered to understand this new paradigm that is increasingly engaging humanity on every continent of the planet, we call home. **Who are they? Where do they come from? Why are they here? Is there alien agenda and if so, what is it? How did they get here?**

It is apropos at this time to mention **Jim Marrs'** book "Alien Agenda" as it covers a lot of the aspects on this topic, a pre-ample in many ways of the subject matters presented in this book that are covered in more detailed. Marrs in his introduction points to the fact that the answers to these questions can be found from examining the bigger picture of the UFO experience which as this author will attest has become a labyrinthine nightmare of sensationalism, contradiction, distortion, convolution and as previously stated a corrupted database of half truths with many leaps of illogic and forced speculations for the sake of selling another UFO book to promote one's ego and status.

"This confusion is compounded by documented government deceit and duplicity aided by the reluctance of conventional science to publicly address the evidence. For few people --- particularly among the smug scientific and political intelligentsia --- are willing to give any public credence to the subject. By failing to publicly to take notice of the phenomenon, these bastions of conformity and conservatism have left the field open to a wide array of private researchers, who range from serious and dedicated investigators to the wildest of charlatans and profiteers. This situation has meant that any serious and unbiased look at UFOs immediately opened the researcher to a barrage ridicule and arrogant dismissal by those who have some reason to ignore the subject." **Alien Agenda: Investigating the Extraterrestrial Presence among Us by Jim Marrs; 1997; published by HarperCollins Publishers; New York, USA; ISBN 0-06-018642-9**

Some of the questions above have already been answered but, most have not. We know that there is upwards of 125 different species of Extraterrestrial Intelligences coming to this planet. Some researchers have said there are 12 civilizations, and some have said 75 different xenotypes, however, the actual figure is speculative based upon eyewitness accounts and those ETs that were unfortunately captured by the military force of this world. It is safe to say that no one knows the exact number but, that they represent potentially a united planetary federation by many interstellar civilizations visiting and interacting with this planet, much like our United Nations, though more unified!

As to who they are is still a harder question to answer and again it is speculation based on ET **close encounters of the fourth kind (CE-4)** and no ETs encountered thus far, have ever stated what planet they call home or what star system they originate from, other than to point up to the stars above. A rather nebulas answer, indeed (pun intended)!

Some UFO researchers have speculated that the little Grey/Gray ETs are from **Zeta Reticuli** based upon the supposed star map of **Barney and Betty Hill's** close encounter and the follow-up 3D modelling by **Marjorie Fish** of the map drawn by Betty where she concluded that the home star system was Reticuli. There is, of course, no confirmation of proof that indeed this is the home star system of the Greys as any and all Grey type ET beings have never confirmed that piece of information.

It is also true that the same non-confirmatory proof can be said of the ET species that are reptilian in appearance known as **Dracos or Draconian** which are allegedly from the constellation of stars known as Dracos (the Dragon). There simply is no proof of this as their home world unless someone knows more than they are saying and are covering up their sources of information.

Again, the **Nordic humanoids** or near-human ET cousins to Earth humans, known as the **Pleiadians,** or the **Lyrans,** or **Sirians,** or **Orions,** or even **Arcturians,** etc, ad nauseum, there is no corroborating evidence to substantiate the claims of their origin. We Earth humans simply do not have as yet, the propulsion systems to venture out into space to these star systems to confirm these claims.

Good people, unless you have corroborating evidence or proof of some kind to back up your claim, it simply means your claim is subjective and /or anecdotal at best!

Author's Rant: I've inserted this next small section into the book to add some humour to an otherwise, rather dry book and in no way is it meant to disparage those who practice

channelling but, it does point out what Jim Marrs refers to in the UFO literature as "contradiction, distortion, convolution" and just plain "old" new age mumble jumble rhetoric.

Pleiadians, also known as **Nordic aliens**, are humanoid aliens that come from the solar systems surrounding the Pleiades stars, and they're really, really, really concerned about Earth and our future. (Since the Pleiades are a pretty group of stars and since they're even more beautiful through telescopes, anything connected with them has to be good and kind, right?) So concerned are they, they've contacted certain special people to channel them and convey their message. Which isn't always consistent. The Pleiadians either came from a group called the Lyrans, or coexisted with the Lyrans, or the distinctions between Lyran, Pleiadian, and Sirians don't really exist. Native Americans are of Pleiadian descent, or white people are of Lyran descent, or humans are of no notable relation to either. They can either switch between the third and the ninth dimension, or they exist solely in the fifth dimension (which is apparently made of love and creativity), or just live in non-specific dimensions higher than three. Basically, the only things generally agreed upon by the Pleiadians themselves is that they want to help Earth and let humanity ascend to higher dimensions.

Pleiadians are supposed to be, well, better-looking versions of us. (**Ashtar** and **Semjase** are both two of the beautiful people of the Pleiades). They're nearly always white and were often blonde/redheaded, and not black, or vaguely Chinese-looking like the grays. *(Sounds racist doesn't it!)* Think that Robinson family from *Lost in Space*, minus the robot and Dr. Smith.

As a rule, Pleiadians don't carry around much excess fat, although female ones are said to have curvy figures. They don't have bad hair days, or have beards usually, *unless, you've been talking to* **"Billy" Eduard Albert Meier**, *lately.*

No particularly elaborate theory is really needed to work out why humans came up with aliens who looked so, human. As such, their beauty is reminiscent of Greek gods, and Christian angels, who are usually pretty good looking.

"Human" aliens have been a staple of science fiction from the beginning. **Erich Von Daniken**

suggests that humans are deliberately designed to look like their makers. In *Star Trek*, it was easier for actors to fit into human type suits.

Pleiadians and Nordic aliens were a staple of the early contactee (as opposed to abductee) trend of people like **George Adamski**.

Pleiadian statements tie in nicely with the **Spiritualist** and **Theosophist** beliefs of the early twentieth century. They're also a bit like the guy out of the original *When the Earth Stood Still* (he was from Venus though). As the hippie thing died down, and the seventies became more cynical (Watergate and all), so Pleiadians took a back seat from contact and moved into channelling. The far right likes them as well, and the whole **Maria Orsic** legend (Nazi Germany) seems tied up with them. (She is said by some to be one herself.)

Most of all, they're the aliens, you'd enjoy a nice quiet cup of tea with down at the yoga centre, rather than being eaten and cruelly ruled over (Reptoids) or having foreign objects lodged in your rectum (Grays).

Pleiadians are somehow connected to **Atlantis** and **Lemuria**. Reiki, ear candling, Shiatsu, reflexology, aromatherapy, and crystal healing all came from the Pleiadians. The Pleiadians brought dolphins to Earth - we know so because **JFK's spirit** has contacted a human through automatic writing to tell us so. **Jesus** was a Pleiadian, as was his father, though his mother was Lyran. They love sex (the Pleiadians do, not Jesus and his family... though they must too if they're Pleiadians). The Pleiadians are helping us fight the evil space reptiles (and you thought we couldn't connect this to Reptoids) as part of the battle against the **Illuminati**. They are old, old, old Earth creationists, and think Earth is 626 billion years old. Pleiadians are apparently truthers.

And the Pleiadians *just might* have been doing something good and kind in the year 2012 to counter gloom and disaster instead of spreading it.

In fairness to the Pleiadians and the other ET beings as described above, it may yet be, that we shall discover in future space explorations that your home worlds are as claimed by the eye witnesses who have interacted with you and *as the channellers have stated but, maybe to a lesser degree!*

There are elements of truth to some of these stories and ET descriptions but, this could also be a clever deception being perpetrated upon the UFO community by military intelligence as part of a continuing disinformation and misinformation campaign.

The real question that needs to be answered that is on the minds of many military, government and religious officials, as well as scientists, Ufologists, and the public, in general, is the **Alien Agenda**, *"Why are they here?"*

Of major interest to the military and intelligentsia is an ancillary question just as important the first which has both these branches of the government spending billions of dollars to get answered, it involves their ultimate motive, the acquisition of alien technology: *"HOW ARE THEY GETTING HERE?"*

Where's the Evidence for Alien Hostility?

When you speak about a possible hidden alien agenda to anyone within the UFO community, you will find researchers polarized into two opposing camps of thought, heatedly debating the motives and agenda of ETs visiting the Earth. One side theorizes that not all Extraterrestrial Intelligences coming to the Earth are benevolent but, that some are actually malevolent with occasional acts of overt aggression; at best, they are possibly neutral or indifferent to the concerns and the affairs of humans.

The opposing camp theorizes that there is no proof of ET hostility or malevolence toward humans with the exception of self-defence from aggressive action taken by humans towards ETI, such as targeting and shooting them down from the skies. The fact that no individual or group or country is being subjugated by an alien force or civilization is, therefore, proof of an ET peaceful intentions or benevolence which has been demonstrated repeatedly throughout recorded history and by their presence at this time; at best, they are possibly neutral or indifferent to the concerns and the affairs of humans.

If ETs are neither hostile nor beneficent then, their agenda may be a neutral one as explorers searching out new life and intelligences or they could simply be galactic vital statistic counters gathering information on the diversity of life in the universe, as well as the types, quantities and locations of particular natural resources that are currently available.

Let's examine all sides of the argument to make sense of this perplexing problem and see what the alien agenda really is, that has everyone who studies this phenomenon so divided.

Are Aliens Malevolent?

Supporters that say ETs are malevolent point do what appears to be solid indisputable evidence, often becoming emotional upset even at times, aggressive in trying to put forth their arguments. Some of their evidence focuses chiefly upon human abductions and cattle mutilations such as the following:

- abductions of humans by little **Grey type ET beings** at night usually when the victim is asleep or alone in some remote location,
- medical examinations and DNA retrieval from humans by these ETs,
- small biopsy scaring on various parts of the body, the removal of ovum, semen, and even fetuses before full term pregnancy is completed,
- the introduction of human-ET hybrids to the abduction victims for bonding purposes or emotion transference with the ultimate goal of introducing hybrids into human societies as eventual human replacements
- implanting of foreign objects into unusual areas of the body for the purpose of mind control or manipulation of actions and thoughts or for tracking and monitoring purposes, etc.
- cattle and other livestock are frequently taken by UFOs sighted in and around cattle ranches and then, the next day or later the cow or bovine carcass is found separated away from the herd having missing body parts as if surgically removed.
- the cow often appears to be dropped from a great height and no human or predator footprints can be found around the carcass
- recent grizzly discoveries of humans have been found with similar excised removal of certain body parts similar to cattle indicating a possible UFO/ET culprit.

These are considered invasive and even hostile actions by ETs often attributed to the little Grey ET beings or the "reptoids" (reptilian ETs) which are so often reported in the UFO literature.

These abduction experiences reported by abductees seem to be very real and it hard to really dispute the claims of the victims, because they experienced a real event, however, the perception and interpretation of the experience may actually be quite different from the true nature of the event and thus, may not necessarily be what the abductee thinks is happening at that moment. Therefore, the interpretation of the experience must be called into question as to the authenticity of the event. In fact, the evidence claimed by abductees is often circumstantial at best usually recalled under hypnotic regression sometimes by trained psychologists but, more frequently by UFO researchers untrained in clinical hypnotherapy.

Chief among the malevolent ETs and the **alien abduction supporters** are **UFO researchers** like the late **Budd Hopkins, David Jacobs MD,** the late **John Mack MD, John Carpenter MD, Yvonne Smith, Richard Boylan Ph.D., Brad Steiger,** and **Edith Fiore Ph.D.** and many others.

Many people report more than one abduction event in their lives, sometimes within hours, days, weeks of the first event with more re-occurring ET visitations spread out over months and years and from one generation to the next generation in the same family. There are two possible scenarios for an abduction event:

1. The person claiming abduction has a genuine ET experience where they are physically removed or more frequently "astrally" removed from their bodies on board a spacecraft. They undergo a set of medical procedures which may be unpleasant and then they are returned to the bodies or homes usually unaware of the experience that has just transpired. They will awake in the morning perhaps with some discomfort physically or emotionally believing perhaps, they had a bad dream maybe about ETs, not really knowing why other than it was upsetting to them. They seek out help from a psychologist or a UFO researcher thinking that there is more to this bad "dream experience". The abductee victim tries to recall as much information consciously as possible, however, hypnotic regression is used by the therapist to dig deeper into the subconscious to extract greater detail of the experience.

   When the information is gained, the assessment of the experience seems to be as described by the abductee, namely that they had an extraterrestrial experience which wasn't pleasant. Conclusions, suppositions, and theories are given that point to a possible ET encounter with perhaps, some discrepancies in the experience that seem to get lost by the wayside in favour of the ET hypothesis. Namely, why so much of what took place often smacks of covert intelligence procedures that are used sometimes in field interrogations and torture, where the victim is made to believe that something else is taking place, other than the reality of the situation, a la **Manchurian Candidate scenario**?

 Bear in mind, that hypnosis and hypnotic regressive as a professional tool to help people suffering from an emotional crisis is not 100% accurate for accessing or assessing information gained by this method. Its reliability and effectiveness are, therefore, in question even among **APA (American Psychological Association)** yet, it has not stopped it use in or outside of the medical profession. Obviously there have been more successful uses of it than failures or misuses of it, otherwise, the APA and other institutions would stop using it, altogether.

2. The person or persons experience a genuine UFO/ETI encounter that is life changing, pleasant, emotionally and spiritually uplifting with no physical harm or side effects. They may experience the wonders of the ETs technology aboard their craft or their common

bonds and similarities that they share through the interaction with ETs. There may be the transfer of information, the contents of which may be informative, reassuring or perhaps as a warning or admonition. The whole experience is not considered an abduction event and the humans are returned to their point of origin (home or car) perhaps, more emotionally charged from the event. No doubt this positive life-altering event will remain with them for the rest of their lives. This is their first encounter with UFOs and ETs and they may feel the need to report it to authorities or to a UFO organization. They recall this event fully awake, sometimes its lies dormant for a period of time before total recall of it.

Unexpectedly, these people then begin to experience black unmarked military helicopter activity either, hovering over their home or following them wherever they go, whether it's driving around town or out into the countryside. They appear to harass these people who have had a close encounter with ETI and sometimes these harassments can last for days or weeks.

They then, experience a second UFO and ET encounter but, this encounter is negative and the complete opposite of the experience of the first ET encounter. Everything that they thought was positive about Extraterrestrial life has now been turned upside down on its head. The second ET encounter is a horrendous experience that falls into the same category as the abduction scenario in #1 above, ala ***Barney and Betty Hill's "Incident at Exeter".*** The Interrupted Journey; 1966 by John G. Fuller; published by Berkley Medallion Books, New York, USA; SBN 425-02572-1

Their memories become a distortion of the reality of events. It becomes a combination of both good aspects with negative aspects with the emphasis on the negativity that alien beings are aggressive, malevolent and maybe even hostile. The takeaway factor from the experience is that these victims now have a fear-based perception toward anything not of this world, particularly ETs such as the diminutive Grey ET beings.

Recall that Barney and Betty Hill had a positive ET experience near Montreal on their way back from vacationing in Canada and then, experienced a **MILAB (Military Abduction)** event hours later when they crossed the border into the US. It seems that a covert arm of the US military had tracked the event in Canada but, did not cross the border to intervene. They waited instead, until the Hills were miles *"safely"* into the US, before intercepting them with another ET encounter of their own creation. One that was scary for both Barney and Betty but, it became a distortion of memories between two separate events, one in Canada and one in the States, one positive, the other negative. The memories became a mishmash confabulation of these two events, of small humanoid ET beings with one wearing a ball cap, the use of *supposed*ly unknown medical procedures at that time and with human officers wearing black Gestapo-type uniforms. The overall effect from these two events was a confused and traumatic experience where ETs were suddenly painted with the paintbrush of fear. The long-term fallout from Barnet and Betty Hill's experience is that there are ongoing abduction scenarios being claimed by millions of people, particularly in the United States and now growing to become a worldwide phenomenon, as an adjunct to the whole UFO/ETI phenomenon.

What is interesting is that there are military bases close to a lot of these abductees usually located within 50 to 1oo miles of their homes and therefore, by air travel, a helicopter or a jet can be over these people's homes literally within minutes to carry out a MILAB.

In other words, most people claiming an abduction experience have been the unwilling victims of a military abduction staged-craft event using reversed engineered alien reproduction vehicles piloted by programmed synthetic or cybernetic life forms designed to look like short Grey extraterrestrial beings! *These are **NOT TRUE** Extraterrestrial Intelligences (Greys); they are* **HUMAN CREATED ET BEINGS (PLFs – Programmable Life Forms)!!!**

We have already discussed in this book that the military is in possession of **Alien Reproduction Vehicles (ARVs)** since the '50s and definitely in full operation during the '60s. The military also has in its possession and control **Programmable Life Forms (PLFs),** genetically created to look and function like little Grey ET beings but, controlled by humans. To all intents and purposes, they are mistaken as actual ET beings but, they are not from another planet or star system but, are "home grown" in underground military bases, here on Earth by covert rogue scientists working within the M.I.C. This is a known fact that is often overlooked by Ufologists and those few inquiring scientists interested in the subject.

Together these two assets are unleashed upon an unsuspecting public for covert campaigns of disinformation, fearmongering, terrorist activity while also, instilling hatred and prejudice toward the actual ETI in which they mimic. These campaigns of fear-mongering inflict a negative mindset toward all Extraterrestrial Intelligences visiting the Earth thereby, setting the stage for an orchestrated **"false flag alien invasion"** of Earth designed to control the human masses upon an already over-populated planet, causing mankind to given up many of its rights and freedoms for protection by the *"Military Mafia"* (aka. M.I.C.).

Look carefully at all the abduction claims which usually follow a carefully scripted set of actions that often perceived by the victims to be against his or her will or beyond their control causing them fear and panic. The individual believes he has become a hapless victim of an *alleged* alien abduction by an alien force or ET entities entering his home for a fulfillment of a nefarious agenda. Search the abduction documents, ask the questions that few people ask, see if there is a human explanation that fits the parameters of the abduction, look for possible human motives, hidden agendas and missions that promote the enslavement of humanity in a fascist or totalitarian regime or a world government plot based upon materialism that benefits only the few wealthy corporate elite and not the global population. If you don't find the answers, then and only then, look above for the ET explanation. It is a simple case of **Occam's Razor**: *"the simplest answer is usually the correct one!"*

Alien abductions are essentially circumstantial evidence not founded on a logical, progressive flow of facts and evidence but, are instead an irrational leap of illogical where the researcher forces the circumstances to fit the evidence, bottom line: the facts don't fit the evidence! The right questions have not been asked, therefore, the **UFO database** is corrupted.

This unsound reasoning is the basis of much of the faulty conclusions often reached in the UFO community which in turn becomes the catalyst by which the **Military Industrial Complex** propagate further disinformation into the UFO community.

The UFO corrupted date base has become a conflagration besetting the **UFO Community** threatening to tear it apart and thus, fulfilling the machinations of the military-industrial complex. Containment of this corrupt data requires a **"Dead Man Zone" (DMZ)** to control the spread of disinformation and misinformation that comes from such things as alien abductions and cattle mutilations, so- called ET-human hybridization programs, false interstellar exchange programs like SERPO, yetis, and sasquatches as ET interdimensional beings, all of these are fuel additives that cause the fires of UFO propaganda to leap beyond its contained boundaries to confuse honest researchers and add negative connotations and mindsets to the UFO/ETI phenomenon.

This corrupted UFO database needs a major overall, an honest introspection of itself, along with a re-analysis of the collected data thus far. Some serious hard questions have to be asked as to how this UFO data became flawed and why no one has come forward to suggest a course correction that redefines the mission statement of UFO and ET research and investigation. Part of the problem has been the infiltration of most national UFO organizations by the military departments and intelligence agencies which have collapsed such organizations as APRO and NICAP. Unfortunately, they continue to operate within the national organizations of CUFOS and MUFON going undetected and *"guiding the research"* into bogus cases that lead to only one inevitable outcome of disinformation or distortion of truth. Most of all, there is a need for the elimination of these agent provocateurs and the egotistical personalities that have upstaged or hindered the progress of unbiased, unfettered research of the UFO/ETI phenomenon.

Are Extraterrestrials Benevolent?

This flip side of the argument is focused on a common sense, more rational approach to investigation by asking the right questions and steering clear of all leaps of illogic which might inadvertently dictate the direction of research. The evidence is based on historical records and documents as well as current UFO and ETI investigative research. The hypothesis of benevolence states if Extraterrestrial Intelligences have always been in our midst since time immemorial, there is at least, over ten thousand years of recorded history that we have never been alone in the universe.

Supporters of the peaceful benevolent ETI hypothesis state that Extraterrestrial Intelligences have been watching, monitoring and occasionally influencing the course of history including the possible evolutionary development of humanity. If we accept the theory that ET beings who have been visiting the Earth are more advanced technologically, socially and spiritually than humans, they as a superior intelligence, if they wish to dominate or enslave humanity or simply wipe us all out, they could have done it tens of thousands of years ago, long before we started down the road to building civilizations and constructing retaliatory weapons of mass destruction.

Not once during all that time has ETI displayed any malevolence toward mankind and they certainly have had more than enough time to eliminate or dominate humanity hundreds of times over. It is simply not a part of their agenda! In fact, they are more appalled and concerned with our current state of development than any other time in our history yet, **we are still breathing the free air of Earth!!**

Primary **supporters of peaceful and benevolent ETI** are **Dr. Steven M. Greer, Dr. Joseph Burkes, astronaut Dr. Edgar Mitchell, Martin Cannon, Jacques Vallee, Barbara LambMFCC, Wendelle Stevens, Leonard Stringfield, James Harder Ph.D., Ralph & Judy Blum, Jim and Coral Lorenzen** and many others, *including this author*.

Careful examination of the UFO literature points to the inescapable conclusion that they are not only concerned for the welfare of mankind but, on many occasions have tried to warn us of our shortcomings and our overarching penchant for "tribal" or global warfare.

There are even numerous reported cases where ETI have interceded unexpectedly in providing cures and healings to many sick and afflicted people. **Preston Dennett** in his book ***"UFO Healings"*** recounts hundreds of these cases where ETs have come to the aid of humans who are sick or dying of some disease and heal them to full restorative health! His book reveals the compelling medical evidence for UFOs, how so-called "abductees" or "experiencers" are taught by extraterrestrials how to perform psychic hands-on healing. Dennett reveals the many parallels between UFO healings and other miraculous healings including cures resulting from lightning strikes and religious miracles.

These are just some of the diseases, injuries and illnesses from more than 100 true firsthand cases of healings of injuries, illnesses and diseases that have had ET healing: AIDs, arthritis, asthma, burns, cancer, colds, diabetes, diphtheria, flesh wounds, infertility, kidney stones, liver disease, multiple sclerosis, muscular dystrophy, myopia, pneumonia, polio, tuberculosis, yeast infections and many more. http://prestondennett.weebly.com/ufo-healings.html

A curious fact is all the research into abduction cases, three of the most prominent alien abduction experts are mystified by these acts of medical benevolence because they don't fit the abduction scenario paradigm of malevolent ET behaviour calling into question as to what is really going on and what is the true alien agenda. For example, **Budd Hopkins** says,

"The question is whether we hear about healing cases. We do sometimes, very rarely, but they do turn up. And we don't know what to make of them. It's kind of a sad thing because I have some abductees who have serious medical problems who wish they were being healed themselves but are not...so we don't know. Incidentally, again, there's no evidence whatsoever that this is a malevolent, evil conspiracy going on in the sky against us; they're going to take us over, or anything else. I'm very optimistic about the outcome because they seem to be most interested in what I consider the most lovely aspects of being human. They're interested in that. There's no evidence, however, that they're here to help us. We wouldn't perhaps have AIDS and the hole in the ozone layer and everything else if they were here to help us. So there's no sense to this."

David Jacobs Ph.D. staying true to form in his research model of abductions (which this author considers is seriously flawed as only some of his assessment are pertinent while the majority are distortions and poorly evaluated conclusions) states,

376

"In extremely rare cases, the aliens will undertake a cure of some ailment troubling the abductee. This is not in any way related to the contactee/Space Brother concepts of benevolent aliens coming to Earth to cure cancer. Rather, in special circumstances, it appears that aliens feel obliged to preserve the specimen for their own purposes. As one abductee said, 'It's

equipment maintenance."

John Mack MD is somewhat more forgiving but perplexed,

"Some encounters are more sinister, traumatizing and mysterious. Others seem to bear a healing and education intent...many abductees have experienced or witnessed healing conditions ranging from minor wounds to pneumonia, childhood leukemia, and even in one case reported to me first-hand, the overcoming of muscular atrophy in a leg related to poliomyelitis."
http://prestondennett.weebly.com/ufo-healings.html

If as Dennett posits that Extraterrestrials are curing people, then we have to seriously re-evaluate the central feature of most abduction accounts, that being the medical examination that is claimed by most abductees. These medical examinations indicate that ETs seem to know a great deal about human anatomy and its afflictions which should not come as a surprise given the obvious advanced technology of their spacecraft and the fact that Extraterrestrials have been around a long time observing and interacting with mankind, it becomes clear that it is well within the capability of ETI to cure a large number of diseases.

Consider two well-known cases that many researchers in the UFO community have always thought of as irrefutable abduction accounts namely, **The Walton Experience** and **The Allagash Abductions.** Recall elsewhere in this book, that **Travis Walton** disappeared for five days after being hit by what his horrified fellow loggers thought was a light beam weapon fired from a flying saucer seen hovering in the wood just off a logging road. Walton is knocked to the ground and his sudden panic-stricken companions drive madly away down the road leaving their logging buddy behind. Walton awakes three or four days later inside strange and unfamiliar surroundings of the ET spacecraft, he sees strange short stature white complexion creatures in uniforms, who are obviously not human standing over him. Dazed and confused, he gets up off a medical table and lashes out at these ET benefactors in typical human fight or flight response. These ETs flee the room realizing that their unappreciative patient wants to "brain" them with one of the medical instruments. Long story short Travis Walton is sedated by human-like ETs and is then dropped back off on Earth near his hometown. The national newspapers and TV networks have "feeding frenzy" over his story and Ufologists conclude he was the victim of an alien weapon and abduction experience on board an alien spaceship for five days.

Even Travis was convinced that he experienced an alien abduction but, recently in 2013, he has recanted his interpretation of the events. Now, after much reflection and after being interviewed by saner rational minds like **Dr. Steven Greer**, is seriously convinced that his stupidity was the cause of his five-day absence off of planet Earth. His close approach to a highly electromagnetically charged spacecraft which discharged its energy accidently at him, knocking him off his feet and into unconsciousness critically injuring him with a probable cardiac arrest like being hit with a bolt of lightning required immediate medical treatment and follow-up observation to ensure his recovery to full health. These strange white skinned ETs were, in fact, saving Walton's life and were falsely accused of being his abductors when in reality they were his medical physicians! Finally, by his own admission, Travis Walton's experience has come full circle with a view of optimism that ETs are not malevolent but poorly misunderstood by humans!

In the second case also discussed previously in this book, four young men, **Jim & Jack Weiner, Charlie Foltz and Chuck Rak,** all good friends and fishing buddies with each other and with a common bond of all being artists head off to Eagle Lake for a few days of fishing in Allagash, Maine. They see a large brilliant globe light one night while fishing on the lake but, it merely moves around in the sky and flies off. The next night, again on the lake for a late-night fishing session, they see what they believe is the same UFO but this time, it hovers directly over their canoe. All four men are levitated or teleported aboard the hovering spacecraft from out of their canoe and are undressed and given medical reproduction examinations. They are then returned to their fishing camp, their canoe is already banked upon the shore and their fire by the campsite has burned down to embers indicating that they are missing some hours of their life with no conscious recall of the events that have just transpired. It is morning and a mysterious man shows up at their cap who claims to be a park ranger or supervisor who is photographed with the men and is never heard of again, no one seems to know who this man is even from those within the parks branch.

Were these men merely experiencing the abduction scenario played out once again, this time upon them as unwilling participants or was it ETs taking an interest in helping these men with corrective medical procedures or was it something else?

Chuck Rak told this author in July 1996 that he was not convinced that the UFO/ETI event that he and the others experienced was a true ET encounter but, a possible military abduction event using **programmable life forms (PLFs)** in a false flag ET encounter designed to be perceived as an **alien abduction**! The men later undergo hypnotic regression by Ufologist, Raymond Fowler who later writes a book about the men's experience called the "The Allagash Abductions".

Stepping back and taking in the bigger picture, one has to admit that some UFO and ET encounters may on the surface appear to be an abduction event particularly in the minds of those who are experiencing such a traumatic event. But, abduction, when instigated by humans against humans, are not only traumatic affairs but most abductions don't end well for the victim, they are usually never returned to their point of origin, murder is the final outcome in these situations! Alien abductions in which humans are always return to their home or wherever they were first picked up from can hardly be said to be an abduction, even if the unwillingness of the experience by the victim was terrifying and the reason was for medical examination or biopsies or DNA retrieval, this alien encounter cannot be said to be an abduction.

A trauma doctor of medicine in any hospital will see mutilations almost on a daily basis from gunshot wounds, knife stabbings or grisly vehicle accidents or he may have a child come into emergency ward who has swallowed something which has become lodged deep down in the throat causing major breathing problems and risking the child's life. To the child's mother or even the semi-conscious child, the doctor may appear to be the incarnation of pure evil intent on

378

harming the child further by before a tracheotomy on the throat of the child, all without anesthesia in order to save the child's life.

In other words, perception and understanding are everything and not everything perceived is actually understood or perceived as it is intended, it may be something else quite entirely! Extraterrestrial healing of human infirmities and diseases hardly sounds like the malevolent acts of aggressive ETs or an agenda of hostility designed in their overthrow of humanity and conquest of our planet!

Another point of consideration that ETs are peaceful and not malevolent is the documented accounts in which Dr. Steven Greer has lead thousands of people out into remote areas to establish contact with Extraterrestrial Intelligences and on no occasion has there been a single case of missing time or alien abduction among the investigative teams. Even, people like Sixto La Paz and his ET contact teams, as well as this author's CSETI Vancouver team and many other teams of CE-5 researchers worldwide, all have never experienced any alien threats or acts of aggression towards us while engaging with ETI, not a single one! So, what are we, chopped liver? Do we not meet the criteria of alien expectations for abductions and therefore, we are simply left in the fields staring up at them in their craft in wonderment?

This really begs the question, what is really going on and is the peaceful initiatives from humans toward ETs the real crux of the situation? Why is it that so many people claim that they have been abducted by aliens and given intrusive medical examinations while a growing body of evidence is indicating that Extraterrestrial encounters are not hostile but peaceful? In fact, these human initiated encounters with ETs is *positively life altering for the human recipients,* as they can now firmly *acknowledge that they are not alone in the universe* and that it is possible to have *a mutual, sustainable and peaceful relationship with visiting Extraterrestrial intelligences!*

Author's Rant: Could this be the real reason behind the hidden agenda of suppression and cover-up of all thing extraterrestrial by the Military Industrial Complex? They simply don't want the public to know about the nature of Extraterrestrial Intelligences and their alien technology. Such alien technology which is light-years ahead of any human technology would have a profound and positive impact upon the course of society which would remove once and for all, the hegemonic controlling forces of the oligarchic wealthy corporate elite that is supported by the M.I.C.!

It should be pointed out, that since 1990, the CSETI Initiative founded by Dr Steven Greer and now operating with more a thousand CSETI CE-5 groups world-wide have gone out to remote locations and made peaceful, sustainable and mutually beneficial contact and communications with Extraterrestrial Intelligences.

There have never been any reports of any malevolent contact, aggression or hostility between humans and visiting ETI. There are no abductions of humans, no invasive medical procedures foreign implants place into people and no loss of psycho-motive ability when contact has been established.

CSETI and the numerous global groups have deliberately put themselves in harms way and yet, the results of the night ET contact work have always been repeatedly a positive experience by all who were involved in the group event. People experience many positive and escalating unusual events during the night's CE-5 encounters.

This flies in complete opposition to all the negative reports of people stating that they have been abducted by aliens or had invasive medical procedures on their person. Why is there this difference in ET- human engagement? Why haven't CSETI groups not experienced this alien negativity? Are we just chopped liver to the whims and agendas of ETI or is there a secret covert agenda being perpetrated by the Military Intelligence Industrial Complex in a false flag alien invasion scenario that has been slowly escalating since the 1960s up to the present time?

Is There an Alien Agenda?

Is there a carefully staged alien invasion of Earth in the near future? There are many websites that would like you to believe that an alien invasion of the Earth has been an ongoing program for thousands of years following a carefully choreographed agenda where there is little or no violence. Instead, the plan of attack is rather benign and very subtle with a good overdosing of manipulation on every level of society via mind control, coercion, threats, and intimidation. The ultimate goal is to bring humanity into submission without hostility, but with ultimate compliance by removing our rights and freedoms thus, forcing humanity to go along with an alien agenda, where we are enslaved to a master race and become a part of their empire. There are too many opposing factors to this hypothesis like humanity's nature of non-confinement or enslavement or the right to exercise free-will and choice as promulgated in all religious writings down through the ages which contradict this absurd and paranoiac belief. This is one more of those pieces of disinformation corrupting the database of Ufology, where people are simply not exercising any common sense and rationality to resolve the nonsense being perpetrated upon them.

When one culture or civilization moves in on another civilization like when Europeans came to North, Central, and South America, the takeover of the indigenous civilization was fairly swift and by the end of the first two hundred years, European culture was not only well established but, by the next couple of hundred years, the original indigenous culture had almost vanished or was segregated onto specified land reserves. This example demonstrated that a more technically advanced civilization can subjugate a less technical society very quickly and thus, establish itself as the ruling power on a continent.

A highly advanced alien civilization could easily have taken over this planet thousands of years if they so desired, when we were still running around in bearskins and our most advanced weapon was the bow and arrow and spear! The fact that we have been able to advance to our current civilization clearly points to the fact that this was never the agenda of ETI visiting this planet. *We are still able to breathe the free air of Earth!* – **Dr. Steven Greer**

Are ETs here to eat us for lunch? This concept makes for good science fiction or horror stories for late night television and movies, but honestly, we need to give our heads a shake! The very fact that we are still breathing the free air of Earth, says that this is a false perception of any alien agenda. In reality, there is no foundation whatsoever, for this belief.

Are they here as our saviours or are they demons from space? Some people believe that ETs represent a sort of divine salvation like some sort of messianic Christ in a flying saucer come to save us from ourselves! This type of question implies a fundamentalist dogmatic point of view often found in the pulpits of churches, synagogues, and mosques of most religions. They are not here to be worshipped like gods, no matter how advanced they may appear to be, nor are they demons or **Darth Vader** along with his evil minions to enslave us in some alien empire. We need to put aside such frightening perceptions and observe with the sound eye of discernment and a dispassion of neutrality until we can make a wise judgment of the ET intent and motives.

There does appear, however, to be a guiding hand in the affairs of mankind but, outright interference is not permitted, it is a kind of *"cosmic prime directive"*, ala Star Trek! Can ET show up on mass as a display of aerial force? Certainly, they can, but that isn't interference, it is a display of presence to invoke a statement capability probably out of genuine concern for our current development and for the future of humanity.

Are ETs here for our natural resources? Zecharia Sitchin wrote a number of interesting alternative history books that stated essentially that humanity was the product of a humanoid race known as the Annunaki who created us as a slave race to mine gold in which these ETs could use in their atmosphere to solve atmospheric and planetary problems of their home, Nibiru. Ever since those successful publications hit the bookshelves globally, there have many UFO researchers and "wannabe Ufologists" who have jumped on the bandwagon to write and published their own or similar versions of Sitchin's work.

Are they here to completely mine out all our mineral resources, to strip the planet bare, leaving the world barren before moving on to the next planet in their search for more resources,

irrespective of its impact on its inhabitants and various life forms? Frankly, we are doing a great job of that all on our own, without the help invading resource hungry aliens. Stripping our planet of all kinds of resources, polluting the environment, killing off whole species of plants, animals, insects, and aquatic life seems to have little impact upon the consciousness of those who are in control of the Earth resources, its political, military and religious systems and financial resources.

To the rest of mankind it an alarming problem of epidemic proportions and maybe, just maybe, the real reason why ETIs are here is perhaps, to tell humanity to take back its power from the fascist kleptocracies that have usurped the world of its power and control and kept it in constant chaos and in an accelerating decline toward global poverty!!!

We simply don't have anything that a technologically advanced interstellar civilization wants because in order to travel among the stars such civilizations would have solved the propulsion problems of space travel, including the engineering of spacecraft and construction of their cities all without mining the resources out of any planet. All advanced civilizations who have reached true interstellar or intergalactic capability in space travel have developed to the evolutionary stage of mind control along with the ability of thought manipulation of the basic building blocks of life by extracting these out of the material matrix, the etheric plane of reality and the quantum flux of space.

Any advanced Extraterrestrial intelligences that have progressed to becoming a **Type I** or **Type II Civilization** or higher will have evolved to a state where impact upon the environment, whether planetary, solar, interstellar or galactic will be minimal or non-existent. All life whether mineral, plant, animal, even planetary or stellar becomes sacred. To extract resources or destroy planetary environments in order to sustain an intelligent, sentient species' evolutionary development is neither intelligent nor sentient in its actions.

Highly advanced civilizations greater than Type I extract resources from the quantum flux of space itself or from the matrix of ideation and causation that if needed, can be materialized into the matrix of the three-dimensional universe. It is essentially, only our limited perception and understanding to grasp the concept and the ability to materialize something out of the state of nothingness! In reality, we and everything in the universe are nothing more than energy, frequency, and resonance!!

In other words from the thought idea to the causation of the thought to the three dimensional matrix of this reality, should an ETI require a physical spaceship to travel in, they could manifest its physical reality for their use. However, any truly advanced interstellar civilization would have no need of physical spaceships, as they could travel in ships constructed of pure thought, ships constructed of light, if necessary. A super advanced civilization could also travel by mere thought alone, a kind of thought teleportation through any dimension to the next.

So, we really don't have anything they want, other than that they want us to live and thrive in peace and if we can't do that simple thing then, they will find a way to ensure that life on this planet doesn't perish, even if we can't all be saved!

Human Hostility Abounds - Land At your Own Risk!

The alien invasion scenario has been used as an allegory for a protest against military hegemony and the societal ills of the time. In H.G. Wells' novel *The War of the Worlds*, he exploits invasion fears by aliens that were common when science fiction was first emerging as a genre. Today, the fear is still with us but, more and more of society is awakening and becoming better educated and better informed; we are simply not just going to accept whatever is being told to us by our elected or appointed officials, without having several sources of confirmation to corroborate our beliefs and our concerns.

We must also consider that if ETI are not malevolent, could it be we, who show signs of aggression and hostility towards those who may come here in peace? Are we the "Klingons" of this sector of the universe?

You would think that as sane, rational human beings which we claim to be in this day and age, that after two self-inflicted world wars among ourselves, along with all previous and post-war conflicts, which have lead to immeasurable misery and suffering, chaos and ruin, and just basically Hell on Earth for our species, these painful and murderous life-lessons should have been more than enough to teach humanity to grow up and seek a more positive approach to the problems that we so frequently encounter.

We have seen this during the Second World War when ET spacecraft showed up in the two major theatres of war, in Europe and in the Pacific as **"foo fighters",** as they were call back then. The L.A. Incident of 1942 where a barrage of US Army artillery was targeted at an Unidentified Flying Object seen flying over Los Angeles may have constituted the first open act of aggression towards an Extraterrestrial spacecraft. They were observing us in global conflict with ourselves and when we exploded the first atomic bombs in 1945 in the American Southwest and again, over the cities of Nagasaki and Hiroshima, Japan, they came and investigated to see what the hell was going on. In no small measure was this just an explosion of a big bomb, it sent ripples through the fabric of the time –space continuum, where other intelligent beings reside. We had sent a signal to all other intelligent life forms in the universe that the *Earth "kids have found the matches to the explosives and seem intent on burning down the planet!"* If you want to get the attention of another intelligent species this is how you do it!!!

Small wonder then, that ETI have come calling to check out the American southwest, particularly the **First Atomic Bomb Wing 509**[th] at Roswell Army Air Force Base near Roswell, New Mexico in 1947. Strange craft (UFOs) were reported flying in the area and the military was prepared for them with a scalar weapon (not with high-powered radar as was initially thought to be the case) to shoot them down. Thus, began the legend of the **Roswell Saucer Crash Incident of 1947** and the growing emergence of hostility towards UFOs and ETI.

Varginha, Brazil Saucer Crash of 1996: Another infamous act of human-initiated aggression towards ETs occurred in 1996 Jan 20 in Varginha, Brazil, when the Brazilian Military used a particle beam weapon to bring down a UFO which had been tracked by the US Military deep space radar. The US then informed the Brazilian Military of the inbound alien craft would come down in their country. The space vehicle crashed to earth on the outskirts of Varginha and seven little brown coloured ET beings escaped for safety only to be rounded by the local fire department and the Brazilian Army. When news about this event reached English-speaking countries, it quickly became a case that ranked second only to the Roswell crash in annals of Ufology.

Spin control was carried out in semi-debunking new articles like those found in the Wall Street Journal in July 1996 which was followed within a few days by similar articles in England's Sunday Times and the Christian Science Monitor. It seemed that there was a coordinated global media damage control in effect.

CSETI had received a confirmed report of particular interest back in Feb. 2000 that Brazil possesses **scalar electromagnetic weapons** *(both longitudinal wave interferometry and quantum potential)* . This program is centered at a facility in the suburbs of Sao Paulo. Not only would this account for much of the ET and UFO interest and activity in Brazil, but also the intense disinformation campaign featuring well reported *"alien abductions",* etc. to cover up the reason for the interest.

It is possible, therefore, that the Varginha craft was, in fact, ***shot down by Brazilian scalar EM weaponry.*** http://www.cseti.org/crashes/089.htm

Seven small brown skinned ET beings with large red eyes and having three bumps or protrusion on their heads had crashed their mini-van sized space vehicle and were caught wondering close to the city of Varginha seeking possibly some safety. They were caught either individually or in small groups of twos or threes; in one situation, it is reported that the local fire dept of the city aided the Brazilian military to catch one of the little ET beings on an embankment off a road outside of town using a net. The firemen snared the creature in the net where they dragged it back up to the road to the waiting military soldiers who immediately placed the frightened little being into a shipping crate and nailed a lid down onto the box to contain the creature inside. ***Welcome to the planet Earth!***

This scenario has been played out numerous times in many different countries and the one overriding factor is that many ET spacecraft have been shot down by the military and ET beings were recovered or caught away from their craft. The retrieval of ET craft and bodies are typically shipped off to parts unknown in the US, but it is presumed that they are sent to the closest military base and then, from there, it's anyone's guess.

Even, the common folk are not exempt from displaying inappropriate or aggressive behaviour toward ETs that have landed on Earth. There are reports of farmers shooting at ET spacecraft that come hovering around their farms and homes; cattle or livestock kick up a fuss and the farmer comes out with a shotgun to deal with these "alien intruders." Another case reports that hunters stumble across an ET craft on the ground. They start throwing stones, rocks and tree branches at the object to see if anything is inside will poke its head out then, when that doesn't happen they begin firing their guns at the landed craft reporting the bullets ricocheting off the hull of the craft. The flying saucer then rises and flies off and is quickly out of site.

CHAPTER 113

HUMAN AGGRESSION AND HOSTILITY - A FEW CLASSIC CE-3 CASES REVISITED

The 1954 José Ponce / Gustavo González Incident

An example of this open irrational aggressive is the classic case occurred in Caracas, Venezuela on November 28, 1954, when two young men, **Gustavo Gonzales** and **Jose Ponce** were driving their flatbed truck to store at 2:00 o'clock in the morning to buy foodstuffs at Petare, 20 minutes away from their home.

Driving along Buena Vista Street, they were surprised to see the street illuminated as though it were 12 noon. Upon exiting the van to see what was afoot, José suddenly ran back toward the vehicle after seeing a strange entity approaching them.

Seconds later, Gustavo also saw the creature and was at first hesitant, but then advanced toward it and wrapped his arms around the being to capture it and drag it back to the van. The small alien, however, was rather strong and managed to break away from the hold. Upon releasing itself, Gustavo fell to the pavement but managed to spring up quickly. According to Gustavo, the entity weighed some 50 kilograms (110 pounds) when he lifted it.

While he followed the small alien, he noticed something even more surprising: two other small aliens were approaching him. One of them flashed him with a "flashlight" – apparently, they had come to assist their comrade. Blinded by the light and unable to see what was happening for a few seconds, he took hold of his Boy Scout knife when his vision was restored and saw that the same diminutive alien was now coming toward him. Instinctively, the man stabbed the creature's shoulder, only to feel the blade slip off its skin, which was as tough as rhinoceros hide. When the extraterrestrial tried to seize him, Gustavo realized it had sharp claws on each of its four fingers. Meanwhile, his assistant, José Ponce, emerged from the right side of the van and headed toward the spherical object. Suddenly, a small hairy extraterrestrial emerged from the right, hurriedly walking up a steep slope with fistfuls of dirt in its hand.

When the tiny alien noticed Ponce, it jumped two meters, entered the hatch and vanished into the object. Seconds later, another entity emerged, armed with a long, shiny tube in its hands, pointing it at both men. They suddenly felt a vibration that encompassed their bodies – Gustavo and José were rendered paralyzed. They later saw the brilliant sphere rising majestically and silently to a point in the night sky before vanishing altogether.

José Ponce and **Gustavo Gonzales** were able to describe the aliens thus: the one Gustavo grappled with was hairy, noseless, with glowing eyes, short stature, barefoot and wore a strange item of clothing resembling a loincloth. It was very agile and strong for its size and was able to break away from his grasp, leaping like a cat.
http://inexplicata.blogspot.ca/2013/12/venezuela-classic-ce3k-revisited-1954.html

Corel Lorenzen reports that the description of the event (slightly different) tells that Gonzales in a state of sudden impulsiveness immediately grabbed at the little hairy man (3 feet tall) intending to take him back to the police. To Gonzales' total surprise the little man was light in weight with stiff bristly hair and a hard body but, more astonishing was that this creature was extremely strong as he pushed Gonzales back with one clawed hand sending flying about 15 feet. Ponce in sheer panic ran off to the police station located a short distance away but, before leaving his friend he noticed two other little hairy creatures emerging from some bushes carrying what looked like chunks of rock or dirt as the hopped aboard an opening in the side of the sphere estimated as 10 feet in diameter.

Gonzales and Ponce arrive at the police station who suspected the men of being drunk but, after some sedatives to calm both men down and after an examination by police that it is discovered that Gonzales has suffered a long red scratch on his side. Several days later a doctor came forward to state that he had witnessed the fracas with Gonzales and the creature but, not wishing undesirable publicity, he left the scene. Flying Saucers: The Startling Evidence of the invasion from Outer Space by Corel E. Lorenzen; 1966; published by Signet Books; New York, USA

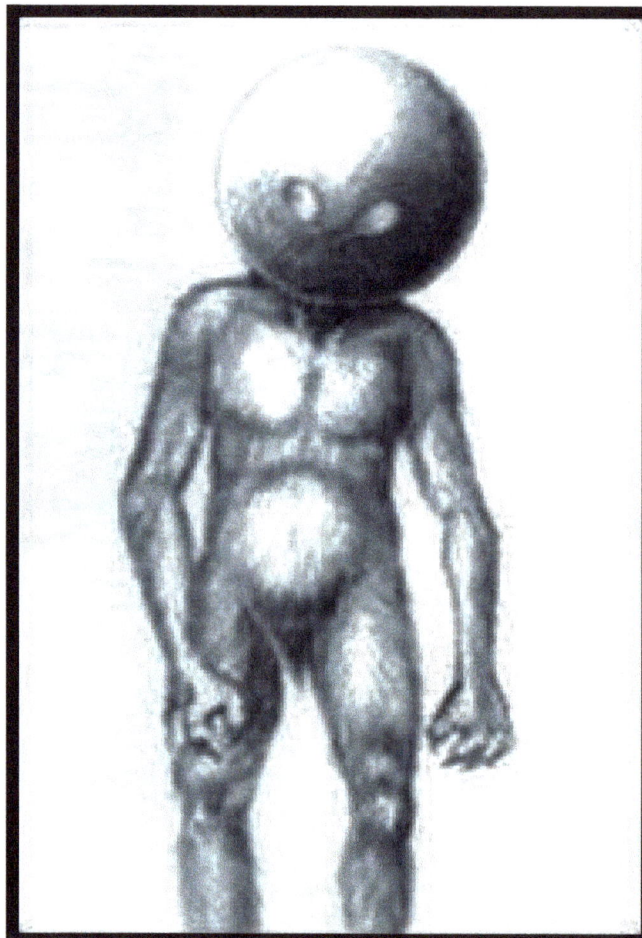

The short hairy creature that Gonzales fought with on a road near Caracas, Venezuela
http://inexplicata.blogspot.ca/2013/12/venezuela-classic-ce3k-revisited-1954.html

**Gonzales and Ponce physically engage some Extraterrestrial
Beings on a road outside of Caracas,**

The two versions of the story both seem to indicate that Gonzales action was provocative and could have been interpreted as aggressive and thus, the ET beings merely defended themselves from larger human beings and potentially a more aggressive attack by paralyzing the two individuals in order to make their escape and fly away from a dangerous situation.

The Kelly–Hopkinsville Encounter

A second even more bizarre case where there was repeated acts open hostility between human and Extraterrestrials is the **Kelly–Hopkinsville Encounter** of August 21, 1955, where a shootout between a family and a bunch of small, very strange looking, unarmed ETs that had landed near the Sutton farm homestead**.**

This most unusual event originated in the rural area of Christian County, Kentucky, this UFO enigma took place in the little town of Kelly, located near the small city of Hopkinsville. The Sutton family lifestyle was typical of a traditional Kentucky rural family which hadn't changed in many decades. "**Lucky" Sutton**, as he was known to friends and neighbors, was the "patriarch" of this bluegrass clan.

Visiting Lucky and his family was a man from Pennsylvania named **Billy Ray Taylor**. Billy left the Sutton house to go for some water from the family well as there was no inside plumbing at the Sutton farmhouse. At the well, he saw an immense, shining object land in a small gully about a quarter of a mile away. Running back to the house, he excitedly reported his sighting to others in the house who only laughed at Billy and his "crazy" tale.

After a short period of time, the family dog began to raise a ruckus outside. Hearing the dog barking, Lucky, and Billy, in customary Kentucky outback fashion grabbed their guns and

387

headed outside, planning to shoot first, and ask questions later.

Only a short distance from the front door, both men were stopped dead in their tracks by the sight of a 3-4 foot tall creature, which was walking towards them with hands up, as if to surrender. Frightened by the **small "goblin-like" entity**, Billy Ray fired a shot with his .22, and Lucky unloaded with his shotgun.

This bizarre-looking creature would be described as having a greenish complexion, "large eyes, a long thin mouth, large ears, thin short legs, and hands ending in claws." (See Sutton family drawings and 3D model). http://www.ufocasebook.com/Kelly-Hopkinsville.html

Both men later admitted that there was no way they missed the creature at close range, but the little being just did a back flip and ran into the woods in fright.

No sooner had the two men re-entered the house, the creature or another like it, appeared at a window. They took a shot at him, leaving a blast hole through the screen. They ran back outside to see if the creature was dead but found no trace of it. Standing at the front of the house, the men were terrified by a clawed hand reaching down from the roof in an attempt to touch them.

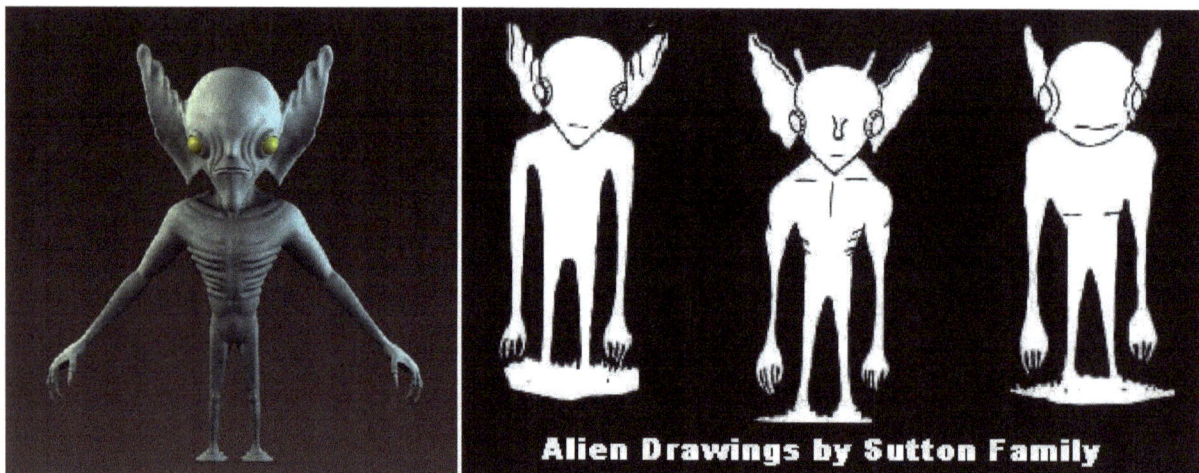

Alien Drawings by Sutton Family

The Sutton family was terrorized by strange looking ET beings described as "Green Goblins" who seemed unaffected by rifle and shotgun fire from the Suttons
https://www.turbosquid.com/3d-models/3d-alien-hopkinsville/670175 and http://www.ufocasebook.com/Kelly-Hopkinsville.html

Again, they shot, but the long-eared being simply floated to the ground and scurried into the cover of the woods. http://www.ufocasebook.com/Kelly-Hopkinsville.html

The two men sought the protection of the house again, only to find themselves under siege from these little men. For a time, the entities seemed to tease the family, appearing from one window to another.

Taking pot shots through the windows and walls, their weapons seemed totally ineffective against the invading creatures. After several hours of fear, the Sutton family decided to make a break from the house and get help at the Police station at Hopkinsville.

388

Family members took two vehicles to the Police Station in Hopkinsville and reported their strange tale to **Sheriff Russell Greenwell**. Finally persuading the policemen that they were not joking, the authorities agreed to visit the Sutton house.

Arriving at the farm, Greenwell, and twenty plus police officers searched the house and the surrounding grounds but, found no trace of the creatures, but did find numerous bullet and rifle holes in the windows and walls.

Police reported that the Suttons seemed sober and were genuinely frightened by something and their reports for that night were entered as the "hearing of shots being fired," and the observation of "lights in the sky." The police left the Sutton place at about 2:15 am after exhausting all efforts to find the origin of this strange report.

No soon did the police leave, the creatures returned to the Sutton home. They began again peeking in the windows, seemingly out of curiosity. More gunfire took place, but again without effect.

Several more hours of cat and mouse antics ensued, finally stopping just before daybreak. The police were finally persuaded to call in Air Force personnel the next morning, but a new search brought no results. After the beings had left, Billy Ray and Lucky had gone into Evansville, Indiana to take care of some business. The other five family members were questioned by Air Force and Police.

The Sutton family encounter with strange little space "critters" although, initially thought of as being a hoax by some of the general public and by **Project Blue Book** was nevertheless, it was determined by independent investigation by **Dr. J. Allen Hynek** and **Isabel Davis** that there were no evidence of a hoax being perpetrated by the Sutton family and no financial gain was made off the story by the family. The case today remains extremely unusual and is considered authentic by many UFO investigators.

When we examine the details of this case, it seems that these strange ET beings may not have had previous experience with humans before but nonetheless, approached the Sutton family with arms raised as a universal gesture of non-hostility or that they carried any weapons. In an all too predictable human "fight of flight" response, "Lucky" Sutton and Billy Ray decide to open fire upon the little ET beings, not realizing that these beings may simply want to communicate. This is an excellent example that indicates that in many ways humanity is not yet ready to engage with visiting Extraterrestrial Intelligences on an equal footing of non-hostility.
http://www.ufocasebook.com/Kelly-Hopkinsville.html

Obviously, gun-toting humans who may not be highly educated or are socially challenged to the world and the universe around them would make any Extraterrestrial Intelligence give serious second thought to landing on this planet in order to make contact with human beings.

This particular ET-human encounter **CE-4** plays right into the **Brookings Institute Report** that essentially states that the revelation of the existence of ETI visiting this planet or when actual ET beings are encountered, it would cause chaos and panic among the general public. For the

Brookings Institute, this was a classic case in point in keeping such discoveries of ET presence hidden from the public until such time that humanity matures or is slowly conditioned to the reality of an Extraterrestrial presence. Problem with this think tank's assessment is that humans respond almost always in the same manner regardless of whether it's a couple of people, a family or as a fully trained professional military army, air force or navy, the only difference is the ultimate agenda or motive of the parties' involved in such close encounters.

Whenever ETs showed their presence in our skies as UFOs or over our towns, cities, or military bases, humanity's response, especially from the military perspective has been to shoot them down first and then, ask questions second. As cited and discussed earlier in this book, we've seen this situation repeatedly played out over every military base and missile site throughout every country around the world and it still continues to this day.

STS-48 Astronauts Video Record a Potential Shoot Down of Orbiting ETI Spacecraft

ETI show up at military or missile sites and then, temporarily disarming these weapons of mass destruction as a demonstration that ETI will not just stand by and allow nuclear arm conflict to

break loose upon this beautiful blue planet. The retaliatory response by the military to this ET action is open hostility towards those ET beings, who take their lives in their hands whenever they fly in orbit about our Earth or whenever they penetrate our atmosphere to land upon terra firma. The major superpowers have on many occasions open fire upon ET craft to shoot them down with missiles, usually with no effect whatsoever to the ETs. Orders are often give by military brass to send jets up to engage ET spacecraft that intruded into a sensitive military area, which often results in chases and "dog fights" with the outcome being that ETS have had to defend themselves with return fire of their weapons or flee to the outreaches of space above the Earth. As if this hostility toward ETI wasn't enough, ET spacecraft are often *"monitored" as they "phase" into our dimensional space* from wherever they originate from as there is a *"distinct magnetic anomaly signature"* given off by these ET craft when *"phase travelling between dimensions"*, they are then *"tracked by radar"*, *"targeted"* and then *"fired upon with particle beam weapons or scalar weapons"!* This was exactly what happened when the shuttle astronauts on mission **STS-48** inadvertently captured on video, back in September 1991.

This is *"Neanderthal" thinking,* a *"Cowboy"* mentality with a *"trigger-happy"* response to those who essentially come here in peace; this is how we greet another intelligent species who travelled a longways to come to Earth! There are numerous examples of this unprovoked militant hostility that comes not from inbound ET spacecraft but, from us humans!!!

The **Cabal,** (aka. the **Military Industrial Complex** or **Majestic 12**) are playing a potentially dangerous war game with alien forces beyond their comprehension. And unless, saner heads prevail and remove this **militaristic kleptocracy** along with its increasing military build-up of space-based weapons around the Earth, its continued provocative actions threatens to instigate a potential global calamity that will thrust humanity into an interplanetary war levelled against this planet with an unknown Extraterrestrial species in which the world knows nothing about its existence nor will want to wage war against it!!!

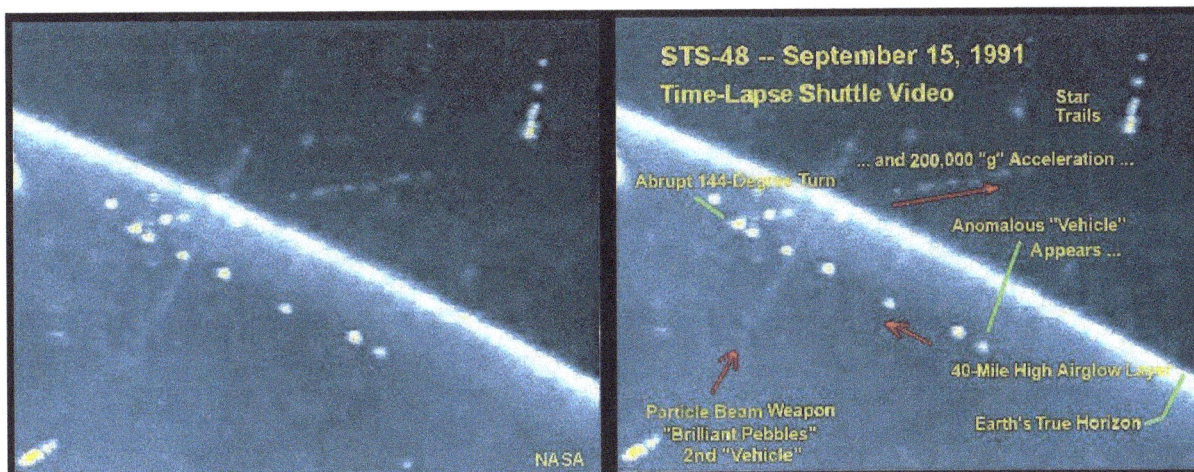

STS-48 (time-lapse shuttle video) showing the firing of a particle beam or scalar weapon upon ETI spacecraft not once, but twice (see image below) by a secret US military base situated in Pine Gap, Australia

http://www.sciences-fictions-histoires.com/blog/ovni-ufo/ovnis-le-film-de-la-mission-sts-48-de-la-navette-spatiale-discovery.html

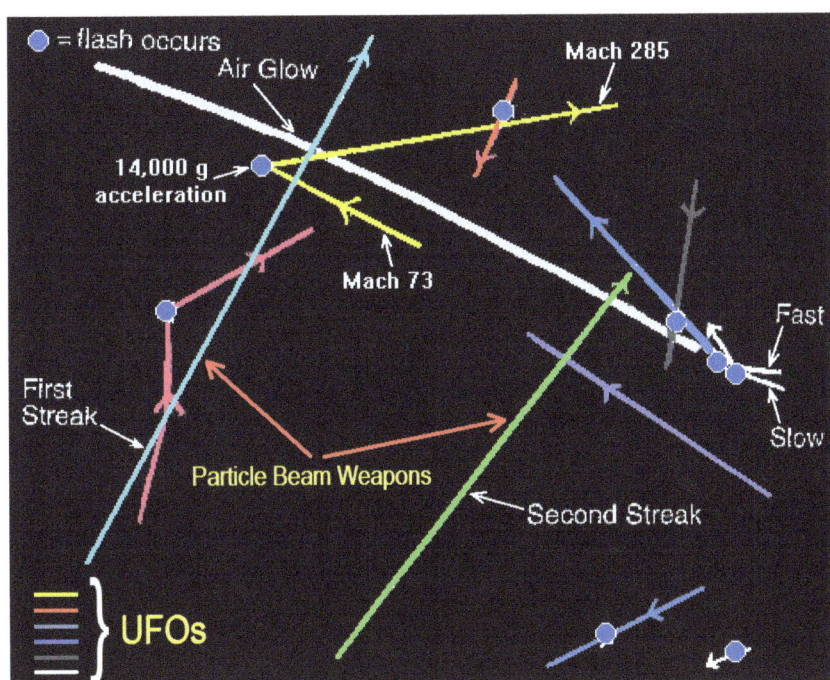

Trajectories of Particle Beam weapons targeted at UFOs – ETI spacecraft (See below explanations of trajectories of UFOs from scalar weapons)

- The distance from the Discovery to the Earth's horizon is 2,757 kilometers.
- The UFO's speed before accelerating into space is 87,000 kph (Mach 73).
- Three seconds after the light flash, the UFO changes direction sharply and accelerates off into space at 340,000 kph (Mach 285) within 2.2 seconds.
- Such an acceleration would produce 14,000 g of force.

http://www.sciences-fictions-histoires.com/blog/ovni-ufo/ovnis-le-film-de-la-mission-sts-48-de-la-navette-spatiale-discovery.html

The **alien invasion** is a common theme used in science fiction stories and film, and most recently in television documentaries where the truth is supposedly contained in "Unsealed Alien Files" which for the most part strongly suggest and re-enforce the concept that Extraterrestrial Intelligences are potentially hostile. Front and centre in all these scary invasion scenarios are the little ET beings known as the **'Greys'** or the **"Zeta Reticulians"** as well as the reptilian ETs known as **"Reptoids"** or **"Dracos"** or **"Draconians".** There is a deliberate fear-based agenda being perpetrated upon the public to create the spread of panic, chaos, and xenophobia, especially toward these two types of ET xeno-types.

This negative mindset toward ETIs is incrementally scaled upward annually by such TV programs and in movies, so that at some near future time, a false flag alien invasion scenario will be unleashed upon an unsuspecting public by the **Military Industrial Complex** and other cooperative covert intelligentsia. The MIC agenda is to seize more power and control over the general public by getting them to surrender more of their rights, freedoms, and privacy for the sake of planetary protection and security against invading aliens, but chiefly to obtain greater financial support from the public for their military and space-based programs. Where have we heard this before? We've seen it during the early 1930s with mobsters, gangs and the mafia taking over major cities across the US with their gangland slayings and killing of innocent bystanders causing societal chaos and fear. Later, as if we still didn't learn our learn back then, we let it happen again with the ruthless and corrupt fascist movement of Nazis Germany throughout Europe during the Second World War. In each time period, the psychological scare tactics and violence were the same and the casualty fallout was always the innocent public sector caught in the middle. Once more the same psychological scare tactics are in play! Are we going to allow this newly disguised malevolence to overcome society again, but this time, on a global scale?

The question here, in all of the above accounts and there, are numerous similar accounts like these, who were the hostile aggressors and who are the ones defending themselves?

It's clear to any rational mind that humanity still has a long way go, to grow up into full maturity. Adulthood requires that we do not bring out the guns like a bunch cowboys from out of the old West in every confrontation with the unknown.

Extraterrestrial Intelligences are not here to invade the Earth, neither are they here to exterminate and supplant human life, enslave it under a colonial system, harvest humans for food, steal the planet's resources, or destroy the planet altogether. This is simply not their agenda.

Planet Earth is Quarantined!

Extraterrestrial Intelligences have shown humanity that they exist, that they have watched mankind's development for the last 6000 to 10,000 years of recorded history. Occasionally, in unobtrusive ways, they guided us in the development of building more advanced societies or civilizations, at other times they found it necessary to intervene in our tribal or nation-state wars and conflicts. However, they have preferred not to interfere in the affairs of men and more often than not, they seem to have completely disappeared out of the physical perception and consciousness of humanity, only to show up in recent times on mass. Their recent appearance has been marked with armadas of disc and globe-shaped craft (UFOs) seen flying over the capitols and cities of every nation on Earth.

Unknowingly at the time, we sent ETs a German TV broadcast of the 1936 Olympic Summer Games in Berlin; the signals are still travelling today, way out beyond our Solar System to other nearby star systems. But what really got their attention to come back and check upon humanity was when humans (the Americans) detonated an atomic bomb in the American southwest and again, twice more over Japan. To all outward appearance, we certainly have a funny way of communicating our presence to the entire universe which via our latter methods were alarming to the Extraterrestrial Intelligences that received the signals.

Civilization on this planet plodded along on a very slow upward curve of development with the exception of few tribal skirmishes and city-state wars. Hostility between city-nations was contained and rarely spilled over into other peaceful nations and again, humanity was left to its own devices for the next 5800 years of recorded history. Then, something happened in the consciousness of mankind which may have been attributable to the increase in the world's population or it may have been political opportunism inspired by imperialistic attitudes for expansionism but, whatever the reason, conflicts along many countries' neighbouring borders began breaking out by the mid-1800s. The outbreaks of nationalist wars around the world in such rapidity kept journalists, historians and military strategists moving around trying to keep up with these outbreaks of human unrest and turmoil. By the 1850s and on into 1900s wars seem to spring up everywhere for little or no reason. The effects on humanity were profound impacting every sector of society to the point that human civilization went from a slow march forward to an exponential thrust upward like a rocket. It seemed that civilization around the globe was in an uncontrolled frenzy of movement and agitation with the world's population doubling over the next hundred years.

The world's first global war must have drawn the attention of ETI as tens of millions of people lost their lives during that four year period between 1914 and 1918 and it is likely that there are recorded accounts in the archives and libraries from which hopefully some resourceful UFO researcher will uncover in the near future. It was also fascinating that many astronomers from the 1850s to the 1950s began reporting mysterious lights on or around the Moon; were ETs gathering for a ringside seat to the biggest social upheaval event in this corner of the galaxy?

The proliferation and testing of atomic and nuclear weapons by the world's superpowers had ETI flying continuously over the global. Whenever and wherever there was movement of radioactive material and weapons, ET spacecraft were there to monitor human activity, whenever and wherever there was war they were there, when we left the Earth to go into orbit about the planet or when we went to the Moon or sent deep space probes to other planets in the /Solar system, ETI were there to track and monitor our explorations.

One could not help but think that an alien invasion was being mounted but, who were the aggressors that started firing the first shots at whom? An aggressive humanity needs to be contained before all hell breaks loose beyond this planet and beyond this Solar System. Saner heads must prevail in order to contain this militaristic madness controlled by a fascist kleptocracy that threatens to spread into space beyond anyone's control. Humanity needs to grow up if it is going to survive this adolescence stage in its evolution!

Most people who have studied and researched the UFO and ETI question or simply channeled ET information agree that the Earth is under some sort of quarantine by intelligent forces or an ETI collective who are working together in a kind of galactic federation, similar to our United Nation, although nowhere near as dysfunctional and impotent of true power. The reasons as to why are many depending on who you talk to but, if we eliminate the new age rhetoric and the pseudo-historical accounts of a former interstellar glory of mankind and focus our attention upon the last one to two hundred years from a minimum ten thousand year recorded history, we will find some very disturbing answers for our quarantine.

It has nothing to do with the notion that we are a backwater part of the galactic rim *(actually and more astronomically correct, we are located in the middle portion of one of the Milky Way Galaxy's spiral arms, not out on its rim. People who make such statements need to do their research or remain silent!)*

It has nothing to do with space litter and debris from spent rockets and lost astronaut toolkits or space-based weapons acting as some sort of shield that interferes with electromagnetic (radio or radar) broadcasts and receptions.

It has nothing to do with NASA or JPL scientists that we are incapable of developing technology beyond rocket power for interstellar travel. The late **Ben Rich** and **Kelly Johnson** have already told the world that we have faster than light speed technology developed from reverse engineering alien technology that has crashed upon this planet, to the point that we are capable of taking ET back home. The optimum word here is *"capable"*. As yet, we have not left the Solar System because of the quarantine in place by ETs to contain humanity's attempts to leave our star system.

It most certainly has nothing to do with interfering with any rocket launches into space or toward Mars or any other planet unless it carries a nuclear-armed device that is not for the purpose of propulsion or may somehow threaten some other life form in our Solar System.

Nor does our quarantine have anything to do with the **Large Hadron Collider** in Switzerland that may inadvertently develop a black hole singularity to swallow the Earth and everything else in the Solar System.

Because we haven't met and been told specifically by ETs of the reasons for quarantine, doesn't

mean that someone in high places or the intelligentsia doesn't know why. People will point to the fact that we have been warned off the Moon by ET transmissions as attested to by **Neil Armstrong** and **Buzz Aldrin** and other NASA astronauts but, that warning hasn't stopped any country from sending spacecraft and lunar satellites to the moon.

How do we know ETs are responsible for the quarantine? Well unless you were bored to death by the massive information download contained in this book and fell asleep between pages 1 and

2000, you'd realized that we have been discussing everything under the Sun about ETI and their UFO craft, so it leaves very little room for any other possibility.

But, the real reason that humanity is in quarantine, an embargo of not being allowed to travel out of the **Solar System** is that we are still immature species. We are a threat not only to ourselves but, potentially to other interstellar civilizations. There are other technological civilizations equally advanced as ours but, they have never experienced war nor do they have a word to express such a concept and God forbid should they encounter our species in its current social or spiritual state of development!

We have seen in the last hundred plus years or so, our evolvement from the horse and carriage, sailing ships and steam trains as our chief modes of transportation and guns and rifle weapons to gasoline powered automobiles, jet aircraft and rocket-powered spacecraft and our most formidable weapons are thermal nuclear, scalar armaments and high-powered lasers. This does not include alien reproduction vehicles and programmable life forms and all the subsidiary alien technology spin-offs.

Are we a potential threat to other star systems? You bet we are! When we started building rockets sending them into space and then exploded the first atomic and nuclear weapons over two cities of Japan, you better believe that the Extraterrestrial Intelligences sat up and took notice of us and our planet. We are bristling and armed to the teeth, for what? Our technological prowess has outstripped our spiritual and social development on this planet. Technology has become a runaway Frankenstein monster set loose upon a gullible and heedless society that has sacrificed their power to a few correct usurpers, forgetful of their God and unaware of the day and age in which they live in. *Technology has become the new religion, the opium of the people!*

This is an opposing counter viewpoint to Karl Marx's statement that *"religion is the opium of the people"*. The pendulum of social development swung from far right to far left, instead of the steady balance of the middle path.

The full quote from **Karl Marx** is: *"Religion is the sigh of the oppressed creature, the heart of a heartless world, and the soul of soulless conditions. It is the opium of the people"*
http://en.wikipedia.org/wiki/Opium_of_the_people

Author's rant: *Personally, I think Karl Marx didn't really understand the concept of religion as a spiritual force that is meant to bind, connect, to unify, to bring people together in spiritual bonds of knowledge, sanctity, and recognition of a divine supreme creator. Religion comes from the Greek word, "regilio" meaning unity. Religion is a set of truths and spiritual values designed by a supreme mind not for control of the masses but, as a source of guidance and affirmations to promote the advancement of civilization toward divinity which is the perfection of God. Marx's perceptions are based upon the non-existence of a supreme being, the negative mindset of conformity and being controlled, in finding meaning in life in a meaningless existence. I wonder, what has so upset him as to cause his perceptions of life to be so negative? Who wronged or did him injustice that he needed to destroy everyone's belief in something good and higher than themselves?*

Can you imagine meeting another Extraterrestrial species who have landed on this planet and immediately, we put them into a heavily military guarded detention building as political prisoners; we interrogate them for information and then, remembering some basic courtesy, we **welcome them to planet Earth** and continue asking them, **by the way, how did you get here and what's the secret to your propulsion system?**

Without a protocol of quarantine placed upon an immature human civilization, we could be travelling out of this Solar System headed toward the nearest star systems, knocking at the home worlds of ETI fully armed with nuclear warheads and other WMD, greeting them with a familiar **hi, remember us? We're here, can we have this piece of dirt, that you call your planet?**

In our current adolescent stage of social development, we are like a cancer or disease that will not be allowed to spread to other star systems infecting other civilizations with our prejudices and negative behaviours, not until we mature as a species. The universe does not need nor will it tolerate another galactic war! Yes, there have been previous interstellar wars that have involved our Solar System eons ago which has become a forgotten memory in the consciousness of mankind but, this is a topic for another book.

Earth Stands at the Crossroads Towards a New Hope

We are at a tipping point in our evolution on this beautiful planet, we call home. We stand at a critical time in mankind's evolutionary development on this planet. It is the pivotal moment foretold by all the world's religions in which all the prophets of former aged long to witness. It is upon this apex of evolutionary change, the process of disintegration of an old world order ending the **Adamic Cycle** or era can be clearly discerned thus, ushering in the beginning process of integration of a whole new era of fulfillment, the **Baha'i Cycle** destined to last for the next 500,000 years, in which the birth pangs of a new world order is currently being born. We see upon the horizon, the dawn of a long-awaited global civilization which will finally be unified by new spiritual values and marked with technological achievements emerging in stark contrast to the decay of an accelerating collapse of an old world order based upon outdated values of excessive materialism, racism, false patriotism and nationalistic pride, all clothed in tired, worn-out religious shibboleths and chaotic political standards. The old world is being rolled up and a new one is being laid out in its stead! God Passes By; by Shoghi Effendi; 1944; published by the Baha'i Publishing Trust; Wilmette, Illinois, USA; ISBN 0-87743-020-9

The world at present is in its final adolescent stage, it has become the night season of the owls and bats of tyranny and chaos, it is, therefore, spiritually bankrupted and moribund in need of divine guidance, only a divine physician can cure the world of its spiritual ills and afflictions.

The values of a material world must be tempered and balanced by a world of moral and spiritual values. What is required at this time is a world government that has the true support and allegiance from all its member nations to employ enforceable checks and balances against the ever-rising political chaos and financial corruption on the world stage.

We must be free from all prejudices, inequalities, and extremes of poverty and wealth; we must eliminate the causes of war, conflict, and hostility. The foundations of peace and unity, of "world stability and order" necessary for a world commonwealth must be built upon and "sustained by"..."justice and wisdom", the "twin pillars of which are reward and punishment", such are the teachings and wisdoms of **Baha'u'llah**. Gleanings from the Writings of Baha'u'llah; translated by Shoghi Effendi NSA of the Baha'is of the US); 1939 and 1952 published by Baha'i Publishing Trust; Wilmette, Illinois, USA; ISBN 52-24896

Can we on this planet evolve to become a space travelling, interstellar civilization exploring new star systems and planets where we become the new ETs to a new race of beings? The answer depends on this particular moment in our history of whether we are able to transition peacefully, to become united as a global civilization or if we wish to suffer further travail and chaos from devastating wars, conflicts, and global financial instability, de-evolving backwards into fighting feudal societies or worse.

Humanity stands at the crossroads of human evolutionary and planetary development, it highest expression is a global civilization marked by ever-increasing advances in higher consciousness, reason, intellect, and spiritual enlightenment. Advancements in the technical aspects of civilization will resolve the current issues of an environmentally polluted planet, distribution of wealth and resources will ameliorate the problems and misery of extreme poverty making such conditions a distant memory in the consciousness of mankind, burdensome, stagnating wealth once formerly held in the hands of a few will be use for the education and advancement of all peoples of the world commonwealth, medicine will make incredible strides to solve many of the diseases and illness that currently afflict humanity thus, extending the longevity of the human race. This is but a foretaste of a hopeful, good future in store for the humanity if we chose wisely the kind of future we want it to be!

There is a story about the late **Col. Philip J. Corso** that echoes this hopeful future sentiment that his son related about his father which occurred in the late 1950s or early 1960s, when he was at **Holloman Air Force Base/White Sands Missile Range** during a landing of an extraterrestrial spacecraft. There was an exchange between the ETs and the military and at one-point Col. Corso, in typical American military fashion, asked, with regard to open ET cooperation, *"What's in it for me?"* The ET responded: *"A new world if you can take it!"*

The question at this moment is do we want a new world order based upon spiritual values that balance our material needs to build an ever-advancing civilization on Earth with the promise and potential of also becoming an interstellar travelling civilization or do we allow the world to be hijacked in its current downwardly spiralling corrupt state to become some false notion of a world order based solely upon materialistic values that serve only the few oligarchic families and their mega wealthy corporations while depriving the mass populace of this planet of any semblance of a reasonable and healthy standard of living?

The people of the Earth, it is this author's belief, will choose the right course of action that will benefit the majority on this planet, however, there will be considerable resistance to the changes that people will want for a better world. These arrogant oligarchical families and their corrupt corporations will become increasingly defiant and even militant creating false charges of public anarchy as people stand their ground. There will be open hostility towards the masses in order to buttress this oligarchy and ensure that their way of life is unaffected by the common will of the people. If it hasn't already happened by then, this oligarchy will unleash a false flag ET invasion scenario upon the people to distract them with a new terror from space thus, taking the focus of the world's attention off these few **wealthy corporate elite (WCEs).** But this Cabal, this small-minded M.I.C. group are inevitably doomed to failure and the presence of Extraterrestrial Intelligence in orbit around Earth and on nearby planets stand ready to assist humanity in its darkest hours.

Author's Rant: Sorry for the doom and gloom scenario and I hope that I am wrong but, my intuitive gut feelings tell me, that I'm not that far off this prediction target!

The positive spin from this gloom and doom cat and dog fight among ourselves is that we will enter into an age of universal peace, whether we want it or not, kicking and screaming, no power on Earth can prevent it! It will happen!!!

The age of fulfillment, of **Universal Peace** on Earth which is the hallmark of this new **Baha'i Era** or **Cycle** is not just here on this planet but, it truly is a universal cycle of peace for all planetary civilizations, star systems, and galaxies throughout the universe. The profound positively charged energy of this time ripples and surges waves upon waves throughout the universe, The synapses, and neurons of the universe are all firing at unprecedented levels because it has been divinely ordained by that **Supreme Extraterrestrial Intelligence (SETI),... God,** Himself**!!!** *(This is by the way... at least in the mind of this author, is the correct usage of that acronym, not that other one that we all know about!)*

Planetary Explorers and the Galactic Commonwealth

Pursuant to that eventual reality (Universal Peace), *we now know the reason why the Extraterrestrial intelligences are here and what is their ultimate agenda*! They are not here start a new religion, each planet, and civilization has their own. They are here not only as **explorers** but, *to assist us in co-creating a time of peace on Earth*, one that holds the promise of us joining them in a greater paradigm of unity with the ability to journey among the stars as an interstellar civilization! As its states in the Bible, "As it is in Heaven, so shall it be on Earth!"

We have stated before, elsewhere in this book that just as we have a **United Nations** upon Earth, (although, somewhat dysfunctional and impotent), Extraterrestrial Intelligences visiting this planet also have a very active and dynamic "Federation of Planets" (pardon the cheesy reference to the Star Trek genre) or "Council of Star Systems" with each interstellar race working together in peace and harmony for the common good of its united planetary members. This is the reason why there is a great variation of eyewitness accounts not only of UFOs sightings but, particularly in the wide variety of different Extraterrestrial beings seen around the world, they are all working together with some having specific fields of specialization on Earth. Contact has been for the most part random among the populace of the world, some deliberate, others by mere coincidence and happenstance. Few deliberate contacts have been made towards any political leaders with a few exceptions, which did not turn out well in an ambassadorial sense.

As previously stated above, we are still immature species which has not formed a fully functional global civilization, but an incipient one. When we can get our proverbial shit together then, there

398

will be great and startling ETI revelations and changes occurring for us on Earth. Until that day, we must all work together (yes, it really does require work and not just some nice fuzzy, warm and loving platitudes bantered about) for the promise of a better future. It not about changing our densities, our vibrations or frequencies to move toward higher levels of consciousness, **it's about spiritual attitude and maturity** which implies a higher state of consciousness.

There is no other way to attain to a greater good or higher consciousness! All the religions of the past that we practice on Earth have been telling us that for millennia but, have we really listened? Did we really understand the message? Or did we arrogantly think that we knew better than our creator and decided on doing something else, which is why we're all in a hell of a mess right now?! Perhaps, if your religion is no long working for you because it has become impotent in one way or another then, it time to change up to something more relevant to the day and age in which we live in. Religion is progressive not stagnate, nor is it mutually exclusive claiming to have all definitive truth. Truth is also progressive not final. My apologies if this is upsetting to the close-minded world of the fundamentalists!

The change necessary to bring about a global civilization is going to require hard work if we want it. It is ***about good old fashion, roll up your sleeves and prepare to get a little dirty kind of work!!!*** This means coming up with viable plans using common sense and logic, not some holier than thou attitudes. Saying something doesn't make it happen, implementing plans through the toil off of one's back and the sweat off of one's brow accomplishes the tasks at hand and obviously having the right attitude gets you further ahead of the game.

When humanity has demonstrated within a reasonable period of time that it not only claims to be a united commonwealth but, is truly at peace with itself and all those around them then, we will find an invitation to join in greater bonds of unity and friendship with other interstellar and galactic civilizations. This is the prerequisite to becoming and interstellar space travelling civilization. Until we lose our predilection for war and demonstrate the prerequisite of peace and harmony among ourselves, we will not be going very far or very fast to anywhere.

There is a good, hopeful future that awaits us if we but can take it!!!

CHAPTER 114

PREPARATION FOR CONTACT AND
SOME "INALIENABLE TRUTHS"!

In recent years we have witnessed many nations releasing secret government-controlled UFO documents into the public domain as a part of an unofficial worldwide disclosure movement. Researchers and news media journalists wonder if these governments know more than they are revealing about the UFO/ETI phenomenon. Some Ufologists suspect that all these releases of UFO documentation are leading up to something bigger, like a worldwide announcement that

finally confirms what most of the public already knows, that we are not alone in the universe nor have we ever been. But more importantly, Extraterrestrial Intelligences are visiting the Earth and that imminent contact with ETIs is near!

Dr. Richard Haines in the preface of his book "***CE-5***", 1998 remarks upon the fascination that mankind has had with the subject of extraterrestrial (E.T.) life for centuries: *"We have pondered what might be true about E.T., always considered through the invisible biases of our own planetary cultures, languages, beliefs, hopes, and numerous unconscious fears. "They must look like us, and think like us,' we may reason, not thinking very logically about the creativity of God, the vastness of the universe, or the awesome span of time of the existence of matter. Why can't E.T. exist? Why can't "they" be different from us?"* CE-5: Close Encounters of the Fifth Kind by Richard F. Haines, Ph.D.; 1998; published by Sourcebooks Inc.; Naperville, Illinois, USA; ISBN 1-57071-427-4

Such a revelation, whether anticipated by the general public or not would nevertheless, still have an impact of initial shock upon people but, contrary to popular official perceptions or think tank reports, no anticipated outbreaks of public panic and chaos in the streets would occur, due in part from the long public conditioning process of expecting to find other intelligent life in the universe. It will become the popular topic of discussion among everyone, almost everywhere for the first few months with many "so-called experts" both in an official and private capacity coming out of the woodwork to be interviewed professing their insight and knowledge as to the nature and agenda of the ET visitors. Such a monumental event in the annals of human history would have many people from within the world's governments, their militaries, and their various intelligence agencies, including leaders from the major world religions, like the Catholic Church, all lining up to express their interest for involvement. In fact, at this particular moment in time, ***"they are already positioning themselves for the biggest game of the millennium".***

There is only one problem with this scenario, from those already in contact with ETI, which seemed to be overlooked by officialdom and that is ETs are not going to make full open contact with the body of mankind until it grows up and matures. We are still in the throes of adolescence; we need to put away childish things, immature attitudes, and behaviours.

As the lyrics in a popular rock song suggests, "The future is so bright that I need sunglasses!", and if we choose the right course of action in our societies and start promoting and defending the God-given rights of all humans then, indeed, the future is bright and hopeful. The question then

400

becomes, how do we achieve this bright hopeful future? How do we bring about a lasting and sustainable **Universal Peace** on Earth that advances human civilization toward a golden age of mankind which will include becoming a member of the universal family of advanced Extraterrestrial Civilizations?

The old Chinese adage that *"journey of a thousand miles begins with the first step"* a means we must identify those obstacles that stagnate and retard the social, spiritual, emotional and intellectual progress of human development. We must acknowledge those human limitations within ourselves, rather than denying them or making excuses for them. If we start down the path to educate every person on this planet of how to remove these limitations and live within a new set or moral values, within a generation (twenty years) or less, humanity will have entered into universal peace and the doors to Extraterrestrial contact will be flung forever open!
https://www.youtube.com/watch?v=8DycG5ZMjTo

Childhood's End and the Path to Universal Peace

We are as yet, nowhere near ready to enter into that stage of universal peace where there will be a mutual and peaceful communication exchanges between ETI and humans; ***they simply don't trust us at this point in our current development!*** And why should they? We haven't exactly shown them that we are as yet, a friendly planet that is why we need to clean house on this planet in order to prepare for the day when we do engage in ETI contact.

There is no point in further clinging to or holding out in the hopes of an old religious prophesy first enunciated two thousand years ago to be fulfilled which will provide some desperately needed miraculous event of rapture to shape our destiny while we all sit back reaping the eternal benefits of its unfoldment. That *"Bus"* has come and gone some time ago and had you recognized its coming, you would have already boarded that *"Bus"* for the ride of your life! Fortunately, for everyone that *"Bus"* makes frequent stops along the way, you still have a chance to board it.

A world that navigates the dark murky waters of civilization towards an unknown future without a moral compass seems inevitably doomed to be shipwrecked upon a rocky shore of materialism and irreligious values only to find itself as the cannibalized victims of that very same civilization.

A world that is being steered by an unscrupulous and corrupt, powerful wealthy corporate elite cares neither about the ship, its passengers nor the destination, save that they journey in comfort and style regardless of the other passengers' living conditions. It is a "Titanic" with incompetent helmsmen at the controls that is about to hit an iceberg and no one seems to be the wiser to avoid inevitable disaster. This is merely one of the many problems that face the world of humanity, namely, who should be piloting this ship called Earth on the murky waters of civilization?

We need to identify and acknowledge those problems that hold humanity back from progressing forward; we need to gather to us those insights and wise counsels that come from the writings of every major religion on Earth. We need to draw upon those new and ancient spiritual teachings for inspiration and to formulate plans to achieve a true world order based upon a balance between material and spiritual values. We should also be prepared to discard those tenets and outdated teachings that are no longer viable, particularly those that are steeped in religious dogma and prejudice.

It will require giving up old concepts and standards for new more progressive ones, a sacrifice of economic resources, political agendas and in some cases even, national sovereignty will need to be addressed in order to resolve political border and boundary issues. We are looking at a total house-cleaning of this planet from top to bottom. It's our "childhood end", the final stage in our adolescence giving way to the maturity of adulthood; it is the coming of age of humanity and the emergence of a world commonwealth.

Before we can examine those social ills and human limitations in a spiritually bankrupted and moribund world which only a "divine physician" can heal, we must first identify if such a "divine physician" exists. This requires searching the world's religious writings to see who this individual may be and then we can examine his teachings for remedies or solutions for the spiritual ills of mankind.

The famous ancient Greek philosopher **Aristotle** once commented on the spiritual condition of mankind during his time and the attainment to a better moral life: *"Moral excellence comes about as a result of habit… we become just by doing just acts, temperate by doing temperate acts, brave by doing brave acts."*

Part of moral excellence, therefore, requires habitual acts of justice, temperance, and bravery or stated another way, the progress of man and society comes from a firm foundation of truth that is grounded in *knowledge, volition, and action.* It is a fundamental prerequisite for advancement of any society or world civilization to emerge upon this planet. Such men, however, receive their inspiration from a higher source of wisdom which can be traced back to the Prophets of God for their particular day and age in which they live. Anyone can arise to become a great man or woman from Aristotle's insight, but human insights and personal wisdoms are limited unless they derive their source of wisdom from the fountainhead of all wisdom that comes from the Manifestation of God. Therefore, divine spiritual insights, wisdom, and right action from the Manifestations affect not only the individual but, has a profound effect on the whole of society.

All the world's religions and holy writings speak of a time in which a particular saviour or avatar or messiah will appear at the end of times ushering in an age of enlightenment and a **Universal Peace.** Within the framework of human experience and understanding, history has benchmarks or touchstone levels in which all things are judged against; they are the pivotal points in which history charts a new course of direction for humanity to travel along. These benchmarks of history are epitomized by the appearance of divine **"Enlightened Ones"** who proclaim teachings of philosophical, moral and spiritual knowledge as well as laws and tenets which lead to the advancement of cultures, societies, and civilizations. They promote through sacrifice of their own lives and comforts noble virtues, goodly character, peace, love and harmony among their nations as a means of inspiring their fellow countrymen to strive for greater ideals that would, in turn, spread to other neighbouring countries.

With this basic understanding, we look towards the **Manifestation of God** and his teachings to educate, guide and inspire the body of humanity to arise to action to advance civilization. It,

402

therefore, stands to reason that God's Prophets are the source of all knowledge and wisdom; their teachings are the healing balm for the afflictions of mankind.

The reader may be asking themselves, so what does religion have to do with the study of UFOs and ETI? Are they not two separate topics of discussion? In many ways, too numerous to recount them all and as already discussed early in this book, the study and research into the UFO/ETI phenomenon is very similar to a religious belief system.

Every religion has in their sacred text, writings that record, either through allegorical accounts or actual statements made by the divine Prophets themselves, that UFOs and Extraterrestrial Intelligences had visited mankind and may have been regarded as angels or simply as strange creatures, who intervened in the course of man's societal development. Therefore, we must take a second look at the world's various religions which point to an age where such questions about ETs will be answered by a **Universal Prophet.**

It is an old standard adage that states, "You never discuss politics and religion with friends, if you wish to keep your friends". To this may be added the subject of UFOs and ETI which whether we like it or not, involves both politics and religion. At times, it is very difficult to discuss the topic of UFOs and ETI as it relates to peace and benevolence, without discussing the religious ramifications of this subject, along with all its philosophical ideals and spiritual values. Most of us in this day and age tread lightly around religious or spiritual matters preferring to just "go over and kick the tires" on the UFO subject and nothing more. However, there is no way to avoid this aspect, as it has a very real spiritual component! Prepare yourselves!!!

WARNING ALERT! THIS NEXT SUB-SECTION COMES WITH A WARNING!

Religion is a touchy subject for almost everyone who expounds upon what they believe to be the truth in life! We are discussing concepts of religious belief systems, which in the heart and mind of this author, are all from God and therefore, are in agreement in their essential truths that they present, their reality is the same, their essence is one and therefore, there are no differences in their message. The "Golden Rule" remains constant and intact! Yet, their seeming apparent differences and interpretations of understanding whether by individuals or whole segments of society in many countries has been the cause of much religious bloodshed, dissension, conflicts, chaos, and wars in the world, all in the name of God. Truth is more often than not, relative to the times in which it arises, and no religion has absolute exclusivity or definitiveness to truth. Truth, however, is renewed approximately every millennium (1000 years) and it is also progressive, not just the same old truth revitalized or showing up again, but new truth for the day and age in which it appears to advance humanity's development! Therefore, there will be no sugar coating of the truth here; no punches will be pulled, just the facts! YOU HAVE BEEN WARNED!!!

The Path to Universal Peace is Filled with Tumult, Grief, and Heartache!
Who is Going to Sort Out the Mess We are In?

In reality, many of our societies and nations have been built upon the spiritual truths, laws and moral beliefs that originate from many of the world's religions. Such societies become the envy of the world and go on to become great civilizations of spiritual enlightenment and material prosperity. In some cases, however, the lack of a religious foundation or its distortion within society has become the cause of that society's collapse and their eventual disappearance from history. But, more often than not and contrary to popular belief, religion has also been the source of strife and war. This particularly true, when a monarch or church leader's limited understanding and interpretation of religious doctrine becomes the source of competitive ideals conflicting with other religions precepts resulting in holy wars and **Jihads**, such as the **Crusader Wars** of the 11th to 13th centuries, between European countries of Christianity and the Islamic **Ottoman Empire**.

Sources of religious prejudice often centered around such Christian concepts as "Christ is God in the flesh" and that "**Christ** is the only begotten **Son of God**" and "no man cometh before God except through Christ" or in Islamic tradition that "**Mohammad** is the **Seal of the Prophets**, there can be no other prophets after Him" and "all those that do not believe in Allah and Islam are infidels", etc, etc. These are statements are highly provocative invoking competitiveness and combativeness because the defenders of the faith are so rigid in their understanding and belief of the sacred texts that they can broach no other understanding or interpretation of the holy writings. This is but one type of religious prejudice.

Another example is Islam's Temple Mount, site of the **Dome of the Rock** mosque believed by the people of Islam to be the site of **Muhammad's Ascension** which just happens to be built upon the ancient site of **Solomon's Temple and Palace** also considered a sacred site to the Jews of Israel. These two sites of worship overlapping each other has led to serious contention and conflict between these two religions, as to who actually has ownership of the sacred site.

Religious prejudice knows no bounds or borders nor does time seem to be able to heal it. Most religions to all outward appearances seem to acknowledge their predecessors but, secretly hold them in contempt and then, in turn, deny any other future religion that supercedes it. The truth of the matter is that each major religion has grown out of a former religion and it flourishes for a brief time (about a 1000 years), then a new faith comes into the world birthed by the decline in efficacy of its mother religion thus, are the generations of religious belief continued and therefore, religion is progressive as the **Word of God** is not limited by the understanding of men.

Think of religion as a school system that has many levels or grades of understanding and disciplines. It begins with the levels of elementary education and progressively moves upward to high school levels and then into university levels. At each level throughout the education system, there are great teachers that the students come to know and love and are educated by the lessons of that teacher. These teachers point to a future time when the student having learned all that can be taught them in this grade level must prepare themselves for graduation to the next grade level and to expect this eventual reality. When the students graduate, they meet a new teacher as great and as loving as their former teacher which they had to leave behind. He teaches them great truths and wisdoms that will guide them for the rest of their lives. He acknowledges the previous teacher and tells the students not to forget the lessons they had learned in the former grade level and to anticipate another new teacher after they have learned all the things in his class.

404

The student is taught to remember all the lessons learn from all former grade levels and to love all their teachers and not to forget them, but to always look forward to the future when many new teachers will come to guide and educate them to higher realities of knowledge and greater worlds of understanding. Thus, the students of the body of humanity grow, develop and advance society and civilization progressively upwards, both spiritually and materially, building upon each former religion's spiritual teachings and always expecting to greet and learn from a new **"divine educator"** of mankind. Thus, has it been and thus, shall it always be!

It should be understood that each **"divine teacher"** or **"educator of mankind"** has a particular mission to accomplish based on the exigencies of the time in which he appears. And do not misperceive these divine educators as being Extraterrestrial in nature or in mission for they are as human as any other human but, they are singled-out by a divine calling beyond their own will to control, their lives are no longer theirs to live but, are in the hands of the Creator of the Universe.

These divine educators never give out more truth or wisdom or guidance than humanity is capable of understanding and implementing into advancing society. It should never be looked upon that one Messenger from God is greater than the former Messenger or the next that will follow him. Each Manifestation of God could easily deliver the whole wisdom and knowledge of God but, man is a finite creature and would not be able to comprehend it let alone contain it and would ultimately pass out of existence. It is because God loves His creation that he sends His divine Messengers and they portion out wisdom and truth according to our ability to absorb it.

Such has been the Divine Will in the past and will be forever more, not just here on this planet, but upon all planets in every star system and every galaxy throughout this universe and even in all worlds of God, both visible and invisible that are beyond this universe!

To this basic understanding, all major religions, both recognized and those not so familiar around the global, follow some form of these teachings, each to some lesser or greater degree. They clothe such understanding in teachings of love, virtue, unity, peace, harmony and the acquisition of knowledge and the advancement of civilization measured out to the receptivity and the potentiality of their followers.

In this section the discussion of religion as it relates to the Extraterrestrial question is not to imply one religion's superiority over another. Religion has never been created to be competitive or combative with any other religious belief system but has been founded upon succession and progression with each religion superseding the former and each former religion supporting its successor. Each religion is, in reality, the same religion, both ancient and new, as Alpha and Omega. It is the reason that we find some form of the **"Golden Rule"** teaching in every religion from God creating the inseparable bond that ties each religion together as the **"Word of God"** or the **"Book of Life"** or the **"Book of Names"** in which there are many *"books"* and *"chapters"*!

Worldwide percentage of Adherents by Religion (mid 2005)

0.04%
0.04% 0.04% 0.02%
0.07%
0.10%
0.12%
0.20%
0.23%
0.39%
1.68%
3.97%
5.87%
6.27%
13.33%
20.28%
11.92%
2.35%
33.06%

- Christians (33.06%)
- Muslims (20.28%)
- Hindus (13.33%)
- Chinese Universists (6.27%)
- Buddhists (5.87%)
- Ethnoreligionists (3.97%)
- Neoreligionists (1.68%)
- Sikhs (0.39%)
- Jews (0.23%)
- Spiritists (0.20%)
- Baha'is (0.12%)
- Confucianists (0.10%)
- Jains (0.07%)
- Shintoists (0.04%)
- Taoists (0.04%)
- Zoroastrians (0.04%)
- Other religionists (0.02%)
- Non-religious (11.92%)
- Atheist (2.35%)

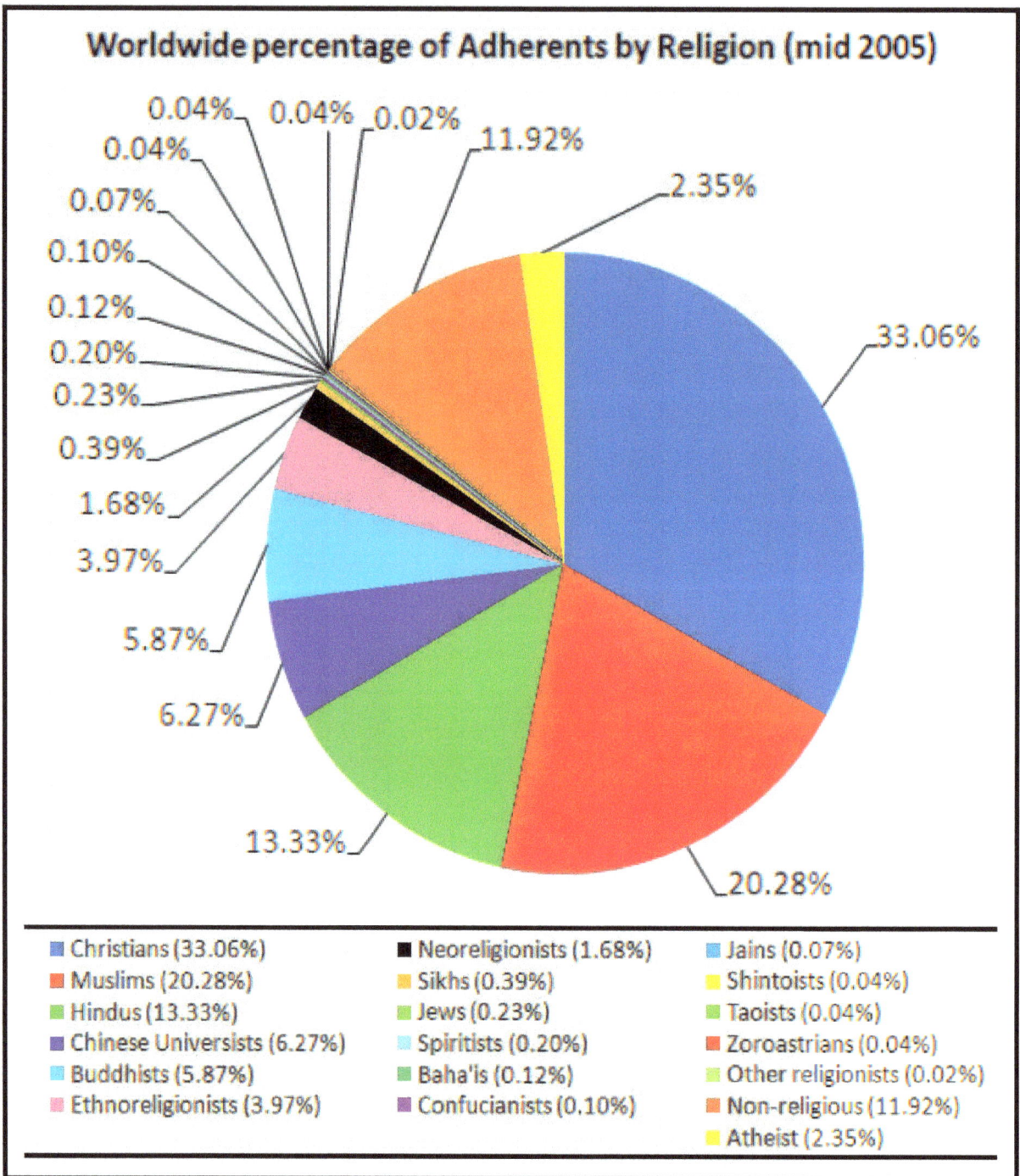

**This pie chart shows the world's major religions as well as
the lesser known and the non-religious orders**

https://commons.wikimedia.org/wiki/File:Worldwide_percentage_of_Adherents_by_Religion.png

The statistical assessment below by Savata indicates the belief by various religions in the existence of extraterrestrial life, although, there is nothing in their holy writings to support this belief! The percentages reflect a number of factors which may include personal belief in the subject, religious writings that support this unique belief, as well as the individual perspective,

regardless of his particular religious affiliation. Unique among the world's major religions, the **Baha'i Faith,** and its holy writings support the premise of the existence of extraterrestrial life which is a small part of their basic beliefs. (From a YouTube interview of Troy Mathew from Savata by **Alejandro Rojas** on *"Open Minds UFO Radio"*).
http://www.youtube.com/watch?v=WfDyr_FDZiI and http://www.openminds.tv/radio

It should be stated that this author has *added* the Baha'i Faith into the bar graph below as Savata was either not aware of the Baha'i Faith as a major world religion at the time it developed this graph or intentionally left it out for reasons unknown to this author.

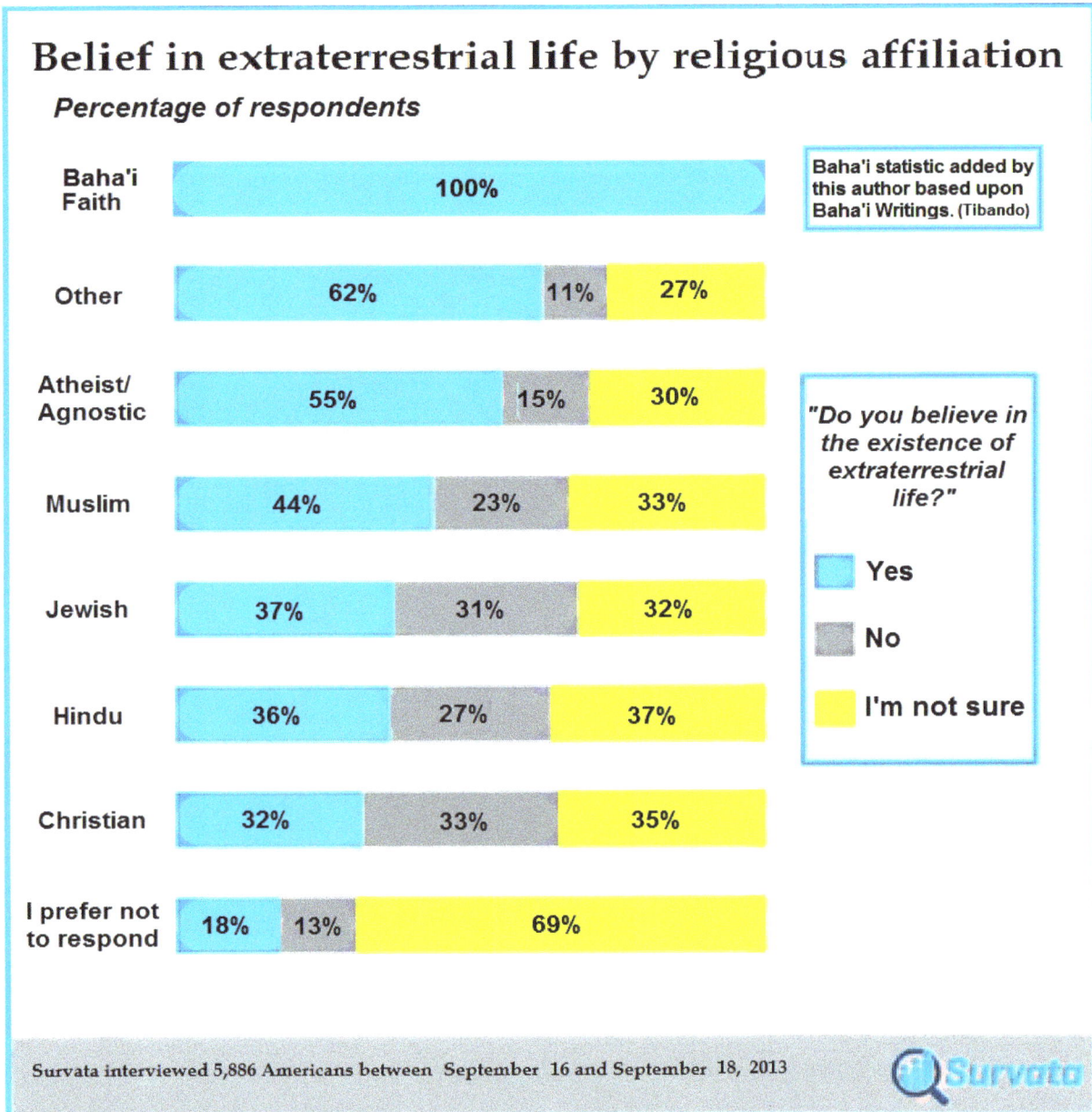

Belief in extraterrestrial life by religious affiliation

Percentage of respondents

Baha'i Faith	100%		Baha'i statistic added by this author based upon Baha'i Writings. (Tibando)
Other	62%	11%	27%
Atheist/ Agnostic	55%	15%	30%
Muslim	44%	23%	33%
Jewish	37%	31%	32%
Hindu	36%	27%	37%
Christian	32%	33%	35%
I prefer not to respond	18%	13%	69%

"Do you believe in the existence of extraterrestrial life?"

- Yes
- No
- I'm not sure

Survata interviewed 5,886 Americans between September 16 and September 18, 2013

Survata

Statistical belief in extraterrestrial life by each major religion. Note that the Baha'i Faith has a 100% belief in this subject because its writings support this premise
http://www.openminds.tv/survey-examines-if-belief-in-god-effects-belief-in-extraterrestrials-video-1151/23878

407

Some of the world's major religions (clockwise from the top) Baha'i Faith, Islam, Christianity, Judaism, Buddhism, Zoroastrianism, Hinduism, Shinto, Native American
(c) Terry Tibando

Religion, therefore, was meant to bind the faithful believers together, not to become a source of dissension and division among the people!

Every religion has its spiritual writings that guide and stress the importance of peace as a form of worship to the divine and as a means of stability and security for society. Religion is, therefore, a means of fellowship, peace, and unity among it follow adherents and thus, among the general populace of humanity. Peace which is the hallmark of a golden age of civilization has been the universal touchstone to which all religions strive to attain through their teachings. An age of

universal peace is often referred to in theological terms as a **Messianic Age**, a future time of universal peace and brotherhood on the earth, without crime, war, and poverty. Many religions believe that there will be such an age; some refer to it as the consummate "kingdom of God", "paradise", "peaceable kingdom", or the "world to come".
http://en.wikipedia.org/wiki/Messianic_Age

A "Golden Age of Mankind" receives its impetus from the teachings of a **Divine Messenger** or **Universal Prophet** or **Manifestation of God**. It is here that we need to examine dispassionately, putting all religious differences, absolutisms, and prejudices aside and see what the world's religions have to say about the **end of days** or the **age of universal peace** which will be established by a **Supreme Manifestation of God**.

Judaism

According to Jewish tradition, the **Messianic Era** will be one of global peace and harmony, an era free of strife and hardship, and one conducive to the furthering of the knowledge of the Creator. The theme of the **Jewish Messiah** ushering in an era of global peace is encapsulated in two of the most famous scriptural passages from the **Book of Isaiah**:

They shall beat their swords into plowshares and their spears into pruning hooks; nation will not lift sword against nation and they will no longer study warfare. **(Isaiah 2:4)**

The wolf will live with the lamb, the leopard will lie down with the goat, the calf and the lion and the yearling together; and a little child will lead them. The cow will feed with the bear, their young will lie down together, and the lion will eat straw like the ox. The infant will play near the hole of the cobra, and the young child put his hand into the viper's nest. They will neither harm nor destroy on all my holy mountain, for the earth will be full of the knowledge of the Lord as the waters cover the sea. **(Isaiah 11:6-9)**

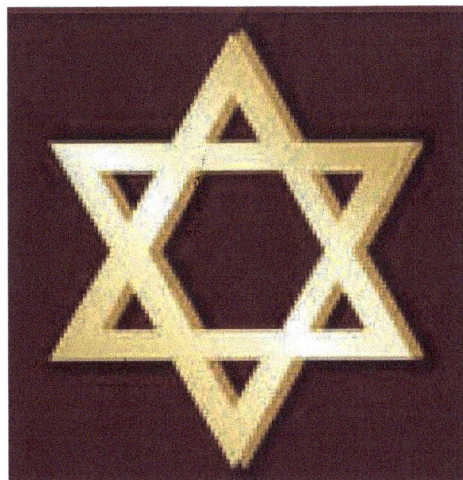

The religious symbol of Judaism
(Google Images)

According to the **Talmud**, the **Midrash**, and the ancient Kabbalistic work, the **Zohar**, the Messiah must arrive before the year 6000 from the time of creation. In Orthodox Jewish belief, the Hebrew calendar dates to the time of creation, making this correspond to the year 2230. http://en.wikipedia.org/wiki/Messianic_Age

Christianity

Christian eschatology includes several views of the Messianic Age.

According to a **realized eschatology** (*Realized eschatology is a Christian eschatological theory popularized by C. H. Dodd (1884–1973) that holds that the eschatological passages in the New Testament do not refer to the future, but instead refer to the ministry of Jesus and his lasting legacy. Eschatology is, therefore, not the end of the world but its rebirth instituted by Jesus and continued by his disciples, a historical (rather than transhistorical) phenomenon. Those holding this view generally dismiss "end times" theories, believing them to be irrelevant. They hold that what Jesus said and did, and told his disciples to do likewise, are of greater significance than any messianic expectations.*), the Messianic Era, a time of universal peace and brotherhood on the earth, without crime, war, and poverty, is already here. With the death of Christ himself, the Messianic Era had begun, but, according to an inaugurated eschatology, it will only be *fulfilled* by the parousia of Christ. Commonly, the **Book of Revelation** is interpreted as referring to the "unveiling" or "revelation" of **Jesus Christ** as **Christian Messiah** in the apocalypse or end of the world. http://en.wikipedia.org/wiki/Realized_eschatology and http://en.wikipedia.org/wiki/Parusia

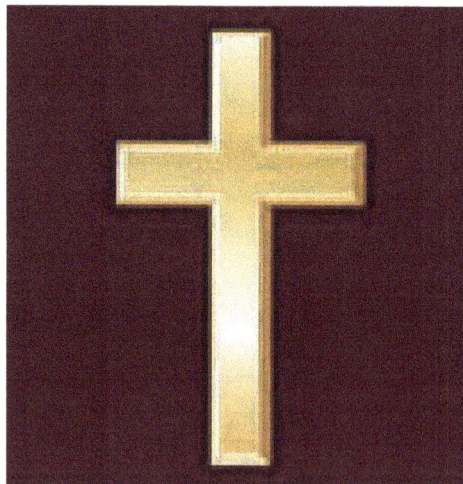

The religious symbol of Christianity
(Google Images)

The **Nicene Creed,** professed by most Christians, expresses the belief that Christ ascended to Heaven, where he now sits at the **Right Hand of God** and will return to earth at the **Second Coming of Christ** to establish the **Kingdom of God** of the **World to Come**. http://en.wikipedia.org/wiki/Parusia

Islam

Qur'an states that Isa ibn Maryam (Jesus the Son of Mary) was the Messiah or "Prophet" sent to the Jews **[Quran 3:45].** Muslims believe he is alive in Heaven, and will return to Earth to defeat the Masih ad-Dajjal, an anti-messiah comparable to the **Christian Antichrist** and the **Jewish Armilus**.

A hadith in Abu Dawud (37:4310) says:

Narrated Abu Hurayrah: The Prophet said: There is no prophet between me and him, that is, Jesus. He will descend (to the earth). When you see him, recognize him: a man of medium height, reddish hair, wearing two light yellow garments, looking as if drops were falling down from his head though it will not be wet. He will fight for the cause of Islam. He will break the cross, kill the swine, and put an end to war (in another Tradition, there is the word Jizyah instead of Harb (war), meaning that he will abolish jizyah); God will perish all religions except Islam. He [Jesus] will destroy the Antichrist who will live on the earth for forty days and then he will die. The Muslims will pray behind him.

The religious symbol of Islam
(Google Images)

Both Sunni and Shia Muslims agree Imam Mahdi will arrive first, and after him, Jesus. Jesus will proclaim that the true leader is al-Mahdi. A war, literally Jihad (Jihade Asghar) will be fought— the Dajjal (evil) against al-Mahdi and Jesus (good). This war will mark the approach of the coming of the **Last Day.** After Jesus slays al-Dajjāl at the Gate of Lud, he will bear witness and reveal that Islam is the true and final word from God to humanity as Yusuf Ali's translation reads: [**Quran 4:159** (Translated by Yusuf Ali)]

And there is none of the People of the Book but must believe in him before his death, and on the Day of Judgment, He will be a witness against them.— (159)

He will live for several years, marry, have children and will be buried in Medina.

A hadith in Sahih Bukhari (Sahih al-Bukhari, 4:55:658) says:

Allah's Apostle said "How will you be when the son of Mary descends amongst you and your Imam is from amongst you.

Very few scholars outside of Orthodox Islam reject all the quotes (Hadith) attributed to Prophet Muhammad that mention the second return of Jesus, the Dajjal, and Imam Mahdi, believing that they have no Qur'anic basis. However, Quran emphatically rejects the implication of termination of Jesus' life when he was allegedly crucified. Yusuf Ali's translation reads:

That they said (in boast), "We killed Christ Jesus the son of Mary, the Messenger of Allah";— but they killed him not, nor crucified him, but so it was made to appear to them and those who differ therein are full of doubts, with no (certain) knowledge, but only conjecture to follow, for of a surety they killed him not. (157) Nay, Allah raised him up unto Himself; and Allah is Exalted in Power, Wise. (158) **[Quran 4:157–158]**

So Peace is on me the day I was born, the day that I die and the day that I shall be raised up to life (again). **[Quran 19:33]**

This implies that Jesus will die someday. The unified opinion of Islam maintains that the bodily death of Jesus will happen after his second coming.

Many classical commentators such as Ibn Kathir, At-Tabari, al-Qurtubi, Suyuti, al-Undlusi (Bahr al-Muhit), Abu al-Fadl al-Alusi (Ruh al-Maani) clearly mention that verse 43:61 of the Qur'an refers to the descent of Jesus before the Day of Resurrection, indicating that Jesus would be the Sign that the Hour is close.

And (Jesus) shall be a Sign (for the coming of) the Hour (of Judgment): therefore have no doubt about the (Hour)... **[Quran 43:61]**

Those that reject the second coming of Jesus argue that the knowledge of the Hour is only with God and that the Hour will come suddenly. They maintain that if the second coming of Jesus were true, whenever it happens, billions of people would then be certain the Hour is about to come. The response given to this is that signs that the Last Hour is near have been foretold and given, including that of the second coming of Jesus, as signs indicating the **Last Hour** is near. They will not clarify when it is to come in any specific sense, and hence do not reveal it.

Now, it is should be obvious to any reasonable inquiring mind that the actual dates, divine dignitaries, prophets, and second coming scenarios is open to interpretation even among the scholars of **Judaism**, **Christianity**, and **Islam** where such truths are asserted in an uncompromising fashion allowing for no alternative insights or versions of truth. These inflexibilities in the understanding of one's own spiritual scriptures foster ***competitiveness*** among differing religions and are the cause of religious prejudices to form and fester leading ultimately into sectarian conflicts and divisions that split the mother religion and which eventually lead to war within or across borders and boundaries.
http://en.wikipedia.org/wiki/Messianic_Age

412

Hinduism

Far Eastern religious beliefs like **Hinduism** and **Buddhism** found in India and it surrounding neighbours believe that there is a hopeful time when the Earth will eventually be united as one people in a golden age.

Hindus believe that there is a return of **Lord Krishna**, a ninth incarnation of Krishna will occur when the world is in turmoil and chaos, whenever the world is in great evil:

"Whenever there is decay of righteousness... and there is exaltation of unrighteousness, then I Myself come forth... for the destruction of evil-doers, for the sake of firmly establishing righteousness, **I am born from age to age."** --- **KRISHNA- Bhagavad Gita- fourth discourse**

The religious symbol of Hinduism
(Google Images)

The Hindu Holy Book **"Manu Smriti" (the Memorandum of Manu)** is traditionally attributed to the mythical first Manu, Svayambhuva. This book describes the **lifespan of the Universe**, specifies calculations of the major and minor **Religious Cycles** comprising various eons, ages, epics and their linking periods. Mankind's appearance is placed at the beginning of a cycle of 12000 years containing within itself 4 minor and uneven periodic cycles. The termination of this 12000-year cycle is described as the beginning of a new **"Krit Yug"** or a new "Cycle of Truth in Deeds".

According to the calculations given in "Manu Smriti", chapter 1, verses 68 to 72... this 12,000-year cycle was to end and the new "Cycle of Truth in Deeds" was to begin in the year 1844 AD.

The following chart lists the calculation of the full-period duration of the 12000-year cycle containing the 4 minor and uneven periodic cycles are as follows:

Sat Yug 100 + 1000 + 100 = 1200

Treta Yug 200 + 2000 + 200 = 2400

Dwapar Yug 300 + 3000 + 300 = 3600

Kali Yug 400 + 4000 + 400 = 4800

Total Years = 12,000

Hindus use both a lunar and a solar calendar. The calendar used 5000 years ago was a lunar calendar. In approximately the year 56 BC, a noble king named **Vikramaditya** appeared who inaugurated the use of a solar calendar.

From Vishnu Puran, Part IV, Chapter 24, Verses 109 and 113, we learn that most Hindus today believe that Kali Yug is an evil age. The word "Kali" has a dual meaning... on one hand, Kali depicts a beautiful bud and a fragrant rose. On the other hand in later stages of its life cycle, it depicts a flower which has died, decayed and dried up. The same is true of the life cycles of religion. In the beginning, religion is like a beautiful bud and a fragrant rose because whenever one of God's Prophets or Manifestations appear on Earth... He renews Religion, and a new era of Righteousness in the world of Humanity is born. Similarly, like the beautiful rose which eventually dies decays and dries up... religion's regenerating force and life-giving vitality are obscured.

Hence, there are clear-cut prophecies in the Hindu Scriptures about the bad events to come and evil occurrence to happen in the end-times of Kali Yug. This is the example provided to us by Shri Krishna's religion. **Krishna** originally did establish a beautifully budding civilization and a spiritualized nobility which flourished like the opening of a "Kali" petal... of a fragrant rose. Krishna claims:

"I am the flowery Spring Season" ("Aham Rutuna Kusumakarah")---Gita X. So, Kali originally was a budding age of goodness, which gradually and progressively declined until it finally died 5065 lunar years later... in **1844 A.D.**

From these 5065 lunar years, we deduct the 126 years of Shri Krishna's life on earth. This leaves us with 4939 lunar years of 355 days each (4939 x 355 = 1753345 days.) If we convert these 4939 lunar years into solar years of 365.25 days each, we get 4800.3969 solar years. (4939 x 355 = 1753345 days divided by 365.25 days in a solar year= we get 4800.3969 solar years.

The year 4800.3969 of the Hindu calendar is the year 1844 A.D. of the Christian calendar. This prophecy says that Kali Yug will end in the year 1844 A.D. It also promises that by the end of Kali Yug age (or 1844 A.D.) there will appear a Manifestation of God who will inaugurate a new "Krit Yug" or a new "age of deeds". http://bci.org/prophecy-fulfilled/hindutim.htm

Buddhism

Maitreya (Sanskrit), **Metteyya** (Pāli), **Maithree** (Sinhala), **Jampa** (Tibetan) or **Di-Lặc** in Vietnamese, is regarded by Buddhists as a future Buddha of this world in Buddhist eschatology. In some Buddhist literature, such as the Amitabha Sutra and the Lotus Sutra, he is referred to as *Ajita* Bodhisattva.

Maitreya is a **bodhisattva** who in the Buddhist tradition is to appear on Earth, achieve complete enlightenment, and teach the pure dharma. According to scriptures, Maitreya will be a successor of the historic Śākyamuni Buddha. The prophecy of the arrival of Maitreya refers to a time when the Dharma will have been forgotten by most on Jambudvipa. It is found in the canonical literature of all major Buddhist schools (Theravāda, Mahāyāna, Vajrayāna), and is accepted by most Buddhists as a statement about an event that will take place when the Dharma will have been mostly forgotten on Earth. http://en.wikipedia.org/wiki/Maitreya

The religious symbol of Buddhism
(Google Images)

"At that period, brethren, there will arise in the world an Exalted One named **Metteyya, Arahat,** Fully Awakened, abounding in wisdom and goodness, happy, with knowledge of the worlds, unsurpassed as a guide to mortals willing to be led, a teacher for gods and men, an Exalted One, a Buddha, even as I am now..."

"...He, by himself, will thoroughly know and see, as it were face to face, this universe, with Its worlds of the spirits, Its Brahmas and Its Maras, and Its world of recluses and Brahmins, of princes and peoples, even as I now, by myself, thoroughly know and see them." --- **Digha Nikaya 1**

"The truth [the Norm, the Dhamma], lovely in origin, lovely in Its progress, lovely in Its consummation, will he **(Metteyya Buddha)** proclaim, both in the spirit and in letter, the higher life will he make known, in all Its fullness and in all Its purity, even as I do know. He will be accompanied by a congregation of some thousands of Brethren, even as I am now accompanied by a congregation of some hundreds of Brethren." --- **Digha Nikaya 2**

This prophecy of **Gautama Buddha** clearly predicts that the teaching of **Mettayya Buddha** will be a worldwide teaching.

The prophecy of the Lord Buddha that the teaching of Metteyya Buddha will be much more extensive than his own is already being fulfilled through the Baha'i Faith, already a worldwide religion with communities in every part of the planet consisting of all the races and nations of humanity.

Buddha replied:

'After my decease first, will occur the *five disappearances*. And what are the five disappearances?' The *disappearance of attainments* [to nibbana], *the disappearance of the method* [inability to practice wisdom, insight and the four purities of moral habit], *the disappearance of learning* [loss of men who follow the Dhamma and the forgetting of the Pitakas and other scriptures], *the disappearance of the symbols* [the loss of the outward forms, the robes, and practices of monkhood], *the disappearance of the relics* [the Dhatu]...

"Then when the Dispensation of the perfect Buddha is 5000 years old, the relics, not receiving reverence and honor will go to places where they can receive them... This, Sariputta, is called the disappearance of relics'.

This passage clearly shows that the Metteyya Buddha will appear 'before this auspicious eon runs to the end of years'. Since Gautama Buddha appeared in India and was speaking to disciples who had also been Hindus and were familiar with the Hindu system of dating cycles.

Sariputta: 'How will it occur?'

Buddha replied:

'After my decease first, will occur the five disappearances. And what are the five disappearances?' The disappearance of attainments [to nibbana], the disappearance of the method [inability to practice wisdom, insight and the four purities of moral habit], the disappearance of learning [loss of men who follow the Dhamma and the forgetting of the Pitakas and other scriptures], the disappearance of the symbols [the loss of the outward forms, the robes, and practices of monkhood], the disappearance of the relics [the Dhatu]...

"Then when the Dispensation of the perfect Buddha is 5000 years old, the relics, not receiving reverence and honor will go to places where they can receive them... This, Sariputta, is called the disappearance of relics".

This passage clearly shows that the Metteyya Buddha will appear *'before this auspicious aeon runs to the end of years'*. Since Gautama Buddha appeared in India and was speaking to disciples who had also been Hindus and were familiar with the Hindu system of dating cycles.

It would seem likely that when Gautama Buddha said 'before this auspicious aeon runs to the end of years', he was speaking of the **Hindu Kali Yuga** in the middle of which he had appeared.

416

This Kali Yuga ended at noon on 1 August 1943, equivalent to the year 2486 of the Buddhist Era. Therefore, according to this prophecy of the Buddha, the Metteyya Buddha should already have appeared sometime *before* 1943.

It is true that this prophecy of the **Five Disappearances** states that the last of the Disappearances will occur when the Dispensation of the Buddha is five thousand years old. Since only 2,500 years have passed, It might seem that it is not yet time for the coming of Metteyya Buddha.

It should, however, be borne in mind that when the *Buddha allowed women to be ordained as nuns*, he then prophesied that because of this time that the Dhamma endured *would be halved. The period during which the full Dhamma is known and Nibbana achieved (the First Disappearance) is halved from one thousand years to five hundred and similarly for the whole process.*

Thus, resulting in 2,500 years during which all of the Five Disappearances must occur. The year 2500 of the Buddhist Era was celebrated in AD 1956. This ties in very well with AD 1943 as the

last possible date for the appearance of the Metteyya Buddha. http://bci.org/prophecy-fulfilled/buddhasa.htm

From these religions, it is clear that each prophet or Manifestation of God had a particular set of teachings in their mission to fulfill which would guide mankind down the road to eventual unity and universal peace. They each acknowledged the former manifestation of God but also spoke of a time of a great universal prophet or manifestation that would not only bring an abundance of new teachings but, would also set in motion the institutions which would birth the new era of universal peace.

In addition to the idea of religion being progressively revealed from the same God through different prophets/messengers, there also exists in Bahá'í literature, the idea of a universal cycle, which represents a series of dispensations, and is used to categorize human history and social evolution in a number of ways. It is viewed as a superset of the sequence of progressive revelations and currently, comprises two cycles.

The **Adamic cycle**, also known as the **Prophetic cycle** is stated to have begun approximately 6,000 years ago with a Manifestation of God referred to in various sacred scriptures as **Adam**, and ended with the dispensation of **Muhammad**. In this cycle, Bahá'í belief is that Manifestations of God continued to advance human civilization at regular intervals through progressive revelation. The **Abrahamic religions** and **Dharmic religions** are partial recognitions of this cycle, from a Bahá'í point of view.

In Bahá'í belief, the **Bahá'í Cycle,** or **Cycle of Fulfillment,** began with the **Báb** and includes **Bahá'u'lláh,** and will last at least five hundred thousand years with numerous Manifestations of God appearing throughout that time. It is stated in Bahá'í literature that the Manifestations of God in the **Adamic cycle**, in addition to bringing their own teachings, foretold of the **Cycle of Fulfillment.**

Baha'i Faith – The Promise and Fulfillment of All the Ages!

This brings us to the **Baha'i Faith,** the youngest of the world's major religions. It's origin of birth was in the ancient land of Persia, now known as Iran with its international administrative centre in Haifa, Israel thus, Israel is no longer just the home to three major world religions (Judaism, Christianity, and Islam) but to a fourth, the Baha'i Faith.

Baha'is recognize **Baha'u'llah** "as the **Judge,** the **Lawgiver** and **Redeemer** of all mankind, as the **Organizer** of the entire planet, as the **Unifier** of the children of men, as the **Inaugurator** of the long-awaited millennium, as the **Originator** of a new **"Universal Cycle,"** as the **Establisher of the Most Great Peace,** as the **Fountain of the Most Great Justice,** as the **Proclaimer** of the coming of age of the entire human race, as the **Creator of a new World Order,** and as the **Inspirer** and **Founder** of a world civilization.

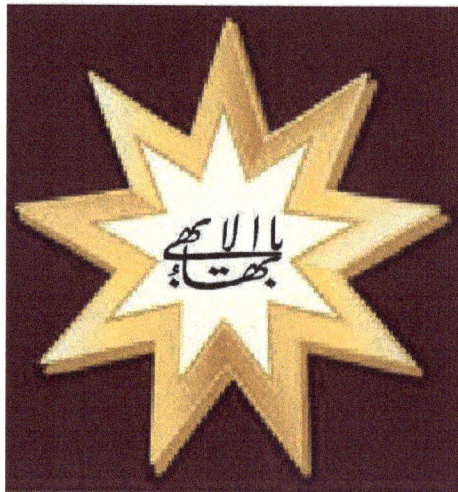

The religious symbol of Baha'i Faith
(Google Images)

To **Judaism** of Israel, He was neither more nor less than the incarnation of the **"Everlasting Father",** the **"Lord of Hosts"** come down **"with ten thousands of saints",** the **"Messiah",** the **"Glory of the Lord"** or the **"Glory of God";**

To **Christendom,** he is the return of the "**Christ** spirit" come **"in the glory of the Father",** the **"Father of the Son",**

To **Shi'ah Islam** the return of the **Imam Husayn;**

To **Sunni Islam** the descent of the **"Spirit of God" (Jesus Christ);**

To the **Zoroastrians** the promised **Shah-Bahram;**

To the **Hindus** the reincarnation of **Krishna;**

To the **Buddhists** the **fifth Buddha**...

"He alone is meant by the prophecy attributed to Gautama Buddha Himself, that "a Buddha named Maitreye, **the Buddha of universal fellowship"** should, in the fullness of time, arise and reveal **"His boundless glory"**...

Baha'u'llah wrote:

"Verily I say, this is the day in which mankind can behold the Face and hear the Voice of the ***Promised One***. *The call of God hath been raised, and the light of His countenance hath been lifted up upon men. It behooveth every man to blot out the trace of every idle word from the tablet of his heart, and to gaze, with an open and unbiased mind, on the signs of His Revelation, the proofs of His Mission, and the tokens of His Glory."* Baha'u'llah , Baha'i World Faith, Selected Writing of Baha'u'llah and Abdu'l-Baha, p.9-11; by the National Spiritual Assembly of the Unites States; 1945, 1956; Baha'i Publishing Trust; ISBN 0-87743—043-8

The Baha'i Faith has existed for nearly 170 hundred years unchanged in its original form as promulgated down by its Manifestation of God, Baha'u'llah. There are no schisms that have divided the Baha'i Faith into sects as there has been within every other world religion that has preceded it. It stands inviolable because of the powerful nature and structure of its holy writings that were penned by the Voice of God, Baha'u'llah!

Now that we have established who Baha'u'llah is, **The Promised One of all the Ages,** we can now look at those human limitations that impede our development on this planet toward a global civilization and **Universal Peace.**

"Among the teachings of Baha'u'llah is, that ***religious, racial, political, economic and patriotic prejudices destroy the edifice of humanity***. *As long as these prejudices prevail, the world of humanity will not have rest. For a period of 6000 years, the world of humanity has not been free of war, strife, murder and blood thirstiness. In every period war has been waged in one country or another and that war was due to religious prejudice, political prejudice or patriotic prejudice. It has therefore been ascertained and proved that all prejudices are destructive of the human edifice. As long as these prejudices persist, the struggle for existence must remain dominant, and bloodthirstiness and rapacity continue. Therefore, even as was the case in the past, the world of humanity cannot be saved from the darkness of nature and cannot attain illumination except through the abandonment of prejudices and the acquisition of the morals of the Kingdom."* The Bahá'í Revelation, Selections from the Bahá'í Holy Writings and Talks by 'Abdu'l-Bahá; first published in 1955; Bahá'í Publishing Trust; London, U.K.; Library Reference 299.15 Bahá'í

Many people have asked why they never heard before about Baha'u'llah and the Baha'i Faith. The problem was not in the proclamation of this religion but, in having it outlawed in the country of its birth, Persia (Iran) and in various areas of the Ottoman empire. People were also, heedless to investigate their own religious prophesies or recognize the signs pertaining to a promised Messiah who would appear at the end of times with a new faith to unite all mankind. There has also been the deliberate suppression and/or persecution of this new faith by certain governments, and by the leading ecclesiastical orders of both Christianity and Islam.

In September 1867, **Bahá'u'lláh**, wrote a series of letters to the world leaders of His time, addressing, among others, **Emperor Napoleon III, Queen Victoria, Kaiser Wilhelm I, Tsar Alexander II of Russia, Emperor Franz Joseph, Pope Pius IX, Sultan Abdul-Aziz,** and the Persian ruler, **Nasiri'd-Din Shah** as well as the **Presidents of the United States** He also sent letters to the leaders of both the Christian **(Pope Pius IIX)** (see letter below**)** and Islamic ecclesiastic orders. He proclaimed that He was the fulfillment of the Christ promise and of all the world's religions, that His Faith was the Cause of God for this day and age.

In these letters, Bahá'u'lláh openly proclaimed His station. He spoke of the dawn of a new age. But first, He warned, there would be catastrophic upheavals in the world's political and social order. To smooth humanity's transition, He urged the world's leaders to pursue justice. He forewarned them of the world's armament build-up which also began in the mid-1800s urging the world's rulers to band together into some form of **Commonwealth of Nations**. Should one nation take up arms against another nation, all nations should arise to overpower the aggressor nation. This says Baha'u'llah was not but manifest justice! Only by acting collectively against war, He said, could a lasting peace be established. He offered solutions to the world's spiritual ills as well as practical solutions to the peace and stability of the world, including its political, financial and religious systems. **The Bahá'í Revelation, Selections from the Bahá'í Holy Writings and Talks by 'Abdu'l-Bahá; first published in 1955; Bahá'í Publishing Trust; London, U.K.; Library Reference 299.15 Bahá'í**

Not one monarch in Europe acknowledged the admonitions of Baha'u'llah or his divine station with the exception of Britain's **Queen Victoria** who responded: ***"If this is of God, it will endure..."* Prisoner and the Kings by William Sears, 1971; General Publishing Co. Ltd., Toronto, Canada; Revised Edition (Jan. 1 2007); U.S. Baha'i Publishing Trust; Wilmette, Illinois, USA**

Shoghi Effendi, in "Promised Day is Come" (p. 51), recounts the following that **Emperor Napoleon III** of France arrogantly threw the Baha'u'llah's letter behind his back."It is reported that upon receipt of this first Message that superficial, tricky, and pride-intoxicated monarch flung down the Tablet saying: ***"If this man is God, I am two gods!"* Promised Day is Come by Shoghi Effendi; 1941; published by Baha'i Publishing Trust; Wilmette, Illinois, USA; ISBN 0-87743-132-9**

Below is a letter from Baha'u'llah to **Pope Pius IX** proclaiming His divine station in fulfillment of biblical prophecy of "***the return of the Son in the Glory of the Father***":

O POPE! Rend the veils asunder. He Who is the Lord of Lords is come overshadowed with clouds, and the decree hath been fulfilled by God, the Almighty, the Unrestrained... He, verily, hath again come down from Heaven even as He came down from it the first time. Beware that thou dispute not with Him even as the Pharisees disputed with Him (Jesus) without a clear token or proof. On His right hand flow the living waters of grace, and on His left the choice Wine of justice, whilst before Him march the angels of Paradise, bearing the banners of His signs.

420

Beware lest any name debar thee from God, the Creator of earth and heaven. Leave thou the world behind thee, and turn towards thy Lord, through Whom the whole earth hath been illumined... Dwellest thou in palaces whilst He Who is the King of Revelation liveth in the most desolate of abodes? Leave them unto such as desire them, and set thy face with joy and delight towards the Kingdom... Arise in the name of thy Lord, the God of Mercy, amidst the peoples of the earth, and seize thou the Cup of Life with the hands of confidence, and first drink thou therefrom, and proffer it then to such as turn towards it amongst the peoples of all faiths...

Call thou to remembrance Him Who was the Spirit (Jesus), Who when He came, the most learned of His age pronounced judgment against Him in His own country, whilst he who was only a fisherman believed in Him. Take heed, then, ye men of understanding heart! Thou, in truth, art one of the suns of the heaven of His names. Guard thyself, lest darkness spread its veils over thee, and fold thee away from His light... Consider those who opposed the Son (Jesus), when He came unto them with sovereignty and power. How many the Pharisees who were waiting to behold Him, and were lamenting over their separation from Him! And yet, when the fragrance of His coming was wafted over them, and His beauty was unveiled, they turned aside from Him and disputed with Him... None save a very few, who were destitute of any power amongst men, turned towards His face. And yet, today, every man endowed with power and invested with sovereignty prideth himself on His Name! In like manner, consider how numerous, in these days, are the monks who, in My Name, have secluded themselves in their churches, and who, when the appointed time was fulfilled, and We unveiled Our beauty, knew Us not, though they call upon Me at eventide and at dawn...

The Word which the Son concealed is made manifest. It hath been sent down in the form of the human temple in this day. Blessed be the Lord Who is the Father! He, verily, is come unto the nations in His most great majesty. Turn your faces towards Him, O concourse of the righteous... This is the day whereon the Rock (Peter) crieth out and shouteth, and celebrateth the praise of its Lord, the All-Possessing, the Most High, saying: 'Lo! The Father is come, and that which ye were promised in the Kingdom is fulfilled!...' My body longeth for the cross, and Mine head waiteth the thrust of the spear, in the path of the All-Merciful, that the world may be purged from its transgressions...

O Supreme Pontiff! *Incline thine ear unto that which the Fashioner of mouldering bones counselleth thee, as voiced by Him Who is His Most Great Name. Sell all the embellished ornaments thou dost possess, and expend them in the path of God, Who causeth the night to return upon the day, and the day to return upon the night. Abandon thy kingdom unto the kings, and emerge from thy habitation, with thy face set towards the Kingdom, and, detached from the world, then speak forth the praises of thy Lord betwixt earth and heaven. Thus hath bidden thee He Who is the Possessor of Names, on the part of thy Lord, the Almighty, the All-Knowing. Exhort thou the kings and say: 'Deal equitably with men. Beware lest ye transgress the bounds fixed in the Book.' This indeed becometh thee. Beware lest thou appropriate unto thyself the things of the world and the riches thereof. Leave them unto such as desire them, and cleave unto that which hath been enjoined upon thee by Him Who is the Lord of creation. Should any one offer thee all the treasures of the earth, refuse to even glance upon them. Be as thy Lord hath been. Thus hath the Tongue of Revelation spoken that which God hath made the ornament of the book of creation... Should the inebriation of the wine of My verses seize thee, and thou* *determinest to present thyself before the throne of thy Lord, the Creator of earth and heaven, make My love thy vesture and thy shield remembrance of Me, and thy provision reliance upon God, the Revealer of all power... Verily, the day of ingathering is come, and all things have been separated from each other. He hath stored away that which He chose in the vessels of justice, and cast into fire that which befitteth it. Thus hath it been decreed by your Lord, the Mighty, the Loving, in this promised Day. He, verily, ordaineth what He pleaseth. There is none other God save He, the Almighty, the All-Compelling.* Proclamation of Baha'u'llah by Baha'u'llah; 1978 reprint; US Bahá'í Publishing Trust, Wilmette, Illinois, USA

As foretold in the bible that in those latter days, kingdoms will be toppled and monarchs and ecclesiastics will lose their thrones and the stars of heaven will fall upon the Earth and the Sun will shine no more. This is not a literal interpretation of prophecy but, an allegorical or spiritual understanding, as stars cannot in actual fact fall and land upon the Earth as this is a scientific impossibility. What we have seen in the last 100 plus years between 1850 to 1950 is that many monarchies and the ecclesiastical orders of various religions in Europe, Russia and in the Middle and the Far East have disappeared into history as forewarned by Baha'u'llah, if they did not heed His warnings, admonitions and acknowledged His divine station. This indeed is that Day of Judgment and they have been weighed and measured by the Divine Judge, the Lord of the Ages... as prophesized in all the holy writings.

Most amazingly in fulfillment of Judeo-Christian biblical prophecy, these monarchs, and their kingdoms were overthrown and lost to history forever, All these monarchies have disappeared and been replaced by governments ruled by the people, the ecclesiastic orders of Christendom, Islam, and Judaism are all now impotent and mere shadows of their former glory. The only exception to this divine creed was that the British monarchy had survived; Queen Victoria's reign became the longest in British royal history due to her acknowledgement of Baha'u'llah's divine station. She set in motion the abolition of slavery in her country and inspire other nations to do the same as per the admonitions of Baha'u'llah; God in return had rewarded her actions and her nation for paying heed to the **Voice of God on Earth**.

Resolving the Societal, Religious, Political and Economic Issues of Humanity

'Abdu'l-Bahá, the eldest son of **Bahá'u'lláh**, the founder of the Bahá'í Faith continued Baha'u'llah's teachings and in 1892, `Abdu'l-Bahá was appointed in his father's will to be his successor and head of the Bahá'í Faith. Abdu'l-Bahá' took up the mantle of teaching by proclaiming the Faith of his Father with extensive travels throughout much of Europe, the Middle East and into America and Canada. Wherever 'Abdu'l-Bahá travelled, he was always greeted with much public enthusiasm, meeting new friends and giving insightful talks on the **Baha'i Faith.** His talks expand and elucidated the teachings of Baha'u'llah about the conditions of the world and the bright and hopeful future of humanity if they but turn to and heed God's Words for this day and age.

Of interest to us with respect to human limitations that hinder our spiritual and material advancement, 'Abdu'l-Baha wrote a letter called **Tablet to The Hague** in which 'Abdu'l-Bahá wrote to the **Central Organization for Durable Peace** in The Hague, the Netherlands on 17

December 1919. In this letter 'Abdu'l-Baha gives some insights into the pressing issues that hinder mankind's progress toward global unity.

'Abdu'l-Bahá

"If this prejudice and enmity are on account of religion (consider that) religion should be the cause of fellowship, otherwise it is fruitless. And if this prejudice be the prejudice of nationality consider that all mankind are of one nation; all have sprung from the tree of Adam, and Adam is the root of the tree. That tree is one and all these nations are like branches, while the individuals of humanity are like leaves, blossoms and fruits thereof. Then the establishment of various nations and the consequent shedding of blood and destruction of the edifice of humanity result from human ignorance and selfish motives.

As to the patriotic prejudice, this is also due to absolute ignorance, for the surface of the earth is one native land. Everyone can live in any spot on the terrestrial globe. Therefore, all the world is man's birthplace. These boundaries and outlets have been devised by man. In the creation, such

423

boundaries and outlets were not assigned. Europe is one continent, Asia is one continent, Africa is one continent, Australia is one continent, but some of the souls, from personal motives and selfish interests, have divided each one of these continents and considered a certain part as their own country. God has set up no frontier between France and Germany; they are continuous. Yea, in the first centuries, selfish souls, for the promotion of their own interests, have assigned boundaries and outlets and have, day by day, attached more importance to these, until this led to intense enmity, bloodshed, and rapacity in subsequent centuries. In the same way, this will continue indefinitely, and if this conception of patriotism remains limited within a certain circle, it will be the primary cause of the world's destruction. No wise and just person will acknowledge these imaginary distinctions. Every limited area which we call our native country we regard as our motherland, whereas the terrestrial globe is the motherland of all, and not any restricted area. In short, for a few days, we live on this earth and eventually we are buried in it, it is our eternal tomb. Is it worthwhile that we should engage in bloodshed and tear one another to pieces for this eternal tomb? Nay, far from it, neither is God pleased with such conduct nor would any sane man approve of it.

 Consider! The blessed animals engage in no patriotic quarrels. They are in the utmost fellowship with one another and live together in harmony. For example, if a dove from the east and a dove from the west, a dove from the north and a dove from the south chance to arrive, at the same time, in one spot, they immediately associate in harmony. So is it with all the blessed animals and birds. But the ferocious animals, as soon as they meet, attack and fight with each other, tear each other to pieces and it is impossible for them to live peaceably together in one spot. They are all unsociable and fierce, savage and combative fighters.

Regarding the economic prejudice, it is apparent that whenever the ties between nations become strengthened and the exchange of commodities accelerated, and any economic principle is established in one country, it will ultimately affect the other countries and universal benefits will result. Then why this prejudice?

As to the political prejudice, the policy of God must be followed, and it is indisputable that the policy of God is greater than human policy. We must follow the Divine policy and that applies alike to all individuals. He treats all individuals alike: no distinction is made, and that is the foundation of the Divine Religions." **The Bahá'í Revelation, Selections from the Bahá'í Holy Writings and Talks by 'Abdu'l-Bahá; first published in 1955; Bahá'í Publishing Trust; London, U.K.; Library Reference 299.15 Bahá'í**

And finally:

"The oneness of the kingdom of humanity will supplant the banner of conquest, and all communities of the earth will gather under its protection. No nation with separate and restricted boundaries—such as Persia, for instance—will exist. The United States of America will be known only as a name. Germany, France, England, Turkey, Arabia—all these various nations will be welded together in unity. When the people of the future are asked, "To which nationality do you belong?" the answer will be, "To the nationality of humanity…. The people of the future will not say, "I belong to the nation of England, France or Persia"; for all of them will be citizens

424

of a universal nationality—the one family, the one country, the one world of humanity—and then these wars, hatreds and strifes will pass away.

The body of the human world is sick. Its remedy and healing will be the oneness of the kingdom of humanity. Its life is the Most Great Peace. Its illumination and quickening is love. Its happiness is the attainment of spiritual perfections…. The divine Jerusalem has come down from heaven. The bride of Zion has appeared. The voice of the Kingdom of God has been raised."

The Promulgation of Universal Peace; p. 18. by 'Abdu'l-Bahá; published in 1982 (originally published 1922-1925) by the National Spiritual Assembly of the United States; BP360.A375 1982 and ISBN 0-87743-172-8

CHAPTER 115

GLOBAL UNITY BEFORE GALACTIC UNITY

Before we can reach out to the stars and communicate with other Extraterrestrial civilizations, we must first establish global unity on Earth! We cannot even begin to consider the remotest possibility of a galactic unity with other intelligences and civilizations, until we get our house in order first, here on Earth. This is the prerequisite for galactic membership in a bigger universe!

There is at this current time a growing awareness among the peoples of the world that technology has shrunk the distances and borders of nations forcing people to acknowledge each other and to co-operate with each other, unlike any other time in its history. It is generally acknowledged by many governments, although, never in any official position or capacity that the world is slowly but, inevitably progressing towards a new level of unity and harmony on the planet that must encompass all mankind into one commonwealth under a banner of universal peace.

Bahá'u'lláh addresses the exigencies of this time for which there are basic truths and prerequisites for the unification of mankind in order for it to become a global civilization. Such truths are inalienable (pun intended) rights of everyone and self-evident to any person of understanding and insight. All nations must come together around the table of consultation to resolve their differences and to share the common bonds that unite them. They must surrender some of their sovereignty in order that the greater peace and stability of the world is secured. All nations must be united under the banner of universal peace and in order to achieve it, Bahá'u'lláh states that there are 12 basic principles from His voluminous teachings that would set the tone and direction toward world peace and unity. Here are but a few drops from the ocean of Bahá'u'lláh teachings:

1. There must be the recognition that there is only **One God.**

2. There must be recognition of the **Oneness of Mankind.**

3. There must be recognition of the **Oneness of all Religions.**

4. There must be a **Declaration of Universal Peace upheld by a World Government.**

5. There must be the **Elimination of Prejudice of all Kinds** (religious, racial, political, economic and patriotic prejudices).

6. There must be an unfettered freedom of **Independent Investigation of Truth.**

7. There must be the recognition of the essential **Harmony of Science and Religion.**

8. The must be the recognition of the basic **Equality of Men between Women.**

9. There must be the promotion of **Universal Compulsory Education** for everyone.

10. There must be the recognition of a **Spiritual Solution to the World's Economic Problems.**

11. There must be the adoption of **One Universal Auxiliary Language.**

12. There must be the adoption of a **Universal currency, weights, and measures.**

These basic principles and hundreds more found in the Bahá'í Faith were established by Bahá'u'lláh in the latter half of the nineteenth century and promulgated by His Son and appointed successor, 'Abdu'l-Bahá during a tireless twenty-nine year ministry. While on an historic teaching trip in 1912 throughout the United States and Canada, 'Abdu'l-Bahá explained Bahá'u'lláh's Teachings to audiences of diverse backgrounds. Many of His talks included discussions of such Bahá'í principles which he elucidates further in greater detail.

The principles of the Teachings of Baha'u'llah should be carefully studied, one by one, until they are realized and understood by mind and heart - so will you become strong followers of the light, truly spiritual, heavenly soldiers of God, acquiring and spreading the true civilization in Persia, in Europe, and in the whole world. This will be the paradise which is to come on earth when all mankind will be gathered together under the tent of unity in the Kingdom of Glory. **'Abdu'l-Baha: Paris Talks, (Page: 22)**

These Holy Words and teachings are the remedy for the body politic, the divine prescription and real cure for the disorders which afflict the world. Therefore, we must accept and partake of this healing remedy in order that complete recovery may be assured. Every soul who lives according to the teachings of Baha'u'llah is free from the ailments and indispositions which prevail throughout the world of humanity; otherwise, selfish disorders, intellectual maladies, spiritual sicknesses, imperfections, and vices will surround him, and he will not receive the life-giving bounties of God. Baha'u'llah is the real Physician. He has diagnosed human conditions and indicated the necessary treatment. The essential principles of His healing remedies are the knowledge and love of God, severance from all else save God, turning our faces in sincerity toward the Kingdom of God, implicit faith, firmness and fidelity, loving-kindness toward all creatures and the acquisition of the divine virtues indicated for the human world. These are the fundamental principles of progress, civilization, international peace and the unity of mankind. These are the essentials of Baha'u'llah's teachings, the secret of everlasting health, the remedy and healing for man. It is my hope that you may assist in healing the sick body of the world through these teachings so that eternal radiance may illumine all the nations of mankind. **'Abdu'l-Baha: Promulgation of Universal Peace, (Pages: 204-205)**

The explanations of these principles in the sections following are excerpts from the public talks of 'Abdu'l-Bahá in America in 1912, published in The Promulgation of Universal Peace:

The Oneness of God

Baha'i teachings focus on unity – the oneness of God, the oneness of religion, and the oneness of the human race. Baha'is believe there is only one God, the Creator of the universe. Although, God may be called by different names in different languages – be it Yahweh, Allah, Brahma, or

God – in actuality, these names all refer to the same singular force and being.

The Bahá'í belief in one God means that the universe and all creatures and forces within it have been created by a single supernatural Being. This Being, Whom we call God, has absolute control over His creation (omnipotence) as well as perfect and complete knowledge of it (omniscience).

Evidence of this Oneness of God can be found in many places, one of which is examining the essential teachings (not interpretations) of all the world's major religions. All religions, although they may at times appear to be separate and the interpretations of their teachings sometimes at odds, they are actually one ever-unfolding religion from a single source. Baha'is call this concept "The Oneness of Religion" and also refer to it as **Progressive Revelation.**

The Bahá'í view of God is essentially monotheistic. God is the imperishable, uncreated being who is the source of all existence. He is described as "a personal God, unknowable, inaccessible, the source of all Revelation, eternal, omniscient, omnipresent and almighty". Though transcendent and inaccessible directly, his image is reflected in his creation. The purpose of creation is for the created to have the capacity to know and love its creator. God communicates his will and purpose to humanity through intermediaries, known as Manifestations of God, who are the prophets and messengers that have founded religions from prehistoric times up to the present day.

While God's essence is inaccessible, a subordinate form of knowledge is available by way of mediation by divine messengers, known as Manifestations of God. The Manifestations of God reflect divine attributes, which are creations of God made for the purpose of spiritual enlightenment, onto the physical plane of existence. **Shoghi Effendi**, the head of the Bahá'í Faith in the first half of the 20th century, described God as inaccessible, omniscient, almighty, personal, and rational, and rejected pantheistic, anthropomorphic and incarnationist beliefs.

Although human cultures and religions differ on their conceptions of God and his nature, Bahá'ís believe they nevertheless refer to one and the same Being. The differences, instead of being regarded as irreconcilable constructs of mutually exclusive cultures, are seen as purposefully reflective of the varying needs of the societies in which the divine messages were revealed. No single faith and associated conception of God is thus considered essentially superior to another from the viewpoint of its original social context; however, more recent religions may teach a more advanced conception of God as called for by the changing needs of local, regional or global civilization. Bahá'ís thus regard the world's religions as chapters in the history of one single faith, revealed by God's Manifestations progressively and in stages.

While the Bahá'í writings teach of a personal god who is a being with a personality (including the capacity to reason and to feel love), they clearly state that this does not imply a human or physical form. **Shoghi Effendi** writes:

What is meant by personal God is a God Who is conscious of His creation, Who has a Mind, a Will, a Purpose, and not, as many scientists and materialists believe, an unconscious and determined force operating in the universe. Such conception of the **Divine Being**, as the Supreme and ever-present Reality in the world, is not anthropomorphic, for it transcends all human limitations and forms, and does by no means attempt to define the essence of Divinity which is obviously beyond any human comprehension. To say that God is a personal Reality does not mean that He has a physical form or does in any way resemble a human being. To entertain such belief would be sheer blasphemy.

Below are just some of the other teachings of the Baha'i Faith:

The Oneness of Mankind

A fundamental teaching of Bahá'u'lláh is the oneness of the world of humanity. Addressing mankind, He says, ***"Ye are all leaves of one tree and the fruits of one branch."*** By this it is meant that the world of humanity is like a tree, the nations or peoples are the different limbs or branches of that tree and the individual human creatures are as the fruits and blossoms thereof.

Although in former centuries and times this subject received some measure of mention and consideration, it has now become the paramount issue and question in the religious and political conditions of the world. History shows that throughout the past there have been continual warfare and strife among the various nations, peoples, and sects, but now.....in this century of illumination, hearts are inclined toward agreement and fellowship and minds are thoughtful upon the question of the unification of mankind.

What incalculable benefits and blessings would descend upon the great human family if unity and brotherhood were established! In this century when the beneficent results of unity and the ill effects of discord are so clearly apparent, the means for the attainment and accomplishment of human fellowship have appeared in the world.

The Common Foundation of All Religions is One

The foundation of all the divine religions is one. All are based upon reality. Reality does not admit plurality, yet amongst mankind, there have arisen differences concerning the manifestations of God. Some have been **Zoroastrians**, some are **Buddhists**, some **Jews**, **Christians**, **Mohammedans** and so on. This has become a source of divergence whereas the teachings of the holy souls who founded the divine religions are one in essence and reality. All these have served the world of humanity... All have guided souls to the attainment of perfections, but among the nations, certain imitations of ancestral forms of worship have arisen. These imitations are not the foundation and essence of the divine religions. Inasmuch as they differ from the reality and the essential teachings of the Manifestations of God, dissensions have arisen and prejudice has developed. Religious prejudice thus becomes the cause of warfare and battle. If we abandon these time-worn imitations and investigate reality all of us will be unified. No discord will remain; antagonism will disappear. All will associate in fellowship. All will enjoy the cordial bonds of friendship. The world of creation will then attain composure. The dark and gloomy clouds of blind imitations and dogmatic variances will be scattered and dispelled; the Sun of Reality will shine most gloriously.

All religions share the concept of the "golden rule" as they are all one from one source, one divine creator and therefore, all religions are progressive in that they acknowledge the religions of former ages but also point to the religions yet to come in fulfillment to prophecy. In this day and age, Progressive Revelation is a major teaching of the Baha'i Faith and the word of God finds it highest attainment and fulfillment in the Manifestation of God, Baha'u'llah. Happy is he who has turned his face his Lord and hear His Divine Words.

Universal Peace Upheld by a World Government

The world is in greatest need of international peace. Until it is established, mankind will not attain composure and tranquility. It is necessary that the nations and governments organize an international tribunal to which all their disputes and differences shall be referred. The decision of that tribunal shall be final.

He **(Bahá'u'lláh)** exhorted them (rulers of the world) to peace and international agreement, making it incumbent upon them to establish a board of international arbitration; that from all nations and governments of the world there should be delegates selected for a congress of nations which should constitute a universal arbitral court of justice to settle international disputes.

Warfare and strife will be uprooted, disagreement and dissension pass away and Universal Peace unite the nations and peoples of the world. All mankind will dwell together as one family, blend as the waves of one sea, shine as stars of one firmament and appear as fruits of the same tree.

True civilization will unfurl its banner in the midmost heart of the world whenever a certain number of its distinguished and high-minded sovereigns--the shining exemplars of devotion and determination--shall for the good and happiness of all mankind, arise, with firm resolve and clear vision, to establish the Cause of Universal Peace. They must make the Cause of Peace the object of general consultation and seek by every means in their power to establish a Union of the nations of the world.

They must conclude a binding treaty and establish a covenant, the provisions of which shall be sound, inviolable and definite. They must proclaim it to all the world and obtain for it the sanction of all the human race. This supreme and noble undertaking--the real source of the peace and well-being of all the world--should be regarded as sacred by all that dwell on earth. All the forces of humanity must be mobilized to ensure the stability and permanence of this Most Great Covenant. In this all--embracing Pact the limits and frontiers of each and every nation should be clearly fixed, the principles underlying the relations of governments towards one another definitely laid down, and all international agreements and obligations ascertained. In like manner, the size of the armaments of every government should be strictly limited, for if the preparations for war and the military forces of any nation should be allowed to increase, they will arouse the suspicion of others. **The fundamental principle underlying this solemn Pact should be so fixed that if any government later violate any one of its provisions, all the governments on earth should arise to reduce it to utter submission, nay the human race as a whole should resolve, with every power at its disposal, to destroy that government.** Should this greatest of all remedies be applied to the sick body of the world, it will assuredly recover from its ills and will remain eternally safe and secure.

430

Elimination of All Prejudices

Prejudice of all kinds, whether religious, racial, patriotic, political or economic is destructive of the divine foundations in man. All the warfare and bloodshed in human history have been the outcome of prejudice. This earth is one home and nativity. God has created mankind with equal endowment and right to live upon the earth. As a city is the home of all its inhabitants, although each may have his individual place or residence therein, so the earth's surface is one wide native land or home for all of humankind.

"The Earth is but one country and mankind its citizens." Such a powerful pronouncement by Baha'u'llah is readily apparent in this age of space exploration, as every astronaut who has gone into space will attest, there are no borders or political boundaries marked upon the Earth; it really is one country!

Independent Investigation of Truth

Everyone must be permitted to **unfettered self-investigation of the truth,** wherever that truth may lie and regardless of the subject matter. God has endowed man with the eye of investigation by which he may see and recognize truth. He has endowed man with ears that he may hear the message of reality and conferred upon him the gift of reason by which he may discover things for himself. This is his endowment and equipment for the investigation of reality. Man is not intended to see through the eyes of another, hear through another's ears nor comprehend with another's brain. Each human creature has individual endowment, power, and responsibility in the creative plan of God.

The Essential Harmony of Science and Religion

Religion must conform to science and reason, otherwise, it is superstition. God has created man in order that he may perceive the verity of existence and endowed him with mind or reason to discover truth. Therefore, scientific knowledge and religious belief must be conformable to the analysis of this divine faculty in man.

Science by itself left unchecked by moral and spiritual laws becomes a runaway Frankenstein monster and civilization becomes debased and develops without a soul. Religion without the checks and balances of reason from science becomes a religion of dogma, superstition and a quagmire of irrational beliefs. Science and religion are the twin pillars of one truth that comes from one God. Science is truth revealed through discovery and investigation and religion is truth manifested through divine inspiration and revelation. So powerful are these twin institutions that mankind discerns all mysteries of this universe and all the spiritual worlds of God; the world of mankind progresses and civilization advances

Equality of Men and Women

There must be an equality of rights between men and women. Women shall receive an equal privilege of education. This will enable them to qualify and progress in all degrees of occupation and accomplishment. For the world of humanity possesses two wings, man, and woman. If one wing remains incapable and defective, it will restrict the power of the other, and full flight of humanity will be impossible. Therefore, the completeness and perfection of the human world is dependent upon the equal development of these two factors.

Universal Compulsory Education

There must be **universal compulsory education** for everyone upon this Earth for it is the bright light that illumines the world of humanity, inasmuch as ignorance and lack of education are barriers of separation among mankind, all must receive training and instruction. Through this provision the lack of mutual understanding will be remedied, and the unity of mankind furthered and advanced. **Universal education** is a universal law. It is, therefore, incumbent upon every father to teach and instruct his children according to his possibilities. If he is unable to educate them, the body politic, the representative of the people, must provide the means for their education.

Girls must be given preference for learning if a family cannot afford to educate all their children for Baha'u'llah has said that they are the mothers of mankind and the first educators of humanity. How lofty is the station of motherhood and how weighty the responsibility that rests upon their shoulders.

A Spiritual Solution to the Economic Problem

Among the results of the manifestation of spiritual forces will be that the human world will adapt itself to a new social form, the justice of God will become manifest throughout human affairs and human equality will be universally established. The poor will receive a great bestowal and the rich attain eternal happiness. For although at the present time the rich enjoy the greatest luxury and comfort, they are nevertheless deprived of eternal happiness; for eternal happiness is contingent upon giving and the poor are everywhere in the state of abject need. Through the manifestation of God's great equity the poor of the world will be rewarded and assisted fully and there will be a readjustment in the economic conditions of mankind so that in the future there will not be the abnormally rich nor the abject poor. The rich will enjoy the privilege of this new economic condition as well as the poor, for owing to certain provision and restriction they will not be able to accumulate so much as to be burdened by its management, while the poor will be relieved from the stress of want and misery.

A beautiful analogy of this **spiritual solution to the world's economic problems** and the extremes of wealth and poverty was stated by "**Abdu'l-Baha** when he compared it to the human body. "Money is the lifeblood of society" that just as the body requires the vitality of blood to nourish and heal the human body; it must circulate to all the limbs and organs and to all parts of the body in order that the body remain healthy, strong and capable of performing great physical and mental feats. If the blood pools or settles into one or two areas of the body, those areas may appear to become very healthy but in reality, the rest of the body suffers. The limbs and vital organs of the body begin to atrophy, weaken and shut down; eventually, death occurs to the body and in doing so, even those areas that were exclusively nourished by the enrichment of blood will also die because they are not independent from the body.

432

Thus, money must flow through the body of society like blood permitting all areas of the internal and external infrastructure of civilization to be nourished, healthy and potent reaping the benefits of economic growth and prosperity.

A Universal Auxiliary Language

Bahá'u'lláh has proclaimed the adoption of a universal language. A language shall be agreed upon by which unity will be established in the world. Each person will require training in two languages, his native tongue and the universal auxiliary form of speech. This will facilitate intercommunication and dispel the misunderstandings which the barriers of language have occasioned in the world.

Currently, a universal language may come from one of the languages already in existence based upon international popularly, international business and commerce usage, population usage and other factors. English is by far the predominant language used by many countries either for business and commerce or as a secondary language, however, China and India have the largest population base of any other countries in the world and therefore, have more people speaking those languages than any other but, even among these countries English is an unofficial secondary language. The Arabic languages are also spoken by a large population base particularly in the Middle East.

It is also possible that a newly invented language like Esperanto or a similarly invented language may gain international popularly and thus become the universal language of the planet. Only time will tell and one interesting fact stated by Shoghi Effendi was that the languages that the Baha'i Writings were originally written in and there were destined to survive for at least a thousand years were **Arabic, Persian,** and **English!**

A Universal Currency, Weights, and Measures

A world script, a world literature, a uniform and universal system of currency, of weights and measures, will simplify and facilitate intercourse and understanding among the nations and races of mankind.

We can already see the universal development implementation of various aspects of this particular teaching of Baha'u'llah's in the use of the decimal or metric system for weights (grams and kilograms) and measures (metres and kilometres) which has virtually replaced the old English standard of ounces, pounds and inches, feet and miles. Almost every country throughout uses the metric system except for the United States which has tenaciously clung to the old English system but, it too is gradually accepting the new metric system.

The decimal currency system based upon units of ten is also becoming popular among more and more countries, however, there is as yet no universal currency adopted as a world currency but, here again, there have been some countries offering prototypes and sooner or later one of these will assuredly be adopted.

A universal language will enable anyone to travel to another country and converse with its inhabitants removing all barriers of differences and prejudices and enabling people to recognize common bonds and similarities which will help to unify the hearts and minds of humanity. (1981) [1904-06]. *Some Answered Questions*. Wilmette, Illinois, USA: Bahá'í Publishing Trust. ISBN 0-87743-190-6. and

'Abdu'l-Bahá (1982) [1912]. *The Promulgation of Universal Peace* (Hardcover ed.). Wilmette, Illinois, USA: Bahá'í Publishing Trust. ISBN 0-87743-172-8. and

Bahá'u'lláh (1976). *Gleanings from the Writings of Bahá'u'lláh*. Wilmette, Illinois, USA: Bahá'í Publishing Trust. ISBN 0-87743-187-6 and

Effendi, Shoghi (1944). *God Passes By*. Wilmette, Illinois, USA: Bahá'í Publishing Trust. ISBN 0-87743-020-9.

'Abdu'l-Bahá declared that these Baha'i principles would find their way around the world and be incorporated into the governments of the world but, the nations of the world must also come together and resolve their differences and work to establishing universal peace. Although the **League of Nations** was a good beginning to bring mankind together, Abdu'l-Baha declares that the **League of Nations** (forerunner to the **United Nations**) is *"incapable of establishing universal peace",* and calls for the establishment of a **Supreme Tribunal**, representing all countries:

"When the Supreme Tribunal gives a ruling on any international question, either unanimously or by majority rule, there will no longer be any pretext for the plaintiff or ground of objection for the defendant. In case any of the governments or nations, in the execution of the irrefutable decision of the Supreme Tribunal, be negligent or dilatory, the rest of the nations will rise up against it because all the governments and nations of the world are the supporters of this Supreme Tribunal."

Keep in mind that the United Nations, the successor to the League of Nations is itself also dysfunctional and impotent to carry out the decisions on behalf of its 193 nation members.

The **United Nations** (UN) is an intergovernmental organization created in 1945 to promote international cooperation. The **UN Headquarters** resides in international territory in New York City, with further main offices in Geneva, Nairobi, and Vienna. The organization is financed by assessed and voluntary contributions from its member states. Its objectives include maintaining international peace and security, promoting human rights, fostering social and economic development, protecting the environment, and providing humanitarian aid in cases of famine, natural disaster, and armed conflict. http://en.wikipedia.org/wiki/United_Nations

The League of Nations and the United Nations were founded upon *some of the principles of the* **Baha'i Faith** but, early in the formation of the League of Nations, a disastrous self-serving decision was made after the first world war to give five very powerful nations (United States, Britain, Russia, France, and China) the **Veto Power** to side track any decision made by the majority of the former League of Nations and now the United Nations members!

In the **United Nations Security Council (UNSC),** the "power of veto" refers to the veto power wielded solely by the five permanent members of the United Nations Security Council (China,

France, Russia, United Kingdom, and United States), *enabling them to prevent the adoption of any "substantive" draft Council resolution, regardless of the level of international support for the draft*. The veto does not apply to procedural votes, which is significant in that the Security Council's permanent membership can vote against a "procedural" draft resolution, without necessarily blocking its adoption by the Council.

The veto is exercised when any permanent member—the so-called **"P5"**—casts a "negative" vote on a "substantive" draft resolution. *Abstention or absence from the vote by a permanent member does not prevent a draft resolution from being adopted.*
http://en.wikipedia.org/wiki/United_Nations

The idea of states having a veto over the actions of international organizations was not new in 1945. From the foundation of the League of Nations in 1920, each member of the League Council, whether permanent or non-permanent, had a veto on any non-procedural issue. From 1920 there were 4 permanent and 4 non-permanent members, but by 1936 the number of non-permanent members had increased to 11. Thus there were in effect 15 vetoes. This was one of several defects of the League that made action on many issues impossible.

The UN Charter provision for unanimity among the **Permanent Members of the Security Council** *(the veto)* was the result of extensive discussion, including at Dumbarton Oaks (August–October 1944) and Yalta (February 1945). The evidence is that the UK, US, USSR, and France all favoured the principle of unanimity and that they were motivated in this not only by a belief in the desirability of the major powers acting together *but also by a hard-headed concern to protect their own sovereign rights and national interest.* **Truman**, who became President of the US in April 1945, went so far as to write in his memoirs: *"All our experts, civil and military, favored it, and without such a veto no arrangement would have passed the Senate."*

The **UNSC** veto system was established in order to *prohibit the UN from taking any future action directly against its principal founding members*. One of the lessons of the League of Nations (1919–46) had been that an international organization cannot work if all the major powers are not members. The expulsion of the Soviet Union from the League of Nations in December 1939, following its November 1939 attack on Finland soon after the outbreak of World War II, was just one of many events in the League's long history of incomplete membership.

It had already been decided at the UN's founding conference in 1944, that Britain, China, the Soviet Union, the United States and, "in due course" France, should be the permanent members of any newly formed Council. France had been defeated and occupied by Germany (1940–44), but its role as a permanent member of the League of Nations, its status as a colonial power and the activities of the Free French forces on the allied side allowed it a place at the table with the other four. http://en.wikipedia.org/wiki/United_Nations_Security_Council_veto_power

From a Baha'i perspective, the problem with the concept of a League of Nations or a United Nations was that not all the principles originating from the Baha'i Faith were implemented and that the original founding members wield more power politically and militarily than any other nations and therefore, are by de facto felt entitled to the power of veto. Among other problematic issues has been dissension arising from other nations historically, Russia had not always gone along with the LON or UNSC decisions and would constantly veto a decision to get its own way but, then all original UNSC members have done this at one time or another. This Veto ploy is nothing more than the immaturity of certain UN members and an abuse of the UN political system.

This is not what Bahá'u'lláh had in mind when he formulated the concept back in the late 1800s and why "Abdu'l-Baha called the League of Nations incapable of establishing Universal Peace and why he suggested a **Supreme Tribunal** to settle the problems of nations, whose decision would be final and absolute!

It is apparent that no nation is willing to give up some of its power and sovereignty for the peace and stability of the world, particularly those powerful nations with the veto power who have their own political agendas that do not align with the commonwealth majority of the world.

Removing the veto power from these original founding nations will probably cause political hostility not only among themselves but with the more peaceful nations of the UN and may also lead to war as 'Abdu'l-Baha has stated that the decision from a Supreme Tribunal (should such a tribunal come into existence) would be final and absolute and failure to comply will result in that dissenting nation's government being overthrown but the rest of the world's nations which may, unfortunately, imply aggressive action or war.

The nations and governments of the world must get it into their collective heads that the people of the world will no longer tolerate, secret government agendas, "bully-boy foreign policy tactics", no matter how justified or sincere the motives, nor do people want to experience any further global conflicts and wars on the planet, just so that some nations may fulfill its nation's interests or hidden agendas.

Such a Supreme Tribunal will assuredly come into being and all nations will support it for the sake of the world's commonwealth, its peace and security, nothing can prevent it eventual existence!

The Baha'i Faith provides the essential modus operandi for this day and age to solve our current material and spiritual issues and problems; it provides divine guidance and a pattern for future society to evolve toward a golden age of civilization. Past religions are but mere shadows of their former glory and in the fullness of time, they must recognize and acknowledge that the Word of God is not exclusive to one particular time or place or with any one religion or people. Their acknowledgement of this basic truth is the beginning of the removal of all religious prejudices, the removal of barriers and boundaries that have for centuries nay, for millennia have kept mankind separated from himself, his kind and even from the other intelligent beings in the universe. It is the beginning of understanding that all religions are one, from one God and are not a multitude of divisive, contentious and competitive religions all vying with each other for power and sway over the masses. Religion is, therefore, progressive; not stagnate and rigid.

Religion is the source of all knowledge and truth and inspires men of understanding to search, investigate and discover new and deeper truths like for example, the quest to search for other life

436

in the universe and if possible, to communicate with other intelligences. Whether we consciously admit it or not, most people who follow a religious belief system communicate with a higher divine intelligence daily, whenever they pray; it is no different than trying to communicate with other beings on other planets or in other star systems.

We must prepare ourselves for the eventuality of Extraterrestrial contact possibly within our lifetime or our children's lifetime and that requires that we be unified as an intelligent species and it requires recognition of a divine plan to get us to that point in our planetary evolution. The Baha'i Faith has for the past 170 years been more than up to the task in guiding us to that age of universal peace on this planet and in preparing humanity for Extraterrestrial contact in the coming years.

CHAPTER 116

GLOBAL UNITY - A PREREQUISITE FOR ADMISSION INTO GALACTIC CIVILIZATION

The time must come when the imperative necessity for the holding of a vast, an all-embracing assemblage of men will be universally realized. The rulers and kings of the earth must needs attend it, and, participating in its deliberations, must consider such ways and means as will lay the foundations of the world's Great Peace amongst men. Such a peace demandeth that the Great Powers should resolve, for the sake of the tranquility of the peoples of the earth, to be fully reconciled among themselves. Should any king take up arms against another, all should unitedly arise and prevent him. If this be done, the nations of the world will no longer require any armaments, except for the purpose of preserving the security of their realms and of maintaining internal order within their territories. This will ensure the peace and composure of every people, government, and nation. http://reference.bahai.org/en/t/b/GWB/gwb-117.html

True civilization will unfurl its banner in the midmost heart of the world whenever a certain number of its distinguished and high-minded sovereigns—the shining exemplars of devotion and determination—shall, for the good and happiness of all mankind, arise, with firm resolve and clear vision, to establish the Cause of Universal Peace. They must make the Cause of Peace the object of general consultation and seek by every means in their power to establish a Union of the nations of the world. They must conclude a binding treaty and establish a covenant, the provisions of which shall be sound, inviolable and definite. They must proclaim it to all the world and obtain for it the sanction of all the human race. This supreme and noble undertaking—the real source of the peace and well-being of all the world—should be regarded as sacred by all that dwell on earth. All the forces of humanity must be mobilized to ensure the stability and permanence of this **Most Great Covenant**. In this all-embracing Pact, the limits and frontiers of each and every nation should be clearly fixed, the principles underlying the relations of governments towards one another definitely laid down, and all international agreements and obligations ascertained. In like manner, the size of the armaments of every government should be strictly limited, for if the preparations for war and the military forces of any nation should be allowed to increase, they will arouse the suspicion of others. The fundamental principle underlying this solemn Pact should be so fixed that if any government later violate any one of its provisions, all the governments on earth should arise to reduce it to utter submission, nay the human race as a whole should resolve, with every power at its disposal, to destroy that government. Should this greatest of all remedies be applied to the sick body of the world, it will assuredly recover from its ills and will remain eternally safe and secure. http://reference.bahai.org/en/t/ab/SDC/sdc-4.html

Although the **League of Nations** has been brought into existence, yet it is incapable of establishing universal peace. But the Supreme Tribunal which Bahá'u'lláh has described will fulfill this sacred task with the utmost might and power. And His plan is this: that the national assemblies of each country and nation—that is to say parliaments—should elect two or three persons who are the choicest men of that nation, and are well informed concerning international laws and the relations between governments and aware of the essential needs of the world of humanity in this day. The number of these representatives should be in proportion to the number of inhabitants of that country. The election of these souls who are chosen by the national assembly, that is, the parliament, must be confirmed by the upper house, the congress and the cabinet and also by the president or monarch so these persons may be the elected ones of all the nation and the government. From among these people, the members of the Supreme Tribunal will be elected, and all mankind will thus have a share therein, for every one of these delegates is fully representative of his nation. When the **Supreme Tribunal** gives a ruling on any international question, either unanimously or by majority rule, there will no longer be any pretext for the plaintiff or ground of objection for the defendant. In case any of the governments or nations, in the execution of the irrefutable decision of the Supreme Tribunal, be negligent or dilatory, the rest of the nations will rise up against it because all the governments and nations of the world are the supporters of this Supreme Tribunal. Consider what a firm foundation this is! But by a limited and restricted League, the purpose will not be realized as it ought and should. http://reference.bahai.org/en/t/ab/SAB/sab-228.html

Unification of the whole of mankind is the hallmark of the stage which human society is now approaching. Unity of family, of tribe, of city-state, and nation have been successively attempted and fully established. World unity is the goal towards which a harassed humanity is striving. Nation-building has come to an end. The anarchy inherent in state sovereignty is moving towards a climax. A world, growing to maturity, must abandon this fetish, recognize the oneness and wholeness of human relationships, and establish once for all the machinery that can best incarnate this fundamental principle of its life. http://reference.bahai.org/en/t/se/WOB/wob-56.html

Some form of a world super-state must needs be evolved, in whose favour all the nations of the world will have willingly ceded every claim to make war, certain rights to impose taxation and all rights to maintain armaments, except for purposes of maintaining internal order within their respective dominions. Such a state will have to include within its orbit an International Executive adequate to enforce supreme and unchallengeable authority on every recalcitrant member of the commonwealth; a World Parliament whose members shall be elected by the people in their respective countries and whose election shall be confirmed by their respective governments; and a Supreme Tribunal whose judgment will have a binding effect even in such cases where the parties concerned did not voluntarily agree to submit their case to its consideration. http://reference.bahai.org/en/t/uhj/PWP/pwp-4.html

The unity of the human race, as envisaged by **Bahá'u'lláh**, implies the establishment of a world commonwealth in which all nations, races, creeds and classes are closely and permanently united, and in which the autonomy of its state members and the personal freedom and initiative of the individuals that compose them are definitely and completely safeguarded. This commonwealth must, as far as we can visualize it, consist of a world legislature, whose members will, as the trustees of the whole of mankind, ultimately control the entire resources of all the component nations, and will enact such laws as shall be required to regulate the life, satisfy the needs and adjust the relationships of all races and peoples. A world executive, backed by an international Force, will carry out the decisions arrived at, and apply the laws enacted by this world legislature, and will safeguard the organic unity of the whole commonwealth. A world tribunal will adjudicate and deliver its compulsory and final verdict in all and any disputes that may arise between the various elements constituting this universal system. A mechanism of world inter-communication will be devised, embracing the whole planet, freed from national hindrances and restrictions, and functioning with marvellous swiftness and perfect regularity. A world metropolis will act as the nerve center of a world civilization, the focus towards which the unifying forces of life will converge and from which its energizing influences will radiate. A world language will either be invented or chosen from among the existing languages and will be taught in the schools of all the federated nations as an auxiliary to their mother tongue. A world script, a world literature, a uniform and universal system of currency, of weights and measures, will simplify and facilitate intercourse and understanding among the nations and races of mankind. In such a world society, science and religion, the two most potent forces in human life, will be reconciled, will cooperate, and will harmoniously develop. The press will, under such a system, while giving full scope to the expression of the diversified views and convictions of mankind, cease to be mischievously manipulated by vested interests, whether private or public, and will be liberated from the influence of contending governments and peoples. The economic resources of the world will be organized, its sources of raw materials will be tapped and fully utilized, its markets will be coordinated and developed, and the distribution of its products will be equitably regulated.

National rivalries, hatreds, and intrigues will cease, and racial animosity and prejudice will be replaced by racial amity, understanding, and cooperation. The causes of religious strife will be permanently removed, economic barriers and restrictions will be completely abolished, and the inordinate distinction between classes will be obliterated. Destitution on the one hand and gross accumulation of ownership on the other will disappear. The enormous energy dissipated and wasted on war, whether economic or political, will be consecrated to such ends as will extend the range of human inventions and technical development, to the increase of the productivity of mankind, to the extermination of disease, to the extension of scientific research, to the raising of the standard of physical health, to the sharpening and refinement of the human brain, to the exploitation of the unused and unsuspected resources of the planet, to the prolongation of human life, and to the furtherance of any other agency that can stimulate the intellectual, the moral, and spiritual life of the entire human race.

A world federal system, ruling the whole earth and exercising unchallengeable authority over its unimaginably vast resources, blending and embodying the ideals of both the East and the West, liberated from the curse of war and its miseries, and bent on the exploitation of all the available sources of energy on the surface of the planet, a system in which Force is made the servant of Justice, whose life is sustained by its universal recognition of one God and by its allegiance to one common Revelation—such is the goal towards which humanity, impelled by the unifying forces of life, is moving.

"In cycles gone by, though harmony was established, yet, owing to the absence of means, the unity of all mankind could not have been achieved. Continents remained widely divided, nay even among the peoples of one and the same continent association and interchange of thought were well nigh impossible. Consequently, intercourse, understanding, and unity amongst all the peoples and kindreds of the earth were unattainable. In this day, however, means of communication have multiplied, and the five continents of the earth have virtually merged into one... In like manner, all the members of the human family, whether peoples or governments, cities or villages, have become increasingly interdependent. For none is self-sufficiency any longer possible, inasmuch as political ties unite all peoples and nations, and the bonds of trade and industry, of agriculture and education, are being strengthened every day. Hence the unity of all mankind can in this day be achieved. Verily this is none other but one of the wonders of this wondrous age, this glorious century. Of this, past ages have been deprived, for this century—the century of light—has been endowed with unique and unprecedented glory, power and illumination. Hence, the miraculous unfolding of a fresh marvel every day. Eventually, it will be seen how bright its candles will burn in the assemblage of man.

"Behold how its light is now dawning upon the world's darkened horizon. The first candle is unity in the political realm, the early glimmerings of which can now be discerned. The second candle is unity of thought in world undertakings, the consummation of which will ere long be witnessed. The third candle is unity in freedom which will surely come to pass. The fourth candle is unity in religion which is the cornerstone of the foundation itself, and which, by the power of God, will be revealed in all its splendor. The fifth candle is the unity of nations—a unity which in this century will be securely established, causing all the peoples of the world to regard themselves as citizens of one common fatherland. The sixth candle is unity of races, making of all that dwell on earth peoples and kindreds of one race. The seventh candle is unity of language, i.e., the choice of a universal tongue in which all peoples will be instructed and converse. Each and every one of these will inevitably come to pass, inasmuch as the power of the Kingdom of God will aid and assist in their realization."
http://reference.bahai.org/en/t/se/WOB/wob-19.html

Let there be no mistake. The principle of the **Oneness of Mankind**—the pivot round which all the teachings of **Bahá'u'lláh** revolve—is no mere outburst of ignorant emotionalism or an expression of vague and pious hope. Its appeal is not to be merely identified with a reawakening of the spirit of brotherhood and good-will among men, nor does it aim solely at the fostering of harmonious cooperation among individual peoples and nations. Its implications are deeper, its claims greater than any which the Prophets of old were allowed to advance. Its message is applicable not only to the individual but concerns itself primarily with the nature of those essential relationships that must bind all the states and nations as members of one human family. It does not constitute merely the enunciation of an ideal but stands inseparably associated with an institution adequate to embody its truth, demonstrate its validity, and perpetuate its influence. It implies an organic change in the structure of present-day society, a change such as the world has not yet experienced. It constitutes a challenge, at once bold and universal, to outworn shibboleths of national creeds—creeds that have had their day and which must, in the ordinary course of events as shaped and controlled by Providence, give way to a new gospel, fundamentally different from, and infinitely superior to, what the world has already conceived. It calls for no less than the reconstruction and the demilitarization of the whole civilized world—a world organically unified in all the essential aspects of its life, its political machinery, its spiritual aspiration, its trade and finance, its script and language, and yet infinite in the diversity of the national characteristics of its federated units. http://reference.bahai.org/en/t/se/WOB/wob-22.html

441

Toward Galactic Unity - One Universe, One People

Dr. Steven Greer takes these Baha'i concepts of global unity one logical step further to the imminent day when ETI contacts humanity and suggest ways to prepare ourselves for that day by understanding the dynamics of unity and the oneness of consciousness. In his 1991 paper, "One Universe, One People", he describes these concepts:

One of the greatest tasks humanity has faced throughout history is the establishment of peace and unity among differing and diverse peoples. Superficial, external and cultural distinctions such as gender, race, ethnic origin, nationality, religion and so forth have long divided humanity and been the cause of much warfare and social turmoil. It is only in the last 100 or so years that humans have seriously begun to explore worldwide our points of unity and begun to overcome the barriers which have separated humanity. Central to this evolutionary process has been the dynamic of at once accepting and celebrating diversity while simultaneously seeing the fundamental oneness which all humans share. This dynamic of unity - seeing with the eye of oneness - is the essential foundation for lasting world peace and prosperity and will be the motivating principle of the next millennium. The long and painful process of overcoming prejudice and embracing humanity's essential oneness, while by no means yet complete, has brought us to the dawn of a true world-encircling community of one people. The recognition that mankind is one, that race, nationality, gender, religion and so on are secondary to our shared humanness, may well be the crowning achievement of the 20th century.

But what does it mean to be human, essentially human, apart from a purely biological definition? Our deepest point of unity transcends race, culture, gender, profession, life roles, even level of intelligence or emotional make- up, since all these attributes vary widely among people. Rather, the foundation of human oneness is consciousness itself, the ability to be conscious, self-aware, intelligent sentient beings. All other human qualities arise from this mother of all attributes. Conscious intelligence is the root essence from which all other human qualities emanate. It is the universal and fundamentally pure canvas on which the dazzling array of human life manifests. The firmest, most enduring and transcendent foundation on which human unity is based then, is consciousness itself, for we are all sentient beings, conscious, self-aware, and intelligent. No matter how diverse two people or two cultures may be, this foundation of consciousness will enable unity to prevail, as it is the simplest yet most profound common ground which all humans share.

As great as the challenges to unity have been and continue to be for humans, how much greater might this be for the emerging and embryonic relationship between humans and extraterrestrial civilizations. The superficial and cultural differences between, say, an American and a Kenyan tribesman may pale before it! If disunity and conflict arise when we look only at the differences between humans, how much greater will the potential disunity and conflict be if we are able only to focus on the points of difference between humans and extraterrestrial beings? The failed and disastrous ways of the past - of seeing only differences and foreign qualities - must give way to a new way of seeing, of seeing with the eye of oneness. This eye of oneness must be directed not only towards our fellow humans but towards extraterrestrial people as well, for the same fundamental basis for unity which exists among humans also exists for the relationship between humans and extraterrestrials.

The term **Extraterrestrial Intelligence (ETI),** so curiously nondescript, wonderfully lends itself to these concepts of unity. Regardless of planet, star system or galaxy of origin, and no matter how diverse, ETIs are, essentially, intelligent, conscious, sentient beings. Humans are essentially intelligent, conscious, sentient beings. We are, essentially, one. On this basis, we may speak of one people inhabiting one universe, just as we now envision one people as children of one planet. Differences are always a matter of degree, but true unity established in consciousness is absolute. The beings currently visiting earth from other planets, while no doubt different from humans in both superficial and more profound ways are nevertheless conscious intelligent beings. Consciousness is the basis for both human and extraterrestrial existence and is, therefore, the foundation for unity and communication between the various people of the universe. Beliefs may vary, biological processes may vary, assorted capacities may vary, social systems and technology may vary - but the simple thread of conscious intelligence which runs through all peoples elegantly weaves our unity. This essential unity is not subject to the trials of diversity, for it is pure, immutable and fundamental to the existence of intelligent life itself.

The challenge of establishing unity among the peoples of the universe is a grand extension of the challenge of establishing unity and peace among the people of the earth. Diversity, distinction, and differences must be met with mutual respect, acceptance, and even celebration, while the deeper foundations of unity are held steadily in view. The eye of oneness does not exclude or reject the diversity among peoples but relates this diversity to a paradigm of universality based in consciousness. The development of this capacity, of this kind of awareness, is the most important prerequisite for not only peace and unity among humans, but also for the peace and unity between humans and other intelligent life in the universe. We must hope and pray that the errors and shortcomings humanity has manifested in its long and, as yet, incomplete march to world unity will serve as well-remembered lessons as we face the task of peacefully interacting with extraterrestrial peoples. The endless diversity which so outstanding a universe can present will only be endured by minds established in the calmness of universal consciousness. In the coming decades, centuries and millennia, it will be increasingly realized that the success of humanity's existence will be dependent on the development of consciousness more than on any outward progress.

As there is one God which manifests one creation, so there is one God which is the source of all conscious beings, whether on earth or elsewhere. The great Universal Intelligence has sent a ray of this light of consciousness throughout all conscious beings, and we are united to God and to one another through its subtle and all-pervading effect. It is for these reasons that I state that the reality of man and the reality of other extraterrestrial peoples are one. Viewed with the eye of differences, we are diverse and unrelated, but viewed with the eye of oneness, we are more alike than dissimilar, more kindred than alien. And so it is that we must look to our inner reality to find not only our oneness with our fellow humans but our oneness with other intelligent life in the universe as well. While ephemeral differences may confound us, our essential oneness in consciousness will never fail us. For there is one universe inhabited by one people, and we are they. http://new.cseti.org/position-papers/14-position-papers/42-art-one-universe-one-people.html

The Universe is Filled with An Abundance of Intelligent Life

Though many religions allude to accounts or allegorical stories about ET visitations which are often thought of as angels or possible demons, there is, however, no major religion that speaks more directly to the existence of other sentient life in the universe than does the Baha'i Faith. (See also the section in this book under **Religious References to Alien Visitations Gleaned from the World's Major Religions**).

Baha'u'llah has stated in Gleanings pg. 163 that ***"Know that every fixed star hath its own planets and every planet its own creatures, whose number no man can compute."*** Gleanings from the Writings of Baha'u'llah; translated by Shoghi Effendi NSA of the Baha'is of the US); 1939 and 1952 published by Baha'i Publishing Trust; Wilmette, Illinois, USA; ISBN 52-24896

Baha'u'llah's writings and prayers are filled with references to other beings and creatures on other planets as well as those astral, celestial and divine entities that are beyond this physical realm of existence and with those that share it with us: *"beings that are invisible and visible," "those that are hidden and those that are manifest", "the kingdoms of the unseen and of the seen", "the Lord of all worlds"* and the loved ones who are in the *"Abha Kingdom!"*

The next set of writings from Baha'u'llah come from "Gleanings" which states that not only are there physical beings on other planets but, there are invisible beings which occupy infinite worlds in the next higher realm of existence known in religious context as the spiritual kingdom. Note that he repeats the fact that there exist invisible beings beyond our known comprehension and that God is the most invisible of the invisibles *or... as this author likes to refer to as the* **Supreme Extraterrestrial Intelligence (SETI)** to whom all other intelligences owe their very their existence to:

"...Thy matchless Beauty have at all times been imprinted upon the realities of all beings, visible and invisible."

Baha'u'llah testifies that god's creation is not limited to mankind or this world but, there are many other worlds and creatures in an infinite universe:

"Know thou of a truth that the worlds of God are countless in number, and infinite in their range."

And again, Baha'u'llah reaffirms:

"Verily I say, the creation of god embraceth worlds beside this world, and creatures apart from these creatures. In each of these worlds, He hath ordained things which none can search except Himself, the All-Searching, the All-Wise. Do thou meditate on that which We have revealed to thee, that thou mayest discover the purpose of God, thy Lord, and the Lord of all the worlds."

Baha'u'llah astutely points out a basic truth for us to ponder:

444

"Wert thou to ponder in thine own heart the behaviour of the Prophets of God thou wouldst assuredly and readily testify that there needs be other worlds besides this world."

Baha'u'llah makes a profound declaration about the station of the Earth for this day and age:

'This is the Day whereon the unseen world crieth out: "Great is thy blessedness, O earth, for thou hast been made the foot-stool of thy God, and been chosen as the seat of His mighty throne."'... and "The realm of glory exclaimeth: "Would that my life could be sacrificed for thee, for He Who is the Beloved of the All-Merciful hath established His sovereignty upon thee, through the power of His Name that hath been promised unto all things, whether of the past or of the future."' Gleanings from the Writings of Baha'u'llah translated by Shoghi Effendi NSA of the Baha'is of the US); 1939 and 1952 published by Baha'i Publishing Trust; Wilmette, Illinois, USA; ISBN 52-24896

In the Kitab-i- Iqan in reference to his own revelation states:

"So potent and universal is this revelation (the revelation of Baha'u'llah), *that it hath encompassed all things visible and invisible."*

And again, Baha'u'llah states that the Divine Messengers of God are the perfect embodiments of God's Word to the world of humanity.

"These Tabernacles of holiness, these primal Mirrors which reflect the light of unfading glory, are but expressions of Him Who is the Invisible of the Invisibles." Kitab-i-Iqan The Book of Certitude; Baha'u'llah translated by Shoghi Effendi ; 1931, rev. edn. 1954; Baha'i Publishing Trust; Wilmette, Illinois, USA; ISBN 0-87743-022-5

Some Inalienable Truths About Aliens

The Baha'i Faith, the most recent of all the religions from God which have preceded it, makes it firmly understood that there is other life in the universe and that it has an awareness of the divine creator:
"...Thy matchless Beauty have at all times been imprinted upon the realities of all beings, visible and invisible."

... which means that they too, as physical beings in this physical universe, as well as beings that are invisible or hidden from the human eye, are under some of the same basic truths and principles that govern humans in this world and in this universe of existence.

"O SON OF MAN! Wert thou to speed through the immensity of space and traverse the expanse of heaven, yet thou wouldst find no rest save in submission to Our command and humbleness before Our Face." (Baha'u'llah, Arabic Hidden Words)

"Gaze upward through immeasurable space to the majestic order of the colossal suns. These luminous bodies are numberless. Behind our solar system, there are unfathomable stellar systems and above those stellar systems are the remote aggregations of the Milky Way. Extend

your vision beyond the fixed stars and again you shall behold many spheres of light. In brief, the creation of the Almighty is beyond the grasp of the human intellect. When this objective creation is unlimited and not subject to suspension, is not the subjective creation of His Majesty the Almighty limitless? When the reflection or physical creation is infinite, how is it possible to circumscribe the reality which is the basis of divine creation? The spiritual world is so much greater than the physical that in comparison with it the physical world is non-existent." (Abdu'l-Baha, Divine Philosophy)

"Consider the creation of the infinite universe. This globe of ours is one of the smallest planets. Those stupendous bodies revolving in yonder immeasurable space, the infinite blue canopy of God, are many times greater than our small earth. To our eyes this globe appears spacious; yet when we look upon it with divine eyes, it is reduced to the tiniest atom. This small planet is not worthy of division. Is it not one home, one native land? Is not all humanity one race? Creationally there is no difference whatsoever between the peoples." (Abdu'l-Baha, Divine Philosophy 178) http://www.inspiremore.com/nasa-just-released-the-largest-photo-ever-taken-what-it-shows-will-shake-you-up/

It is, therefore, possible to deduce some basic inalienable truths that govern other intelligent beings in the universe, based upon the Baha'i religious perspective, common sense, and logic, although, admittedly scientific discovery of such theories and hypothesis will need to be confirmed through space exploration and contact with such diverse creatures.

Based upon the millions of eye witness accounts and the hundreds of thousands of released government files from many nations around the world, we have discovered that the Extraterrestrial Intelligences that have come to Earth appear to have some common universal characteristics and standards. There are universal constants in play within the universe, while at the same time still allowing for a rich diversity of uniqueness that reflects the infinite creativity of the omnipotent mind of the **Divine Creator**.

It is to be expected that every planet in a stable orbit around a stable star will have unique environmental and atmospheric conditions which will determine and shape the type of life forms that each planet will develop. Uniqueness should be the rule, however, if we accept all the reports of ET encounters with humans, we find to date that the rule of thumb is that uniqueness in xeno-morphology is bounded by universal conformity. The star shape geometric form or the humanoid form as in **Leonardo da Vinci's Vitruvian man** appears to be the best model for intelligence however, intelligent life can and does assume any shape or form.

Avatars, Divine Educators and Manifestations of God

As we have divine messengers or avatars or ascended masters or prophets or manifestations of God that appear frequently throughout our history to move us toward an ever-advancing civilization on this planet, so it is, according to Baha'i writings, that all intelligent beings in the universe have an awareness of their creator. Therefore, it is no stretch of the imagination that Extraterrestrial Intelligences likewise, also have manifestations of God who come regularly to them every millennium to give them divine guidance.

446

This pre-supposes that if we are divinely educated then, ETI are also divinely educated and depending on the acceptance of the divine guidance, some Extraterrestrial Intelligences advance very quickly spiritually and their civilizations always appear in the state of a **"golden age"** of great enlightenment and understanding. For some interstellar civilizations, they recognize immediately the *"Voice of God"* through His **Divine Educators**; there is no opposition to the divine message. There are even, civilizations that have never known or experienced internal conflicts or wars, the concept is completely *"alien"* to them (pun intended). Dare it be said but, there are interstellar civilizations so divine and celestial in nature that they receive the **"Word of God"** directly from the **Creator** without the need of an *"Intermediary"* because their higher consciousness is so developed and highly attuned to the **Will of God** that they exist only to receive his message and to pass it on to other celestial and divine beings who relay it to lower levels of interstellar civilizations. These beings are distinct from the divine Messengers that appear to humanity or to other planetary civilizations. Without sounding like a *"new age guru"*, *it is the reason why we are being visited* by Extraterrestrial civilizations, to assist us to develop beyond the material state of existence, to help us attain greater consciousness that is traditionally known by the mystics and shaman as **Cosmic Consciousness, God Consciousness or Unity Consciousness.**

For this reason, emerging planetary civilizations like ours with a similar technological and spiritual development are closely monitored, perhaps, even guided in their planet's cultural and societal development and are even, temporarily quarantined until we pass through the adolescence phase to full mature adulthood, eventually emerging as a full **Type One Civilization**. At that point, the planetary quarantine is lifted and we are offered the opportunity to join a greater bond of unity as part of the larger **Community of Advanced Galactic Intelligences (CAGI).**

These then, are just a few of the many similarities and common bonds we will discover in the coming years, decades and centuries that lie ahead that we share with other Extraterrestrial civilizations.

This would also explain why such so-called inter-species exchange programs like **SERPO** are in reality disinformation programs designed to confuse the public, and the private UFO investigators or merely the wishful thinking on the part of deluded Ufologists who haven't really thought the whole concept through, having not considered the quarantine placed upon this planet.

There is a pecking order in the universe based upon technical and spiritual development of a planet's civilization and currently at this particular time on Earth, humanity is not near the top or even in the middle of advanced interstellar or intergalactic civilizations, we are as yet, still near the bottom of the ladder in planetary social development.

To allow an immature civilization like humanity to go gallivanting around the galaxy with under-developed spiritual attitudes and behaviours, armed to the teeth with nuclear weapons of mass destruction and with stolen or hijacked alien technology is simply too dangerous to permit. The universe can ill afford interplanetary or interstellar wars to be brought to the doorsteps of non-aggressive, non-hostile planets.

As we enter into the tumultuous time when humanity finally matures and we have full open contact with other interstellar civilizations, we will discover many bonds of commonality, friendship and other similarities in which we share. We will wonder why it took us so long to get our house in order, until now but, maybe, it's all about the right timing in the universe!

The light at the end of the tunnel is becoming larger and brighter, as we rapidly begin to emerge from out of the darkness of heedlessness and corruption and into the full daylight of true reasoning and spiritual understanding. *__The future is so bright that we will all need sunglasses__*!

A Rich ET Bio-Diversity Including Humanoid Similarities

For example, from the numerous UFO and ET reports and accounts, we know that many ET beings that have been encountered appear in a humanoid form suggesting as many scientists have been stating over the years that the basic human form is a marvelous precisely created bio-mechanism to carry out a myriad of movements and functions.

We have described earlier in this book the similarities and differences between humans and humanoid ET beings as well as the diversity of other intelligent beings with xeno-morphologies that are truly alien or bizarre to the human perspective, therefore, no further discussion need be repeated here at this point, save one. The commonality in the human or star shape morphology does appear to be a universal standard, a constant or template that seems to show up time and again through numerous ET encounters particularly CE-3s through to CE-5s. The diversity of ET species from the many UFO reports indicate that unusual body types may also be a common universal standard, however, the fact that we do not see many of these unusual ETIs with morphologies resembling body-like organs on Earth may be, due to certain environmental factors of adaptability which make the Earth inhospitable to these ETs beings.

Humans consider themselves a noble being of creation. We have two eyes in the front of the face allow for stereoscopic vision necessary for a predatory mammal and for a highly evolve intelligence to be developed. Cranial bone plates allow for the expansion of the brain to grow and develop over a sustained evolutionary period of time to reflect the development of intelligence. The nose is situated in the centre of the face that permits olfactory detection of all types of smells before us. Our ears are situated on the central side of the head giving us the ability to hear in stereo all sounds around us a full spherical 360 degrees. The mouth is appropriately located below the nose to allow us to smell the foods we eat, to exhale, to permit the emanation of speech, to display emotions of happiness, contentment, frustration, sadness and anger, etc. The head is elevated to permit us the ability to see things over a long distance or to turn our heads 90 degrees to the right or left or to look down and up to observe what is beneath or above us.

We have to arms and two hands with multiple fingers and an opposable thumb to manipulate our environment and to operate tools which are also necessary for a highly evolve sentient intelligent creature. A body containing the organs necessary for the complete function of the being encase in a protective rib cage and of course two legs and feet to allow the intelligent being to walk, run,

448

climb or swim through diverse terrain surfaces and liquids (water). There are also the sex organs in which the intelligent being may reproduce itself to continue the generations of its species.

Such is the fascinating creation of human beings; however, because we are not alone in the universe, the **Divine Creator** has seen fit to duplicate this marvelous biology with unique differences among interstellar beings that are also humanoid in appearance. There are interstellar beings that are so close to being human in appearance that they could be our *cosmic cousins* thus, no further explanation is necessary to describe them. There is, however, the example of the often reported small gray-like beings that are frequently encountered.

Their morphology is similar to humans: A head with a large cranium denoting high intelligence which makes up about one-quarter to one-third of the body proportions, two large black eyes suitable for both day and particularly nighttime vision. Two small ears or orifices for stereo sound, small nose or nasal slits for olfactory sense and a small mouth for eating, speaking, etc. The organs of the head may indicate that these ears, nose, and mouth are either not as highly evolved as in humans or are becoming vestigial due to an increase of brain development and possible telepathic ability.

They have thin long arms and hands long fingers (three fingers with an opposable thumb which may have long claws or nails, may have webbing between fingers and may have minute suction cups on tips of fingers instead of fingerprints). A small narrow rib-caged body supported by long thin legs and feet with three or four toes on each foot that may also be clawed.

Other alien xeno-types are also described as humanoid in appearance, be they reptilian or insect-like or mammalian in appearance but each having their own uniqueness of morphology distinction. Of course there are intelligent beings that defy logic or reason whose bodies resemble extremely large body organs shaped like an eyeball or a brain or a sponge-like mass, there are even beings that look like the family pet cat or dog but humanoid in shape, some appears as avian or bird-like, marine like dolphins, manatees or squids or octopi but again, humanoid in shape, there are metal or stone-like beings or robotic or cybernetic creatures. These too may also conform to a set universal standard for their xenomorphic types.

These biological similarities are bonds of acceptability between two different intelligent species who meet for the first time in contact as it gives each other common grounds of familiarity and yet, allow for unique distinctions to also be apparent.

In order to develop relationships with Extraterrestrial Intelligences visiting the Earth, it would be natural and very probable that relationship will begin with beings that appear most human-like or with morphologies that are most familiar to us. This pre-assumes that ETI having been monitoring us and know something about us as a species. It is of course, just as likely that the first formal contact with another Extraterrestrial civilization may be with a species that is completely foreign and bizarre in appearance. Intelligence knows no bounds nor is limited to a particular morphology and as humans, we must not project anthropocentric attitudes familiar to us upon any visiting ETI.

The point being is that there seems to be a universal standard in creation where intelligence is created in particular physical forms to maximize its intellectual functionality which separates it from the rest of creation. This then is a common trait among sentient intelligence that we have thus far determine for this corner of the universe.

Physical uniqueness is, however, superficial and loses its charm or unusualness over a period of time as two interacting intelligent species become more familiar with each other. Apart from the physical attributes that we may share with other interstellar civilizations, we share much more. The focus of any relationship is mutually supported in an atmosphere of amity and goodwill by establishing the commonalities that we share and which will inevitably bind us in close bonds of affinity and camaraderie. Friendship is after all the motivation behind any relationship that is mutual, sustained and peaceful. Everything else derived from such a relationship is secondary and thus a bonus to each civilization.

The most obvious attribute we share is **sentience, consciousness or being aware of awareness**; it is the hallmark of all intelligent species in the universe.

Now, many scientists have debated endlessly upon consciousness, as a set of reductionist values that can be measured, quantified and be reproducibly wrapped in nice little packets of familiar conformity.

Even, medical professionals, doctors, and psychologists have weighed in on the matter viewing the consciousness as a set of existential experiences derived from the electromagnetic currents within the neuron synapses of the brain that produce "neurological correlates of consciousness" as postulated by Nobel Prize-winning neuroscientist, **Gerald Edelman.**
http://www.bibliotecapleyades.net/ciencia/ciencia_consciousuniverse43.htm

He determined that there is no one place in the brain where consciousness takes place. No command center.

He agrees with **William James** that, *"thoughts don't necessarily need a thinker."*

His research points to the possibility that our working brain was not designed, but evolved, as he postulates a "neural Darwinism" **(O'Reilly).**

"The basic idea is that consciousness involves brain activities compiled to self-organizing ripples in fundamental reality. Brain stimulates reality based on sensory input and is also intimately connected to that reality at the quantum level". **(Huff)**

*"In **Panpsychism theory**, mind is fundamental in the universe. All matter has associated mental aspects or properties... Everything in the universe is seen as conscious".* **(Blackmore11)**

David Chalmers claimed that consciousness was a fundamental constituent of reality. It may be a building block of the universe, as photons are to light. Consciousness may be an inherent requirement of all that surrounds and composes us **(Huff).**

The physicists, neurosurgeons, philosophers and mathematicians substitute the less threatening term proto-consciousness to indicate that consciousness may be a fundamental constituent of reality, a building block.

Is this a spiritual force? **Danah Zohar** merges religion and science with proto-consciousness.

Zohar quotes David Chalmers: *"suggests that something called proto-consciousness is a fundamental property of all matter, just like mass, charge, spin and location. In this view, proto-consciousness is a natural part of the fundamental physical laws of the universe and has been present since the beginning of time".*

"Everything that exists ... fundamental particles like mesons and quarks, atoms, stones, tree trunks...possess proto-consciousness". **(Zohar 81)"**
http://www.bibliotecapleyades.net/ciencia/ciencia_consciousuniverse43.htm

Erwin Schrodinger, Nobel Prize Physicist in 1933 probably said it best of all that, *"the total minds in the universe is one."*
http://www.gestaltreality.com/articles/favorite-quotations/

There are not pieces of minds. The fact that we perceive individuality existing in many bodies is a delusional state. Such a delusional perception is the cause of separation among the societies of mankind. The reality is that there is but one mind in many places at the same time and we are all part of the **One Mind.**

Baha'u'llah writes *"...that all men shall be regarded as one soul..."* Gleanings from the Writings of Baha'u'llah; translated by Shoghi Effendi NSA of the Baha'is of the US); 1939 and 1952 published by Baha'i Publishing Trust; Wilmette, Illinois, USA; ISBN 52-24896

Abdu'l-Baha, son of Baha'u'llah says that in order to live the life: we must *"... be as one soul in many bodies, for the more we love each other, the nearer we shall be to God;"*

It is fascinating that this is the basis of all religious belief, that God is the One Mind in the universe that is conscious at all times and at all places, thus the attributes of divinity extolled by all religions that God is *omnipotent* (all-powerful), *omnipresent* (present everywhere), *omniscient* (infinite knowledge), and *omnibenevolent* (perfect goodness).

Consciousness is at the heart of mankind's existence and is the one thing we share with Extraterrestrial beings who are also sentient and conscious; they too are creations of a supreme creator. Consciousness is, therefore, a vital attribute that we share with all sentient intelligences in the universe. It also implies that there is in reality, only one consciousness in the universe and not pieces of consciousness according to Schrodinger and the writings of the Baha'i Faith.

The most obvious attribute about **ETI** is their **advanced intelligence** in the realm of technology and their ability to control their emotional response to the hostility that humans seem to project towards them. The fact that ETI have travelled vast distances from other star systems and even from other galaxies to reach the Earth in advanced spacecraft is a testament to their understanding of quantum and hyperdimensional physics which is perhaps, many hundreds or thousands and possibly millions of years beyond human understanding and capability.

The ability to control one's emotional response to aggressive and hostile situations encountered on this planet implies an advance understanding of spiritual concepts that govern their actions in everyday life and a greater understanding of the divine purpose of the Creator.

Extraterrestrial beings appear to be able to work and get along with the rich bio-diversity of other interstellar civilizations that have come to this planet and in exploring the universe. It is NOT just the "Greys" that are visiting this planet for they are in full co-operation with other interstellar civilizations and not here on a separate agenda.

There is a basic fundamental principle of acceptance of differences and an acknowledgement of similarities in order to work and cooperate with each other. This is a quality that still seems to elude much of mankind at this time, even though the Baha'i writings state that "*unity through diversity*" is not only possible but, as an eventual destiny that mankind will inevitably attain to in the near future.

These Extraterrestrial Intelligences have evolved to become **Type I, Type II, Type III Civilizations** and even higher levels which means that with technical advancement there is a commensurate level of spiritual development. Or is that vice versa: with a greater level of spiritual development there is a commensurate level of technical advancement? According to many eyewitness accounts and encounters of **CE- 4s** and particularly **CE-5s** (human initiated contact), many of these ETI operate both physically and interdimensionally but, prefer to interact with humans at this time interdimensionally for reasons of personal safety. Besides the potential aggression of humans who seem to act as if they were barely out of the trees toward anything that seems to be threatening to them, it is a safe method of interaction where contagion from unknown bacteria, viruses, and diseases are not freely transmitted between species. This is not to say that open physical contact between humans and ETI hasn't happened as there are accounts of this also occurring and most likely it will also continue to take place, during those times when it is appropriate to do so.

452

CHAPTER 117

CAN YOU PUT THAT IN WRITING? XENOLINGUISTICS - EXTRATERRESTRIAL SCRIPTS AND SYMBOLS

Humanity is on the verge of adopting a universal language, script, weights and measures some of which are already developed and in full implementation while others like language and script are still in need of development and universal acceptance by all nations. We see primarily certain languages being more commonly used than other languages, like English used extensively in international business and trade. Chinese is spoken by more people than any other language but, is not universally recognized as an international language. Arabic is also a major language particularly in Middle Eastern or Muslim countries but again, it is not a universally adopted secondary language like English. A universal language will need to be selected and adopted by all nations from one of the existing languages or a new language will need to be invented.

As the world shrinks via global business, commerce and trades, along with technological developments, the increased cross-cultural exchanges in education; the adoption and acceptance of multi- culturalism with regard to social, political and religious belief systems and the burgeoning global travel, transportation, and tourism has made the necessity of a universal language in which to converse with anyone, in any country and be understood by one's fellow man, mandatory. In the light of this understanding in human communications could there also be *"a true Universal Language and Script"* that most advanced Extraterrestrial civilizations have adopted to communicate between ETI species and is this what we are seeing on their spacecraft?

Intelligent beings that are sentient are able to communicate verbally or telepathically and most likely have developed a written script to conceptualize higher thoughts and ideas like humans and no doubt have also a sense of community that has developed technology. With the development of social values and concepts, ETs are very probably spiritual beings with an understanding of a supreme power or greater intelligence or basic oneness in the universe which humans call God that creates, guides and directs the flow of creation.

In view of ever-evolving concepts of unity upon this planet and there would be a natural expectation that unity is a universal concept within an interstellar community or in an intergalactic community among the myriads of sentient intelligent civilizations. With this infinite growth in the concept of unity, there is in all likelihood vast interstellar and intergalactic communities with a common language or method of communication and a universal script!

A similarity that we share with Extraterrestrial Intelligences is the concept of written languages which is a clear outward sign of intelligence. The UFO literature is filled with eyewitness accounts of strange markings seen on ET craft such as those found in such well documented cases, like the **Utsuro Bune UFO** or **USO (Unidentified Submerged Object)** of the 1800s in Japan, the **Roswell Saucer Crash of 1947**, **George Adamski's** ET contacts of 1950s, the 1964 **Zamora UFO Incident**, the **Kecksburg Crash of 1965**, the **Barney and Betty Hill Incident of 1969**, all in the USA, the **Bentwaters/ Rendlesham Incident of 1981** in the U.K. and the **Shaitan Mazar UFO Crash Retrieval of 1991** in Russia.

In all these cases, eyewitnesses to these crash site areas reported seeing unusual writings, pictoglyphs or hieroglyphics of unknown characters that were found on pieces of metal or seen either, inside the craft or on the hulls of the spacecraft. Some of these eyewitnesses were even able to touch the surface and run their hand over the letters on the ship's hull as in the Bentwaters/ Rendlesham Incident.

These may be symbols denoting a military, mathematical or a scientific emblem or possibly a place of origin or organization. So far no one seems to have been able to interpret their meaning.

In the case of George Adamski's contact with blonde **Nordic type ETs** in the 1950s; the ETs left footprints in the ground with strange writings that were from the bottom of their boots. The reader is encouraged to compare alien script and symbols with each other.
https://www.youtube.com/watch?v=QnSGE_plJVM

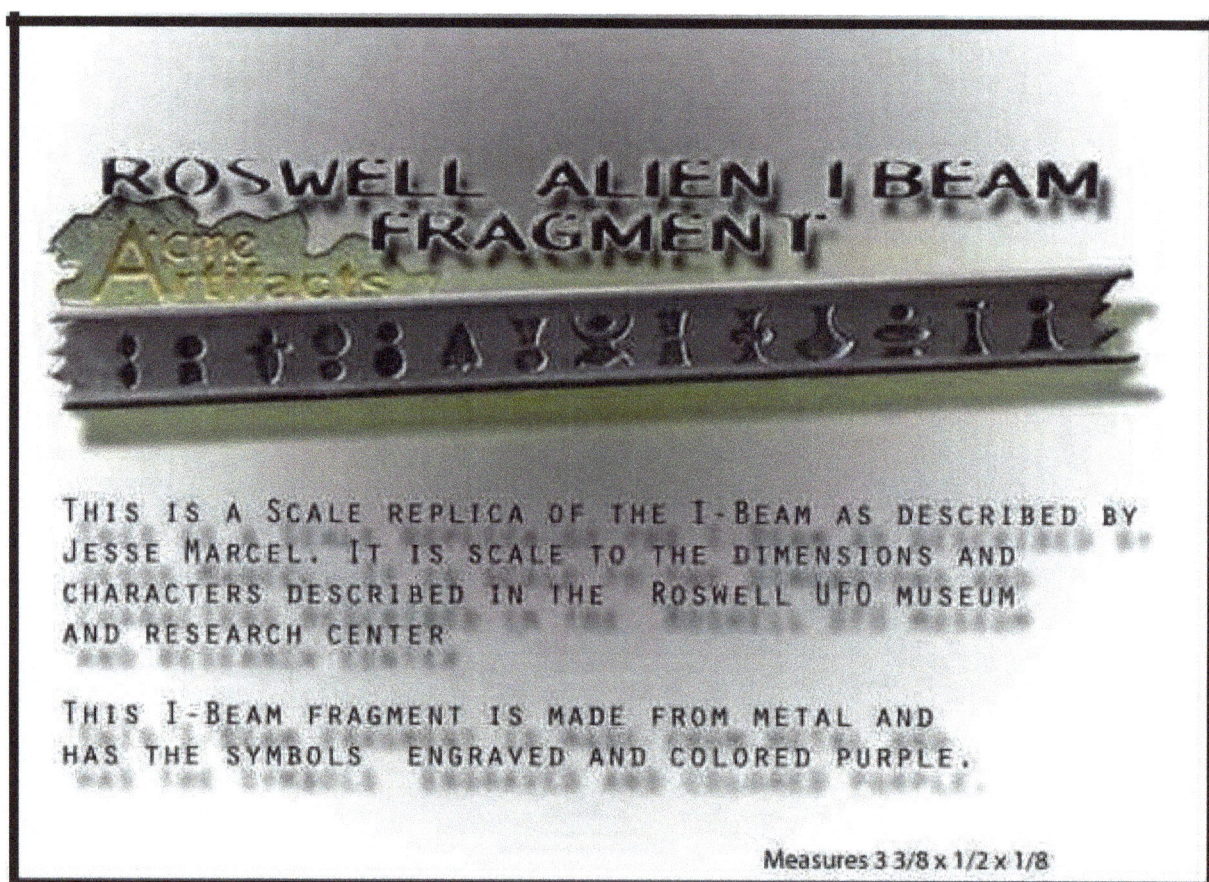

A replica of the Roswell I-beam
http://www.replicaprops.com/Roswell-I-Beam-Fragment-from-crash-site-Jesse-Marcel_p_56.html

What becomes readily apparent in comparing alien scripts is that some are hieroglyphic or pictorial in nature like the Roswell I-beam fragment or what was found on the Rendlesham ET spacecraft, while others seem symbolic or mathematic-like in appearance as in the Barney and Betty Hill, the George Adamski and the Sgt. Lenny Zamora alien scripts. The ET scripts on the spacecraft at the Shaitan Mazar UFO Crash Retrieval site, the Utsuro Bune UFO/USO and on the David Adair's recollection of alien writing that came from a symbiotic alien engine are a combination of mathematical and pictographic symbols.

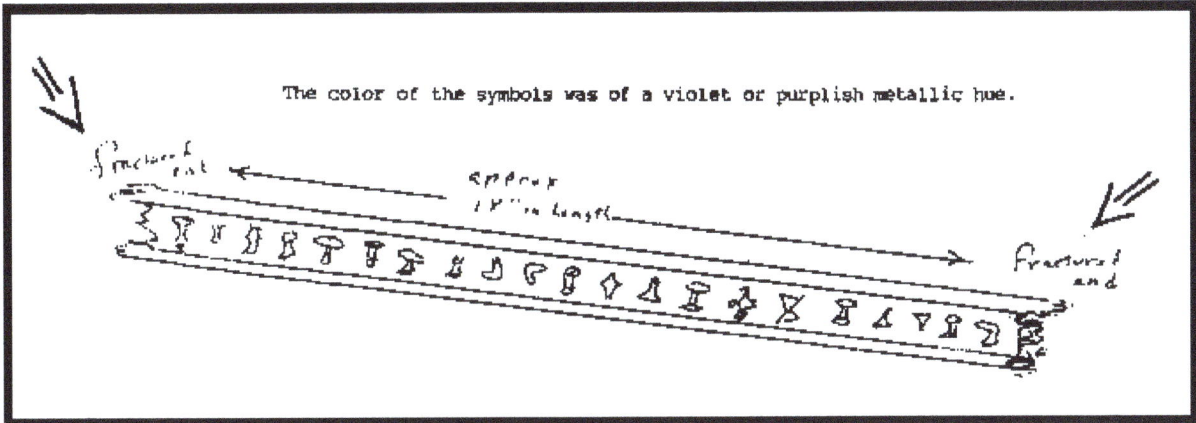

The sketch of the Roswell I-beam drawn by Jesse Marcel Jr.
http://ufologie.patrickgross.org/rw/w/jessemarceljr.htm

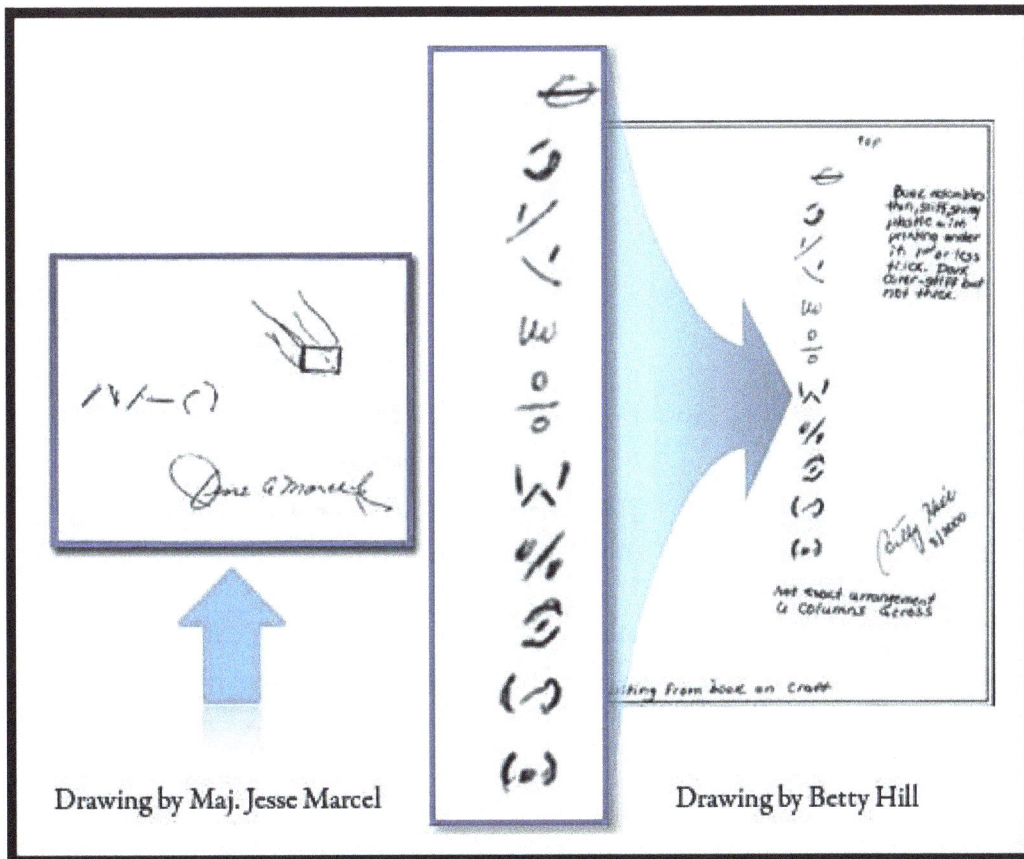

Drawing by Maj. Jesse Marcel

Drawing by Betty Hill

Comparison between Maj. Jesse Marcel alien writings and Betty Hill's alien script
http://www.truthseekeratroswell.com/60-years.html

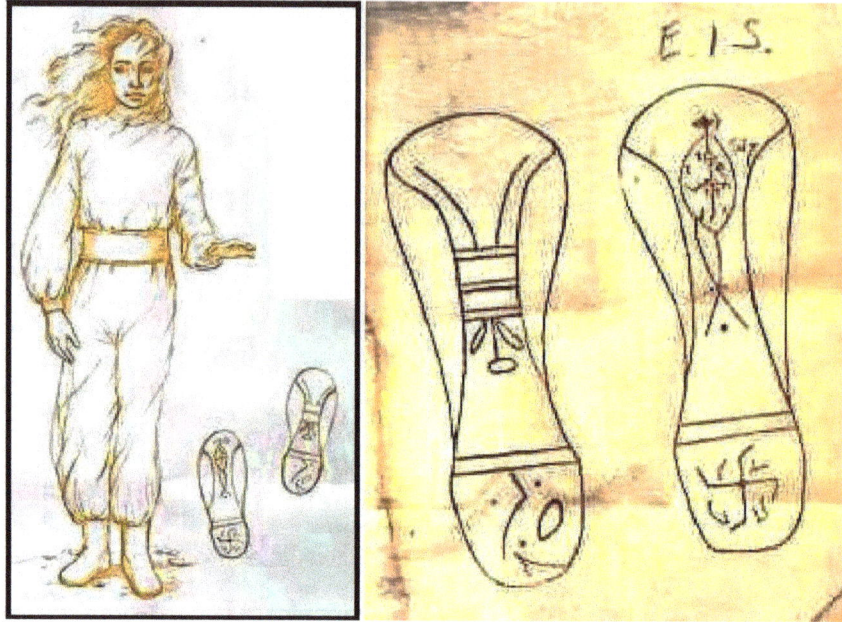

George Adamski's ET "footprint script"

Oil painting by Chris Lambright based on photographs taken of the actual landing site
Sgt. Lonnie Zamora stated that it is a good representation of what he observed.

THE C.I.A. AND THE SAUCER

Part of the evidence that the C.I.A., under Allen Dulles, was the builder and operator of the flying saucers seen in close encounters.

ANALYSIS OF THE INSIGNIA SEEN BY PATROLMAN LONNIE ZAMORA ON A LANDED FLYING SAUCER AT SOCORRO, N. MEX. ON APRIL 24, 1964

A.   B.

Step 1: Exact copy of Insignia as drawn at time of sighting. It was about 2 x 2 ft. in area, in red lettering, consisting of an arc, a line, and an "arrow-like" figure, with "stem" and "point" disconnected.

A. Reprinted from p. 67 of Emenegger, "UFO's Past, Present, and Future", Ballantine Books #25036, 1974 ($1.75).

B. Reprinted from p. 107 of Steiger, "Project Blue Book", Ballantine Books, #26091, 1976 ($1.95)

Step 2: Rotate the entire Insignia, counterclockwise or clockwise.

Step 3: Rotate the "arrowhead" only, to form the letter "A" with the "stem".

Step 4: Interchange the second and   Shift the vertical line to the
third letters, forming the   right to meet the arc, forming
initials of the "C I A" ex-   the letter "D", thus giving the
actly!   initials of Allen Dulles, "A D"!

Entire analysis copyright © 1977 by Dr. Leon Davidson
Blue-Book Publishers, Room 6, 64 Prospect St., White Plains, NY 10606

Project Blue Book makes an absurd attempt to say the glyph Lonny Zamora witnessed on the craft of unknown origin on April 24, 1964 at Socorro, New Mexico was the initials "CIA"
www.therendleshamforestincident.com/Zamora_UFO_Incident.html

The strange hieroglyphics seen on the ET spacecraft that had landed in the Rendlesham Forest which Sgt. Penniston had touched and then drew on paper.

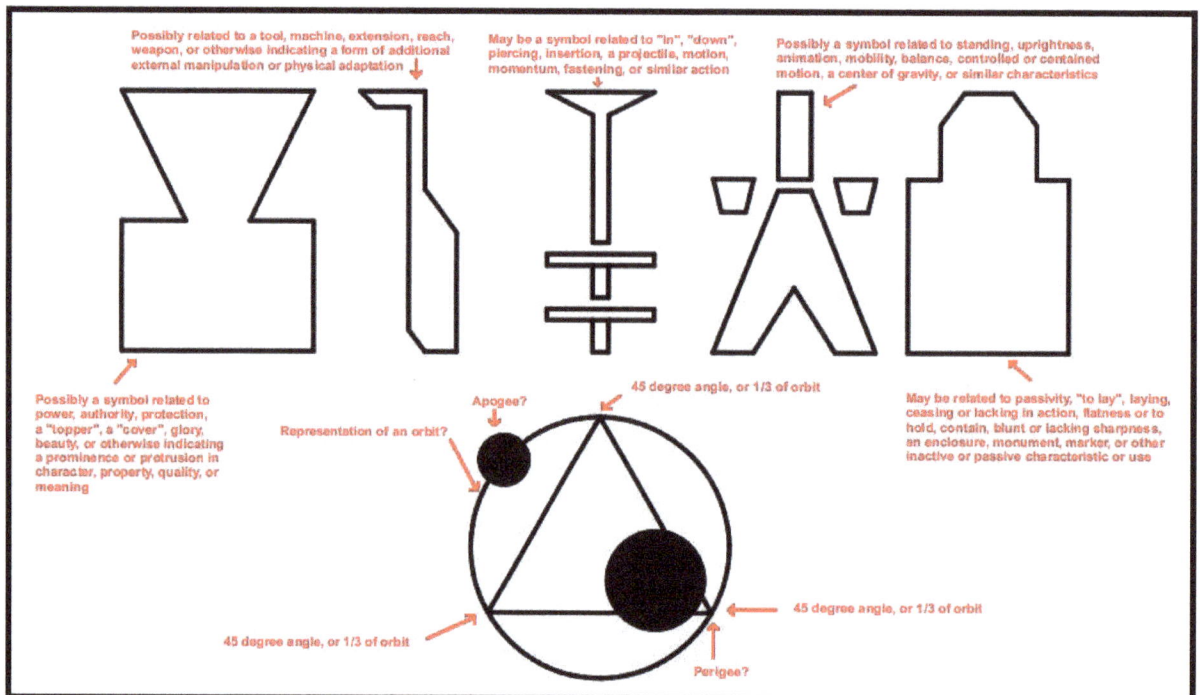

Possible interpretation of the Bentwaters/Rendlesham UFO symbols

"Дирижабль" в Тянь-Шане.
"Урочище „Ш. – Мазар." Вост. ущел.
Невозможно снять. Жертвы. Г.Э. Св.
обн. 1991, (?) 1992г.
МИ-8 – 1995г.

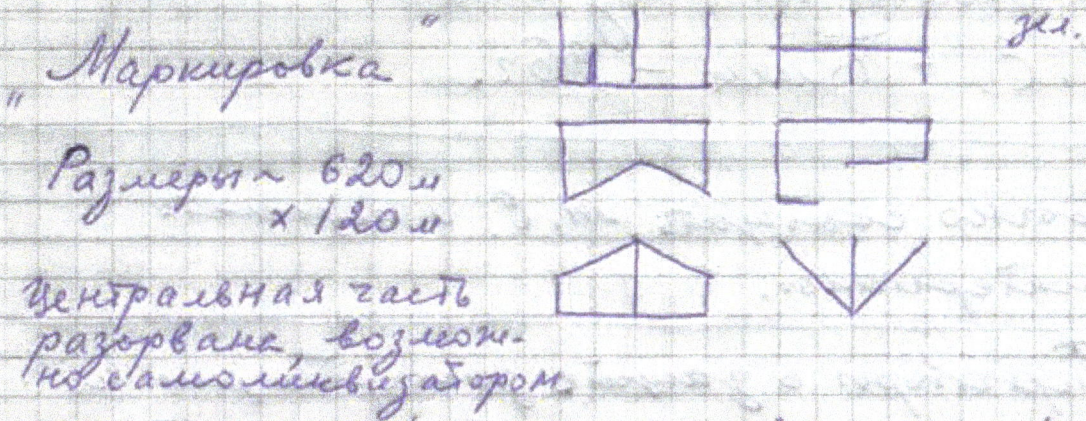

Маркировка "

зл.

"Размеры ≈ 620м × 120м

Центральная часть разорвана, возможно самолётным забором.

Поражающий эффект вызван сильнейшим отрицательным хрональным полем, до полной остановки всех часов на удалении 600м от объекта.
Кроме того отмечается почти полное отсут-ствие магнитного поля на том же расстоянии даже в образцах скальных пород (диориты).

обвод шпангоут
след взрыва
окед скольжения
И=У. "210м

The drawing from a witness at the Shaitan Mazar UFO Crash Retrieval site in Russia
http://www.therendleshamforestincident.com/Shaitan_Mazar_UFO_Crash.html

An artist concept of the Shaitan Mazar UFO Crash Retrieval site
http://www.therendleshamforestincident.com/Shaitan_Mazar_UFO_Crash.html

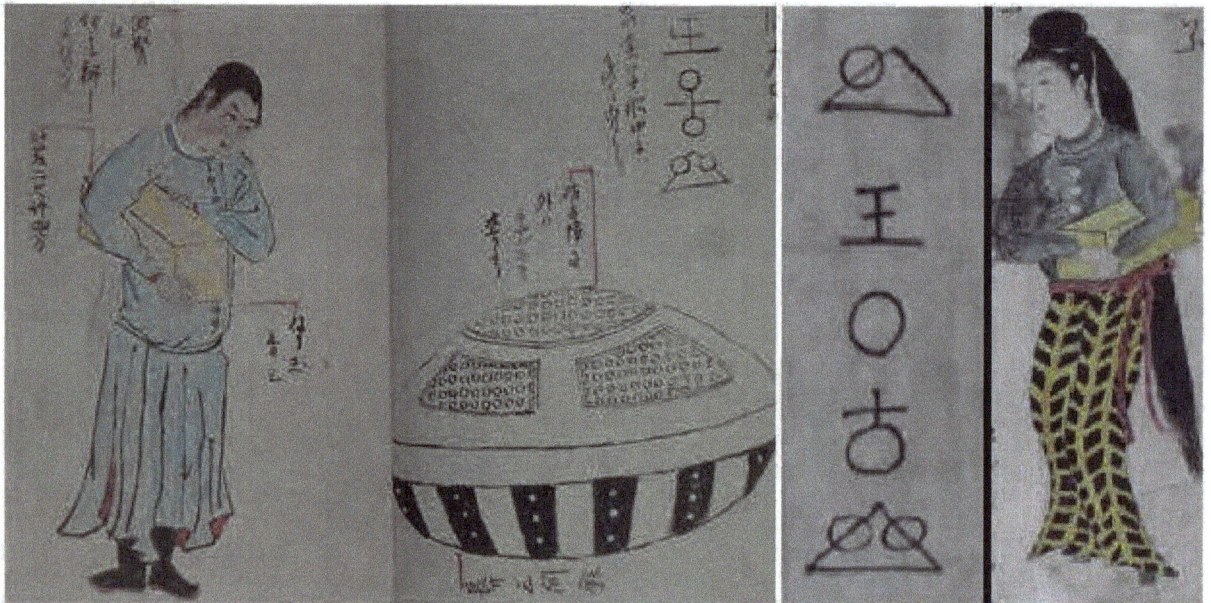

Two images of the Utsuro Bune UFO or USO (Unidentified Submerged Object) seen in a Japanese fishing village in the 1800s. Both images depict a "stylized Japanese" alien female tightly holding a box who had emerged from the saucer-like craft (centre). The right image also shows an alien script seen on the side of the USO
http://thelivingmoon.com/49ufo_files/03files2/1803_Japan_Utsuro_Bune.html

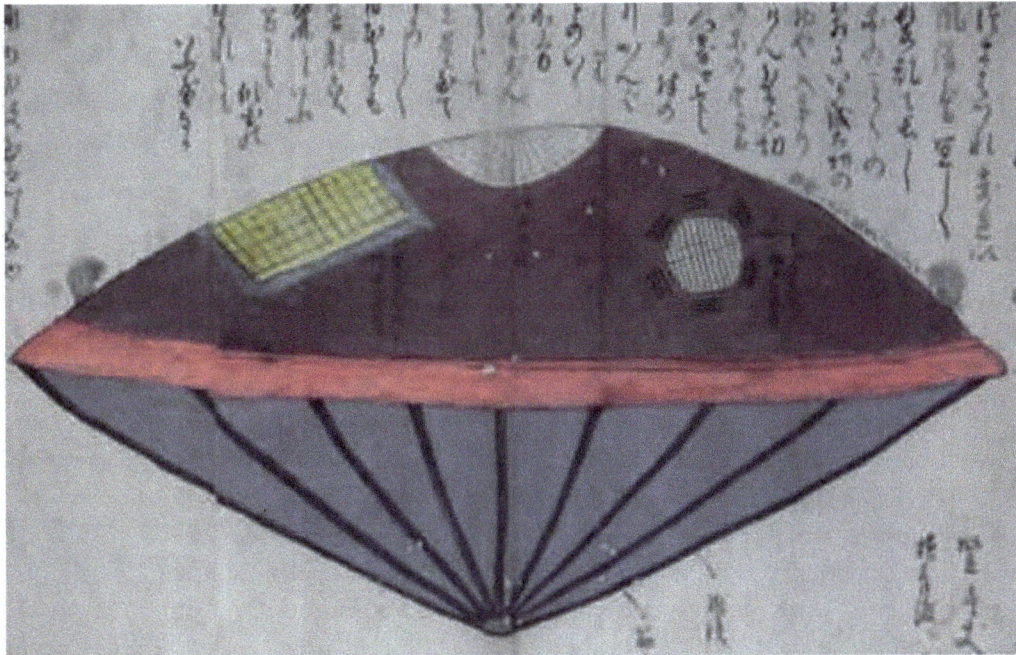

The Utsuro Bune UFO /USO seen by fishing villagers in the 1800s

http://thelivingmoon.com/49ufo_files/03files2/1803_Japan_Utsuro_Bune.html

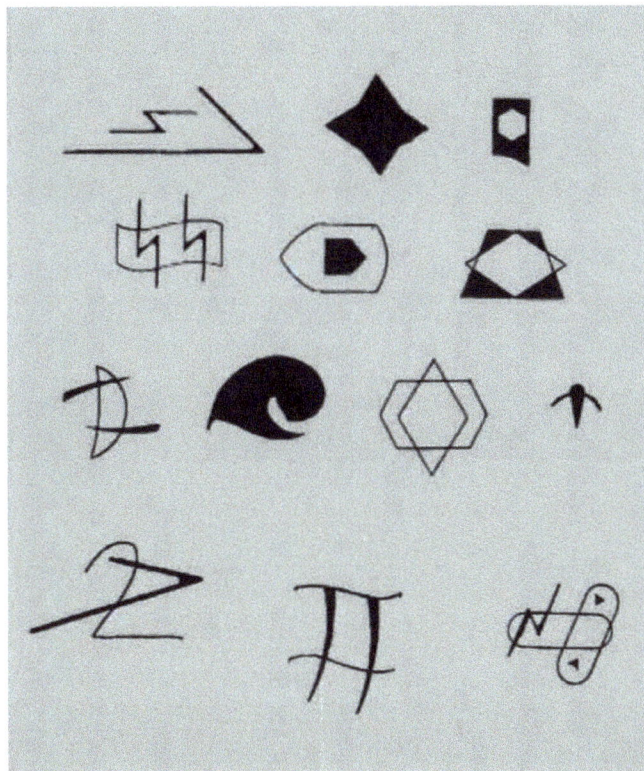

David Adair's recollection of alien writing that came from a symbiotic alien engine

https://www.youtube.com/watch?v=jqhLO4RZgdo

Obviously, what we can conclude from these various alien scripts is that many Extraterrestrial civilizations use symbolic script to convey various concepts of thoughts and ideas to communicate with each other of their kind. This would also imply that institutes of learning both primary levels through advanced institutions of higher learning are taught to a younger generation of ET beings. Education, it would seem is universal if an intelligent species is to advance and civilization is to progress!

As David Adair points out that certain symbols may be universal in their concept such as the symbol for Pi (see the above picture and note the bottom middle symbol and compare it to our mathematical symbol for Pi $(\pi\pi)$.

Having a written script also implies a verbal language where sounds are made to communicate thoughts and ideas, perhaps, emotions and physical feeling of pain or laughter.

Interrupting these extraterrestrial scripts may be the first step to understanding another interstellar civilization and hopefully removing the barriers of ignorance and allowing the discovery of similarities as well as the obvious differences between two intelligent species.

We are after all more alike than different as humans and extraterrestrial being are both sentient, aware and intelligent and no doubt, may even share similar concepts of morality and a special reverence for a divine power or supreme being!

CHAPTER 118

WHO BEST SPEAKS FOR THE PEOPLE OF EARTH WHEN CONTACT WITH EXTRATERRESTRIAL INTELLIGENCES IS IMMINENT?

The universe is full of life, Extraterrestrials Intelligences abound everywhere and many have found their way to our little blue planet, we called home...Earth! We have previously discussed the various attempts to reach out and communicate with these advanced Extraterrestrial Intelligences and we examined the methodologies in which that communication may take form.

Now, with a view of optimism, we can logically assume that ETI has reached the point in their final observation and surveillance of this planet and its people that they have decided that humanity is ready for open and direct contact with them. A signal of some sort is picked up from a radio telescope or a ham radio station or a military base indicating that ETI wish to establish contact and communications with us. Naturally, this would be the greatest event in human history and it would have a major impact on our lives and the way we perceive ourselves as a species in this vast, beautiful Universe.

Scientific literature and science fiction put forward various models of the ways in which extraterrestrial and human civilizations might interact. Their predictions range widely, from sophisticated civilizations that could advance human civilization in many areas to imperial powers that might draw upon the forces necessary to subjugate humanity.

The implications of discovery depend very much on the level of aggressiveness of the civilization interacting with humanity, its ethics, and how much human and extraterrestrial biologies have in common. These factors will govern the quantity and type of dialogue that can take place. The question of whether contact is physical or through electromagnetic signals will also govern the magnitude of the long-term implications of contact. In the case of communication using electromagnetic signals, the long silence between the reception of one message and another would mean that the content of any message would particularly affect the consequences of contact, as would the extent of mutual comprehension.
http://en.wikipedia.org/wiki/Potential_cultural_impact_of_extraterrestrial_contact

The Benevolent ETIs vs. The Malevolent ETIs (Human Advancement or Destruction)

Futurist **Allen Tough** suggests that an extremely advanced extraterrestrial civilization, recalling its own past of war and plunder and knowing that it possesses super weapons that could destroy it, would be likely to try to help humans rather than to destroy them. He identifies three approaches that a friendly civilization might take to help humanity:

- **Intervention only to avert catastrophe:** this would involve occasional limited intervention to stop events that could destroy human civilization completely, such as nuclear war or asteroid impact.
- **Advice and action with consent:** under this approach, the extraterrestrials would be more closely involved in terrestrial affairs, advising world leaders and acting with their consent to protect against danger.

- **Forcible corrective action:** the extraterrestrials could require humanity to reduce major risks against its will, intending to help humans advance to the next stage of civilization.

Tough considers advising and acting only with consent to be a more likely choice than the forceful option. While coercive aid may be possible, and advanced extraterrestrials would recognize their own practices as superior to those of humanity, it may be unlikely that this method would be used in cultural cooperation. **Lemarchand** suggests that instruction of a civilization in its *"technological adolescence",* such as humanity, would probably focus on morality and ethics rather than on science and technology, to ensure that the civilization did not destroy itself with technology it was not yet ready to use.

According to Tough, it is unlikely that the avoidance of immediate dangers and prevention of future catastrophes would be conducted through radio, as these tasks would demand constant surveillance and quick action. However, cultural cooperation might take place through radio or a space probe in the Solar System, as radio waves could be used to communicate information about advanced technologies and cultures to humanity.

Even if an ancient and advanced extraterrestrial civilization wished to help humanity, humans could suffer from a loss of identity and confidence due to the technological and cultural prowess of the extraterrestrial civilization. However, a friendly civilization may calibrate its contact with humanity in such a way as to minimize unintended consequences. Michael A. G. Michaud suggests that a friendly and advanced extraterrestrial civilization may even avoid all contact with an emerging intelligent species like humanity, to ensure that the less advanced civilization can develop naturally at its own pace. "What Role will Extraterrestrials Play in Humanity's Future?"; by Allen Tough; (1986); Journal of the British Interplanetary Society 39: 491–498. Bibcode:1986JBIS...39..491T.

A point of interest is that all of these hypotheses: **"Intervention only to avert catastrophe, Advice, and action with consent and Forcible corrective action"** as stated above, have already been realized or enforced according to **Dr. Steven Greer**, who has had Extraterrestrial contact on numerous occasions and has received precisely these particular messages from ETI!

Science fiction films often depict invading aliens as coming to Earth to wipe out humankind or eats us for lunch or enslave mankind while pillaging and raping the Earth of its resources! Scientists, however, take the view that an Extraterrestrial civilization capable of reaching the Earth would have more than enough power to destroy human civilization with minimal effort. Extraterrestrial totalitarian civilizations with a hostile **Klingon mentality** traveling to another planetary system and using unimaginable power just to obliterate all population centres on a planet, thereby reducing the planet to a cinder, seems like a waste of natural resources.

Deardorff speculates that only a small proportion of the intelligent life forms in the galaxy may be aggressive, but the actual aggressiveness or benevolence of the civilizations would cover a wide spectrum, with some civilizations "policing" others. **Dr. Steven Greer** points to planets like ours which at an adolescent stage in their planetary development and which are still aggressive by nature are quarantined until they reach a state of adult maturity. **Tough** also indicates the same position with his theory that *ETs will intercede to avert catastrophic war, the possible extinction of an intelligent species and if need be forcible corrective action!*

According to **Harrison** and **Dick**, hostile extraterrestrial life may indeed be rare in the Universe, *just as belligerent and autocratic nations on Earth* have been the ones that lasted for the shortest periods of time, and humanity is seeing a shift away from these characteristics in its own sociopolitical systems. In addition, the causes of war may be diminished greatly for a civilization with access to the galaxy, as there are prodigious quantities of natural resources in space accessible without resort to violence.

SETI researcher **Carl Sagan** believed that a civilization with the technological prowess needed to reach the stars and come to Earth must have transcended war to be able to avoid self-destruction. These would be minimum Type 1 Civilizations whose spiritual development has kept paced with and balances its technological achievements. Representatives of such a civilization would treat humanity with dignity and respect, and humanity, with its relatively backward technology, would have no choice but to reciprocate. **Seth Shostak**, an astronomer at the SETI Institute, disagrees, stating that the finite quantity of resources in the galaxy would cultivate aggression in any intelligent species and that an explorer civilization that would want to contact humanity would be aggressive. **(One has to wonder if Shostak has his head up a "black hole" as contrary to his thinking, the infinite universe is infinite in its resources and not finite)!** Similarly, **Ragbir Bhathal** claims that since the laws of evolution would be the same on another habitable planet as they are on Earth, an extremely advanced extraterrestrial civilization may have the motivation to colonize humanity, much as British colonizers did to the aboriginal peoples of Australia. *(Another scientist with anthropocentric thinking filling his head)!* (Bold italics added by author for emphasis).

Disputing these analyses, **David Brin** states that while an Extraterrestrial civilization may have an imperative to act for no benefit to itself; it would be naïve to suggest that such a trait would be prevalent throughout the galaxy. Brin points to the fact that in many moral systems in ancient Earth civilizations, "exalted non-military killings" were considered justifiable and acceptable by society, such acts are even found throughout the animal kingdom. Animals, however, are without rational reasoning and react to their environment by instinct.

Baum *et al.* speculate that highly advanced civilizations are unlikely to come to Earth to enslave humans, as the achievement of their level of advancement would have required them to solve the problems of labor and resources by other means, such as creating a sustainable environment and using mechanized labor. Moreover, humans as a food source for Extraterrestrials may be off the ET menu because of marked differences in biochemistry. For example, the **chirality of molecules** used by terrestrial biota may differ from those used by extraterrestrial beings.

Politicians have also commented on the likely human reaction to contact with hostile species. In his 1987 speech to the United Nations General Assembly, **Ronald Reagan** said, *"I occasionally think how quickly our differences worldwide would vanish if we were facing an alien threat from outside this world."*
http://en.wikipedia.org/wiki/Potential_cultural_impact_of_extraterrestrial_contact

Candidates for Possible Positions as Cosmic Diplomats

Well, get ready as the leaders of officialdom bump and jostle each other for front seat positioning in the biggest show in human history. There will undoubtedly be many aspects of this first official encounter with ETs that will look much like something out of the movie "Contact"! Who exactly will be participants in the greatest show on Earth?

Military brass will try to cordon off the areas of concern from all the non-essential personnel meaning the public, most government officials, etc. and try to place the whole event under eminent domain for reasons of *"National Security"*!

The Intelligence community will want to follow similar lines of action as the Dept of Defence as there may be highly sensitive information that will need to be kept from the public and from all other nations who are enemies of the country, even those who may be friendly nations will be left out in the cold!

Science community will state that this **first contact** will require the most brilliant scientific minds, who are highly specialized in related fields of science as most people will most likely miss things without examining all situations, the finer details, and nuisances that may be easily missed and since it is more than likely that the first signals from ET are going to be in the language of mathematics and physics, that give scientists a leg up on everyone else!

Of course religious leaders will want to jump in on the fray and state their position that thesebeings are more of *"God's children"* who will need to be spoken in a more moral or spiritual fashion of welcome and may require the special holy ministrations of Christian Catholicism, right from the **Pope of Roman**, himself or from the world's leading **Sunni and Shia Islamic leaders,** the **Grand Imam**, the **Grand Mufti** and the **Grand Ayatollah**! http://en.wikipedia.org/wiki/Islamic_religious_leaders

We can also expect the radical elements in society to want to participate in order to have their voices heard and their personal agendas met or they may even decide to disrupt the historic proceedings of humanity first official contact with Extraterrestrial intelligences.

When the Extraterrestrial Galactic Federation or whatever ETI cosmic council shows up on Earth at our doorstep, who would be our cosmic ambassador, our alien liaison?

There is no real world government that can speak on our behalf, particularly, as Stanton Friedman has often stated that there is no unified cooperation among governments, the military scientists or even among religions: *"Nationalism is the only game in town."* The UN is impotent in its efforts to exercise total world governance. Citizens are still blindly encouraged to owe their allegiance to outdated monarchies, iron-fisted military leaders, or to premiers and prime ministers of their countries who truly do not represent them, but are the lackey servants to the wealthy corporate elite regardless of which political systems, be they democracy, communism, socialism or dictatorship, all political systems are founded in a narrow nationalism that precludes anything resembling true altruism. Making Contact, A Serious Handbook for Locating and Communicating with Extraterrestrials" edited by Bill Fawcett; 1997; published by William Morrow and Co. Inc. New York, NY. USA; ISBN 0-688-14486-1

Is it possible that the ETI have already pre-selected who they will converse with from the vast populace of humanity? Maybe, the ETs have decided that common folk are the best candidates to communicate with, thereby, bypassing all the usual lines of officialdom and choose people from everyday walks of life, people who are the common salt of the Earth, who may actually be less hostile and perhaps, more open and receptive to a cross-species communication exchange?

Who do you think would best represent humanity? What qualifications should that person or persons should have? What language would they need to communicate with to the visiting ETI, would we have to learn a new language, an invented one or perhaps, an ET language, one of theirs or would the language simply be telepathy and/or sign language?

But, what if this scenario has already played without the knowledge of the general public being aware of these historic events and all of it was covered up and suppressed, so that another agenda could be implemented and a new destiny already pre- determined for the people of Earth without the consent from the rest of us.

People in the know, like the militaries of the most powerful nations, know that ETs are real and have determined beyond a shadow of a doubt that we are not alone in the universe. The Intelligence community of governments are also in on the **BIG SECRET**, a few politicians and a few of the most powerful, wealthiest businessmen of the world are also in on the secret. Has first contact already occurred but, the knowledge of that first encounter has been covered up and suppressed from the public awareness? This book is evidence that it has indeed occurred however, we are not likely to hear about it in the near future.

Did this hidden and mysterious cabal of the powerful and the privileged had that historic meeting and what communications were exchanged? What agreements were made? What treaties were signed or made binding upon an unsuspecting global populace? Should such covert contacts and communications between ETS and humans be even considered binding upon the rest of humanity, when they weren't present at this historic meeting nor had they even appointed or elected representatives to speak on their behalf?

The **Brooking Institute** in their commissioned report for NASA stated emphatically that the two most powerful institutions on Earth to be profoundly and adversely affected by the implications of an ET presence would be the science community and most of the world's fundamentalist and orthodox religions! The reason why these institutions would be the most affected, which is not taught in any school system, is that Science and Religion are in really one and the same thing; which teaches the same basic truth!

Members of the **Baha'i Faith** understand this concept very well, as it is a major tenant of their teachings. Science is after all the investigation of truth by discovery and religion is the understanding of truth by inspiration and revelation. Truth is one and comes from one source, the wellspring of all truth, God!

467

One of the basic problems for the science community and the institutions of religion is not understanding this simple concept and promoting it to the general public. It is a basic truth not only on Earth but, in the whole universe. A second problem is that most people have lost respect because of trust issues, lies, and deceptions on the part of these institutions.

This is particularly true for scientists who hold positions of influential power who work in covert black projects and programs that work against the welfare of humanity. Spiritual leaders and their institutions from the two most powerful world religions have become corrupted from within. Spiritual leaders through abusive powers and positions have taken advantage of their adherents or have suppressed the knowledge of hidden transgressions of their priesthood or failed to disclose pertinent religious information to their masses.

If there is a negative fallout to the people of Earth through contact with ETI, it will more than likely be these institutions that will be greatly affected. Most of the common populace of the planet will be accepting of this historic revelation with little to none in the way of chaos and panic. In fact, everyone will be looking at this big, new opportunity to make a buck! Some would say, there is no one who best speaks for humanity! Nevertheless, let's examine the pros and cons of those who think that they are the best to speak on behalf of us.

Is the UN Preparing to Appoint an Extraterrestrial Ambassador for First Contact?

The question: "Who best speaks for the people of Earth when contact with Extraterrestrial Intelligences is imminent?" has come up more frequently in these days of heighten awareness with the knowledge that we are being visited and engaged by ETI. This subject matter has increasingly become the topic of discussion privately amongst politicians within governments, it is a constant concern within the militaries of the US, UK, Russian and Chinese, etc. and within the **General Assembly of the United Nations** . What is odd at this time is that no government, military or any agency has come out to publicly disclose the existence of the Extraterrestrial presence visiting our planet or in the Solar System, so even if contact had occurred behind secret closed doors, we would never hear about it.

The only inkling that such an event has happened (besides the possibility that one or two insiders going public in the near future with such information and who undoubtedly would be seeking political asylum much like an **Edward Snowden,** will be revealed with the careful conditioning of the public with the slow release of pertinent information on the UFO/ETI matter. Knowledge regarding the ET presence on Earth would be distilled down over a period of time into the public domain, so that by the time the public figures it all out, the infrastructure on any ET/human agreements will be securely in place and hard to dismantle or those involved in controlling contact would have died off.

Is this one possible near-future scenario that awaits humanity, where we have no say in the matter or is there a more hopeful, glorious future that awaits us, one which will require everyone's full participation in order to shape our common destiny and allows us to establish a full open diplomatic relationship with Extraterrestrials from the stars?

Most of the public will look at the United Nations or their world leaders for the leadership in diplomatic relations with ETs but, Greer does not give that much hope. In Dr. Steven Greer's book, "Hidden Truth, Forbidden Knowledge" he recounts the severity and the lengths to which the shadow government operating through a covert paramilitary group will go to suppress the release of UFO and ET information or create disinformation into the public domain.

He tells of how he and one of his daughters were invited to the four Seasons Pierre Hotel in New York to meet with a well-known **Crown Prince (S.A.)** regarding Greer's disclosure of the UFO phenomenon. The Prince told Greer that he won't be able to do the disclosure because *"The aliens won't let you."*

As it turns out the Prince's brother had been abducted by what he thought were aliens. But what the Prince didn't know was that such abductions were carried out by a covert paramilitary operation, an important detail learned by Greer from a NASA researcher who investigated this case. Unbeknownst to the NASA researcher, Greer was also aware that this NASA researcher was part of the shadow government and is an eschatological, fundamentalist, end-of-the-world, Christian.

The role of this NASA scientist was to appear to be a legitimate UFO researcher seriously investigating the UFO phenomenon and this NASA scientist told Greer that the Prince's brother was abducted from the castle in his country!

Now, Dr. Greer knew that from an insider within the paramilitary operations who does abductions routinely who told him that, *"Yes, of course, we abducted him so that this particular powerful family and the banking empire would be on board with our program to fight the aliens."*

The Prince believes incorrectly, that human abductions by aliens had been going on since **Adam** and **Eve**, yet he could not explain why some people who go out to deliberately contact ETI, never experience the abduction phenomena or why UFOs would appear over cities worldwide if they wanted to remain secret. Greer politely disagreed with the Prince's assessment and pointed out, ***"I think it is a very human group that wants to keep this secret because to disclose it would mean the end of the entire centralized power system that exists, all fossil fuels, and the entire paradigm of anthropocentric religiosity."*** "Hidden Truth, Forbidden Knowledge: It's time for You to Know" by Steven M. Greer M.D.; 2006; published by Crossing Point, Inc. Publication; Crozet, Va., USA; ISBN 0-9673238-2-7

To say that the Prince's perceptions of what was real and factual and those that were disinformation and hoaxing were seriously called into question, as his fear-base belief and hate supported military action against aliens.

Now, just when we think we heard everything, the whole matter becomes even, more sinister and complicated as the Prince tells of his relationship with the Bushes particularly President Bush Sr. who wanted to do disclosure. The Prince explained that he was involved with a group after the Cold War back in 1989 that included **Mikail Gorbachev**, **George President Bush**, **UN Secretary General Perez de Cuellar** who met to plan the release of this information to the public. He said, "But one night, during a late night planning session in New York, the Secretary-

General of the United Nations, Perez de Cuellar was coming back from a meeting where the plans were being put into place to make this announcement about ETs, and a UFO appeared, stopped his motorcade, and abducted him out of the limousine! They took him on board the craft, and told him that if they didn't stop this plan to disclose the extraterrestrial presence, that every world leader involved, including the United States President, would be abducted and taken off this planet and that they would stop the process!"

People may recall this famous case in which the late abduction researcher, Budd Hopkins talks about a major international figure being abducted along with a civilian who witnessed it. Greer knows of people in the cell group operating the **alien reproduction vehicles** and the **psychotronic weapon systems** that conducted this abduction. Dr. Greer states that a relative of the infamous red-headed sergeant at Roswell who threatened everyone, was there to coordinate the event. He was inserted as part of the security detail for the Secretary-General, and he set up the electronics at the site of the Secretary General's motorcade so that this late night pseudo-abduction of the UN Secretary could take place.

Greer says that the goal of this operation was to stop the entire attempt by the world's power elite, including Gorbachev, to disclose the truth to the world.

These paramilitary abduction scenarios like the one the Prince's brother experienced are run by humans using craft that look like **UFOs (flying saucers)** and humans disguised as aliens or **PLFs (Programmable Life Forms)** that resemble small grey ETs that are manmade. It's all "**Stagecraft**" and the whole thing is a **MASSIVE BIG LIE!**

It is a Big Lie carried out against the general public as well as the world's leaders to get them to hate Extraterrestrials and support **Star Wars (weapons of mass destruction in space)** and to convince world leaders to shut down any effort to disclose information to the public!!! **"Hidden Truth, Forbidden Knowledge: It's time for You to Know" by Steven M. Greer M.D.; 2006; published by Crossing Point, Inc. Publication; Crozet, Va., USA; ISBN 0-9673238-2-7**

On the flip side of this reality, the perceptions that most people have is based upon blissful ignorance to the cold harsh reality that most world leaders really have no power as such, except for things that maintain the status quo within their nation and any foreign policies that do the same. In other words, it should be business as usual and don't rock the boat. Any world leader or nation that steps out of line and goes down a different path will feel the full wrath of the secret power elite, pseudo-government. But, it doesn't stop people from hoping!

According to a story printed on 26th Sept. 2010 by journalist, **Heidi Blake** in the British newspaper, "The Telegraph," it was believed that the UN was set to appoint Earth's first ambassador to visiting Extraterrestrial Intelligences. Earth's first ET representative as stated in the British newspaper was an unknown Malaysian astrophysicist, **Mazlan Othman**, who is set to be tasked with co-ordinating humanity's response if and when Extraterrestrial Intelligences make contact. Aliens who landed on Earth and asked: *"Take me to your leader" would be directed to Mrs. Othman,"* at least that is what is reported by Ms. Blake.
http://www.telegraph.co.uk/science/space/8025832/UN-to-appoint-space-ambassador-to-greet-alien-visitors.html

470

**Mazlan Othman, Director of UNOOSA but, she is unfortunately,
not Earth's first ambassador for extraterrestrial contact**

The story as it turns out is ***bogus***, without a shred of credibility or authenticity. Even Mazlan Othman, herself has officially denied media reports that she was selected by the United Nations to represent earthlings in their future dealings with aliens.

It seems Ms. Blake may have jumped to erroneous conclusions by not understanding her information source. In September 2010, several news sources reported the United Nations would soon appoint Mazlan to be the ambassador for extraterrestrial contact, apparently basing their claims on remarks she made suggesting that the UN coordinate any international response to such contact, and her scheduled appearance on a Royal Society panel that October, "Towards a scientific and societal agenda on extra-terrestrial life." However, a UN spokesperson dismissed the reports as "nonsense", dismissing any plan to expand the mandate of **United Nations Office for Outer Space Affairs (UNOOSA).** She later explained that her talk would illustrate how extraterrestrial affairs could become a topic of discussion at the UN using as an example, the advocacy that led to UN discussion of "***near-Earth objects and space debris.***"
http://en.wikipedia.org/wiki/Mazlan_Othman

"It sounds really cool," Othman told The Guardian, another U.K. newspaper, *"but I have to deny it."*

The United Nations has also denied the story, *"The article in the Sunday times is nonsense,"* a U.N. spokesman said.

The idea, according to the original erroneous piece, was that the planet would have a person in place when the eventual arrival of an alien ship led to the inevitable *"take me to your leader"* moment. http://www.aolnews.com/2010/09/27/mazland-othman-is-not-earths-alien-ambassador-after-all/

In November 1999, **Kofi Annan, UN Secretary-General,** appointed Mazlan as Director of the Office for Outer Space Affairs (UNOOSA) in Vienna. She was also reappointed again, as UNOOSA director in 2007 by **Secretary-General Ban Ki-moon**.

At **UNOOSA** she deals with issues of international cooperation in space, prevention of collisions and space debris, use of space-based remote sensing platforms for sustainable development, coordination of space law between countries, and the risks posed by near-earth asteroids, among other topics.

It remains a debated point, however, whether the world's governmental bodies should take the subject of possible alien encounters more seriously. As AOL News contributor Lee Speigel wrote in June, some European politicians are urging the U.N. to bring the topic of UFO sightings out into the open and urge world governments to declassify documents that chronicle human encounters with alien space ships. http://www.aolnews.com/2010/09/27/mazland-othman-is-not-earths-alien-ambassador-after-all/

The Times article claimed that Othman would tell delegates that the recent discovery of hundreds of planets around other stars has made the detection of extraterrestrial life more likely than ever before -- and that means the U.N. must be ready to coordinate humanity's response to any **"first contact."**

The Times article cited a talk Othman gave recently to fellow scientists, where she said: *"The continued search for extraterrestrial communication, by several entities, sustains the hope that someday humankind will receive signals from extraterrestrials. When we do, we should have in place a coordinated response that takes into account all the sensitivities related to the subject. The U.N. is a ready-made mechanism for such coordination."*

Professor Richard Crowther, an expert in space law and governance at the U.K. Space Agency and who leads British delegations to the U.N. on such matters, said: *"Othman is absolutely the nearest thing we have to a 'take me to your leader' person."*

However, he thinks humanity's first encounter with any intelligent aliens is more likely to be via radio or light signals from a distant planet than by beings arriving on Earth. And, he suggests, even if we do encounter aliens in the flesh, they are more likely to be microbes than anything intelligent. http://www.foxnews.com/scitech/2010/09/27/appoints-contact-visiting-space-aliens/

If first contact with an Extraterrestrial civilization were to occur, the United Nations, as many people believe, would be the logical first international organization of choice to represent humanity along with a duly appointed UN representative who has diplomatic credentials and experience. But to date, no UN official has been designated to this position.

The question then at this moment, does the UN have a person(s) in mind for the role of Earth's **ambassador to the universe** along with a set of protocols in place for the eventuality of such a human- ET **"first contact"** encounter? Until we know for sure the UN's response to ETI, we are left pondering who would be the best person to represent humanity when ETs arrive on Earth at our doorstep.

"We must recognize the limitations of pursuing this idea through the United Nations. Although the U.N. is the most comprehensively inclusive intergovernmental body, that organization's operating style has disadvantages. The United Nations, not known for its efficiency, may be slow to react to contact.

That organization eventually would convene a meeting of the General Assembly to hear general statements from national representatives. The issue would be referred to **COPUOS (Committee on the Peaceful Uses of Outer Space)** and possibly to other organizations such as the **International Council of Scientific Unions**. These bodies would conduct assessment studies that could take years. In Doyle's view, there would be no reason to hurry.

COPUOS is in session for only a few weeks a year. As the committee operates on the basis of consensus, a single nation could block the approval of a proposed policy statement. If the U.N. process proves too slow and unwieldy, some governments may turn to other options. There may be pressure for a quicker response, particularly if we find alien technology in our own solar system. This has led to proposals that the matter be referred to the U.N. Security Council, which is always in session. Referring contact with extraterrestrials to that body makes some people uncomfortable because it implies that they are a threat to human security

What would stir the United Nations to action, other than a confirmed detection? Related discoveries such as fossils on Mars or an Earth-like planet near another star might provoke a discussion, possibly leading to consideration of something like the proposed **Declaration of Principles** on communicating with extraterrestrial civilizations. Although the United Nations is unlikely to take policy action in advance of contact, its members could call for a study of that event's implications.

What if there is no agreed international procedure when contact occurs? Nations with the necessary technical capabilities might act pre-emptively in an uncoordinated way—for example, by sending separate messages to the detected civilization. One or more governments could head this off by quickly proposing a coordinated set of actions, within or outside of the U.N. system."

Contact with Alien Civilizations: *Our Hopes and Fears about Encountering Extraterrestrials* by Michael A. G. Michaud; 2007; published by Copernicus Books; New York, NY, United States; ISBN 978-0-387-28598-6. Archived from the original on 24 December 2012.

For the United Nations to be taken seriously in any matters of foreign policies and negotiations, we have already stated elsewhere in this book that the original five founding member nations of the **UN Security Council** must give up their **VETO powers** and then, all nations will have equal footing and voting powers as these five powerful nations. When this happens the UN will finally have the power to truly represent the world and effect powerful changes that will benefit all mankind and in turn, they would truly represent the world in first contact with Extraterrestrials. Until that day happens, we will all have to hold our breath for our global destiny to unfold.

In the meantime we should at least prepare for that day when open contact occurs and we begin to communicate with ETI, so let's look at a few ambassadorial prospects of people we want or not want to represent us as spokespeople on behalf of humanity.

There are of course as many thoughts, ideas, and suggestions as to who should represent us as there are people interested in this UFO subject. Some people will say emphatically that it should be the military, while others support the Prime Minister or the President of their country. Still, others will want a more truly representative political body such as the **Secretary-General of the UN** and the religious minded will want to support or promote the **Pope** or the **Grand Imam** of Islam, also the **Dali Llama** of Tibet would no doubt be a good candidate. Maybe people will look towards royalty like the **Queen of England** or an international political statesman as ambassador.

The issue may come down to which nation the ETs decide to land in and if this happens, it a safe bet that the military and the political leaders of that country will quickly move toward controlling the whole historic situation and its outcome.

If this contact event took place in a small country where their military might is minimal would that government invite the assistance of other countries to be involved or go it alone? What if the event took place in a country where the foreign relationship is non-existent, minimum, in cold-war political tension or even worst, adversely hostile?

The overriding factor will, of course, will be the nature of how communications will take place (spoken word, mathematics, sign language, pictographic, telepathy, etc) as this will be a key issue in the success of interspecies relationship so, communicating effectively and understanding correctly will be of primary importance.

The Military and Intelligence Representatives – It's all in the Name of National Security

We've seen that the UN is basically impotent on this matter or least they are not saying much at this time so could the military with all its might and incredible resources at its disposable be the right people as Earth representative? Certainly, if push came to shove, we can be assured that the military would not back down and would hold their ground in a worst case scenario.

If history as taught one thing, violence has never really solved any situation in an amicable fashion, where one side becomes the victor and the losing side is left resenting the conditions imposed upon it by the victor. In fact history, for the most part, is always written from the perspective of the victor, rarely is it ever written from the viewpoint of the loser! It is often the task left to the historians to figure out the details and accuracy of the reason why conflict or war arises in the first place.

Militaries, for the most part, are trained to defend themselves, their allies, their country and ultimately their people and at the same time to attack the opposing enemy, bring chaos and ruin upon their military, their country and their people in order to bring about conditional surrender. Militaries are always in a state of readiness, always monitoring, tracking, spying, building, repairing, constructing, upgrading, researching and developing new and more improved weapons arsenals and machinery. These are the militaristic capabilities of diplomacy. Bottom line: National Security and protection supported by a shoot first and ask questions second mentality!

When war comes, we are also happy to have our military well equipped to protect us and defend our way of life and in this sense, militaries will always have a place in society as long as they serve our purpose and follow no hidden agenda of their own.

How does this type of diplomacy translate into the history of the UFO and ETI phenomenon?

Shoot first, retrieve the alien spacecraft, the alien technology and the scattered debris from the crash site, pick up and body-bag any dead Extraterrestrials, capture and arrest any surviving ETs, reverse-engine the ET spacecraft, autopsy the dead aliens, perform DNA experiments, interrogate the surviving EBEs then, place them in a comfortable, very secure dwelling under 24 hour guarded protection with remote surveillance to monitor their every movement as interstellar house guest/prisoners!

"Welcome to the planet Earth! Enjoy your long stay at one of our finest, all inclusive military's bases in the United States of America"!!!

This has historically been the procedure and the military mindset, since day one when it was determined before the end of the Second World War that alien spacecraft were monitoring our global conflicts.

Dr. Steven Greer tells how one former military officer who was on the board of directors of a company whose founder and inventor had developed a **"neutrino light detector"**, a device that detects subtle electromagnetic emissions associated with **Extraterrestrial Vehicles (ETVs)** when they are in a dematerialized form. This detector was stolen by the **NRO, National Reconnaissance Office**, for use in their detection satellites so they could target ET vehicles prior to their full materialization. Dr. Greer knows the inventor whose is a very accomplished scientist. "Contact" Countdown to Transformation – The CSETI Experience from 1992 - 2009 by Steven M. Greer, M.D. 2009; publish by 123PrintFinder, Inc.; Ladera Ranch, VA, USA; ISBN 9780967323831

Therefore, are these really the type of people you want to represent you as peaceful, non-threatening **"Universal Ambassadors"** to visiting Extraterrestrial Intelligences coming to our planet? Definitely, positively, absolutely **NOT!!!**

These are the very people who would callously jeopardize the very lives of everyone on this planet by engaging in a secret interstellar war with ETs which would become in a short time, a publicly recognized war with an alien species. War could break out with Extraterrestrial Intelligences who do not know the customs of humanity (intrusive entries into a nation's airspace without prior permission) which would make them immediate targets in a shoot down campaign, a policy that has been in effect since the early '50s.

However, the real agenda that permits such aggressive military action which is supported by MJ-12 and the Military industrial Complex, is the deliberate tracking, targeting and shooting down of

ET spacecraft that appear either in our atmosphere or in near orbit purely for the acquisition of their alien technology!

Is there a role to be played by the military in first contact? It would be naive to think that they won't but since this event requires diplomacy their participation may be more in securing the venue site and the ET landing site from unwanted human intruders and staying alert for any unforeseen eventualities.

Needless to say, military or intelligence people are absolutely and unquestionably, the last people on Earth who should represent mankind in our attempt to communicate with Extraterrestrial Intelligences.

Government Representatives – Will Political Hot Air on the Hill Speak Exopolitics to ETI?

When Extraterrestrial Intelligences do land either in **New York's Central Park** or on the **White House Lawn** or in **Moscow's Red Square** or **China's Tiananmen Square** or in front of **Ottawa's Parliament Hill**, will our government officials be ready and are they the right people as **Cosmic Diplomats**?

Would news of radio contact with an extraterrestrial civilization prove impossible to suppress and travel rapidly, or could it quickly contained by the military or intelligence communities? Media coverage of the discovery would probably die down quickly, though, as scientists began to decipher the message and learn its true impact. Different branches of government (for example legislative, executive, and judiciary) may pursue their own policies, potentially giving rise to power struggles. Even in the event of a single contact with no follow-up, radio contact may prompt fierce disagreements as to which bodies have the authority to represent humanity as a whole. **Michaud** hypothesizes that the fear arising from direct contact may cause nation-states to put aside their conflicts and work together for the common defense of humanity.

Apart from the question of who would represent the Earth as a whole, contact could create other international problems, such as the degree of involvement of governments foreign to the one whose radio astronomers received the signal. The United Nations discussed various issues of foreign relations immediately before the launch of the Voyager probes, which in 2012 are leaving the Solar System carrying a golden record in case they are found by extraterrestrial intelligence. Among the issues discussed were what messages would best represent humanity, what format they should take, how to convey the cultural history of the Earth, and what international groups should be formed to study extraterrestrial intelligence in greater detail.

Even in the absence of close contact between humanity and extraterrestrials, high-information messages from an extraterrestrial civilization to humanity have the potential to cause a great cultural shock. Sociologist Donald Tarter has conjectured that knowledge of extraterrestrial culture and theology has the potential to compromise human allegiance to existing organizational structures and institutions. The cultural shock of meeting an extraterrestrial civilization may be spread over decades or even centuries if an extraterrestrial message to humanity is extremely difficult to decipher.
http://en.wikipedia.org/wiki/Potential_cultural_impact_of_extraterrestrial_contact

Governments, for the most part, serve the people that why every country has one to ensure the welfare, growth, and prosperity of their nation. Government officials represent us through the due voting process of national and regional elections and thus, give voice to the public's concerns and issues. But of late, there are many Governments but particularly in the US that do not seem to be listening to their nation's people.

Consequently, over a period of time bad bills are made and passed on the government's exorbitant spending on military programs, supporting financially draining wars in other countries in the name of protecting national interests abroad, bailing out the major banking cartels and automotive corporations which have essentially ruined the economy in America, failure to bolster and support a national medic-care program, failure to prevent the national decline in economic growth that has primarily affected the middle income earners, cutting back on essential programs that would lead to improve education, proper housing and eliminating the growing poverty that is becoming all to prevalent in major cities and towns and the list goes on.

The only ones that seem to benefit from this type of government are the *"Fat Cats"* in Washington (insert your own country's government, here) and the wealthy corporate elite who ensure that nothing ruffles the feathers of their wealth, power, and control.

Also keep in mind that when it comes to diplomacy, governments are the first ones to create political dissension in foreign affairs with other nations usually through over-inflated male egos which often comes through disagreements, threats, provocative and intimidating language and generally through bully-boy tactics and flexing of military muscle and strength. When wars come, they blame the other side and when war ends, they take all the credit of victory or they blame the other side, if the loose!

These government officials will always tell you with a smile on their faces that everything is fine, everything is under control, that the top, best people are looking after things and everything is going as planned. The question is for who mis it benefitting?

More often than not when you want to know something in detail and the subject is exquisitely sensitive, you'll find your side getting the runaround or the *"Washington Two Step"* on your concerns or you are simply just blown off as if you never existed.

In matters of UFO and ETI subject matter from the current US government which ran on the election campaign of promising to be the government of the people and for the people and as a government that was about open transparency and disclosure, it has become a dismal failure! To this can be attested the serious public efforts of various UFO disclosure programs and events on **Washington's Capitol Hill** like Dr. Greer's **Disclosure Project** campaign at the **National Press Club** and the Bassett's **Citizen Hearing on Disclosure.** Both were great citizen efforts to bring the public attention to a subject for which the American government has done nothing to investigation, research or offer official acknowledgement as a legitimate matter of public concern that needs to be seriously addressed.

Most other countries governments have gotten on board the *"UFO Express Train of Disclosure"* by releasing hundreds of thousands of UFO documents and files but the US has stubbornly resisted doing the same; going so far as to site bogus military investigations in the matter that there is nothing to the so-called UFO phenomenon.

No wonder the public is pissed with their government and are ready *for "some good old fashion American western justice that comes from the end of a rope?"* The advice here to the US government and its various arms and branches is **BE CAREFUL,** the citizenry is watching intently and they're telling you, *enough is enough already!*

Bottom line here is do you trust anyone currently in politics to speak on your behalf when the ETs arrive to make contact with us? The government representatives are one more group of people you can write off your list as trusted servants to the people, simply because, they've never really been there for you in the past!

Scientists as Cosmic Diplomats –The Rational, Logical Language of Mathematics and Physics Or Does E Still Equal MCC^{22}?

Most of us as the movie viewing public will recall the movie called *"Contact"* with **Jodi Foster** as the astrophysicist Ellie Arrowway who as one of the final ET diplomat selectees is being interviewed by an international body of selectors. In the interview she is asked why she should be the one to go on the interstellar trip to meet and communicate with ETs to which she replies that since the original message from ETs was in the language of mathematics and science and as an astrophysicist, she would be the logical choice to travel to the stars.

This fictional situation is based upon reasons of logic and initially it would seem picking a scientist would be a good choice as the person to speak on behalf of humanity as they are some of the smartest people on the planet, very focused on matters of detail with highly developed analytical minds but, that does not necessarily equate to being adept in the ways of a diplomat.

It has been suggested that when ET calls on us, the person who speaks for humankind may be **Paul Davies**, chairman of the **SETI Post-Detection Task group**, as a likely ambassador. The mission of SETI Post-Detection Task group, chaired by Davies, a theoretical physicist and cosmologist at Arizona State University, is to prepare, reflect on, manage, advise and consult in preparation for and upon the discovery of a putative signal of Extraterrestrial Intelligent origin.

In much the same vain as his fellow scientist, **Seth Shostak** who is also outspoken on the possibility of **SETI** receiving that history making signal from ET. In fact, Shostak feels that if ET calls, it will more than likely be he that will receive that famous signal and inform the rest of the world.

SETI has even graded the impact assessment of Extraterrestrial contact depending on the method of discovery, the nature of the Extraterrestrial beings, and their location relative to the Earth using the **Rio Scale.** The concept was first proposed in Rio de Janeiro, Brazil (hence its name) by **Iván Almár** and **Jill Tarter** presented to the 51st **International Astronautical Congress**, on the Search for Extraterrestrial Intelligence, in October 2000.
https://www.youtube.com/watch?v=hx9i-KRMCCc and
https://www.youtube.com/watch?v=IVV4zRuE1mw

Rio Importance

Rio	Importance
10	Extraordinary
9	Outstanding
8	Far-reaching
7	High
6	Noteworthy
5	Intermediate
4	Moderate
3	Minor
2	Low
1	Insignificant
0	None

Interpreting Rio Scale Values
http://avsport.org/IAA/rioscale.htm

Considering these factors, the Rio Scale has been devised in order to provide a more quantitative picture of the results of Extraterrestrial contact. More specifically, the scale gauges whether *communication was conducted through radio, the information content of any messages,* and *whether discovery arose from a deliberately beamed message* (and if so, whether the detection was the result of a specialized SETI effort or through general astronomical observations) or by the *detection of occurrences such as radiation leakage* from astroengineering installations. It has also been expanded to include *"technosignatures",* including all indications of intelligent Extraterrestrial life other than the interstellar radio messages sought by traditional SETI programs.

The Rio Scale is an attempt to quantify the importance of a candidate SETI signal. It is an ordinal scale between zero and ten, used to quantify the impact of any public announcement regarding evidence of extraterrestrial intelligence. The importance factor comes from a list of questions that determines the response to possible contact or the lack of action to no ET event.

There isn't any greater potential threat to the status quo than the discovery of extraterrestrial life, which is why some people would prefer we didn't try.

A poll was administered to American and Chinese university students and their responses to questions about the participants' belief that extraterrestrial life exists in the Universe, that such

479

life may be intelligent, and that humans will eventually make contact with it. The study showed significant weighted correlations between participants' belief that *extraterrestrial contact may either conflict with or enrich their personal religious beliefs* and *how conservative such religious beliefs are*. Not surprising, the more conservative the respondents, the more harmful they considered extraterrestrial contact to be. Other significant correlation patterns indicate that participants took the view that the search for extraterrestrial intelligence may be futile or even harmful. http://en.wikipedia.org/wiki/Potential_cultural_impact_of_extraterrestrial_contact

Post-Detection Protocols Are they Binding on Everyone?

Government officials do not want the general public to establish contact and communications with ETI because you may unwittingly create an interplanetary conflict or start an interstellar war.

Various protocols have been drawn up detailing a course of action for scientists and governments after extraterrestrial contact. Post-detection protocols must address three issues: what to do in the first weeks after receiving a message from an extraterrestrial source; whether or not to send a reply; and analyzing the long-term consequences of the message received. No post-detection protocol, however, is binding under national or international law, and some scientists believe that such protocols are likely to be ignored if contact occurs.

One of the first post-detection protocols, the "Declaration of Principles for Activities Following the Detection of Extraterrestrial Intelligence", was created by the SETI Permanent Committee of the **International Academy of Astronautics (IAA).** It was later approved by the Board of Trustees of the IAA and by the **International Institute of Space Law**, and still later by the **International Astronomical Union (IAU),** the **Committee on Space Research, the International Union of Radio Science,** and others. It was subsequently endorsed by most researchers involved in the search for extraterrestrial intelligence, including the SETI Institute.

The Declaration of Principles contains the following broad provisions:

1. Any person or organization detecting a signal should try to verify that it is likely to be of intelligent origin before announcing it.
2. The discoverer of a signal should, for the purposes of independent verification, communicate with other signatories of the Declaration before making a public announcement, and should also inform their national authorities.
3. Once a given astronomical observation has been determined to be a credible extraterrestrial signal, the astronomical community should be informed through the **Central Bureau for Astronomical Telegrams of the IAU. The Secretary-General of the United Nations** and various other global scientific unions should also be informed.
4. Following confirmation of an observation's extraterrestrial origin, news of the discovery should be made public. The discoverer has the right to make the first public announcement.
5. All data confirming the discovery should be published to the international scientific community and stored in an accessible form as permanently as possible.

6. Should evidence for extraterrestrial intelligence take the form of electromagnetic signals, the Secretary-General of the **International Telecommunications Union (ITU)** should be contacted, and may request in the next ITU Weekly Circular to minimize terrestrial use of the **electromagnetic frequency bands** in which the signal was detected.
7. Neither the discoverer nor anyone else should respond to an observed extraterrestrial intelligence; doing so requires international agreement under separate procedures.
8. The SETI Permanent Committee of the IAA and Commission 51 of the IAU should continually review procedures regarding detection of extraterrestrial intelligence and management of data related to such discoveries. A committee comprising members from various international scientific unions, and other bodies designated by the committee, should regulate continued SETI research.

http://en.wikipedia.org/wiki/Potential_cultural_impact_of_extraterrestrial_contact

A separate "Proposed Agreement on the Sending of Communications to Extraterrestrial Intelligence" was subsequently created. It proposes an international commission, membership of which would be open to all interested nations, to be constituted on detection of extraterrestrial intelligence. This commission would decide whether to send a message to the extraterrestrial intelligence and if so, would determine the contents of the message on the basis of principles such as *justice, respect for cultural diversity, honesty, and respect for property and territory*. The draft proposes to forbid the sending of any message by an individual nation or organization without the permission of the commission and suggests that, if the detected intelligence poses a danger to human civilization, the **United Nations Security Council** should authorize any message to extraterrestrial intelligence. However, this proposal, like all others, has not been incorporated into national or international law.

Paul Davies, a member of the **SETI Post-Detection Task group,** has stated that post-detection protocols, calling for international consultation before taking any major steps regarding the detection, are unlikely to be followed by astronomers, who would put the advancement of their careers over the word of a protocol that is not part of national or international law.
http://en.wikipedia.org/wiki/Potential_cultural_impact_of_extraterrestrial_contact

There has been some outrage recently over attempts to contact intelligent aliens, where instead of hiding in the corner and listening real hard some astronomers beamed intense directional messages up and away. Critics like Stephen Hawking decried these actions as dangerous, though these critics' fears reveal more about us than any eventual ETs.

A big problem for scientists is they do not all agree as to whether other life exists in the universe and secondly if they can agree, they are still divided as to whether we should even contact other ET civilizations. As **Stephen Hawking** has stated advertising our presence to inhabitants of other planets may go adversely for us. It is possible that some ETS are hostile and are merely after resources. This shows the limited thinking among scientists who do not seem to consider other more positive outcomes that may result from "**first contact**".

They assume that they would be similar to humanity, so their first response to finding a more primitive culture would be to exploit the hell out of it. It may seem deservedly like just deserts for humanity whose history is filled with war and conquest, while others contend that any species that can journey the long interstellar distance to here have probably advanced along a more peaceful and spiritual evolution where their goals are rather higher-minded than "Target and destroy all competing life forms".

They have bought into the negative doomsday tactics of the military industrial sector and thus, the scary ET invasion scenario *"has a powerful effect upon the weak-minded!"* Recalling how first contact went between medieval European explorers and indigenous Native Americans is typical anthropocentric thinking. Because humans may behave in this manner does not translate to an Extraterrestrial species that would start acting like a hoard of invading barbarians! They are alien by nature not human that means they will not think or necessarily behave like humans!!

Needless to say, having the news media and press covering such profoundly impactful events upon human civilization and referring to it in a giggle journalistic piece with a tongue in cheek attitude does nothing to instill scientific confident or credibility by referring to ETI as little green men!

It has already become a problem for scientists that have chosen to silo themselves away to work within the **black world of science**, just so that they can be involved in unique cutting edge alien technology while turning their backs on the rest of mankind when their intellect could be used to better serve all of mankind instead of just a few the elite.

The scientific and technological implications of extraterrestrial contact through electromagnetic radio waves would initially be most probably small. However, if the message contains a large amount of information, deciphering it could give humans access to a galactic heritage perhaps predating the human race itself, which may greatly advance our technology and science. A possible negative effect could be to demoralize research scientists as they come to know that what they are researching may already be known to another civilization.

For humans to respond to an ET message by politely asking them not to send us access to the *"Encyclopædia Galactica"* until they have reached a suitable level of technological maturity advancement to prevent harmful impacts of extraterrestrial technology is like put candy in front of children and telling them not to eat any of it! In fact, if we have accepted most of the evidence in this book that the military industrial complex already has it hands on alien technology busily reverse engineering most of it, then we must conclude that up to this point no harm has befallen mankind and some of it in small measure has already entered into the public domain.

Extraterrestrial technology could have profound impacts on the nature of human culture and civilization. Just as television provided a new outlet for a wide variety of political, religious, and social groups, and as the printing press made the Bible available to the common people of Europe, allowing them to interpret it for themselves, so an extraterrestrial technology might change humanity in ways not immediately apparent.

Harrison speculates that a knowledge of extraterrestrial technologies could increase the gap between scientific and cultural progress, leading to societal shock and an inability to compensate for negative effects of technology. He gives the example of improvements in agricultural technology during the **Industrial Revolution,** which displaced thousands of farm laborers until society could retrain them for jobs suited to the new social order. Contact with an Extraterrestrial civilization far more advanced than humanity could cause a much greater shock than the Industrial Revolution, or anything previously experienced by humanity.

The discovery of extraterrestrial intelligence would have various impacts on biology and astrobiology. The discovery of extraterrestrial life in any form, intelligent or non-intelligent, would give humanity greater insight into the nature of life on Earth and would improve the conception of how the tree of life is organized. Human biologists could learn about extraterrestrial biochemistry and observe how it differs from that found on Earth. This knowledge could help human civilization to learn which aspects of life are common throughout the universe and which are specific to Earth.
http://en.wikipedia.org/wiki/Potential_cultural_impact_of_extraterrestrial_contact

Under close examination, most of the objections to contacting aliens are feeble. After nearly 75 years when the first TV signals were sent out from the pre-war Nazi Germany at the Olympic games in their country, we can't suddenly decide to hide without rhyme or reason or maybe, because we thought better of our initial actions. In fact, we'd better hope that an advanced civilization has watched an episode of "America's Got Talent" and just vaporize us outright for a lack of talent and intelligence!

But there is a basic assumption by astronomers like Davies and Shostak that aliens would have the same kind of technology as humans - despite the extremely obvious fact that our microwave technology pulsed or otherwise is too damn slow for interstellar communications. Any neighbouring star that is a thousand light years away just to say, "Hello, how are you?" using radio telescope transmissions would take 1000 years to reach an intelligent life form. For them to respond back and say, "We are fine, how you're doing?" would take another 1000years! That's two thousand years, the length of time since Christ to the present time!

But we are led to believe that we're in good hands, if we should make contact because the Post-Detection Task group includes some of the planet's finest minds from the Konkoly Observatory, Hungary; the British Interplanetary Society; Leeds University, UK; the International Academy of Astronautics, Italy; the Max Planck Institut für Radioastronomie, German; Jodrell Bank Observatory, UK; the Australian Centre for Astrobiology, Australia; the Raman Research Institute, India; the Vatican Observatory; the University of California at Berkeley and SETI.
http://www.dailygalaxy.com/my_weblog/2010/03/if-et-calls-who-speaks-for-humanity-.html

Some scientists still believe that eventual contact with Extraterrestrials is at least a century away from reality but, they said the same thing in the mid to late 20th Century so, it appears that even they really don't know and are guessing like the rest of us. Even if we do find signs of alien life, and we're planning to talk to it, it is likely to take decades to learn and understand them from Earth via radio or lasers before open contact occurs.

What is most amazing is that most scientists like SETI scientists don't consider or respect the idea that ETs have already arrived on Earth and are currently engaged in contact and communications with human beings. They still believe that Earth scientist need to discover them as coming from some point out in space, rather than consider that they have already found us first thus, saving us the time and effort to go searching for them among the stars.

As long as astrophysicists and scientists believe that nothing can travel or exceed beyond the physical limits of the speed of light, how can they possibly accept the concept that ETI can travel the vast distances of space in very short periods of time? But, that is precisely the unknown factor that needs to be explored: *"how are they getting here?"*

Although Scientists will no doubt be called upon to share their considerable expertise and participate in the implications of first contact, scientists are, however, for the reasons outlined above, probably not the people that we want to represent humanity in a close encounter of the communication kind.

Religious Representatives – Converting the Alien Heathens to Christ or Allah

The confirmation of extraterrestrial intelligence could have a profound impact on religious doctrines, potentially causing theologians to reinterpret scriptures to accommodate the new discoveries. Surveys of religious leaders indicate that only a small percentage are concerned that the existence of extraterrestrial intelligence might fundamentally contradict the views of the adherents of their religion. **Gabriel Funes**, the chief astronomer of the **Vatican Observatory** and a papal adviser on science, has stated that the Roman Catholic Church would be likely to welcome extraterrestrial visitors warmly.

Contact with extraterrestrial intelligence would not be completely inconsequential for religion. Most non-religious people and a significant majority of religious people believe that the world could face a religious crisis, even if their own beliefs were unaffected. Contact with Extraterrestrial Intelligence would be most likely to *cause a problem for western religions, in particular, traditionalist Christianity, because of the geocentric nature of western faiths.* The discovery of extraterrestrial life would not contradict basic conceptions of God, however, and seeing that science has challenged established dogma in the past, for example with the theory of evolution and the teachings of **Giordano Bruno**, it is likely that existing religions will adapt similarly to the new circumstances. **(In other words, traditional religious bigotry and ignorance to learn something new will continue to thrive!)** In the view of **Paolo Musso** however, a global religious crisis would be unlikely even for **Abrahamic faiths**, as the studies by Musso and others have shown on Christianity, the most *"anthropocentric"* of all religions, see no conflict between that religion and the existence of extraterrestrial intelligence. **(How this is not possible is beyond this author's understanding, unless, the interpretation of the study results, fanatically maintains false Christian perceptions of reality)**. In addition, the cultural and religious values of extraterrestrial species would likely be shared over centuries if contact is to occur by radio, meaning that rather than causing a huge shock to humanity, such information would be viewed much as archaeologists and historians view ancient artifacts and texts. (Bold text added by author for emphasis).

Funes speculates that a decipherable message from extraterrestrial intelligence could initiate an interstellar exchange of knowledge in various disciplines, including whatever religions an Extraterrestrial civilization may host. **Billingham** further suggests that an extremely advanced

and friendly Extraterrestrial civilization might put an end to present-day religious conflicts and lead to greater religious toleration worldwide. On the other hand, **Jill Tarter** puts forward the view that contact with extraterrestrial intelligence might eliminate religion as we know it and introduce humanity to an all-encompassing faith. We need to first acknowledge all Earth-based religions that are recognized globally and consider if one or more have the ability to bring unification to humanity before considering an Extraterrestrial based religion which many scientists warn against at this time in our global development.
http://en.wikipedia.org/wiki/Potential_cultural_impact_of_extraterrestrial_contact

Approaching the question of first contact may be perceived as either an event that will unfold in a peaceful positive manner or in a disturbingly negative and hostile manner depending upon one's religious values which an ambassador from Earth judges the Extraterrestrial Intelligence that he is trying to communicate with. If our ambassador is a representative from one of the world's major religions like **Christian Catholicism** or from **Sunni or Shi'a Islam** or from the **Hindu, Buddhist** or **Jewish** persuasion then, the outcome may go in almost any direction based upon religious perception and understanding of what is the nature of the alien being and its race. Most religions have aspects within their teachings that make them open and receptive as long as it does not destabilize one's core beliefs. Unfortunately, fundamentalists in any religion will definitely have problems in maintaining their core beliefs particularly when *"the central figure of their faith is reduced to just another messenger of God!"*

Paolo Musso, a member of the **SETI Permanent Study Group of the International Academy of Astronautics (IAA)** and the **Pontifical Academy of Sciences**, took the view that extraterrestrial civilizations possess, like humans, a morality, driven not entirely by altruism but for individual benefit as well, thus leaving open the possibility that at least *some* extraterrestrial civilizations are hostile.

Quite frankly, none of the world's religious speak with any definitive understanding or authority as to what is extraterrestrial in nature other than, what may be an angel, demon, Jinn or a beast. Everything in the **Bible**, both in the new and old Testaments is based on interpretation, conjecture, and spiritual allegory. The same holds true, (what little there is), in the **Koran**, also in the scriptures of the **Vedas (Vedanta),** the **Upanishads**, and the **Bhagavad-Gita.** There are descriptions in some of these religious texts which do allude or describe flying machines and strange beasts but, these are interpretative and do not necessarily mean that they are Extraterrestrial in nature but rather, symbols of spiritual matters and conditions.

In modern times these particular passages from the world's holy texts, because of their unusual spiritual nature have been interpreted into modern context and terminology based upon modern day events and experiences which are similar to the Biblical, Koranic and Hindu texts.

The only religious writings and texts that have definitive statements about life on other planets is the world's youngest major religion, the **Baha'i Faith** which we have already discussed earlier in this book. A **Cosmic Diplomat** from the Baha'i Faith may actually be the only viable alternative as a representative whose world views are all encompassing, whose belief system includes the concepts of other life in the universe and someone who truly speaks for all mankind without any hidden agendas or personal gain for himself or his religion!

Religion is a highly competitive spiritual sport not for the spiritually faint-hearted or the weak morally minded! Without, becoming prejudiced or spitefully aggressive but, by mere history alone, **Christianity** and **Islam** are by far the two most competitive warring religions on the planet today judging also, by the numbers of their followers who claim distinction in these two faiths. This not to say, that other major religions don't have some warring conflicts with one another, but these two religions stand out more than any other belief systems on this planet. In many ways, their somewhat intolerant behaviour and attitudes stem from their follower's misunderstandings of their own religious teachings and not from the actual teachings as originally promulgated by their divine messengers or prophets!

The real issue for every religion with perhaps, the exception of the Baha'i Faith, will be the realization that the Almighty Creator of the Universe has more than one prophet or Manifestation of God appearing on Earth at different times. The eye-opening shock will also come from the knowledge that is eventually communicated to humanity is that God's **Divine Word** or His **Holy Messengers** are present everywhere else, throughout the universe and on every planet, teaching and providing spiritual guidance to all intelligent beings!!! Our only duty in response to these Divine Messengers is recognition of who they are and obedience to their teachings, a divine teaching that is imparted in every age, throughout our history on this Earth but, which is blindly ignored by a heedless humanity!

At best, religious leaders will be unable to convert any ET to Christian Catholicism and letting them know that **Christ** also, "died for their sins"; nor will Islam persuade them to follow in the footsteps of the "Seal of the Prophets", **Muhammad.** Why? Because, all ETs already will have their own unique spiritual understanding of God, life, and the universe. The best that any religious leader will be able to do is to share the similarities in Earth's spiritual diversity with that understanding of the ETI spiritual insights. It is more probable that they will be able to teach us a thing or two in understanding the true nature of the supreme Creator and the universe that we all share.

Ted Peters, a professor of systematic theology at the Pacific Lutheran Theological Seminary in California, considered what might happen to the world's religions in the event of ET making contact. Conventional wisdom suggests that terrestrial religion would collapse if the existence of extraterrestrial intelligence (ETI) were confirmed, he wrote.

"Because our religious traditions formulated their key beliefs within an ancient world view now out of date, would shocking new knowledge dislodge our pre-modern dogmas? Are religious believers Earth-centric, so that contact with ET would de-centre and marginalize our sense of self-importance? Do our traditional religions rank us human beings on top of life's hierarchy, so if we meet ETI who are smarter than us will we lose our superior rank? If we are created in God's image, as the biblical traditions teach, will we have to share that divine image with our new neighbours?"

His conclusion, however, is that faith in Earth's major religions would survive intact. "Theologians will not find themselves out of a job. In fact, theologians might relish the new

486

challenges to reformulate classical religious commitments in light of the new and wider vision of God's creation."

"Traditional theologians must then become **astrotheologians** ... What I forecast is this: contact with extraterrestrial intelligence will expand the existing religious vision that all of creation – including the 13.7bn-year history of the universe replete with all of God's creatures – is the gift of a loving and gracious God," he speculated.
http://www.theguardian.com/science/2011/jan/10/earth-close-encounter-aliens-extraterrestrials

Will we want a religious leader to act as a spokesperson on behalf of mankind? Probably not, at least not at this time, as we are not all spiritually united and religious intolerance and prejudices still exist throughout the world. The world's major religions have yet to recognize the concept of **"Progressive Revelation"**, that all religions are one, from one source coming at various times in mankind's development which will allow them to come together in harmony and acceptance of each other!

Doctors, Teachers and Artistic Representatives - The Common Folk as ET Ambassadors

Let's consider teachers, professors, caregivers and medical doctors, musicians, actors, artists, poets and those of sublime thought as spokespersons who could engage ETI on behalf of humanity. These are people who are both smart, knowledgeable, creative, often intuitive, insightful, open-minded, receptive, perceptive and responsive. They are talented in related disciplines and highly trained in specific areas of their profession and in addition, most of them have excellent social and communicative skills allowing them to interact with a variety of people from every sector of society. Best of all these people are the some of the best-educated people in society and they are a natural bridge between the multiple economic and social levels in society.

These are the educators of mankind, the ones that inspire society to look beyond the everyday conventionalities of life, to reach for what seems to be impossible, they are the health-care givers and professionals who genuinely care about people, their children, and the elderly. They are people persons who will often sacrifice their time and comfort in selfless acts. They ensure that society advances ever forward. They are the "salt of the Earth!"

When the Extraterrestrial Intelligence arrive and communicate with us, it could very well be that these are the people that may be able to set the tone of communications in a comfortable, peaceful atmosphere of friendship. It has been suggested that collaboration with such an ET civilization could initially be in the arts and "humanities" before moving to the hard sciences. Artists and poets may not necessarily be the primary diplomats but, they may still be involved in the collaboration process by given artistic expression to the proceedings in the form of music, plays or theatre, dance or displays of artwork.

In the final analysis, Earth's **Universal Ambassador** or **Cosmic Diplomat** may be a collaboration of people from all walks of life, from the professional to the artist and the common labourer, from the most intelligent scientist and academician to the young inquiring minds of the youth from the military officer to the mother who raise the children. Their backgrounds, every one of them may be essential in first contact speaking with one voice.

Most people involved in this debate have supported a collective message, believing that the human end of the communications process should speak for all humankind, not for a particular political or occupational subdivision or faction of society. Astronomer, **Dr. Frank Drake**, for one, believed that any reply should be crafted on a ***worldwide basis***.

Others argue that anyone with access to a transmitter should have the right to send separate messages. Some see this as freedom of speech; others believe that many individual messages would more correctly reflect human diversity.

Having humankind speak with many voices may be congruent with individual rights and cultural diversity, but may be bad policy. Imagine yourself in the place of extraterrestrials who receive thousands of uncoordinated messages from Earth. How could you conduct a rational dialogue with such mixed signals? Which ones matter the most? Which most accurately reflect human policy? Who would you believe those humans who seek an exchange of scientific information, those who want to convert you to the true faith, or those who announce their intent to exterminate you?

Donald Tarter proposed that we send a brief initial reply that says "we have received your message and will communicate with you in the near future." The intent would be to establish an "official" channel of communication. If this signal were sent with the most powerful transmitter available, the power and priority of this transmission would differentiate it from others. Contact with Alien Civilizations: *Our Hopes and Fears about Encountering Extraterrestrials* by Michael A. G. Michaud; 2007; published by Copernicus Books; New York, NY, United States; ISBN 978-0-387-28598-6. Archived from the original on 24 December 2012.

In debating the merits and the qualifications as to who humanity should select to represent us in first contact with an Extraterrestrial civilization from the stars, the whole exercise has been rendered at this time, purely academic in nature. We seem to have forgotten or have overlooked in this equation of human cosmic diplomats, the fact that the ***Extraterrestrial Intelligences may have ALREADY PRE-SELECTED the person or persons they wish to act as Earth's first Ambassadors to the Universe!*** As we have discovered, there already exists **Cosmic Ambassadors** representing the Earth, chosen not by humanity but, by Extraterrestrial Intelligence, themselves!

By a priori, the decision has by default been taken out of the hands of mankind at this time based upon our inability to be unified as a global commonwealth and to decide collectively on a course of action!

How is that possible you say?

Since the early '40s when people first reported strange unidentified flying objects in the skies, the ETs have had the capability to access the minds of an individual singularly or collectively

488

with their technology previously described as **Conscious Assisted Technology (CAT)** or **Technology Assisted Consciousness (TAC).**

In much the same way we turn a dial or push a button (or with the use of the remote control) on a radio or a television set to move from one station or channel to another, ETs have the same ability to tune from one individual's conscious thoughts to another. If we accept this premise to be true, then it goes a long way in explaining how their spacecraft can out-maneuver our military jets and be in other places before they can react. To be able to bank simultaneously when a jet banks or dives or ascends or accelerates, most times always beyond the military jet reach.

It also explains how they may have already pre-selected individual experiencers to speak to and they, in turn, speak for our planet. In which case, there is really no such thing as a *"random UFO sighting"* or a *"random ETI encounter"! All contact experiences are carefully staged events, guided by the often unseen hands of Extraterrestrial intelligences!!!*

Not surprisingly, ETI have been *eavesdropping* on our telecommunications, when we use our computers, our cell phones, our TV broadcasting signals and yes, they can also *monitor our thoughts!*

This might be a good time for all of us to clean out our minds of all that fluff we've accumulated over the years! The new science that has appeared on the horizon is the science of consciousness; we should start to prepare ourselves for the possibility of telepathic communications in society and with ETI!

Keep in mind that ETI have been engaging in contact with humanity on multiple levels throughout the history of mankind but, more so in recent times they have escalated that contact as they observe that humanity is quickly evolving technologically, particularly towards space exploration and also, with the development of weapons of mass destruction.

ETI have been contacting us through their appearance in flying saucers and other strange craft which essentially is a statement of here we are, we exist, you are not alone in the universe! They have communicated to us through crop circles in cereal crop particularly in Britain and other places in Europe and now in many places around the world. Britain may have been the first chosen area for this type of contact - communications because of its ancient historic influence, even in modern times with its former global empire that was the largest in history up until the last remnants of that empire fell in 1945 to 1997. http://en.wikipedia.org/wiki/British_Empire

Crop circles are one form of communication modality engaging us on multiple levels of higher learning, in math, physics, geometry, biology, astronomy, music, even in horticulture!

ETI have communicated with people randomly at first in open contact, but in secluded or remote locations imparting messages of warning, admonition, and hope. These engagements were non-hostile demonstrating the intentions of the Extraterrestrial agenda to be somewhat friendly. With military aggression from many powerful nations, the ETs have moved back from their openness of contact and have resorted to more self-preservation actions, however, contact and communications continue. Their methods had to be refined to transdimensional engagement for the time being, as there have been losses of ET spacecraft and capture of ETI due to aggressive military tracking, targeting and shoot downs.

So, where does this leave us, if the ETs have already decided to speak to pre-selected individuals of humanity, thus bypassing governments, militaries, intelligence officials, the scientific and religious communities, even the UN? If first contact by ETI is pre-determined, are there still elements of randomness in which communications can still take place giving humans some measure of control of the contact engagement?

The evidence strongly indicates there is!!!

CHAPTER 119

THE ROSETTA STONE OF EXTRATERRESTRIAL COMMUNICATIONS

As it turns out, some very clever people have discovered the **"Rosetta Stone of Extraterrestrial Communications "**! This is a method by which humans may make contact and establish communications with Extraterrestrial civilizations at almost any time of their choosing! This has created all sorts of headaches and upset to the carefully laid plans within the US Military and the covert organization of MJ 12, causing them to occasionally lash out irrationally to those who would dare cross into their exclusive domain!

Now, it has become possible for anyone in society to be able to make contact with Extraterrestrial Intelligences directly, like picking up a phone and simply calling, thereby, bypassing the tight national security controls place upon this subject by the authorities of officialdom and the Military Industrial Complex!

When you are able to out think or out maneuver a tightly controlled situation, when you are no longer willing to be passive, when you no longer want to *"play ball in their game"*, then the rules of the game change and they change forever! When you try to tighten your grip upon something to control it, the more it will slip through your hands and the more you realize that you were never really in control. Those who try to control other people's lives simply do not have any control in their own lives!

Extraterrestrial intelligences are one of those proponents that refuse to be controlled or manipulated by the powers that be on Earth. Now, there are smart people who have figured out how to contact and communicate with ETs that are no longer accepting the rules of engagement that MJ-12 has been trying to force upon them and the public in all matters of UFOs and ETI. This has caused things within the covert MJ-12 group to be in free-fall as they try to figure out how to get things back under their control. At least 70% of Majestic believes that the veil of secrecy should be lifted from the UFO/ET subject, unfortunately; the remaining 30% which is the *"old guard"* refuse to budge from their entrenched positions of cover up and suppression. They do not believe that power should be shared with the public.

In the forefront of this people's movement for a **Citizen Initiative** to establish ET contact and communications, is none other than **Dr. Steven M. Greer**, the one who has discovered the **"Rosetta Stone of ET Communications"**!

His discovery was predicted to occur within the military and intelligence think tanks and when he publicly came out with it, he raised major red flags within those communities. When those departments or agencies approach him about his contact protocols and asked him (paraphrasing):

"What the hell do you think you are doing and who the hell gave you permission?

Greer replied that *"I'm just a country doctor and quite frankly, I don't need your permission as this isn't under government control and there ain't a damn thing you can do about it"*.

The proverbial "cat was already out of the bag"!

Greer has since gone on to ensure that human-initiated ET contact and communications protocols, **CE-5 (Close Encounters of the Fifth kind)** message was spread both nationally and globally. He has successfully endeavoured to shine a large public spotlight on the UFO/ETI subject and has affirmed repeatedly that ET contact was not only possible but, reproducible by anyone who arises to employ his CE-5 protocols. This author can vouch that the methods are reproducible and that they work in establishing contact and communications with ETI!!!

Earlier, we discussed the successful work of **Sixto La Paz Wells** of Lima, Peru, who has been using certain contact methods and protocols to communicate with ETI where they would show up in a particular area. In this section, we will look at the UFO/ET contact work of **Dr. Steven Greer** who has developed similar reproducible methods and protocols that he teaches worldwide known globally as the **CE-5 Initiative** organized under the banner of **CSETI (Center for the Study of Extraterrestrial Intelligences)** .

Like everything in life, there correct ways to do things and there are incorrect ways of doing the same thing and contact with ETs is no different. First and foremost is having the right mental state and proper moral attitude toward diplomatic relations which is essentially the establishment of a mutually beneficial friendship between two parties.

The Don't Panic Approach to Establishing ET Contact and Communications

Let's look at one method of contact protocol that leaves considerable room for improvement which was developed by **Bill Fawcett**, an author, editor, a professor, teacher, corporate executive, and college dean. His entire life has been spent in the creative fields and managing other creative individuals. He is one of the founders of Mayfair Games, a board, and role-plays gaming company. As an author, Fawcett has written or coauthored over a dozen books and dozens of articles and short stories. As a book packager, a person who prepares series of books from concept to production for major publishers, his company, Bill Fawcett & Associates, has packaged more than 250 titles for virtually every major publisher. He founded, and later sold, what is now the largest hobby shop in Northern Illinois.

Fawcett's first commercial writing appeared as articles in the **Dragon magazine** and include some of the earliest appearances of classes and monster types for **Dungeons & Dragons**. With Mayfair Games he created, wrote, and edited many of the Role Aides role-playing game modules and supplements released in the 1970s and 1980s. During this period, he also designed almost a dozen board games, including several **Charles Roberts Award** (gaming's Emmy) winners, such as Empire.

By gathering information from top UFO researchers and acclaimed writers, editor Bill Fawcett has created the only serious handbook for locating and communicating with interplanetary visitors (at least for its time). From examining the levels of contact to detailing the potential hazards involved in a meeting, this accessible guide is invaluable when you come face-to-face with the unexpected.

In his book, "Making Contact, A Serious Handbook for Locating and Communicating with Extraterrestrials", reads initially like a fear-based guide to **"The Hitchhikers Guide to the Galaxy"... _DON' PANIC!_**

Each set of protocols depends upon the type of UFO/ET encounter that you may experience. The book states it is a "serious handbook", it's protocols appear guanine, but are untried and untested as the author gives no examples of people using his protocols or anything remotely similar to them in their own UFO/ ET encounters, unlike **Dr. Steven Greer's CE-5 protocols** which are tested, proven and reproducible.

The **Military Industrial Complex** probably love this book, mainly because it's negative fear-base approach that basically is designed to scare the common person away from engaging in contact with Extraterrestrials. At the outset of the book, there are certain protocols that come into play depending upon the type of close encounter that one experiences whether that be a **Close Encounter of the First Kind (CE-1)**, a relatively safe sighting encounter up to a **Close Encounter of the Fourth Kind (CE-4)**, a possible abduction event in which you have no control of the situation! Making Contact, A Serious Handbook for Locating and Communicating with Extraterrestrials" edited by Bill Fawcett ; 1997; published by William morrow and Co. Inc. New York, NY. USA; ISBN 0-688-14486-1

In almost every type of encounter, it is strongly suggested that the observer-witness/participant run away from the area of possible escalating events **_TO AVOID CAPTURE!_** Now, allowances are made for very brave individuals, who may want to participate in a close encounter and therefore, if one can retain a certain amount of calmness then, one may be able to control the outcome of unfolding events, however, according to these protocols, there's no guarantee! This is not the author's fault as, like so many people in Ufology and the general public, he has bought into and drank the **_CE-4 Abduction "Kool-Aide"_** manufactured and propagated by the M.I.C. therefore, his book reflects this mindset.

In a CE-1or CE-2 event, he suggests you **"do not panic**!" "**_Stay put._**" "**_Don't run away or try to hide_** except to protect yourself from debris and other flying objects" (UFOs?)!"

In a CE-3 event, Fawcett again reiterates, **_"DON'T PANIC!"_** Remain **_CALM!_** If ETs approach you, in this case: **_"RUN AWAY!" "DON'T WAIT", "DON'T PERMIT YOURSELF to be CAPTURED by the BEINGS!" "RUN AWAY!"_** "Unfortunately, this may be less up to you and your response and more up to **_"THEM"!_**

It seems we are getting a mixed message here from Fawcett! Do we stay or do we run away when encountering a UFO or an ET being? His advice in a CE-4 situation doesn't really get any better, as this close encounter according to most Ufologists is an abduction event!

In a CE-4, it's pretty much a lost cause at this point in staying calm and **_NOT PANICKING,_** since you've more than likely will be abducted and will probably be going through an **_alien medical procedure that employs an ANAL PROBING DEVICE or the removal of DNA and reproduction specimens!!!_**

Needless to say, "DON'T PANIC" will seem like useless advice at this point so, go ahead and let it all out in a good screaming session!!! You owe it to yourself for trying to stay so F#@%ing CALM!!!

You'll have the reassuring knowledge that after your unique ordeal is over that you will have undergone a ***MAJOR SHIFT*** in your ***DEEPEST BELIEFS*** and that ***ATTITUDE ADJUSTMENT*** that your friends have been pleading for you to get ***WILL FINALLY have TAKEN PERMANENT ROOT*** in your ***NEW AND IMPROVED PSYCHE*** and your ***PERSONALITY WILL BE IRREVOCABLY ALTERED! YOU WILL BE A NEW PERSON!***

But, ***DON'T WORRY!*** Those humans in the white coats will take good care of you as they try to get you to come down off the walls as you do your impersonation of a cat climbing up the curtains!

Fawcett suggests you keep ***a written record*** of those bizarre experiences but, how you are going to do that when you've got that funny, little cold thing up your butt? But who knows, one of those little alien beings may have ***a writing instrument*** that you can burrow while they continue their medical procedures on you!

This written account of events may be useful in "helping you work through your ***TRAUMATIC AFTERMATH.*** It may require ***MONTHS or YEARS*** of living with ***TURMOIL INSIDE YOURSELF!*** Sometimes you may even come to think that you are ***LOSING YOUR MIND!!!!!"*** **Making Contact, A Serious Handbook for Locating and Communicating with Extraterrestrials" edited by Bill Fawcett; 1997; published by William morrow and Co. Inc. New York, NY. USA; ISBN 0-688-14486-1**

"Don't Panic" seems to be the mantra and the number one rule that Fawcett subscribes to and this, in reality, is good advice but, do we need the scare tactics and the scary scenarios of abduction? It strongly indicates that Fawcett buys into the alien abduction scenarios as real Extraterrestrial events which occur to hapless victims worldwide. He doesn't consider that it may be in reality, a **MILAB (Military Abduction)** experience perpetrated by humans upon humans, as a psychological warfare program with a hidden covert agenda. But, it is his book and his perceptions, after all.

To be fair, Fawcett does give useful advice in recording UFO and ET encounters and he sites many corroborating accounts to support his hypothesis and a grocery list of items (lengthy piece of string, magnets, specific number of coins in various denominations, paper, pencil or pen, flashlight and a good imagination) in following a set of procedures utilizing the items to communicate with the ETs who may not verbally communicate or if there is a language barrier of understanding. Sign language may be the only effective method left in order to communicate ideas and this small grocery list may come in handy assuming you are prepared for such a human – ETI encounter!

Fawcett says that when you see a distant UFO which appears to be landing and an ET is approaching you, it may be best to simply run away in self-preservation, rather than engaging them in any friendly welcome or meaningful communications. However, Fawcett allows for that

494

possibility and thus, his grocery list of items and his set of protocols of questions and answers to achieve some degree of mutual beneficial communications.

Author's Rant: I can tell you from personal experiences that communications with Extraterrestrial Intelligences is a lot easier, very friendly, mutually satisfying and best of all, you're in control. It is better than having a negative mindset or carrying a pocket or a purse full of specific items as outlined by Fawcett in his book! But, his book is one methodology of how communications with ETs may occur.

In **Richard F. Haines'** book: "CE-5 Close Encounters of the Fifth Kind" points to an overlooked factor in alleged alien aggression against humans which few writers or Ufologists even consider and that is whether UFO or alien beings were provoked into a response by humans!

Haines quotes from various sources to illustrate his point of view that includes cases of personal; injury allegedly caused by UFOs and/or Extraterrestrials. He states *"If our visitors were provoked by acts of human aggression and **still did not react in kind, it suggests that the alleged intelligence behind the UFO is highly self-controlled."***

Haines says that J. U. Pereira's 1974 study of 198 cases involving human-alien contact, he found:

Twenty-four cases (12.2%**)** in which *aliens approached humans*, another
twenty-seven (13.6%**)** in which the *aliens departed*. The other cases did not make any distinction in this regard.

Of the **twenty-seven cases** which Pereira classed as being ***"hostile":***
twelve (**44.4%**) involved *alien aggression shown towards humans*,
eight (29.6%**)** involved *human aggression toward the alien(s)*.
Three cases (11.1%**)** involved an *actual fight of some kind*, and another
four (14.8%**)** involved *some kind of "accidental violence."*

Within the **thirty-six cases** classed as ***"friendly behavior":***
seven cases (**19.4%**) the *alien smiled*, in
six cases (16.6%**),** *put his hand on the shoulder of the human as a friendly gesture*, in
Twenty-two cases (61.1%**)** *extended his hand or gestured toward the human*, in and
one case *involved conversation*.
"CE-5 Close Encounters of the Fifth Kind" by Richard F. Haines ; 1999; published by Sourcebooks, Inc.; Naperville, Illinois, USA; ISBN 1-57071-427-4

Most importantly, in the percentages of "hostile" cases as Haines points out, how many aliens were provoked? To which could be added, how many cases were based upon a personal interpretation of what constitutes violence, as some people view violence as a smack on the face, where most people would consider a punch to the face as violence and some consider any intention of aggression as violence. It is a matter of perspective based upon individual interpretation.

There are the New age concepts of ET contact and communication all with varying degrees of success which we will not get into at this point but suffice it to say, that most of these new age groups take their lead and example from the contact protocol work of Dr. Greer.

CSETI (Center for the Study of Extraterrestrial Intelligence)

Like many people during the '50s and '60s who were seeing unusual craft in the skies, some people in their youth *(like this author)* were experiencing actual ETI contact, who would on occasion engage in communications with ET beings. Dr. Steven M. Greer was one such individual when back in the early 1960s, at the age of eight or nine, he and some neighborhood boys saw a silver, disc-shaped, windowless, seamless craft that hovered, silently, then simply vanished.

It was not an airplane or a helicopter but an ETV (Extraterrestrial Vehicle), a name often used by the **National Security Agency** also commonly known as a UFO. Steven ran back home to tell his parents about what he and his friends witnessed and his parents in typical response was, "That's very nice" and ignored it, but Steven knew what he'd seen, and it was life-changing.

The next few weeks re-enforced his experience with lucid dreams and night encounters with beings who were not from this Earth. Believing it was all a natural part of his first experience, Greer felt that the ETs were focused on instilling him with an awareness and acceptance of things beyond the world, unprejudiced with the youthful view of innocence.

In his mid-teens in high school, young Steven began studying the Vedas and the Sanskrit writings learning prayer and meditation not participating in the drug and alcohol pop craze of the '60s and '70s. At 17 years old as a cyclist he developed a septic infection in his leg after a 200-mile bicycle ride which lead to a near-death, out of body experience. This, in turn, led to God consciousness awareness and the state of perfect oneness on unbounded Mind which later became useful when he became a teacher of **TM (Transcendental Meditation).** Greer was now able to enter into cosmic awareness through the clear, pure channel of transcendental consciousness – full *"samadhi"* at will!

On a clear late afternoon in October 1973 in Steven's 18th year, he walked up to the top of the 5000 foot Mtn. Rich overlooking the township of Boone to see the sunset and to meditate. At the end of his meditation while walking back down the mountain road he senses a presence and saw a luminous glow off to his right. It was an extraterrestrial biological life form which reached out and touched him firmly on the shoulder. He soon found himself transported onto a translucent craft out in space which Greer described as if there was nothing around him but space; it was as if the craft was made of fiber optics.

Greer said that there were small ETs three or four feet tall with deer-like eyes and they want to contact a human being experiencing cosmic consciousness and to participate with them. Greer taught them what that state was like for humans and they meditated together. This was an encounter unlike any of the commonly circulated accounts of extraterrestrial contacts. The meditation session with the ETs was a non-local experience beyond all known realms of space, time and relativity.

496

It was during his time with the ETs that they co-created the human communication contact protocols, the birth of the **CE-5 (Close Encounter of the Fifth Kind) Initiative**. This human initiated contact involves the modalities of sound, light and non-local consciousness and directed coherent thought to communicate with Extraterrestrials and interface with their electronic devices.

At that point, it became clear to Greer that humans needed to move away from a mutually assured destruction and into a peaceful civilization that could co-exist with itself and with space in harmony. It was clear to Greer that he needed to teach his fellow humans to do the same. He suddenly found himself back in full awareness on the gravel road near the top of the mountain near the fire tower. The sky had become a starry night, it was late, about 1:00 AM; he had been away for about 3 or 4 hours and was feeling elated by the experience!

At that moment, Steven had an epiphany from his time with the ETs: "the conscious mind we are awake with at this moment is the same as that of the divine Being, *and of all beings."*

This epiphany was the same as **Erwin Schrodinger's proposition** that *the total number of minds in the universe was one.* "There is one conscious mind and we are it. So there are only one people in the universe and we are they. No "alien" or human, just an unbroken, perfect, seamless conscious life in the universe and we are all a part of it."

This was the occasion when the initial CSETI concept of one universe and one people was conceived. **"Hidden Truth, Forbidden Knowledge: It's time for You to Know" by Steven M. Greer M.D.; 2006; published by Crossing Point, Inc. Publication; Crozet, Va., USA; ISBN 0-9673238-2-7**

Every free moment that Greer, as a medical student, not spent in study or when not with his family was spent honing his meditative skills, even at night just before he would fall asleep. He would sometimes have interactive connections with ETs during his lucid dreaming or in higher states of consciousness, to the point that he could communicate in the language of the ETs! The ETs would at times show up even in the daytime searching for Greer's whereabouts, and in encountering other people by mistake, a somewhat embarrassing situation for Greer as the ET presence was a verification that his coherent thoughts and higher consciousness were been received and heard.

After medical school, Greer studied T M at the Maharishi International University in Iowa where his meditative skill was further refined and expanded. He learned the psychic skill often reported in the '70s by the news media in association with transcendental meditation were true that through training people could perceive the future through dreams or lucid states of awareness, that physical bodies could levitate, or dematerialize or re-appear in another location (bi-location and teleportation). All abilities that were formally ridiculed could be taught and through training, anyone could achieve higher states of interaction with the environment around them.

Even, such abilities as remote viewing and remote sensing as has been previously discussed with the military's **SRI remote viewing program "Stargate"** was also possible. It was a time of great experimentation in higher consciousness and in **TM** abilities like levitation for Greer. On two separate occasions with a witness in attendance with Greer, he beckoned ETs to show up at a retreat in the Maritime Alps of France, whereupon a large tetrahedral craft appeared. The second time was in North Carolina, another blue-white ET craft appeared near a home where he and a friend were staying which unnerved his friend. The occupants of the craft sensing his friend's state of mind left. The next morning the local newspapers wrote that police officers in a helicopter had been buzzed by UFOs. One flew off into the direction of Greer's location and the other that had buzzed the helicopter simply disappeared; both objects were picked up on radar.

By this point in time, it was becoming obvious to Dr. Greer that remote vectoring through meditation of ET craft was not only possible but, the responding ETs would also want to interact with people! By now, a plan of action was crystallizing in Greer's mind that eventually would result in the formation of the **CSETI Initiative.** "Hidden Truth, Forbidden Knowledge: It's time for You to Know" by Steven M. Greer M.D.; 2006; published by Crossing Point, Inc. Publication; Crozet, Va., USA; ISBN 0-9673238-2-7

A necessary point of explanation needs to be made here: **CSETI (Center for the Study of Extraterrestrial Intelligence)** a small public UFO organization is in no way affiliated with the government organization known as **SETI (The Search for Extraterrestrial Intelligence).**

SETI operates on a billion dollar budget with acres upon acres of radio telescopic equipment, paid worldwide travel and is routinely financed by the government, yearly. Whereas CSETI operates on a "shoe-string" budget through public funding, members bring their equipment (camp chair, binoculars, camera, flashlights, stereo ghettoblaster, walkie-talkie, laser pointers, radar and magnetometer detectors) and travel around the world is at each member's own expense! SETI usually works indoors in a temperature controlled environment and CSETI works out in open air remote locations under the cool starry night skies!

SETI is ***still searching*** for signs of distant ET communications (**WOW!**) and CSETI ***knows that the ETs are already here*** and ***routinely communicates with them*** through human-initiated CE-5 protocols!!!

CSETI History - Over Two Decades of Disclosing the ET Presence

1990

By 1990, Greer learned that **NATO** and the Belgian Air Force were tracking an enormous triangular craft -- about 800 feet in diameter -- and it reignited his interest in UFOs. He decided to go to Belgium and actually saw the F-16 fighter jet radar tape of a massive UFO hovering, then accelerating at several thousand miles per hour -- way outside the envelope of what any man-made object could do.

During 1990, Dr. Steven Greer living in Asheville, North Carolina at the time, establishes CSETI, calling it a scientific and visionary project of 60 to 100 years to make assessments about ETI and, ultimately, to establish peaceful interplanetary relationships.

498

Later in the year, small **Working Groups** are established during a University of North Carolina conference after hearing Greer speak of his visionary plan in the next evolutionary step in UFO and ETI research. A temporary Executive Committee is also formed along with an **RMIT (Rapid Mobilize Investigation Team)** to respond to any major UFO "hotspots" anywhere in the world.

1991

Working Group Protocols are established in April for real-time research and diplomatic projects. After a thorough review of existing evidence about so-called **Unidentified Flying Objects (UFOs).** CSETI believes that the Earth is being regularly visited by Extraterrestrial Intelligences – with non-hostile intent – in Extraterrestrial spacecraft (aka. **Extraterrestrial Vehicles (ETV)).** The ultimate goal of the CSETI organization is to establish mutually sustainable relationships with ETI through peaceful, systematic means and the good will of global citizens.

To maintain and foster the growth of its bilateral mission between all human beings and any and all ETs, CSETI has developed a system of research-driven communication protocols and public education programs. The principle research platform is the CE-5 (Close encounters of the Fifth Kind) Initiative.

On March 15, a CE-5 event occurs in Dandridge, Tennessee, near the site of a CSETI research operation. The incident occurred during multiple sightings in that area, including one UFO appearance which the local sheriff videotaped. These UFO activities culminated in May with the reported landing of one UFO which left a circular imprint approximately 28 feet in diameter. The first out of state CSETI Working Group is formed in Denver. Subsequent groups are also established in other American, Canadian and European locations.

1992

February 9, 1992, in Belgium, a CE-5 occurs with a CSETI team and Belgium authorities and scientists present during a wave of UFO activity. More than 3,500 official sightings have occurred in this European capital since November 1989.

A dramatic CE-5 occurs in Gulf Breeze, Florida, initiated by CSETI. Three UFOs break into a triangulated pattern, mimicking the light formation projected into the sky by the local Working Group. One UFO responds in kind to repeated light flashes.

The event is witnessed from seven different locations by more than fifty people and captured on video, audio, and photographic equipment. http://www.siriusdisclosure.com/ce-5-initiative/past-events/cseti-history/

May. In Denver, Dr. Greer addresses the **International Association of New Science (IANS)** co-founded by former Apollo astronaut-scientist **Brian O'Leary**.

June 28. Laramie, Wyoming is the site of another CE-5, during which two UFOs vectored to within 2-4 miles of a CSETI Working Group.

July 20-30. Southern England. Alton Barnes region. A CSETI RMIT team joins **Colin Andrews' Crop Circle Phenomena Research (CPR)** International to study a possible link between UFO activity and the crop circle creations. One of the most significant events in the history of Ufology transpires. Using the **Contact Trilogy**, the two teams visualize and illuminate in the sky an equilateral triangle with circles at each point. The next day such a formation is found within a direct line from the CSETI-CPR research site. During the ten-day period, many anomalous lights are seen. An unusual CE-5 is multiply witnessed by Greer and others on July 27 when a structured craft made a near landing only 400 yards from the team.

During this time, a BBC recording crew picked up a series of beeping tones [from a UFO]. We recorded that, and we fed it back out. Greer says: *"For all we know, it's their trash compactor, but we figure that if they see humans sending their signal back to them, they'll identify it as communication!" "In Pensacola, Florida, we signaled [a UFO craft] with high- powered lights and lasers -- and they signaled back to us! We have this on videotape. We go to areas where these objects have been seen; so we're not just sitting in our backyard."*

In the Autumn, CSETI formally organizes its Executive Council and opens its main office headquarters in Asheville, NC. Dr. Greer semi-retires from his hospital position to devote more time to CSETI's operations and ongoing research.

1993

A lot of events and activity begin to unfold for the new year. In January, CSETI begins its official media campaign and public education programs to begin informational briefings about CSETI activities and ETI research.

Later in the month and the beginning of January, a CSETI team is dispatched to Mexico's volcano zone (Popocatepetl) where several UFOs are spotted during day and night time, including a large triangular craft that interacts with the team in early February by signaling and making a close approach near to the ground.

CSETI is now moving into high gear and conducts its first extensive Working Group seminar in Arizona during the Spring month of March.

By July, Dr. Greer returns to England and addresses the International MENSA Society at Cambridge University.

Come October, **Project Starlight** – a public education forum about the disclosure of extraterrestrial life is launched with a core strategy team. Dr. Greer starts identifying and contacting high profile "insiders" in various levels of government, science, military, intelligence, foreign government and military officials, astronauts and cosmonauts, etc. that will eventually become a part of the **Disclosure Project**.

1994

The following year of 1994 CSETI's RMIT team returns to Mexico's volcano zone and interacts

with several anomalous objects over a week's period and begins a liaison with citizens in the area to form a CSETI Working Group. There are subsequent **RMITs** trips to England during the summer, New Mexico in the fall and Monterrey, Mexico in December, all yielding results, including a dramatic 2.25 hour interaction between a craft and the team.

In October, Dr. Greer begins a very noticeable profile of his CSETI Initiative when he appears on Larry King Live Special on TNT and later learns that his interview was the most popular show Larry King has ever aired.

Later in the year before the Christmas season, Dr. Greer addresses a group at the United Nations.

1995

Events are rapidly unfolding on multiple levels for CSETI and Dr. Greer personally as the UFO phenomenon escalates worldwide much to the chagrin of the **M.I. C./MJ-12** .January.

Once again, an RMIT team journeys again to the volcano zone of Mexico to conduct a lecture and training workshop for the people of the area. Several significant daylight and night time sightings of craft take place.

Dr. Greer gives lectures at the **NASA Welcome Center** and Virginia Air and Space Museum in Norfolk, Virginia. NASA officials are extremely amazed at the extent of Greer's insights and experiences.

In April at Denver, Colorado, Dr. Greer conducts an intensive week-long course on the Foundations for Planetary and Interplanetary Unity which is later published in one of his books.

In late May at Crestone, Colorado CSETI holds its first Ambassadors to the Universe retreat and training. It is a week-long program where interested citizens spend time learning techniques of meditation, remote viewing, coherent thought sequencing and spending days in the San Luis Valley and nights under the stars making contact with ET Intelligences and their craft.

An RMIT is activated to Southern England in July. During that time, multiple witnesses observe **'sky quakes"** (which are reported on in the worldwide press with national space agencies unable to explain), and several sightings of possible craft (a phenomena that continues to this day globally). A joint project with researchers from Germany and including participants from many nations results in a small, brilliantly lit craft appearing within 50 feet of three CSETI team members. Dr. Greer personally briefs a former head of the British military forces (**Admiral Lord Hill-Norton**) and in November, Dr. Greer addresses the International UFO Congress in Mesquite, Nevada. http://www.siriusdisclosure.com/ce-5-initiative/past-events/cseti-history/

1996

1996 is marked by many historic UFO encounters and ET interaction for CSETI teams not only in the US but, with continuous activity down in Mexico, in Canada, England and Scotland with

reports appear on several national network television programs.

An RMIT team is activated to Tepotzlan, Mexico where structured craft and brilliant light formations are observed nightly. A talk and brief training are given to an international community in the mountains.

In late June, CSETI's Ambassadors to the Universe training retreat in Crestone, Colorado draws 33 participants. Large structured craft are seen and two by participants. Corroborating reports reach local researcher from independent observers. **(This author and a friend from Vancouver, B.C. attended this retreat and can attest to many unusual UFO and ET events)**

Further CSETI field work for a tour group conducted in England, witnessed that **Shari Adamiak** (Dr. Greer's assistant) had tones returned through a walkie-talkie as the CSETI tones are being broadcast which were briefly recorded the second time it occurred.

Another RMIT group is activated to the Bonnybridge area of Scotland. Videotape of a string of rapidly moving objects is obtained. Three team members observe a low-flying triangular craft. Multiple distant anomalous objects are sighted by team members. A daylight disc appears shortly after making its presence felt to two team members.

There is another "Ambassadors to the Universe" training in Desert Hot Springs, California draws over 30 participants. A very fast, slow moving golden disc is observed by all participants on the second night. In Joshua Tree National Monument, twelve members see a brilliantly lit, structured craft descend from the sky and vanish into the ground. Electrical anomalies accompanied this sighting.

Dr. Greer lectures as a keynote speaker at the first ever UFO conference in Helsinki, Finland and at presentations in Denver, Colorado. He also continues to brief government and military representatives individually on the CSETI and Project Starlight initiatives.

1997

Sometimes, CSETI's RMIT team encounter incursions and hostility at some location where things do not always go as plan, such as the time RMIT team journeys to Metepec, near Mexico City, and Metepec, near Mt. Popocatepetl, in Mexico to conduct research. The CSETI team is placed in jeopardy when they are ambushed by paramilitary/police at their previously secure field site on public land. A number of sightings, however, occur, including possible craft on the ground.

In April, CSETI's **Project Starlight finally** accomplishes a goal they had been working towards for three years. Briefings in Washington, D.C. were given to congressional offices, embassy and

White House members, and high-level military figures in the Pentagon. Dr. Greer's team brought nearly 20 of our first-hand, military and government witnesses, all with top secret clearances, to speak at these invitational briefings. This historic event resulted in interest among congressional offices. All of the witnesses signed oaths stating they were willing to testify to congress during hearings, even without guarantees of protection or release from national security oaths.

Dr. Steven Greer gives talks and workshops in Atlanta, Georgia, and members of the Denver Working Group and Shari Adamiak are filmed by the BBC during field work in Crestone, Colorado.

Ambassadors to the Universe training retreats routinely continue in Crestone, Colorado, in Southern England and in Desert Hot Springs, California during summer and fall months conducted by Dr. Greer. Anomalous sightings of UFOs become almost commonplace as witnessed by foreign diplomats at Alton Barnes, UK. Video recordings of UFO events are now a standard practice and at times military jets fly over to intervene between Dr. Greer's CSETI groups and the anomalous objects which are being communicated with.

Throughout 1997 CSETI Working Groups continue to proliferate, with significant CE-5 encounters from the Dorchester UK, Santa Barbara, Ca, and from British Columbia and South Ontario (Canada) groups.

1998

In 1998 Dr. Steven M. Greer resigned his post as Chairman of Emergency Medicine at Caldwell Memorial Hospital in North Carolina to spend all his time and energy on exposing a massive corporate-government cover-up of extraterrestrial contact that's been kept under wraps for decades.

In January, Dr. Greer appears on Art Bell radio show and provides revelatory details of his historic 3-hour briefing of sitting CIA director Woolsey, plus recounts how Laurance Rockefeller gave UFO/ET briefing package to the Clintons. Public are requested to contact key Congressional committees to request open hearings on the issue. Dr. Greer becomes Art Bell's favourite and most listened to guest whenever he is on the air.

The continued public education program of Ambassador to the Universe training grows steadily with first-time locations like in Hawaii, in the UK near Stonehenge and in Sedona, Arizona. There are numerous anomalous sightings and CE-5s, and experiences of entities on the ground at these location sites. http://www.siriusdisclosure.com/ce-5-initiative/past-events/cseti-history/

1999

Ambassador to the Universe Training moves to CSETI Headquarters, near Charlottesville, Virginia. And this is followed by the first 2-week back-to-back trainings for Senior Researcher and Instructors training in Crestone, Colorado. Mt. Shasta, California becomes the next location for Ambassador to the Universe Training, where 3 craft fly over the group in a rigid triangular formation at orbital altitude 3 of the 7 nights, each time within minutes of an intense group meditation for interplanetary and world peace.

2000

CSETI has witnessed a decade of phenomenal growth and propagation of its core ideals being adopted globally by people who desire to see governments become transparent disclosing what they know about the UFO and ETI phenomenon. This has been due largely in part to Dr. Greer's global CSETI CE-5 Initiative.

Ambassador to the Universe Training continues at Pine Bush, NY, in Crestone, Colorado, in Wiltshire, England, at Mt. Shasta, California.

The **Disclosure Project** is established and videotaping of witnesses begins leading toward the day when in absence of a full government disclosure due to an impotence in government duty, responsibility and courage will witness nonetheless, a full citizen's disclosure the likes of which has never been seen before in history.

2001

Apart from the retreats for Ambassador to the Universe Training, this year stands out from all other years because on May 9, 2001, the **Disclosure Project** Press Briefing was held at the **National Press Club** in Washington DC. Twenty Witnesses spoke before a crowded room of National Media coverage including CNN Headline News. These heroes of the country gave testimony of their first-hand UFO and ETI involvement and experiences which they would willingly do again, under oath before congress. The Disclosure Project went on tour around the country in America and came up to western Canada first at Vancouver on September 9, 2011, and then to eastern Canada and then, into the U.K.

2002 to 2008

Ambassadors to the Universe and Cosmic Consciousness Training retreats are a sustained integral part of CSETI's education of the public to the ET presence visiting the Earth.

2009

This year witnessed many more training sessions and retreats but two training events stand out for their historic significance. The first is the **"Contact: Countdown to Transformation"** retreat that was held on October 23 in Rio Rico, New Mexico. It was by far the largest of its kind with over 200+ people from around the world attending. Almost immediately ET craft started showing up in the clear skies with Dr. Greer's arrival to the resort along with military drones and a helicopter that tried to interfere with the communications and contact that took place during that time period between ETs and humans. Many ET craft were sighted and filmed including a transdimensional landing of a large ET spacecraft that interacted with electronic radar and magnetometers devices and with all of the 200+ people in attendance! **(This author was there and recalls many of these experiences.).** http://www.siriusdisclosure.com/ce-5-initiative/past-events/cseti-history/

504

The second retreat was the Ambassador to the Universe Training in at Joshua Tree National Park in California, November 17, 2009. There was a lot of UFO and high strangeness activity in this area as there usually has been from years before UFO, however, at this time **Dr. Raven Nabulsi**; a long trusted member of the senior CSETI team had the presence of mind to have her camera with her. She had asked the ETs permission if she could photograph them, even if she could not see them with the naked eye.

Dr. Nabulsi pointed her camera in a direction which she felt compelled to do and snapped a photograph. To everyone's amazing she had captured the CSETI expedition's first ever photograph of an Extraterrestrial being! (See below).

You will note that the ET is suspended in a cone of light which is originating from a small orb to the left of the bush. This is precisely the orb that the CSETI team had seen that followed them up the path, and is in the location where the ET voices were heard just before the photo was taken.

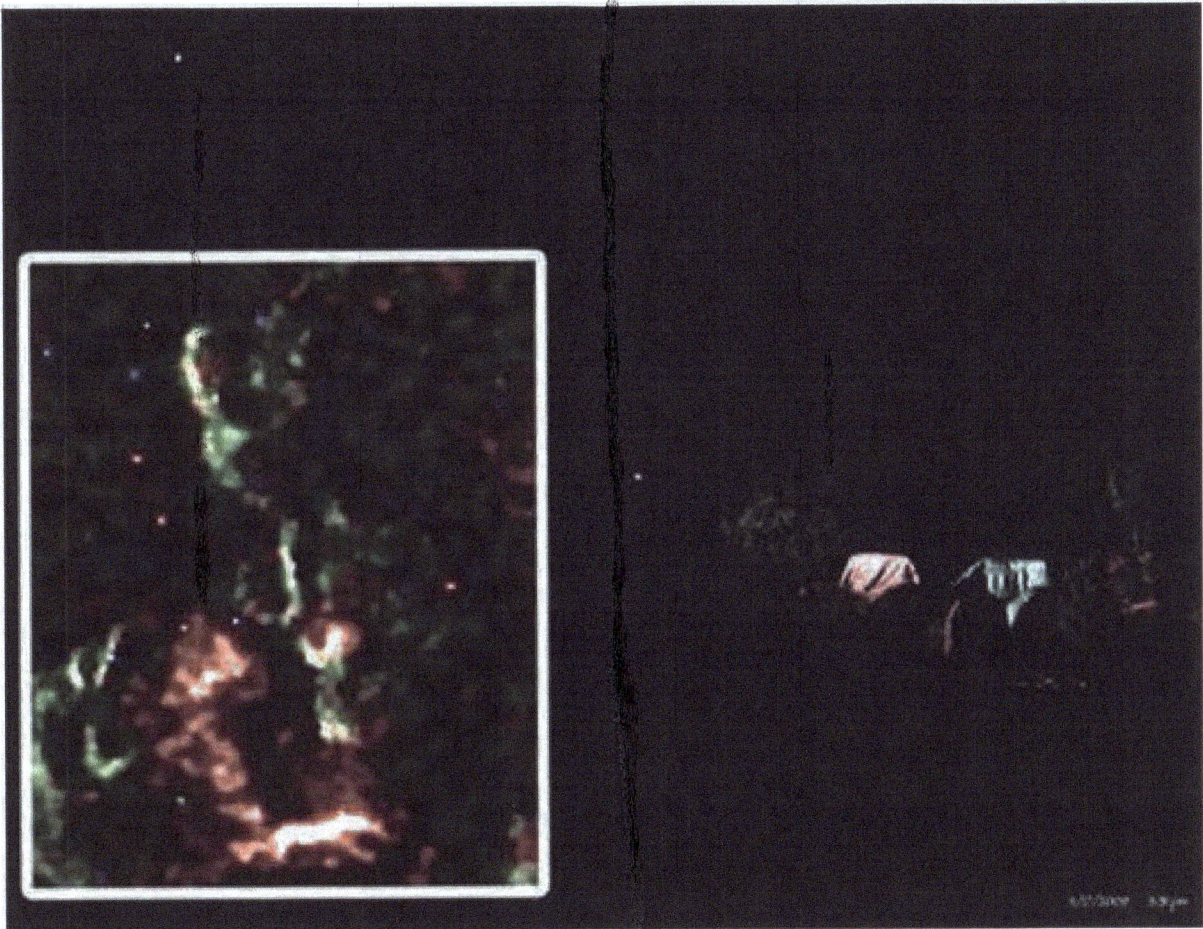

Photo of ET Being taken by Dr. Raven Nabulsi at Joshua Tree National Park.
Barely discernible above the chairs (red and blue) at the right and
shown in insert at left, enhanced and enlarged to show detail
http://siriusdisclosure.com/all-evidence/light-being-boca-grande-fl-february-2010/?portfolioCats=67%2C66%2C71%2C72 (Credit:
CSETI and Dr. Raven Nabulsi)

The ET appears to be a male, wearing a type of vision augmenting goggles, with a very large head that is demarcated with an area of indented ridges in the forehead. The hairline, ears, eyes, mouth and chin are clearly visible. Both arms can be discerned, as well as a torso and both legs, with boots on the feet. He appears to be hovering about a foot or two above the chairs that make up team's contact circle and is just east-southeast of the circle. His size is estimated at 3-5 feet in height. Note that he is leaning forward (possibly, leaning out of his craft?), with his torso and head twisted to look directly at the camera. His right leg is bent behind him.

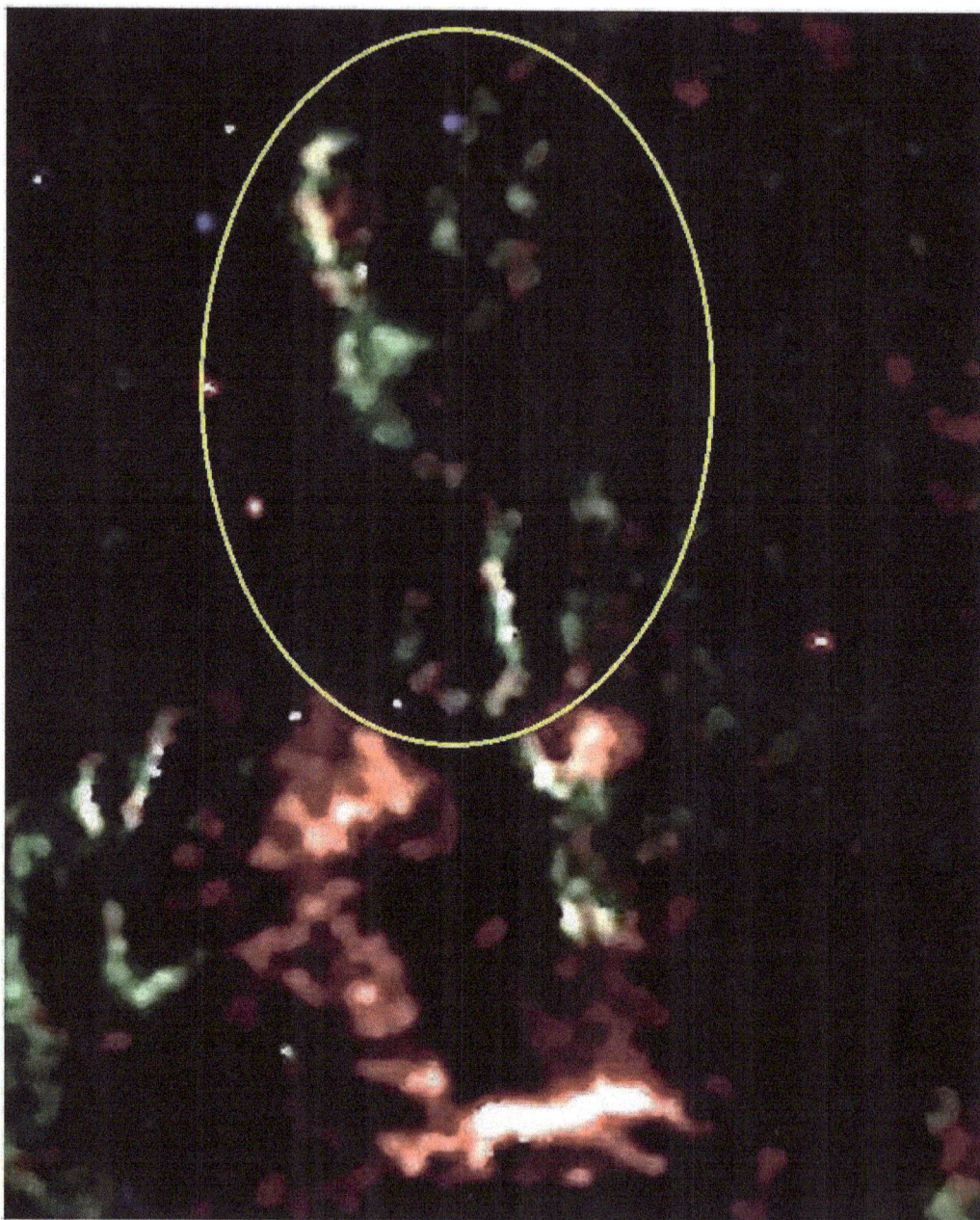

Enlarged image of ET Being (3 to 5 feet tall). Note high forehead, wearing possibly a skull cap and augmenting vision goggles; leaning out of the spacecraft door hatch

(Credit: CSETI and Dr. Raven Nabulsi)

2010 to 2013

In February 2010 the CSETI holds an Ambassador training session in Boca Grande, Florida and once again, a team member photographs another ET being, a "**Light**" **Being**! (See image below).

The last four years has seen "the message" of an ET presence on Earth been further ramped up on the radar of public consciousness. On April 22, 2013, the Los Angeles release of the **"Sirius"** film documentary that has become the largest crowd funded movie of its type and an another step forward for CSETI in the UFO disclosure movement. This movie has awakened more people to investigate the UFO/ETI disclosure movement because of CSETI's efforts and the efforts of many other Ufologists. The movie follows Greer's efforts to reveal information about top secret energy projects and propulsion systems.

Light Being – Boca Grande, FL, February 2010
http://siriusdisclosure.com/all-evidence/light-being-boca-grande-fl-february-2010/?portfolioCats=67%2C66%2C71%2C72

Sirius features interviews with former officials from the government and military as well as images and a DNA analysis of the six-inch humanoid skeleton known as "**Ata**" that was found in the Atacama desert in northern Chile in 2003.

Late April and early May 2013 Dr. Greer participates in the **Citizen Hearing on Disclosure** giving testimony along with 40 other witnesses of the cover up and suppression of the subject. Researchers, activists, and military/agency/political witnesses representing ten countries gave testimony in Washington, DC to six former members of the United States Congress about events and evidence indicating an extraterrestrial presence engaging the human race.
http://www.siriusdisclosure.com/ce-5-initiative/past-events/cseti-history/

Sirius – The Next Phase of Disclosure

With the Disclosure Project Press Briefing held on May 22, 2001, at the National Press Club in Washington, DC well behind them, Dr. Greer and the CSETI team were being questioned what next? Would there be another disclosure event to follow up on the first event? When will it be?

Dr. Greer plays his cards close to himself as he follows a plan that only he and perhaps the ETs of the universe seem to know. But, somehow, we in the public knew that something more was in the offering, once that *"barrel of explosives"* known as the UFO/ETI phenomenon was lit, that it would be almost impossible to stop it from exploding!

We weren't disappointed when the next step in UFO and ETI disclosure premiered on April 22, 2013, in Los Angeles, California.

Following a traditional enigmatic pattern of naming new projects after stars and constellations, CSETI and Disclosure Project Director Dr. Steven Greer, film director Amardeep Kaleka, and producer J.D. Seraphine offer the world the next step in UFO disclosure with a documentary film called **"Sirius"**! *Sirius* shines a brilliant spotlight on the cutting edge perspective of one man and on the whole issue of unidentified flying objects and extraterrestrial disclosure and the web of secrecy that surrounds it. It's a story based on the life work of Dr. Steven Greer, a man who left behind a successful career in medicine to walk a path that has already demonstrated significant results in raising the extraterrestrial subject out from the black vaults of secrecy and tabloid journalism.

Under the umbrella of the CSETI Initiative, Dr. Greer has for over 20 plus years has lead the charge in the Disclosure movement, due largely in part from the identification of 450 witnesses and the video testimony of 200+ former military, government and intelligence witnesses that had first-hand military and government involvement with UFO events, projects and programs. From this large body of heroes of the country, twenty of those witnesses stepped forward to give their testimony for **The Disclosure Project** stating their involvement in UFO and ET matters, as well as other related evidence to be used in a public disclosure.

The Disclosure Project witnesses revealed landmark presentations to the media press in attendance and the online internet public, revelations of UFO encounters, radar tracking, shoot down, retrievals of spacecraft and ET beings, and reversed engineering of alien technology. The witnesses made it abundantly clear that a nefarious UFO cover-up exists, and the public was given an alternative but plausible, down to earth explanation for an out-of-this-world subject.

Perhaps the greatest focus that came from the extraterrestrial revelations was the issue of how the ETs were getting here and the energy propulsion systems that originated from these alien space vehicles in our skies. The existence of alien spacecraft and their technology indicated conclusively that not only do overunity and free energy systems exist but are now in the possession of tightly compartmentalized projects of the US military's black programs. The UFO cover-up is in reality, an energy cover-up.

The goal of the Disclosure Project was an audience on Capitol Hill, and the level of acceptance Dr. Greer had brought to the UFO topic was unprecedented. By mid-2001, it seemed possible the momentum Dr. Greer was driving forward may have resulted in making Disclosure a reality.

But just after 11 September of that same year, people were understandably less inclined to put the UFO topic at the front of the agenda.

I do think the reason for secrecy... [has] much more to do with the economic order, and the fact that the technologies related to these extraterrestrial vehicles – as well as human breakthroughs unrelated to these extraterrestrial vehicles – have given us energy and propulsion systems within the black budget, compartmented, unacknowledged world which would have a significant beneficial effect for the world, but would have a deleterious short-term effect for certain special interests. I believe these special interests would include the conventional energy part of our economy – the internal combustion engines, the fossil fuel plants, gas, oil, coal, and ionizing nuclear technology, technology related to nuclear power plants. We know that those technologies are already obsolete.

I want to emphasize here that this is not a theory. We have testimony from people who have worked on these projects, within these projects, that can establish the fact that we have fully operational technologies which would enable the world to have no need for these polluting forms of energy. Dr. Greer, from The Disclosure Project's *4 hr. Witness Testimony*

Without a doubt, Dr. Greer is recognized as the **"Father of Disclosure"** for his initiative and efforts to bring this subject into greater public awareness.

In the year 2012, a time of superstitious belief in the end of humanity's existence or the possibly the beginning of a new age of enlightenment, Dr. Greer announces the next phase in disclosure will be a publicly crowd-funded documentary called *Sirius* and appeals to the public for funding to support the project. By July, a goal of $250,000 has not only been raised through word of mouth and the internet but, by the time the movie premiered in April 2013, half a million dollars has been raised, making *Sirius* the most crowd-funded documentary yet.

Sirius director **Amardeep Kaleka** proclaims *"Dr. Greer is a lightning rod."* It now looks promising that Dr. Greer will do for free energy systems what he has done UFOs and ETs by spearheading it into mainstream consciousness. *Sirius* will be a big step forward in manifesting that dream into a reality that will not only establish Dr. Greer's legacy for mankind but inevitably will go down in history as a legendary journey by a modern mystical man with practical feet!

Now, 2001 Disclosure Project event, the DVDs of the witness testimony as well as the YouTube Disclosure videos all went viral worldwide as people acknowledged the reality of the phenomenon but, it did not force the secret government to confess its involvement in a UFO cover-up and free energy technology was never disclosed. It was becoming obvious that there was a tight control on the news media as breaking news was spin-doctored and sanitized before public consumption.

This time, Dr. Greer is demonstrating we don't need our government agencies to make official statements heralding the era of Disclosure. Disclosure is already here. All it takes is a desire to make contact and the willingness to work on the spiritual nature of that cosmic connection.

Mr. Kaleka and producer **J.D. Seraphine** had recently been contracted to do a narrative feature relating to the topic of UFOs, ancient aliens and the current interest surrounding those topics. *"Once you start cracking that research as a writer, it takes you down a deep, deep rabbit hole,"* said Mr. Kaleka. http://www.newdawnmagazine.com/articles/sirius-the-film-disclosures-next-step

Out of the blue Mr. Kaleka cold-calls Dr. Greer with a proposition to do a documentary on the UFO subject to which Dr. Greer recognizing a synchronistic opportunity agrees and this was before suggesting they meet for dinner. Dr. Greer offered to be Mr. Kaleka's "White Rabbit" guide down the long rabbit hole into the world of UFOs and ETI!

Once they found out they could really control gravity, in 1954, it all went black.
– **Dr. Greer,** *Sirius*

At the outset of *Sirius*, Dr. Greer sits alone, deep in thought, waiting to take the stage to deliver a presentation while the ominously deep voice of narrator **Thomas Jane** explains the meaning of *"dead man's trigger."* Dr. Greer is one of those rare men with a dead man's trigger, and those on whom he is shining the light must know this because no one has put the free energy topic into the public discussion with as much success as Dr. Greer. That's a dangerous job. And he's still doing it.

"Dead man's trigger is a safety valve," Mr. Jane explains. *"For reasons of security, a person prepares a recourse of such severe action that if harmed, they will release a cache of damaging evidence against those enemies."*

In *Sirius*, Dr. Greer exposes the four primary members of the **banking cartel – Bank of America, Wells Fargo, JPMorgan Chase and Citibank** – as the controlling interests in the four **largest oil companies – Exxon, Chevron, Shell, and BP.** These four banks *are among the controlling shareholders* of the **Federal Reserve**, a private corporation *that controls the currency economy of the United States, and arguably the world*. With the value of the American dollar so closely tied to the price of oil, it's clear who stands to lose the most when free energy becomes public.

Sirius takes the conversation well beyond whether or not ET exists. The documentary makes it clear that we've not only been working on free energy and anti-gravity technology since Tesla, but *we've also had demonstrable success in this field since at least the 1950s*. It seems new *energy scientists* have always had two roads from which to choose: *Do their research under the auspices of the American military, or get bullied out of the game.*

Yet there is one additional thread weaved into this exopolitical tapestry – the **Atacama Humanoid.** http://www.newdawnmagazine.com/articles/sirius-the-film-disclosures-next-step

510

Sirius - The Atacama Humanoid

Throughout the *Sirius* documentary, the camera returns time and again to a subplot within the film which focuses on an unusual six-inch fossilized skeleton known as the **Atacama humanoid, "Ata"**. Is it an Extraterrestrial entity that died upon our planet in the high desert area of Chile or is it something more natural to Earth, a possible new undiscovered human cousin? Besides exposing the machinery of UFO and ET cover-up and suppression in America and the rest of the world, the little being may be the proof that the UFO community has been searching for over many decades. At the very least, the Atacama humanoid is an anomalous example from humanity's fossil record that suggests perhaps we haven't yet figured out the entire story.

The Atacama humanoid was first discovered in 2003 in the remote Atacama Desert region of Chile. Dr. Greer found out about it in 2009 and, apparently, has had his eye on it ever since.

**A size comparison of the Atacama humanoid (6 inches tall)
in the hand of a full grown adult male**
http://www.criptozoologia.net/El-extrano-ser-de-Atacama-El-extraterrestre-de-la-Noria-Chile/166

Dr. Garry Nolan (Rachford and Carlota A. Harris Professor in the Department of Microbiology and Immunology at Stanford University School of Medicine) headed the team investigating the Atacama specimen. Dr. Nolan suggested a protocol of extracting DNA samples, along with X-Rays and CT scans in order to determine a proper evaluation of the skeletal anomalies. The genetic results, though incomplete, could not account for the abnormalities of the humanoid.

Dr. Ralph Lachman, also of Stanford University, is one of the leading experts in the world on skeletal dysplasia and abnormalities. He also studied the skeleton and concluded that the *"humanoid's appearance is NOT the result of any known deformity, genetic defect, skeletal dysplasia or any other known human abnormality."*

According to Dr. Lachman, the *little humanoid lived to between six to eight years old* and Dr. Nolan confirmed that *the specimen is not a non-human primate*, yet *nothing in its genes or biology suggests it is a deformed human.* http://www.newdawnmagazine.com/articles/sirius-the-film-disclosures-next-step

The Atacama Humanoid has a 13 centimeter or 6 inch body that is very desiccated but completely intact. The CAT scan clearly shows internal chest organs (lungs and what appears to be the remains of a heart structure). There is absolutely no doubt that the specimen is an actual organism and that it is not a hoax of any kind. This fact has been confirmed by Dr. Nolan and Dr. Lachman at Stanford. See more at http://www.siriusdisclosure.com/evidence/atacama-humanoid/#sthash.bWe240nZ.dpuf

The specimen has only 10 ribs, a finding not yet found in humans, and a very unusual cranium. It is noted that the cranial vault is, proportionally, much larger than is found in normal humans. The bones are quite well developed and are not those of a fetus (see below). There are multiple skeletal anomalies seen throughout the specimen. Importantly a mature, not fetal, tooth is seen in the mandible (jaw bone). A fracture of the right humerus (upper arm) is seen as is a concave fracture of the right posterior-lateral skull, which was most likely the cause of death.

Importantly, Dr. Lachman has concluded that the humanoid's appearance is **NOT** the result of any known deformity, genetic defect, skeletal dysplasia or any other known human abnormality. However, the most startling conclusion to date is that Dr. Lachman concluded the humanoid lived to be 6- 8 years of age! (See Dr. Lachman's full report here…). This was assessed by examining the epiphyseal plates in the knees and comparing these to normal humans of various ages.

He noted that there is no known form of human dwarfism that has this presentation and set of findings. No human has been known to be able to live for 6-8 years and remain only 6 inches in length.

It should be noted also that **Dr. Manchon**, of the Manchon Radiology Center in Barcelona, also examined the X-Rays and concluded that the specimen was most certainly not a fetus and had lived for a year or more and probably several years.

Previously, there had been false reports published elsewhere that the humanoid was a fetus.

This is clearly not the case as can be concluded from the research of Dr. Lachman and the examination by Dr. Manchon. A comparison of fetus X-Rays shows a remarkable difference between human fetus skeletal development and the X-Rays of the Atacama humanoid. http://www.siriusdisclosure.com/evidence/atacama-humanoid/

Side view of the little 6 inch Atacama humanoid
http://siriusdisclosure.com/evidence/atacama-humanoid/

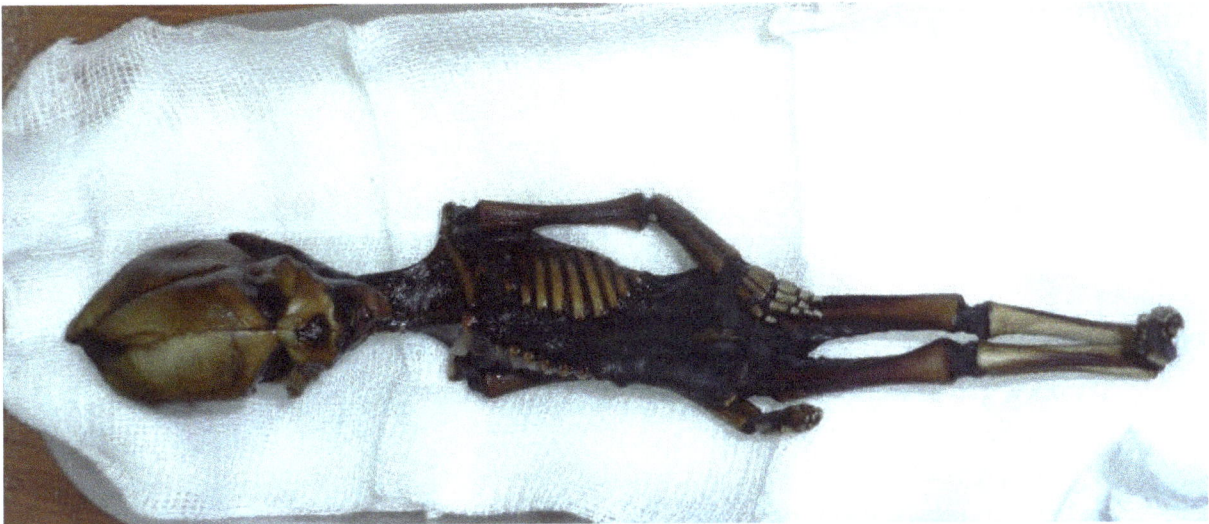

**Overhead view of the Atacama Humanoid, "Ata" is a 13 centimeter
or 6 inch body that is very desiccated but completely intact.**
http://siriusdisclosure.com/evidence/atacama-humanoid/

Importantly, Dr. Nolan has found that: "Preliminary results demonstrate no statistically relevant alterations of genes encoding proteins commonly associated with known genes for primordial dwarfism or other forms of dwarfism. Therefore, if there is a genetic basis for the symptoms observed in the specimen, the casual mutation(s) are not apparent at this level of resolution and at this stage of the analysis."

It should be noted that Neanderthals are 99.5% genetically identical to humans, and chimps and apes are 96-97% identical.

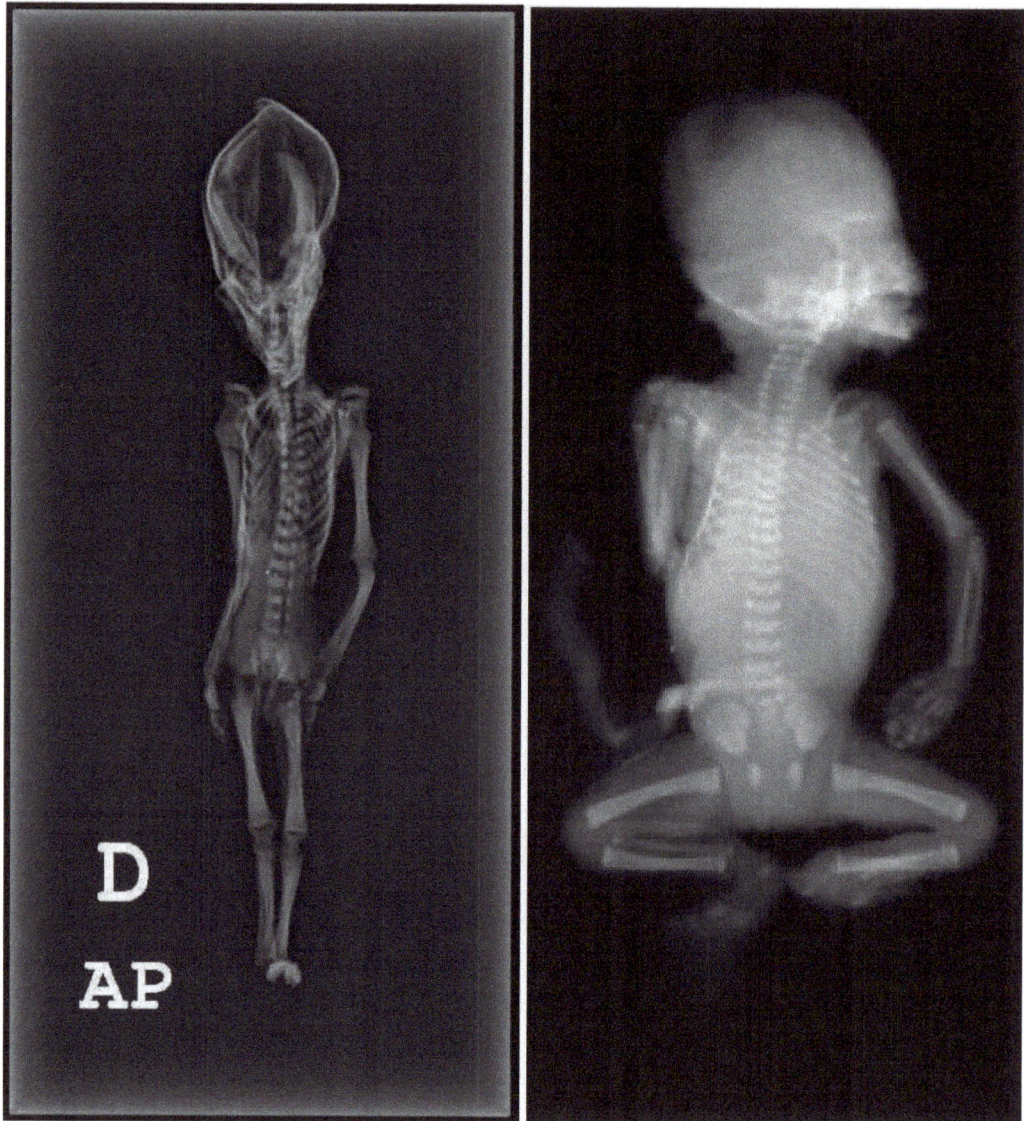

A comparison of the skeletal features of the Atacama humanoid left and a normal human fetus right
http://siriusdisclosure.com/evidence/atacama-humanoid/

As of this date, the genotype does not seem to match the phenotype (meaning physically expressed form of the genetics). The answer to this mystery will require further analysis of the DNA and confirmation of the findings through the peer review process.

 Dr. Greer ponders how long ago this little being lived and how primitive that area would have been- totally lacking in modern medical technologies and facilities- how would this child have lived? And with whom? http://www.siriusdisclosure.com/evidence/atacama-humanoid/

The mystery is further compounded by the reports from Ramon Navia-Osorio Villar and his associates who traveled to the region and obtained information from local native peoples of sightings of UFOs and very small living creatures fitting the general description of this

514

humanoid. There are also reports that other intact humanoids may be stored in various remote sites and locations. These reports have not been confirmed, however.

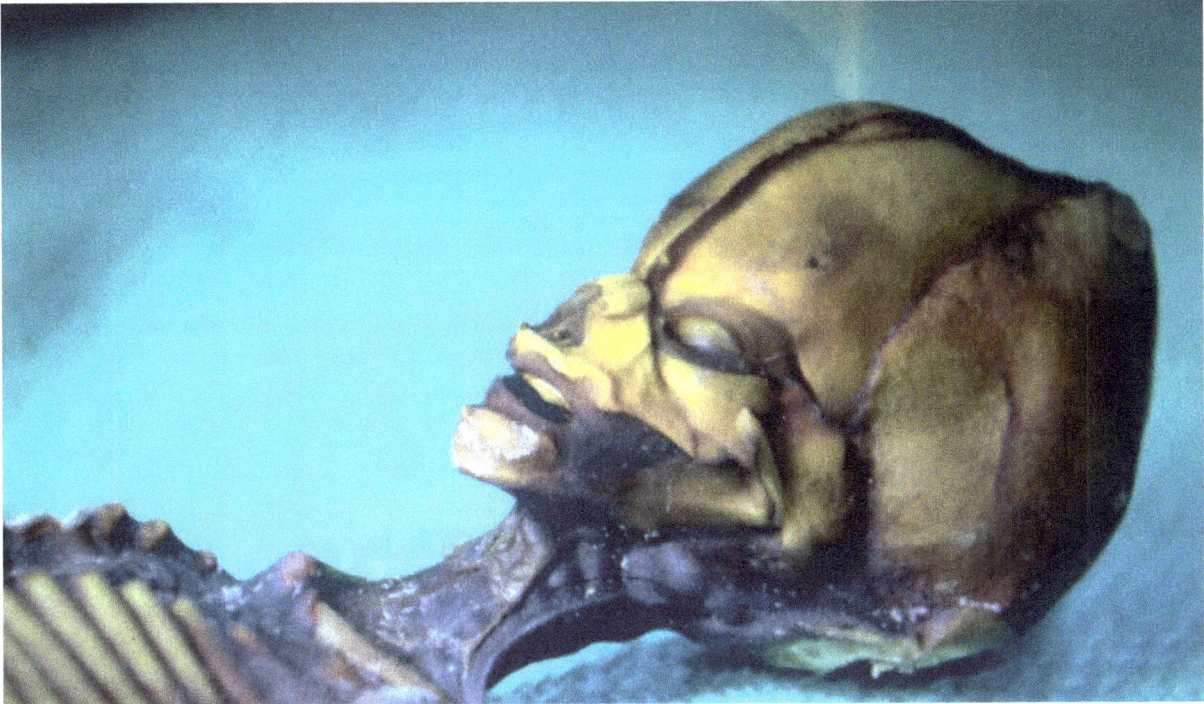

A concave fracture of the right posterior-lateral skull, was most likely the cause of death to "Ata"
http://siriusdisclosure.com/evidence/atacama-humanoid/

It is necessary to do much more thorough research into this case. The DNA work is really in its early stages, and we need to take a scientific expedition to the Atacama Desert to see if there are more examples of this humanoid- and to see if indeed there is on-going UFO/ ET activity in that region as reported.

Recently, some scientists working on DNA and computer analysis found that DNA has been around for over 10 billion years- but Earth has been here for less than half that time. Perhaps life is indeed universal, and contact spreads life from world to world…

I have discussed with other scientists the possibility of epigenetic augmentation of the human genome. Is the Atacama Humanoid a so-called hybrid? Are we all some type of hybrid? Could this have occurred via contact with other extraterrestrial civilizations over millions of years? A source who refuses to be identified stated to me several years ago that he had seen a **National Security Agency** document that concluded that there had been 64 epigenetic augmentations of the human genome in the past that have resulted in modern humans. Could this be possible?

A **Jet Propulsion Laboratory (JPL)** scientist once told me that the reason the objects found on and near Mars – like the obelisks that Astronaut **Buzz Aldrin** wants us to go back to Mars to examine - would show an ancient connection between ETs and humans - and that this is why that information is being kept classified. When I asked why he said: ***"Because the foundations of every fundamentalist orthodox belief system on earth would be up-ended".***

To pursue science is to pursue the truth of matters. What is needed from this point going forward is an open mind so that, together, we may discover the truth about many things yet hidden. http://www.siriusdisclosure.com/evidence/atacama-humanoid/

Following the release of the movie, the story of the Atacama humanoid hit some major news outlets, like NBC. And one week later on 29 April, the well-timed **Citizen Hearing on Disclosure** began. For the hearing, the main ballroom of the **National Press Club** was configured to resemble a Senate hearing room and for five days, dozens of experts and whistleblowers testified in front of a committee of eight former politicians and professionals hoping to ***"accomplish what the US Congress has failed to do for forty-five years – seek out the facts surrounding the most important issue of this or any other time."***

Perhaps the most important result of *Sirius* will be the new energy research facility that the creators of this movie hope to fund with the movie's revenue.

It is hoped that the documentary becomes a huge success and Dr. Greer gets a research facility. It could become a game changer in much the same way Greer's efforts to disclose government secret files on UFOs and ETI, as Dr. Greer says in *Sirius*, *"It isn't about overthrowing the Pentagon. It's about leaving them behind."*
http://www.newdawnmagazine.com/articles/sirius-the-film-disclosures-next-step

To learn more about Dr. Steven Greer's work visit www.disclosureproject.org.
To download or purchase the DVD of the film Sirius, go to www.siriusdisclosure.com.

CHAPTER 120

BECOMING AN AMBASSADOR TO THE UNIVERSE
OR A COSMIC DIPLOMAT

A rare opportunity unlike any other in history now stands before every individual upon this planet. It is the opportunity to advance the human race to the next level on this planet by becoming a global civilization engaged in peaceful contact and communications with other visiting Extraterrestrial Civilizations, as ambassadors of Earth. At present, there are very few people (a few thousand out of 7 billion people) doing what CSETI is doing which is, in reality, the next step in the investigation of **Unidentified Flying Objects** and **Extraterrestrial Intelligences**. This time, however, the research is no longer passive, that is to say, waiting for UFOs or ETs to show up and reveal themselves to us. The new research is one that is pro-active, human initiated, inviting ETI to show up in peaceful engagement with the purpose of bilateral communications between humans and Extraterrestrial Intelligences!

Because our elected and appointed officials have failed miserably to lead in good faith with the due responsibility of their office and to disclose the existence of Extraterrestrial Intelligences, that responsibility, and opportunity is now by default, forfeited and given to the common people. This is the time for citizens to exercise their God-given rights and powers to bypass their governments, their militaries, their intelligence communities, their religious officials and their scientists, et al and reach out to all visiting Extraterrestrial Intelligences coming to the Earth with open hearts, minds, and hands of friendship. **"If the people lead, the leaders will follow!"**

This has been the **CSETI (Center for the Study of Extraterrestrial Intelligence)** philosophy and it is the driving force behind Dr. Greer's push for ETI contact and communications, for public disclosure and for release of hidden and suppressed ET technology held by the military industrial complex. This three-prong approach will build upon itself and will become an unstoppable force that will gather the support of the public that will eventually propel humanity toward a new advanced global civilization.

The **CE-5 (Close encounters of the Fifth Kind) Initiative** is the principle platform for a system of research-driven communication protocols and public education programs. The CE-5 Initiative involves scientists, public investigators and others who voluntarily initiate human contact and/or interaction between Extraterrestrial spacecraft and its occupants. Close Encounter types 1-4 are essentially passive, reactive and ETI initiated. A CE-5 is distinguished from these by conscious, voluntary and proactive human-initiated or cooperative contacts with ETI. CE-5s are active, real-time research activities conducted on-site by trained **Working Group** members, using scientific and diplomatic methods to establish a non-aggressive and evolutionary relationship between humans and ETIs. Evidence exists indicating that CE-5s have successfully occurred in the past, and the inevitable maturing of the human/ETI relationship requires greater research and outreach efforts into this possibility. While ultimate control of such contact and exchange will (and probably should) remain with the technologically more advanced intelligent life forms (i.e., ETI), this does not lessen the importance of conscientious, voluntary human initiatives, contact and follow-up to conventional CEs 1 to 4.

CSETI is the only worldwide effort to concentrate on putting trained teams of investigators into the field where active waves of UFO activity are occurring, or in an attempt to vector UFOs into a specific area for the purposes of initiating communication. Contact protocols include the use of light, sound, and thought. Thought (consciousness) is the primary mode of initiating contact.

More specifically, CSETI utilizes the **Contact Trilogy** to initiate a CE-5 by using a tri-modal communication process developed by Dr. Greer. These modalities include:

- High powered lights, lasers, and visual ground formations which convey intentional, intelligent activity in the sky.
- Specific auditory tones recorded at previous UFO sightings and Close encounters that serve as audio beacons.
- **Coherent Thought Sequencing (CTS)**, an experimental psi component *(remote viewing)* to enhance the CE-5 setting and research staging area, using the tenants of consciousness *(meditation).* http://cseti.org/ (updated, see: http://siriusdisclosure.org)

Qualifications of an Ambassador to the Universe

People have often asked Dr. Greer what type of qualifications, a person should bring if he or she wants to be a part of the CSETI Initiative. Perhaps a simple answer would be the attribute of open-mindedness but, certainly, there are many more qualities that should be considered, here are some of them:

- People should be open to the strong evidence that ETs are currently visiting the Earth, they should accept that intelligence and sentience may come in many different forms with some being viewed as unusual or even bizarre by human standards. It appears probable that more than one extraterrestrial civilization is responsible for the ETI/ETs contact so far observed. It is likely that this represents a cooperative effort.
- The recognition that consciousness, sentience, and intelligence are the hallmarks of all intelligent life in the universe and therefore, both human and ETI are more alike than dissimilar; CSETI is dedicated to the study of both our shared and unique characteristics.
- This in turns requires the elimination of any pre-judgments that humans may have toward stereotyping what Extraterrestrial Intelligences are good or bad. The old adage: "actions speak louder than words" would be appropriate in any contact and communications with ETI. Needless to say, negative attitudes and fear-based minds are not the prerequisites for cosmic diplomacy.
- CSETI operates on the premise that ETI motives and ultimate intentions are peaceful and non-hostile, contrary to the alleged stories military hostility or reports of citizen abductions by ETs, therefore people are expected to conduct themselves in a manner the reflect positive, peaceful intentions.
- People who want to be future goodwill ambassadors of Earth should be mature enough to function in any and all relationships, whether with fellow humans and particularly with Extraterrestrials with co-operative, peaceful, non-harmful intentions and procedures.
- Practicing and living the moral qualities of a high functioning, mature human being particularly in advancing a lasting world peace on Earth is essential to the full development of the ETI-human relationship.

518

- No one should engage in ETI contact or communications with some personal hidden agenda to acquire something of value whether it be pictures, video or alien technology. It is not the goal of CSETI to acquire ET advanced technologies which may have a harmful or military application if disclosed prematurely.

- These qualifications are also the core principles of CSETI and without a doubt this could be an endless list filled with lots of do's and don'ts but, its purpose is to serve to the reader as an example, that an "**Ambassador to the Universe**" functions at a high moral level, is desirably highly educated and is willing to engage with other intelligent ET species or civilizations with a positive attitude that allows for a mutually peaceful and sustainable relationship. Once that relationship of trust and friendship has been established then, all the things that we enjoy about being human can be shared and exchanged with all the things that individual ET civilizations enjoy and wish to share with us. It's a whole new universe that has opened up to humanity at this point!!! http://cseti.org/ (updated, see: http://siriusdisclosure.org)

A Typical CSETI CE-5 Field Trip Experience

Those who are fortunate enough to experience a week-long **"Ambassador to the Universe"** training seminar with **Dr. Steven Greer** and the senior **CSETI** team will inevitably come away from the session with their life forever changed. One thing for sure, they will have a whole new perception for life in the universe, that they are not alone in the universe.

Typically, people start arriving at their accommodation site, usually a hotel or motel in the city area where the field trip to the work site will be organized. The first evening is usually meeting with Dr. Greer who welcomes everyone and introduces the senior CSETI team members to the new people followed a small presentation talk.

The next morning, after breakfast, people gather in a conference room to introduce themselves with some personal background and/or related UFO experiences. Dr. Greer then gives CSETI updates and briefings on the history of CSETI, UFOs, and ETI as well as the **CE-5 Protocols** and what expectations one may encounter in the field work site. This is followed by a training session in **Meditation, Remote Viewing** and **Coherent Thought Sequencing (CTS), (**a somewhat controversial method used in communications) to get people used to the concepts and this becomes the learning platform for day time and night time field trips.

People are encouraged to practice meditation, remote viewing, and CTS on their own throughout as well as in collective gatherings during the training sessions. The goal is to develop a cohesive group fully synchronized with each other so that the group dynamics is finely honed, and everyone begins to feel as if there is one soul with one consciousness when out in the field site.

A daytime field trip expedition to the work site allows everyone to gather in a circle to implement what they have learned in the morning and to practice remote viewing places, objects, and events. This is also a time for people to familiarize themselves with the surroundings and the overall environment in which they will gather again, later in the evening to do the actual vectoring of ET spacecraft into the work area.

Meditation and coherent thought sequencing is an important part of the CE-5 Protocols as it permits ETI to locate the CSETI Working Group in order to engage them in contact and communications. Essentially CTS is established by knowing your locations right down to the square meter or foot and through the process of meditation and expanding consciousness, one visually expands their consciousness to contain their location, their immediate surroundings, the valley or mountains or desert area that they are situated in or which may be nearby. The person consciousness expands to visualize the country and the continent; whether it's the west or east coast they are on, etc., whether it is nighttime or day ime.

They then perceive the whole Earth and the Moon as they move further out into space. Through their consciousness, they perceive the Sun and the stars and moving further away, they may see Mars, the Asteroid Belt, Jupiter, Saturn and the outer planets. Moving further away their consciousness expands to contain the whole solar System with the sun far in the distance and now, moving out of the spiral arm of the Milky Way Galaxy, one perceives the galaxy in its entirety. At this point, one can continue to move out into very deep space and perceive clusters of galaxies or they may stay where they are and remain in a state meditative bliss.

At some point, the person then remote views if possible, rather than merely visualizing, any ETI or a nearby spacecraft or an ET civilization on a planet. The person asks politely if they can board their spacecraft and invite the ETs to engage in contact and communications with their team back on Earth. It is at this point, that CTS is used to show them where you are located. In the reverse order of where you are located in space, you show the ETs the Milky Way spiral arm where our Solar System and Sun is located; you guide them through the Solar System passed the outer planets, Saturn, Jupiter, and Mars towards the Earth - Moon system. You show the ETs the continent and that part of the continent where you are located, right down to valley or mountain or desert area you are situated, right down to the square foot! You show them how many are in your working group and that you will be using high powered lights, lasers, and audio tones to pinpoint your location!

This then is the basic CE-5 Protocols which anyone can learn, practice and teach to others who are also interested in doing more than being just a passive observer of unfolding UFO/ ETI events. This is the **"Rosetta Stone of ET Communications!" This is what the military doesn't want you to know!!!** By employing the **CE-5 Protocol Initiative** you are essentially bypassing, the government, the military and the **Military Industrial Complex** and informing the visiting Extraterrestrial Intelligences that there are other more peaceful people they may wish to communicate without the fear of hostility or harm.

During your ambassador training, every morning becomes a debriefing session and people are once again, encouraged to practice meditation and remote viewing after breakfast and before debriefing sessions. By the end of the week of training, most people are becoming proficient in the CE-5 protocols.

As the days go by, ET activity will begin to increase even during the day time but, certainly during the night time and it may start to get the attention of the military who may show up to chase the ET spacecraft around in the skies that you are trying to interact with using your high powered lights and lasers, etc.

520

Such unwarranted attention by the US Military is an unfortunate aspect of communicating with ETI and must be dealt with but, always within a sphere of safety for your team and for the visiting ETs. At times it may be wiser to say goodbye to the ETI and just simply pack up your equipment and move to a more remote location or call it a night. Safety always takes priority over any attempts at ET contact. **This author has seen military interventions during many CE-5 events, chasing UFOs and even firing light-enhancement flares at ET craft on the ground in which our CSETI team were engaged with, in photon communications!** This is obviously, not an exercise for the faint of heart!

CHAPTER 121

UFO AND ETI PHENOMENON OBSERVED BY CSETI FIELD TEAMS

CSETI field teams can expect a plethora of UFO/ETI phenomena to unfold, regardless of where in the world they gather, as long as the **CE-5 Contact Protocols** are done with pure intent and the Working Group intention is positive, open and receptive. Although ETs and their spacecraft are physical by nature and many researchers would love the chance to *"kick the tires"* on the ET vehicle, for the most part, ETs deem it a safer modality of travel if they move beyond the crossing point of light and appear in the state of transdimensionality. Therefore, a CSETI Working Group should expect that interaction with ETI and their craft will be in this "out of phase" state of reality. This is the ***"high strangeness"*** that the late **J. Allen Hynek** often referred to in his books.

When observing in the field, it is important for a CSETI field team to be alert for a significant variety of phenomena that are associated with ET activity. These include phenomena that impact on the full range of senses including visual, remote viewing, hearing, touch, smell, and emotional. It is sometimes possible to determine the efficacy of a suspected craft by using CE-5 protocols including light/laser signals or thought/mind requests for a response signal indicative of a true ET craft. The following list includes phenomena and responses that have been observed by many CSETI working groups in the field at many locations throughout the world.

Night-Time Aerial Sightings: - Alleged Meteorites

Incoming ET craft can often appear to look like a meteor, however, sometimes they act very differently than a normal meteor. The reader may want to refer to the historical UFO accounts by meteor expert **Lincoln LaPaz,** who led an investigation into the **"Green Fireballs"** that were ***often seen clustered around sensitive research and military installations***, such as **Los Alamos** and **Sandia National Laboratory**, and the Sandia base. The strange green balls of light appeared suddenly and were reported many times per month near such New Mexico installations, but hardly anywhere else. **(This author has seen these "green fireballs" in Crestone, Colorado).**

- They may move more slowly than a normal meteor.
- A second meteor follows the same path through the sky within seconds.
- They move across the sky in a horizontal manner.
- Their flight path changes directions sometimes by as much as 90 degrees or they zigzag in flight.
- A number of meteors fall along the same path during the evening.
- They respond to thought command to change direction
- They just feel different (hard to really describe or nail down)
- They are larger, brighter, and more spherical.
- They don't have a tail.

- They flash bright or get dimmer on their own or in response to being signaled at with a spotlight or laser light.

- They may enter a building through a window. This phenomenon was seen by a CSETI team in England in July 1997. A distinct blue-white sphere moved through the sky in an arc from behind a large tree and entered the window of the house where Dr. Greer and Shari Adamiak, inside, saw it appear as a small, shimmering ET.

- They may be huge, brightly colored, and streak directly down from the apex of the sky and go into the ground with no explosion, disturbance of the ground, etc. A CSETI team witnessed a bright teal object do exactly this in Joshua Tree National Park in November 1996. http://cseti.org/ (updated, see: http://siriusdisclosure.org)

-

Example of Alleged Meteorites
https://thumbs.dreamstime.com/b/%D0%BF%D0%B5%D1%87%D0%B0%D1%82%D1%8C-shooting-stars-galaxy-sky-background-falling-meteorite-comet-glowing-light-vector-illustration-174285639.jpg

Alleged Satellites

Satellites move at a constant speed across the sky and are usually seen during the first few hours after sunset. They may appear with a constant brightness or may pulse in a regular fashion as the satellite rotates reflecting sunlight. Some satellites (i.e. **Iridium type satellites**) may flash very brightly as the sun glints from their solar panels and reflects to the ground. Constant speed and course characterize real satellites. Low level satellites usually traverse the sky from zenith to horizon in 2-4 (?) minutes. ET craft or **ARVs (Alien Reproduction Vehicles** - made on earth) can travel at speeds ranging from virtually standing still (hovering) to speeds allowing them to traverse the width of the sky in a fraction of a second to several seconds. They also can change speed and direction more rapidly than conventional aircraft. These are sometimes referred to as **"Fast Walkers"**. Alleged satellites may appear to come out of a particular constellation and then 10 or 15 minutes later it is followed by another satellite coming from the same direction or constellation and yet again, within another 15 minutes, one more satellite appears travelling in the same trajectory from the same spatial coordinates. **(This author has also seen these types of ET spacecraft)**.

- Satellites are visible until they gradually fade from view. Anomalous "satellites" suddenly disappear, or if viewed through powerful night-vision binoculars, can be seen to dart swiftly into space at an angle perpendicular to their earlier trajectory.

- ET craft may change direction or speed or may change brightness sometimes in response to a directed thought or signal such as a powerful flashlight or laser.

- ET craft may also appear at any time of the night, unlike normal satellites.
  http://cseti.org/ (updated, see: http://siriusdisclosure.org)

Example of Alleged Satellites

Star-like UFOs

Here we refer to objects that appear to be stars at first observation but act differently in the following manners.

- They blink off and on, sometimes randomly, sometimes moving slightly between the blinks. We observed a whole "squadron" of craft one evening in Sedona that blinked off and on for 10-15 minutes in one area of the sky. When a laser was pointed at one of the craft it glinted off the craft. Night vision scopes can be useful in determining the number of craft if they are far away.

- A star-like object appears on or near the horizon (most common though they can be most anywhere) and remains there motionless for a long period of time (minutes to hours). These are often dismissed as a star until it suddenly flies off, changes brightness, or fades out in a cloudless sky. It may also change its appearance in response to signals.

Examples of Star-like UFOs

- Some stars twinkle actively and change colors, twinkling white, green, and red, especially when they are near the horizon. So do some UFOs. A star-like object may be seen in the sky, and over a period of a few hours, may be observed to move east while all the other stars in the sky move west. This phenomenon was observed by a CSETI team in England in 1997. A large, twinkling "star" was seen to move from behind one large tree and travel 30 degrees across the sky and disappear behind another tree, while all the other stars moved in the opposite direction.

- Some ET craft can zoom off into space from the low atmospheric altitude and appear to be just another "twinkling star" in the night sky. **(The author has seen this type of evasive maneuver in 1996 in Crestone, Co. when a large triangular ET craft was being pursued by military jets which suddenly moved off into space to become a just another multi-coloured twinkling star).**

- ET craft may be travelling in space in an unusual manner then, suddenly change their speed to match a geosynchronous Earth orbit, moving at the same rate as the background stars. **(The author has also seen this type of UFO behaviour when a massive (several miles in diameter) "white" flying saucer appeared over the Sangre de Cristo Mountains at Crestone and took up a geosynchronous orbit).** http://cseti.org/ (updated, see: http://siriusdisclosure.org)

526

(Air) Plane-like UFOs

Here we refer to airplane-like objects that are lighted and fly like planes but are not planes. They may be either ET craft or ARVs (Alien Reproduction Vehicles - made on Earth).

- Although all planes are required by the FAA to flash or strobe lights at night, some planes and some UFOs do not. (Is this true?) In Sedona, AZ, a CSETI team watched a probable ARV fly by silently, with no flashing lights in the company of military reconnaissance jets. Its speed, odd lighting, and silent behavior made it appear very anomalous, although at first glance it looked like another plane in the night sky.

Example of an (Air) Plane-like UFOs
https://line.17qq.com/articles/ssugecruax_p3.html

- Usually flashing or strobing planes that sound like a jet or prop engines are normal planes, however, totally silent planes may be UFOs (or ARVs), note the sound of other "planes" in the same area of the sky.

- Any plane that appears to be silent should be watched very carefully even though it may appear to be a perfectly normal plane. ET craft can be cloaked to appear perfectly normal. In Crestone, CO, several years ago, a group of CSETI workers observed a small private plane fly very low right over the group in total silence, and then proceed into a mountain canyon and just disappear. Many of the witnesses later reported seeing the plane in different colors. http://cseti.org/ (updated, see: http://siriusdisclosure.org)

Orbs and Lights:- Distant orbs

- Depending on their size and proximity to the observer, these objects may appear as single star-like objects up to round glowing objects of varying sizes. They are often a uniform amber or gold in color though they can appear in various colors. Airplane landing lights are often mistaken for orbs, but can usually be differentiated by noting the light's location (such as proximity to an airport), strobing lights and/or associated red and green navigation lights.

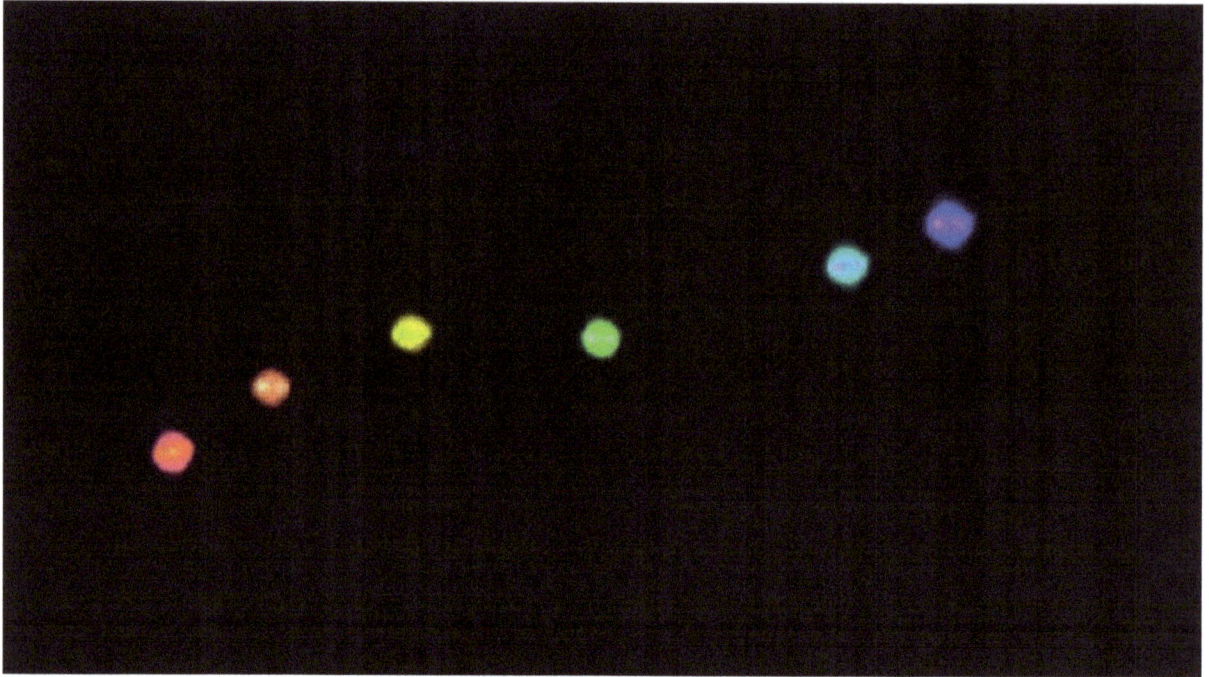

Example of Orbs and Lights:- Distant orbs
https://unsettlingstories.com/2016/12/12/long-fingers/

- **Orbs** often will remain stationary for a period of time though they do move about as well. Lights may be any color and can appear singly or in groups. Group lights may be either individual craft flying together or a single large craft with lights on the outer edges such as observed over the Santa Barbara channel or Phoenix, AZ. Flares, often dropped by the military to confuse observers after a genuine sighting has occurred, float downwards at different rates and give off smoke seen above the floating lights. Since they fall at different rates, a line of flares will often have a jagged appearance after several minutes. **(The author has also seen these types of orbs of various colours in Crestone, Co. and has interacted with them in "photon communications" using high powered lights).**

Small Close-Up Orbs

- These often appear as small spots of light (like a laser spot) or small globes that may appear within several 100 meters or as close as touching a working group member. These are often considered to be probes and either contain intelligence or are under intelligent control (the difference may be academic). They can move about, change intensity and just appear and disappear. **(This type of orb (light gray coloured) has also been seen within 10 to 15 feet from this author in Stave Lake near Mission, B.C.).**

528

- They can also appear as amber-colored "street lights", and may even be mistaken for street lights until they suddenly disappear, or are seen in rural fields where there are no streetlights (e.g. the "golden orbs" often seen in Wiltshire, England).

Photo of the author and his nephew and wife in the backyard when small Orbs appears above (the other Orbs require brightness enhancement to see them – see photo below)
© Terry Tibando

There are three more orbs in the photo plus a wavy line at the left which can be seen through photo enhancement and brightness of the above photo.
© Terry Tibando

- Close orbs can also be very misleading when they are initially observed and mistaken for people with flashlights or other simple explanations. For example, a small group of very experienced CSETI team members described the following observations in Sedona one night in November 1998. ".....I saw four lights behind J. which I thought at the time were shining from the road. There were three lights, light amber in color, roughly the size of an average flashlight, in an uneven row. To the right of the 3 amber lights was a dull red light at least twice the size of the others. In the red light there seemed to be a small grid-like pattern. There were no beams coming from the lights. The four lights then moved rather irregularly to the right and out of my sight. I then commented to the others that I had seen some lights and perhaps there were people on the road. Everyone commented that they hadn't heard anything and there certainly was dead silence on the site." ..." I kept thinking about the lights that I had seen and finally asked D. to shine the big light in the direction I had seen the lights. When she shined the light it did not illuminate the road - the road was too far down the ramp to be seen - it illuminated the edge of the mesa at least 20 feet behind the truck. That's where I had seen the lights, no more than 40 feet (roughly) from where we were sitting! The lights had moved to the right behind the truck out of sight and had not reappeared." http://cseti.org/

Very Large Orbs

- Observers (CSETI and others) in England in 1998 twice saw a very large (3X full moon) orange globe rise above the horizon, then dip back below, then rise again before it suddenly disappeared. This object was observed on two nights, in different directions each night. The second night, after it rose above the horizon for the second time, it "dissolved" as it disappeared, and several British military jets and helicopters appeared in the area within 30 seconds, even dropping a flare in the vicinity of where the object had been seen.

- Although orb appearances may be for extended time periods (minutes to hours) sometimes they appear so quickly, one has to be looking in just the right direction to see one. In **Joshua Tree** in 1997, 5 to 6 CSETI team members observed a globe-like object 1/4 to 1/3 the diameter of the nearly full moon appear to the left and below the moon and travel to the right and below the moon and then just vanish all in just a few seconds.

Very Large Orb over Montreal, Quebec
https://www.mtlblog.com/en-ca/life/9-montreal-ufo-sightings-that-can-never-be-explained

Dark objects

These objects (craft) are often observed under starlit skies with minimum moonlight and may be moving or still. They may appear as rapidly moving black objects (unlit) that stand out against the star field and can range in size from small to apparently huge.

- For example, in 1997, a group on Mount Blanca, CO watched a rectangular cloud form over the top of the mountain on a cloudless night. A few minutes later a black object was seen to fly up, out of the cloud, which then disappeared.

- More examples of these darting black objects were seen by a CSETI team in Hawaii in 1998 through a foggy mist.

Dark objects that can blot out portions of the sky and stars

Distorted Sky

There are times when a craft hovering just beyond the crossing point of light will cause a distortion in the star field. It will appear as though there are heat waves or shimmering even though no object can be seen. There are other times when a small portion of the sky may appear darker than the surrounding sky. This may be indicative of a hovering cloaked craft. These distortions have been observed both at a distance and close to an observing group when a craft was either just above a group or surrounding a group.

> In England in 1998, the beam of a powerful laser directed at an area approximately 25 feet from the group was distorted/bent when it hit the edges of a cloaked object on the ground (and partly on a river). http://cseti.org/ (updated, see http://siriusdisclosure.org)

Grids and Energy Fields

Some observers have reported seeing a manifestation of **energy grids in the sky** which appears as lines of light which may sparkle or fade in and out. These can often be very subtle and therefore not seen by every member of the team. However, on Mt. Blanca several years ago nearly everyone in the group was able to see energy sparkles all over the mountain, appearing like lightening bugs from a distance.

Some Grid and Energy Fields are similar to auroras
http://www.lovethesepics.com/2011/02/24-amazing-auroras-aurora-borealis-aurora-australis/

Flashlight or Camera-like Flashes

Flashes of light have appeared in mid-air, with no apparent source. These flashes can also occur in space and some may actually be Iridium satellites which is a good reason to have satellite charts to determine satellite types and trajectories or orbits.

- In England in 1997, two flashlight-like streams of light burst horizontally amid a group of people, but others surrounding the group were not using their flashlights.

- In Hawaii in 1998, four CSETI observers witnessed two flashes within a few feet of the group, as though flash photos had been taken, but there were no cameras nearby. (**The author has seen these flashes of light both in space and close on the ground as if someone set off a camera flash**).

533

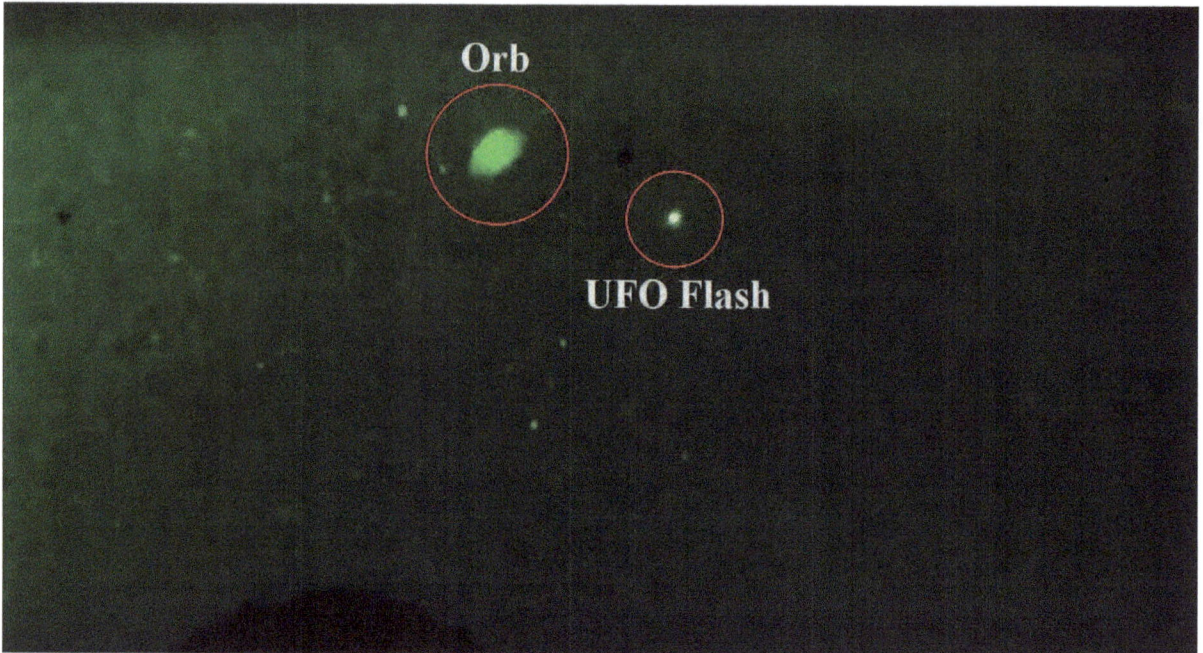

A night vision camera captures a UFO Flash (powering up) similar
To a camera flash and an Orb in the same photo
https://www.youtube.com/watch?v=dIi9ndSowks

Close Proximity Events with a Craft Beyond the Crossing Point of Light

There are a number of signs to look for in the event that a craft has approached a group in the field and is just beyond the crossing point of light. We will assume that there may also be ETs on the ground nearby or among the observing group. Although a craft or ETs may be observed at any time by someone skilled at remote viewing techniques, we will deal here with phenomena observed more typically with the "usual" senses. The use of slightly out-of-focus or **"soft eyes"** will often aid in seeing. Different observers will observe some, none, or all of the following. Seeing with both your physical eyes and the mind requires practice.

If the craft is close to the **crossing point of light** there may be **"bleed through"** and the craft may be partially visible, or perhaps just a sparkling is apparent to some. This sometimes appears as a faint glow or even a scintillating full form of a ship. Craft may appear either whitish or in soft colors. Keep in mind that craft may also become totally visible to everyone on the team. Both types of phenomena have been observed numerous times by CSETI field teams. http://cseti.org/ (updated, see: http://siriusdisclosure.org)

Sensory Awareness of High Strangeness Phenomena

During an event, members of the group may sense or observe some, none, or all of the following; these are some of the elements of high strangeness: **(This author has also experienced most of these high strangeness elements).**

- A sense of body warming.

534

- A sense there is an increase in the surrounding temperature.

- An apparent change in atmospheric pressure, which can be felt in observers' ears (**Sound Compression**).

- A decrease in wind - a stillness or quietness feeling perhaps leading to the feeling of warming mentioned above.

- Body vibrations from barely detectable to full-out shaking.

- Hair on head, arms or legs stands up (pilo erection)

- Sounds including: buzzing, humming, clicking, or strange, otherworldly screeching (sometimes heard by only some of the observers).

- Radar detectors setting off for no obvious reasons.

- Animals in the area will respond with howling, barking, etc. Animals will often respond to ET presence before humans are aware of it.

- Scents or smells including ozone and flowers scents such as violets, roses, carnations, sage, etc.

- Emotional feelings especially that of warmth and love, sometimes so strong that people are moved to tears. http://cseti.org/ (updated, see: http://siriusdisclosure.org)

If ETs are on the ground nearby or among the CSETI team, which has happened many times, the following additional phenomena have been observed:

- Shuffling sounds on gravel or rustling branches, leaves, or grasses. **(The author has experienced this phenomenon in a profound way as if the ETs were trying to get out of my way while I was setting up strobe light beacons).**

- Strange breathing or coughing sounds.

- Soft and gentle touches. **(The author has experienced this "ET touching" on several occasions).**

- Sparkling lights moving around and within the group. The lights can appear as small probes or as vertical forms or shapes of light often greenish or white.

- When a bare hand is moved through the light forms the hand will sparkle as though there was an electrical discharge

- The appearance of the ETs may range from just areas of faint sparkly light to indistinct shapes to fully visible entities with clothing, facial features, and hands, etc. all fully distinguishable. It should be noted that different people will see the ETs very differently, even when the people are standing beside each other so that some report seeing nothing, others a slight shimmering or shadows, and others can describe the ET in detail. Often, however, what is observed is just dark or fuzzy forms near the group as described by this team member observing in a small group in Sedona. "We had been visiting for about an hour when I noticed a fairly tall form to the right of the tree . It was upright and the same shape as the tree. I kept having a running conversation with myself that went, "Is that part of the tree or is that a life form" and "It really looks familiar but it can't be a life form, so it must be part of the tree." It didn't move, but I kept an eye on it. I have no idea why I didn't tell the others what I was seeing." She later described: "We decided to call it a night and started to pack up at which point I looked to see if the form by the tree was still there - it was gone. I stood in the spot where I had seen the form and estimated that it must have been at least six to seven feet tall." She added that she has seen this type of very tall form at least four other times. **(The author has also experienced these tall beings which were 6½ to 7 feet tall wearing dark hooded robes in 1996 at Crestone, Co. using "soft eyes" or the periphery corners of the eyes).**

- If the ETs are being "projected" into the area and a person walks into the projection area or "energy field", there is a distinct feeling of soft pleasant warmth as one passes into the "field".

- Interactions between ETs and individual group members range from just a sense of presence to loving personal acknowledgement to full telepathic conversations. The sense of love is almost always present, no matter the level of the interaction, and is truly wonderful and unforgettable. The conversations are typically non-verbal. Field observers have reported shimmering-light ETs that have stood in front of them or sat at (on) their feet for prolonged times.

- The ground around the group may appear colored such as emerald green or red. Some observers have noted complex geometric shapes and forms or beautiful unusual pictures while interacting with the ETs. These colors, forms, pictures may occur with the observer's eyes both open in the "soft eyes" mode or closed. One observer described the experience as though the ETs were sharing beautiful art with him.

- If the field team has been surrounded by a craft, partially in the ground and partially above, members can sometimes see structural parts of the interior of the ship depending on how visible the ETs make it and the person's ability to *'see'*. This is when there is a feeling of quiet, warmth, the star fields may appear to distort and there is often a strange sense of time distortion. Often when the craft leaves there is a distinct brightening around the group.

- Some or all the above interactions may also occur in this situation and when they do, it is truly a wonderful *"out of this world"* experience. http://cseti.org/ (updated, see: http://siriusdisclosure.org)

Group Members May Bi-locate

- In England in 1998, seven people sitting in a circle in a pasture were engulfed by a craft. The air became still, as though the wind were blocked, the temperature rose 10-15 degrees, and many ETs were perceived among the group of people. While remaining ever conscious of sitting in the pasture, the group was also conscious of looking down at Earth from space (through a window in the bottom of the craft) and joining the ETs in a meditation for the healing of the planet. **((This author has experienced this bi-location in Crestone in 1996, along with one other person which involved remote viewing, bi-location (being on board a ET spacecraft and viewing through some window or hull of the craft at the Earth below), and clairvoyance of future events which later**

unfolded the following day with a full unfoldment of foreseen events. The CSETI team responded to this subjectively acquired information as a real ET message)).

Other "Strangeness" Happenings added by Tony Craddock

These happenings by themselves appear as isolated "funny things that happened to me the other day while out observing or at home" yet taken as a group they may indicate some sort of interaction or communication by someone with perhaps a sense of humor. This is just a sample of some of the types of things that have happened.

- Randy's flashlight being disassembled in his lap during CTS and the pieces scattered around him and his sleeping bag. **(The author has experienced a similar occurrence which he refers to as "Hard" contact - see below).**

- The ETs insistence in locking Pat's door on my car. Being a Metro, it is 100% mechanical. We were at the jobsite a couple of weeks ago, and she heard a click as it was locked once again while she was standing next to it. Doors were locked and interior lights were turned on SG's car while we were in Hawaii.

- Smells. After the last night at Sedona had an overpowering smell of sage in my room all night (no, it was not toilet freshener!).

- C. being on-line the day after a field trip, her phone rings, and she carries on a conversation while still logged on to the Internet .... she only has one phone line, and it is not dedicated.

- Washing machines, dryers turning themselves on and their doors being left open (C.)

- Microwaves turning on. VCRs turning on and off and locking up. Squawking noises being broadcast thru pocket Dictaphone, and machine locking up.

- Logged on to the Internet, and while in Quicken, the program skips to an entry two years back and highlights a check written to someone whose name Pat was trying to remember ...of course this could also have been a mental feat.

- Pieces of toilet paper appearing to flutter around the sky in the dark of night (of course this could have been toilet paper if the night was windy!)

- Being touched onsite or having a sleeve tugged.

- Having pictograms beamed into your head - half inside, half outside, the bedroom pulsating with light visible with eyes open and closed. .... AJC

- SG sees UFO while we are chatting on the phone. http://cseti.org/ (updated, see: http://siriusdisclosure.org)

- **This author has experienced small things being moved around in his home or go missing entirely like remote control devices. Sometimes, there are sounds of movement in the upstairs living room, kitchen or bedrooms when everyone is downstairs in the family room. Other times voices or a single unintelligible word may be heard in one spot in the family room or coming from upstairs.**

- **Small ETs have come at least on one occasion to visit the author late at night right after a nighttime CSETI field trip just when he is in bed about to fall asleep. On such occasions, the author has asked the ETs to communicate with him in his lucid dreams as he physically needs to sleep!**

- **Some CSETI team members will have personal UFO sightings away from the team as validation of their efforts to communicate with them.**

- **ETs may on occasion, follow you around to monitor you and your activities and at times some people may be able to perceive them around you with "soft eyes".**

The above examples of UFO/ET phenomena were compiled by T. Loder (Dec. 4, 1998) and updated May 11, 1999, with help from L. Willetts, T. Guyker, T. Craddock, D. Foch and others Copyright 1998 CSETI
(Additional information in brackets and in bold text was included by the author) and "Contact" Countdown to Transformation – The CSETI Experience from 1992 - 2009 by Steven M. Greer, M.D. 2009; publish by 123PrintFinder, Inc.; Ladera Ranch, VA, USA; ISBN 9780967323831

**High Strangeness and ETI Interactions Experienced by this Author
and the CSETI Vancouver Team**

"Hard" physical interactions with ETs may take the form of ET spacecraft showing up near the field team as a close encounter of the first kind or it may engage the team by moving or manipulating objects in the CSETI worksite environment.

This occurred when this author and the CSETI Vancouver team were being interviewed by **BCTV John Valle Rao** and his news crew in the early Fall for a new piece on CSETI Vancouver organization and our CE-5 work. As we proceeded with the interview and the CE-5 demonstration, I couldn't help, but think of the last time a TV crew from the **Discovery Channel** interviewed myself and our CSETI team, that two ET craft appeared and there was an ET presence in amongst our group; I wondered then, if it would be the same again with this BCTV crew.

By the end of the interview, the news crew and our team began packing up our equipment to load back into our cars and vans. We had set up six strobe light beacons as part of our CE-5 demonstration and as I went to pick up these lights which were still strobing to pack away, I noticed that one light was missing. Thinking it had malfunctioned or had been knocked over causing the sensitive light to burn out, I began searching for it with my flashlight in the dark. I soon realized that it was no longer in the area as a part of the circle of strobe lights that I had set out earlier. One of my team members helped to search for it but, we could not locate it. At this point, I told my teammate that I would come back in the morning and look for it in the daylight.

As we got the rest of the equipment loaded into the car and everyone was about to depart the area, one of the TV crew people pointed in a southern direction and said, *"What's that light over there?"* Looking in the direction he was pointing at, near some small bushes, my teammate and I ran over toward the area of the light and sure enough, there was my missing light, 250 feet from its original position and still strobing!

My natural impulse was to think that some animal, like a coyote or raccoon, purloined the light thinking it was some type of "bright" food. However, realizing that it was not food, the animal dropped it and ran off to parts unknown. But, there was no evidence of salvia or smell associated with any wild animal on the light. It was quite dry without moisture on it, so how could the strobe light that was still functioning have travelled across our path between the cars and the work site with people were walking back and forth, without anyone seeing it being moved? No one on our team or the TV crew admitted to moving the light or trying to create hoax situation as a practical joke. What is curious was the fact that the light was one foot away from falling into a small creek gulley, which if it had been carried there by a wild animal and dropped into the small ravine, I would never have found it, nor would the TV crew have seen the light off in the distance.

In the back of my mind, I knew that something like this would occur as I had silently asked the ETs several months earlier to give me undeniable proof of their presence in this particular worksite location and therefore, was expecting something like this incident to unfold at some point in the near future. It seems the ETs did not deny me this proof of their presence and we have had other experiences with ETs at this same location, since that time.

A "Very Close" ET encounter incident which occurred at this location was when this same team member from the above incident and I went out together to do a CE-5 on our own. The other CSETI Vancouver members could not make it out that evening. The ETs showed their presence with a small flash of light that moved around near the ground of the work site. This was a small and brief encounter, so we thank the ETs for showing up and left the site just about midnight.

Upon reaching my home and then preparing for bed, I soon fell asleep around 12:30 AM. My daughter was out with some friends and she came home at 1:00 AM and as was her custom at that time, she came into our bedroom and kissed me and my wife good night.

Approximately an hour later, my wife suddenly turns over towards me and punches me in the shoulder!

I immediately woke up with some pain in my left shoulder wondering what was going on. I ask her why she hit me thinking it was probably because of my snoring but, she said with some consternation in her voice, *"It's one thing for you to go out and make contact with ETs, but do you have to bring them home!!"* I asked her, *"What do you mean?"*

She said that she suddenly awoke for some unknown reason and ***opened her eyes to see a face staring back at her, less than a foot away from her own face!*** In that instant, my wife said that the being with large wide eyes was also surprised by her reaction to its presence, as she was, to it starring at her. Then, it immediately disappeared from sight, which is when she rolled over and punched me in the arm.

From the rational and skeptical side of my brain that had somewhat been rudely awakened, I told my wife that she may have been half asleep when our daughter came home from being out late and having kissed each of us, she may have confused our daughter's face with the face of an ET being in her half sleep-dream state. My wife was adamant that it was not our daughter that she saw which caused her to awaken from a dead sleep but the unfamiliar face of an ethereal presence not of this world!

It is not unusual for people who become involved with CSETI to suddenly find themselves seeing UFOs or ET beings on their own away from their CSETI team and this was becoming almost habitual for me personally. People would often tell me, particularly those who are intuitive or seeing with "soft eyes" that they could see small ET beings hanging around me most of the time they were in my company, for which at times, I had become accustomed to and somewhat oblivious to their presence.

A "Remote ET" interaction incident occurred on a more subtle level, yet was nonetheless, a physical manifestation of the ET presence and their willingness to interact with me. I had read that some people can get ETs to communicate with them through their computers whether intentionally asked for or by mere serendipity. I put out my intention to the ETI that if they could "hear" me or read my mind to please provide me with another indication of their presence by interfacing and controlling my computer. People may remember that Microsoft's operating system **Windows XP** had a built-in desktop image of "green grassy rolling hills with some distant mountains in the background all under blue skies with some fluffy clouds".

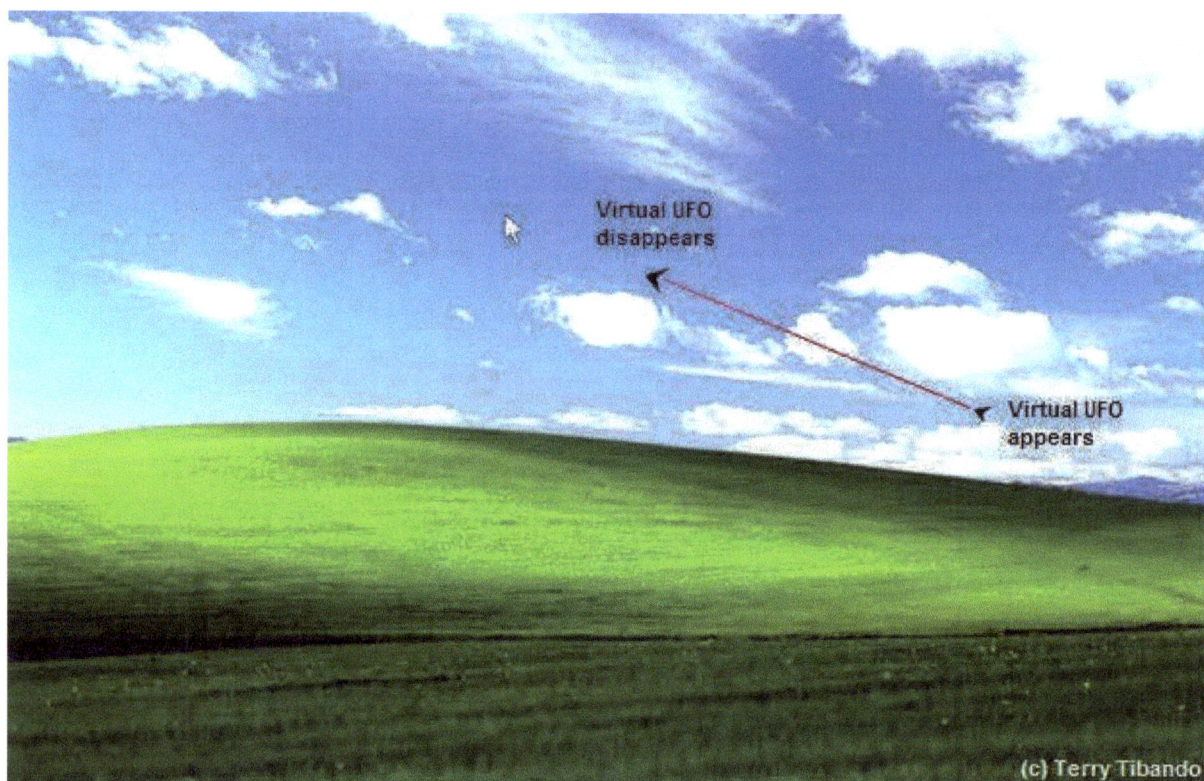

ETI interfaced their technology with my computer to communicate by placing a "black boomerang UFO icon on my desktop screen. It would slowly "fly" From right to left as indicated by the red line "flight path" over a period of a few days and then it mysteriously disappeared.
(c) Terry Tibando

540

In the late '90s, I came home one afternoon from work and jumped onto the computer and to my amazement I had two mouse cursor arrows on by screen. One was the typical small white arrow that I could move about and the other was a ***small black boomerang object*** which I had no control over and which appeared to be stationary on the screen "floating" above the hills close to the mountains. It was not immediately noticeable and I thought at first, it was a piece of dirt or dust on the computer monitor and when I tried to "wipe" it off the screen, I realized it was a part of the desktop image and it was not the mouse cursor arrow. (See photo image above). Now, it is possible to photoshop this image but, what was unusual, was that it moved very slowly across the monitor screen on its own growing slightly larger over a period of a few day before disappearing completely when it reached the centre of the screen.

I remembered telling a dear friend who has since passed away about what I was experiencing and as a computer programmer, he had not heard of such a thing before and was very curious how it could appear and move on its own over a period of a few days and then disappear on its own.

This was the proof I needed that the ET presence was still around me or at the very least would still interact and engage with me as long as my intention were pure and genuine. ETI can through their advanced technology interface with our simple electronics like radar detectors and magnetometers and now computers to control them for the purposes of communications as has been the experience numerous times by many **CSETI** teams worldwide.

I am also of the conviction, that it may be possible to record ET conversation and their language using iPhone and Android cell phones. This may give us a breakthrough in ETI-human communications whereby they could interface with cell phones and record their spoken and possibly written language for us to decipher and learn!!! Unless, of course, they have learn to speak English or some other recognizable language in which to communicate with humans!

Dr. Greer's latest book, "Contact: Countdown to Transformation", deals exclusively with numerous examples of UFO/ET phenomena that the CSETI field teams have observed or encountered globally, far more than can be recounted here, but the reader is encouraged to read Dr. Greer's book as it gives excellent accounts of the types of **Close Encounters of the Fifth Kind** that people can expect when engaging in the **CE-5 Initiative.**

The photographing of ETI presence on camera and video is beginning to be more common with CSETI groups that show up during a CE-5 event or around the homes of some researchers either as ETI or orbs and craft. It appears that some ETI have an intimate interest in some people for reasons only they seem to know.

Recently, this author had security cameras installed in the front and back of his home as there were reports in the neighbourhood that some homes were being broken into through the garage doors of the homes.

In the morning of Feb. 15, 2021, I was checking to see if any activity had been caught by these security cameras as my wife and I have sometimes hear noises in and around our home at night, so I was curious to see if the cameras had video recorded an animal (raccoons are common in the area) or an unwanted intruder.

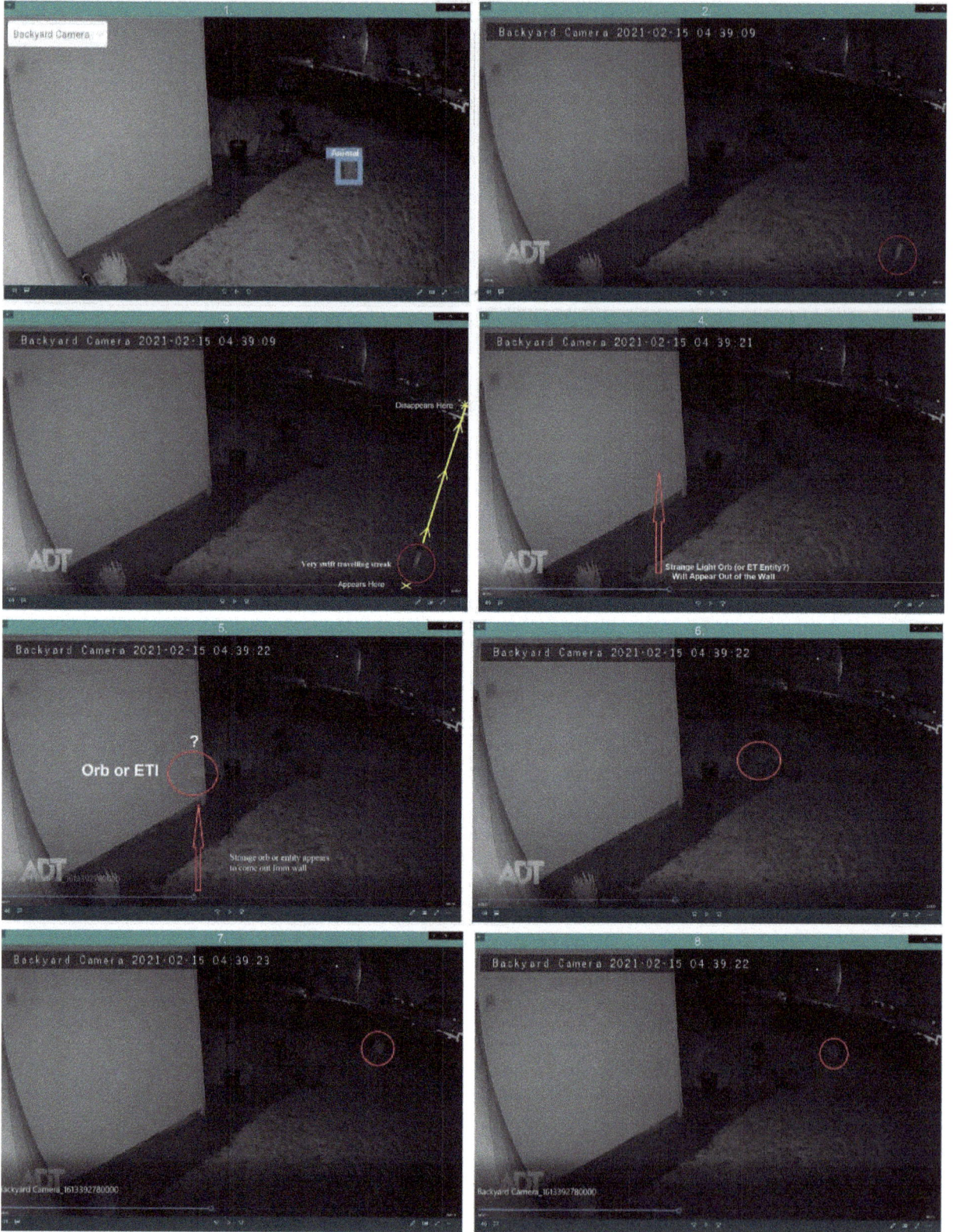

In these sequences of photos taken from my security camera video, it can be seen that several ETI have emerged from my home through the wall inter-dimensionally.

© Terry Tibando

542

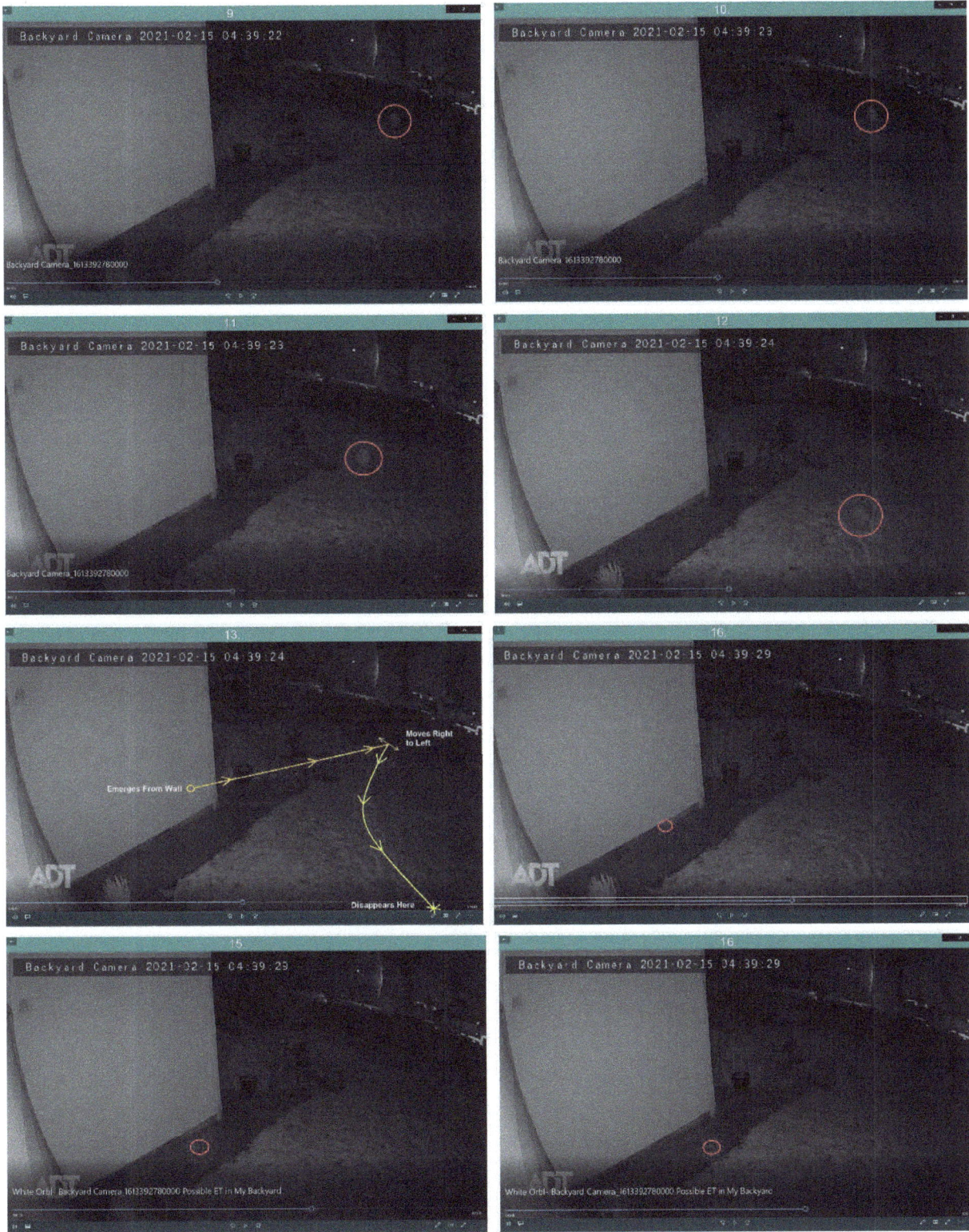

In the second sequence of photos can be seen the continued movement of the large
amorphous entity and the small entity emerging through the wall.

© Terry Tibando

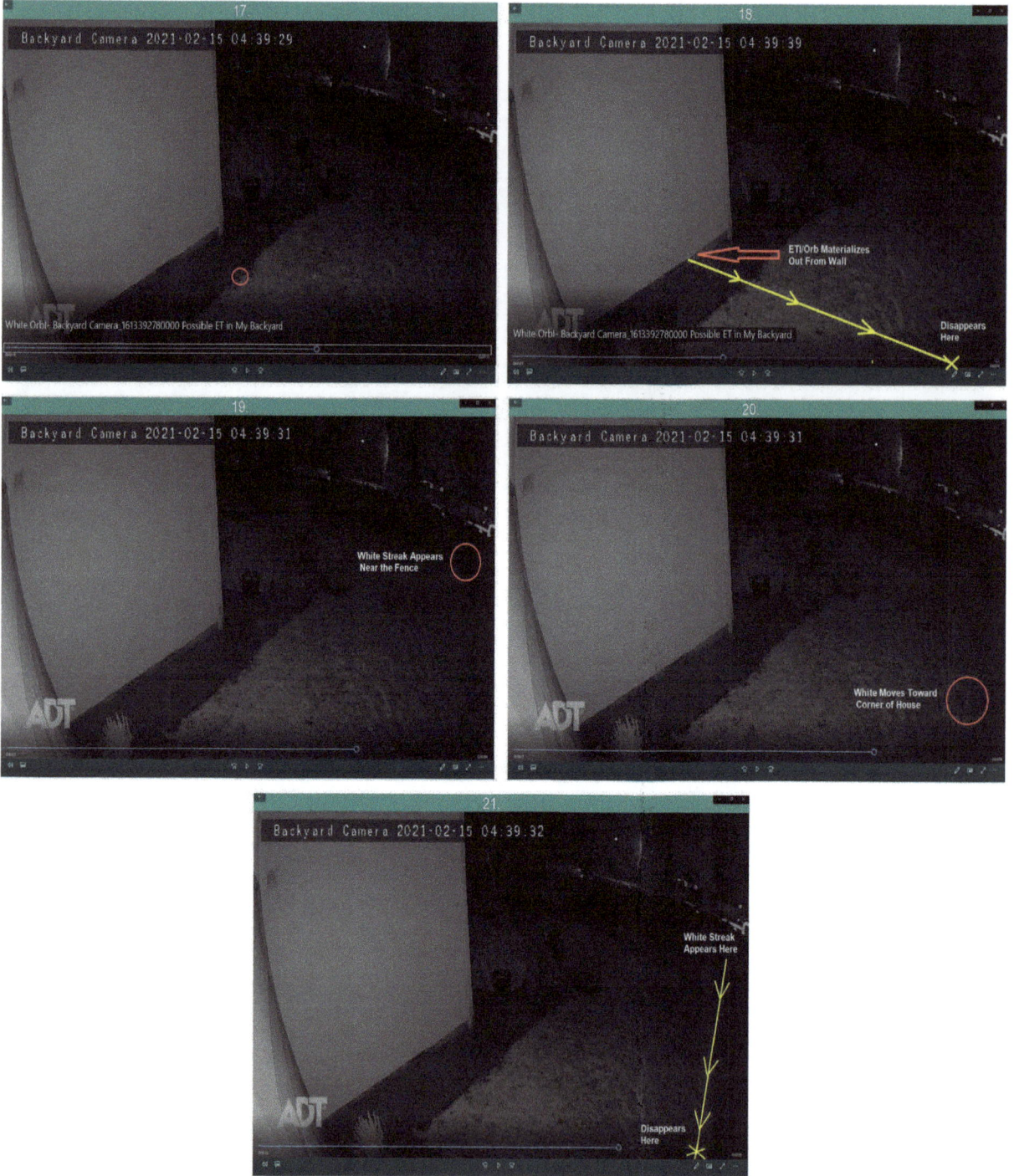

In this third sequence of photos the movements of the small amorphous entity and a streaking light appear to be gathering at the corner of the house.
© Terry Tibando

I soon realized that the backyard camera had recorded what it thought was an *"animal"*, however upon closer inspection, the *"animal"* was not a recognizable and in fact, it wasn't even an animal, but an amorphous blob or orb!

Initially, a white object came streaking out of the corner of the house where the camera is located and quickly disappeared near the fence.

A few seconds later, the larger object which the camera labelled as an animal materialized out through the wall of the house and moved quickly to the back gate; it shifted right and left and then moved toward the corner of the house where the first object had appeared and then this large amorphous object disappeared at this corner.

This was followed by a much smaller amorphous orb materialized in approximately the same spot as the first entity, but lower to the ground. This smaller entity/object moved swiftly toward the corner of the house and it too disappeared.

Almost immediately, a third object streaked from the fence toward the same corner of the house where the other entities/objects had disappeared. Was this a gathering of anomalies congregating in one location, the corner of my house?

The fact that the camera has an infrared setting and it captured what it thought was an *"animal"* makes me suspect that what were recorded were living entities employing some aspect of inter-dimensionality (i.e. moving through the house wall), that was either natural or technological.

With the odd sounds and noises in my home, I intuitively felt that ETs have been hanging around the house perhaps, monitoring us or simply paying the occasional a friendly visit, late at night! (see photos).

The CSETI End Game – Becoming an Interstellar Civilization

What does the future hold for CSETI and for humanity with regards to Extraterrestrial contact; this is the question that constantly occupies the mind of Dr. Greer as well as for many Ufologists. His hope has always been that there will be an about face in the position held by the US government and the military with an open transparency and disclosure of the UFO/ET phenomenon. He feels with great conviction that one day there will be televised landing of an ET craft in a desert like area, where a very wise Extraterrestrial being representing a coalition of Extraterrestrial civilizations, will emerge from his egg-shaped craft, who will be greeted by major world leaders and wise elders of humanity with the full support of the military including the air force! Dr. Greer hopes to be there when that day comes, as does this author. It is a dream of hope and peace represented by an ETI-human contact event of great historic importance for the whole of humanity.

The hurdles to overcome in order to reach this future destiny will no doubt require a massive social re-alignment to the way humans think and behave towards one another. One of the biggest struggles Greer has had to deal with since 1990 has been not letting people hijack the CSETI efforts with their own agenda. Greer says that many people have childishly adopted racist archetypes, and as a result are just recapitulating the tribalism and racism of Earth. This is not what we need to do, as we step from Earth into space. We need to leave that primitive thinking behind. That is the key principle of the **CE-5 Initiative**.

One of the hardest tasks for **Dr. Greer** was to ensure that people who wanted to be diplomats when attending the **CSETI expeditions** had the appropriate altruistic motive and consciousness – a place of higher consciousness of unprejudiced interplanetary diplomacy. **"Contact" Countdown to Transformation – The CSETI Experience from 1992 - 2009 by Steven M. Greer, M.D. 2009; publish by 123PrintFinder, Inc.; Ladera Ranch, VA, USA; ISBN 9780967323831**

This means being peaceful rather than wanting to shoot a vehicle down to acquire the technology or wanting to meet one type of Extraterrestrial, but not another or wanting to make an alliance with one group only, etc. Such tribal nonsense has been the bane of human civilization for thousands of years. That divisive thinking needs to be transformed into a new vision of universal Oneness.

Greer believes that it is possible for a few dedicated souls to make a difference in changing the world if they but arise and know that they have the capability to act and affect change.

"One fruitful tree is conducive to the life of society." -- **'Abdu'l-Bahá**, son of **Bahá'u'lláh**

"Never doubt that a small group of thoughtful committed people can change the world; indeed, it is the only thing that ever has." -- ***Margaret Mead, anthropologist***

"Few men are willing to brave the disapproval of their fellows, the censure of their colleagues, the wrath of their society. Moral courage is a rarer commodity than bravery in battle or great intelligence. Yet it is the one essential, vital quality for those who seek to change a world which yields most painfully to change." -- ***Robert F. Kennedy 1966 Speech, US Democratic Politician***

"Here's to the crazy ones, the misfits, the rebels, the troublemakers, the round pegs in the square holes... the ones who see things differently -- they're not fond of rules... You can quote them, disagree with them, glorify or vilify them, but the only thing you can't do is ignore them because they change things... they push the human race forward, and while some may see them as the crazy ones, we see genius, because the ones who are crazy enough to think that they can change the world, are the ones who do." -- ***Steve Jobs, US computer engineer & industrialist (1955 - 2011)*** http://disclosureproject.com/

Greer realizes that any project that benefits mankind may not be easy and that a herculean effort is often required to achieve success, particularly when there are enormous powers working to stop CSETI that represent the most powerful interests on Earth. When Presidents, senators, congressmen and high-ranking military officers and a sitting CIA Director, people who serve the US government in some of the highest offices and posts in the nation are cut out of the loop of information on the UFO subject by a subversive, rogue government within a government, then you know that things have gotten dangerously out of hand. **"Contact" Countdown to Transformation – The CSETI Experience from 1992 - 2009 by Steven M. Greer, M.D. 2009; publish by 123PrintFinder, Inc.; Ladera Ranch, VA, USA; ISBN 9780967323831**

However, there is "no force" or power "on Earth" that can prevent or stop "the inevitable transformation of life" on this planet from moving to "the next level of evolution."

546

There is a universal process and purpose at work here and it is beyond the ken and control of men.

Of course, **CSETI** is doing its work on a relatively low-tech level. One thing I need to emphasize is that this entire project from its inception to the present has been an unfunded, volunteer, un-staffed, citizens' diplomacy effort. We have gone all over the world doing this work without an office, without staff, without equipment, and without a budget. And yet, we have not been limited at all in what we can achieve in actual contact. This is a very important point. We do not have access to the secret arsenal of the vast military industrial consortium dealing with the ET presence and free energy subject. They have been developing electromagnetic systems, anti-gravity and free energy systems since the 1940s. Nor do we have the hundreds of billions of dollar of funding they have illegally obtained over the last few decades. But what we do have is what the extraterrestrial civilizations are looking for – an unprejudiced mind and heart, clear intent and peaceful purpose. This all you need, and if you have it, you can do the **CE-5** work and make contact wherever you are.

A question that has been constantly asked of **Dr. Steven Greer** from aerospace engineers and scientists is that they knew that there was a nexus between ET technologies and consciousness because they had seen the reports but, had not figured out how it worked and they knew that CSETI protocols had successfully worked as they are based in consciousness and because ET communications use technologies that interface with thought.

One scientist asked Greer how people experience contact with extraterrestrial civilization and was told by Greer that the number one way that actual and real contact occurred between humans and ETI was in the lucid dream state. ET technologies that interface with consciousness are seamless, even their spacecraft are conscious because their technologies are already in the critical mass of artificial intelligence. When they travel faster than the speed of light through interstellar space, they pass through the astral domain – the etheric realm of the lucid dream state, it is at that point that humans and Extraterrestrials can communicate. This aspect of ET technology has always baffled western scientists who always approach things with a reductionist point of view which limiting. Consciousness is, however, at the core of ET science. The synthesis of technology/consciousness is so advanced it is believed to be impossible by mainstream science and thus, is view as even beyond magic to most people. And yet, it will be the future of science on Earth.

The next big breakthrough in science according to Greer will be in the area of consciousness and no doubt the superpowers will jump all over this new science to sequester it, claiming it as the imminent private domain of the military and the intelligence community. This may be something which they may already have done, but consciousness is the God-given right of everyone on Earth and in the universe and therefore, cannot be confined, limited, sequestered or suppressed from the public domain. Communications between humans and extraterrestrials have always been a traditional domain of all cultures and societies as far back in recorded time, particularly among shamanistic tribes, monks, and clerical orders. *"Contact" Countdown to Transformation – The CSETI Experience from 1992 - 2009 by Steven M. Greer, M.D. 2009; publish by 123PrintFinder, Inc.; Ladera Ranch, VA, USA; ISBN 9780967323831*

Part of the new science of consciousness will be the development of consciousness- assisted technologies and technology-assisted consciousness the central core science of these technologies is the consciousness and mind interface with the cosmos, space, time, matter and electromagnetism. Greer has written an excellent paper on this topic back in 1991 titled, "A Comprehensive Assessment of the UFO/ ET Subject" in which ET civilizations that have strong sensing capabilities, and which are thousands to millions of years more technologically and socially developed than we are, use such extremely advanced technologies to mediate their capabilities.

One of the goals of CSETI is to describe and explain how extraterrestrial technologies are manifested and are experienced among CSETI trainings and expeditions and the role that consciousness plays in these ET manifestations.

This brings up the question of why are Extraterrestrial Intelligences here now, visiting the Earth? Again, Greer explains that there is no one simple answer. There appears to be thousands of reasons why there are here and that they are carrying out specific tasks which would account for the reports by witnesses of seeing many different kinds of ETs and different ET spacecraft. This is also indicative that a large number of ET civilizations are represented and are working together much like the **United Nations,** except with more solidarity and unity of conviction. Some ETs may be here to monitor Earth's tectonic plates while others have weapons storage bases under surveillance like the Los Alamos weapons site area. The extraterrestrials may also be here to mitigate other Earth problems and they are definitely monitoring our global war conflicts.

Lord Mountbatten, a member of the British Royalty stated before he was murdered, *"When we started detonating thermonuclear weapons in the atmosphere of Earth, we kicked a cosmic hornet's nest!"* **British intelligence agent interviewed by Dr. Greer.**

Colonel Ross Dedrickson, senior official of the **Atomic Energy Commission (AEC)** was in charge of inspecting all our nuclear weapons storage facilities informed Greer that every single one of them was being monitored by extraterrestrial vehicles due to their deep concern over nuclear weapons. Such concern has been repeatedly and independently reported as well.

Here is a prime example of ET technology interfacing with and actually taking control of human terrestrial based technology, not just manipulation of small home computers but, missile base computer technology that control the launch of weapons of mass destruction anywhere on the planet. As indicated below by Greer, this is a major wake-up call to humanity. The real question is how long do we have to get control of our destiny and start developing a peaceful, united civilization before that decision is made for us by off world visitors?

Arnie Arneson, one of Dr. Greer's **Disclosure Project Witnesses**, who was in the **Malstrom Air Force Base** nuclear silos incident in 1967 when an ET craft took multiple ICBMs offline, rendering them un-launchable. The same morning another ten were neutralized at a squadron 30 miles away. About the same time, the Soviets also lost launch control of their missiles at the control centre. There was a clear message to these events that the Extraterrestrial civilizations were warnings us and that was not to blow up this beautiful planet, and this loss of missile launch control was a demonstration that they could intervene. Destroying the Earth would have effects on deep levels of consciousness and reality elsewhere because all life is connected. Earth is not as separated in space as many of us would like to think. It is a very deep issue to contemplate. **"Contact" Countdown to Transformation – The CSETI Experience from 1992 - 2009 by Steven M. Greer, M.D. 2009; publish by 123PrintFinder, Inc.; Ladera Ranch, VA, USA; ISBN 9780967323831**

One reason they are here, Greer points out is that CSETI teams are inviting them to cultivate a relationship with us for the higher purpose of interplanetary, universal peace, to help heal the burdens and stresses placed upon the Earth by humanity and to help humanity to evolve spiritually to higher levels of consciousness. CSETI recognizes that the source of conscious intelligent life in the universe is **One Awake Being**, an awake mind within us. Greer says that unity consciousness is at the heart of all spiritual traditions, **Christian, Jewish, Buddhist, Hindu, Muslim, Baha'i, Shamanic, Native American,** etc.

ETs also seem to have an interest in humans who understand the interconnectedness of all things which may explain why some people are singled out with a stronger ET attention than others because these individuals are more consciously developed. Truly wise people see the interconnectedness that exists rather than falling for the illusion of separation. Modern science is excellent in the *"male"* linear explosive, destructive process and pursuit of reductionism that focuses on the parts by eliminating the whole. In this day and age, the new science of consciousness or spirit will help us to focus on the integrative *"feminine"* function that is implosive and creative, where reality is non-dualistic.

Contrary to the traditional concept that we might expect, that "the whole is greater than the sum of its parts", we should consider the **holographic universe** concept where *"the part is the whole!"* **"The Holographic Universe" by Michael Talbot; 1991; published by Harper Collins Publishers, Inc.; New York, N.Y., USA; ISBN 0-06-092258-3**

An integrated, conscious reality affects all of relative existence at all strata from the finest celestial manifestation to other planets. This is **Rupert Sheldrake's** cosmic extrapolation of the **morphogenic fields** which operates at all levels. This might be the reason why ETs have an interest and concern for what is happening on Earth.

In reality, all matter is consciousness resonating as a specific object. On one level you can see it as a "discrete object", and it is. On another level of consciousness – unity consciousness – it is actually consciousness manifesting as a discrete body, or stars, plants, wood, minerals, etc. so from the point of a cosmically awake person who is aware of **Cosmic Being**, everything is emanating continuously from mind, and thus we create form and events from Mind.

Each person, then, is the **Universal Being** – and can manifest from within this state of Oneness. But this emerges from beyond the ego – now we are speaking of the **Great Being** standing within all of us as one... From that station in awareness, we can manifest all.

The CSETI philosophy is to give people the tools, articulate the techniques, enhance understanding and encourage individual exploration. That is why CSETI is so decentralized and lacking a bureaucracy. You just go out, form your own Contact group, and explore. **"Contact" Countdown to Transformation – The CSETI Experience from 1992 - 2009 by Steven M. Greer, M.D. 2009; publish by 123PrintFinder, Inc.; Ladera Ranch, VA, USA; ISBN 9780967323831**

The CSETI Vancouver team coordinates a mountain top field expedition with the Vancouver UFO Meetup Group to establish ET contact and communications using the CE-5 protocols
(c) Terry Tibando

550

CHAPTER 122

SOME ANSWERED QUESTIONS ABOUT MILITARY COVER UPS, UFOS, ETI, ZERO POINT ENERGY

"There is no doubt that we are dealing with beings who are capable of reading, by both telepathic and conventional means, our true intentions and motivations, and who can sense the "spirit of our endeavor". Success requires that the spirit of our endeavor be one of scientific openness, the search for truth, altruism, selflessness, harmlessness and non-covetousness. A desire for the peaceful furtherance of the ETI- Human relationship is paramount. For these reasons, "purity of motive" on the part of human researchers and investigators is a primary requisite while specific skills, expertise, and technology are important but secondary considerations. The breadth and clarity of our consciousness is imperative and transcends all other considerations. Our obsession with technology and outward things tends to obscure the Big Picture of the ETI/Human relationship, and all that it entails. While competence and knowledge cannot be slighted, we must insist on the primary importance of consciousness. In this regard, it is likely that a novice possessed of noble intentions and equipped with only a flashlight would meet (has met?) with greater success than a governmental agency motivated by lesser intentions, even though it has advanced technology, personnel and billions of dollars at its disposal. Indeed, an aboriginal with only a bonfire may go further in the establishment of communication and the discovery of truth!"
Extraterrestrial Contact: The Evidence and Implications by Steven M. Greer, MD p. 178 – 180 and http://www.siriusdisclosure.com/cseti-papers/the-imperative-of-consciousness/

The next 500,000-year cycle is the cycle of the fulfillment of the potential of humanity. One of the cornerstones of that fulfillment is for individuals to take full responsibility for their spiritual development, experience, and knowledge. They must not allow themselves to be passive and infantilized by a priesthood, leaders or authority figures. A certain amount of true (and not feigned) humility needs to accompany a spiritual teacher. Let us share knowledge and wisdom, open the door and point the way with some techniques, then let people find their *own* path.

In this new cycle of fulfillment destined to last 5000 centuries, the Extraterrestrial Intelligences are here, now, to understand us, to assist us and to eventually build a bridge between humanity and other peaceful civilizations. Humanity is still in a state of adolescent immaturity and chaos, but our adolescence is nearing an end and one day we will go into space among the stars and planets as a peaceful civilization. At that glorious stage in our evolution, we will become the new Extraterrestrials to other planets! This is the heart of what Extraterrestrials are interested in - **the establishment of universal peace!!!** **"Contact" Countdown to Transformation – The CSETI Experience from 1992 - 2009 by Steven M. Greer, M.D. 2009; publish by 123PrintFinder, Inc.; Ladera Ranch, VA, USA; ISBN 9780967323831**

S.G.: *There's a baseline level of energy that's in all the space around us -- not outer space, but the space in this room. It's estimated that every cubic centimeter of space here has enough energy to run the entire Earth for a day. This can be tapped. Some scientists are trying to fabricate motors that extract energy from the quantum vacuum, the energy that surrounds us, the* **zero point energy***.*

I've actually seen one of these [devices] working. Of course, the inventor has been threatened. **Dr. Eugene Mallove,** *who was murdered in May [2004], and I were working on this. I'm carrying it forward, but it's high-risk. [For more on Dr. Mallove and cold-fusion technology, check out* **PureEnergySystems.com**

S.G.: *We're fairly close. There are people with "proof of principle" things, but no products for sale yet. Right now we're working with an inventor who has created energy from this quantum vacuum in the hundreds of watts range -- enough to run several things in your home. We can also demonstrate the antigravity effects of high-voltage systems under certain controlled experiments.*

An electronic field makes something fairly weightless. We're working with a man who's done very advanced antigravity work. The colonel who classified it is willing to declassify it for us to take forward.

We've seen this done in a lab. The techniques for tapping this energy have already been fully developed within rogue and covert programs. **Lockheed Skunkworks** *has enormous ships zipping around that are [powered by] antigravity. A lot of the UFOs sighted in the high desert of California and Utah are actually manmade prototypes.*

Who else is making these phony UFOs?

S.G.: *The companies involved are* **SAIC (Science Applications International Corporation), TRW, Northrop, Raytheon,** *and* **EG&G.** *We have enormous intelligence on this. I know the buildings where this stuff is going on. This needs to come out so people know the truth.*

Does our tax money support these secret programs, or are they privately funded?

S.G.: *Both. There's private, corporate funding as well as what's called "black budget" sources. I met with* **Senator Robert C. Byrd's** *staff in 1994. His senior investigator and chief counsel for the* **Senate Appropriations Committee***, of which he was chairman at the time, told me that this stuff was real, [but that] they could not penetrate this black world dealing with UFOs. At that time between $40 billion and $100 billion, a year was going into these projects, and they could not trace the money.*

Why are these programs kept secret?

S.G.: *Fossil fuel and the nuclear power industries would be made redundant by these technologies that very elegantly extract energy from the quantum vacuum, or "zero point" energy field. We have a $7 trillion part of the world economy dealing with fossil fuels and conventional*

552

transportation. If this information comes out - aside from people realizing that we're not alone in the universe - they'll quickly see that we don't need oil, coal or central utilities. It's all about maintaining the homeostasis and status quo of the world macro-economic and power dynamic.

How would zero-point energy transform the global economy?

S.G.: *It would replace everything. You wouldn't need oil, but the $30 trillion-a-year global economy would quickly grow to $200 trillion because there'd be clean, sustainable energy, and manufacturing and transportation would be very inexpensive. Eighty percent of the world's population lives in amazing poverty, and it would lift that. It would revolutionize the planet. People talk about the "peace dividend," but it's time for a "space dividend."*

Who's hiding these advanced energy systems from the public?

S.G.: *The entity that runs this stuff is the world's largest **RICO (Racketeer Influenced and Corrupt Organization).** They used to be called **MJ-12 or Majestic - 12** but, the last term I heard was PI-40. It's not one society. There are sweeping conspiracy theories about the **Masons, Bilderbergers, Trilateral Commission** and **Counsel of Foreign Relations**. I know people in all these entities, and most of them couldn't find their ass in a well-lighted room. It's much more prosaic and nuanced than that.* Witness Testimony Overview - "The Greatest Secret in Modern History" - excerpts from *Disclosure: Military and Government Witnesses Reveal the Greatest Secrets in Modern History* by Steven M. Greer, MD.

Why do they feel threatened?

S.G.: *It would decentralize power. Right now. the centralized financial and oil system is so integrated with the way the world runs. We have testimony about what the agenda is from people who've been on the inside.*

Who's in this covert group?

S.G.: *There's a committee of 200 to 300 people who are on the policy board for this issue. Admiral Bobby Ray Inman, who went from head of the National Security Agency to the board of SAIC - which is one of the crown jewels of this covert entity - is a member. So is **Admiral Harry Trane, George Bush Sr., Dick Cheney** and **Donald Rumsfeld** are involved, as is the **Liechtenstein banking family**. The **Mormon corporate empire** has an enormous interest in this subject; they have much more power than the **White house** or the **Pentagon** over this issue. And there are secret cells within the **Vatican**.*

Have you met with any of these people?

S.G.: *There are factions within this group, and I've met with some of the "good guys." People think that it's a monolithic conspiracy, but they're wrong. About 40% to 50% of people involved in these super secret projects want this stuff out. They know we're running out of oil, China is industrializing, and the polar ice caps are melting. And they know that if this [advanced E.T. technology] was announced today, it would take ten to 20 years to get it into widespread* *application to avert an economic, strategic, geopolitical and environmental catastrophe. They see that and want to fix it, but they're still a minority. They're more enlightened, but it's the ruthless ones who rule.*

Was CIA Director William Colby involved at some point? Was his "accidental" death in 1996 connected to your work?

S.G.: *Bill Colby was defecting from the super secret group, and he was assassinated because he was going to transfer some hard technologies - operating devices - and $50 million in funding to us. He knew that with my kind of willpower and connections, I would have gotten that out to the world. He was found floating down the Potomac River the week he was going to meet with my closest friend. They made an example of him.*

*In fact, Colby's best friend, a colonel who set up the meeting, said it was absolutely a hit. Even his wife said on **CNN**, "You know, it was strange because he would never go out canoeing in a flooded, rain-swollen Potomac River at night and leave the house open and the coffeemaker and the computer on. That's not like Bill Colby at all."*

Colby's widow stopped short of calling it murder.

Someone may have said, "Play along, or your children are next." I mean, these people are thugs.
Witness Testimony Overview - "The Greatest Secret in Modern History" - excerpts from *Disclosure: Military and Government Witnesses Reveal the Greatest Secrets in Modern History* by Steven M. Greer, MD.

Has your life been threatened?

S.G.: *[Around the time] Colby was killed, my right-hand assistant, a member of Congress who was working very closely with us and myself all got a deadly type of cancer in the same month -- different kinds. I don't talk about this publicly much because people say, "Come on!" But it absolutely can be done, it was done, and everyone died but me. I was devastated, and it took me 18 months to recover.*

You're saying the UFO-coverup conspirators can target people from a distance and produce cancer?

S.G.: Yes. ***Dr. Tom Bearden's*** *books about the **scalar electromagnetic weapons** systems explain how longitudinal waveforms can be a carrier wave for this kind of thing. They can get it down to a cell, to your DNA. It's all resonance. We've had this technology since the 1950's.*

Do you have witnesses willing to step forward and reveal what they know about this massive coverup?

554

S.G.: *We have about 450 military and government "insiders" who have been present during rather undeniable events, including the study of these energy and propulsion systems.*
Witness Testimony Overview - "The Greatest Secret in Modern History" - excerpts from Disclosure: Military and Government Witnesses Reveal the Greatest Secrets in Modern History by Steven M. Greer, MD.

*"We have a number of smoking gun documents, including a wiretap of **Marilyn Monroe** the day before she died, which has never been declassified. She was threatening to hold a press conference to tell the world what Jack Kennedy had told her during pillow talk about having seen debris from an extraterrestrial vehicle at what the document calls a "secret air base." She was murdered for this.*

The biggest problem is the media. If a congressman starts looking into this, they get pilloried and ridiculed by the media. There's a CIA document from 1991 that clearly states that the Agency had contacts or resources at every major media outlet to kill, spin or stop stories. It's a canard that we have a free press. Also, people can't wrap their minds around it. A former editor at the Boston Globe told me, "This is just too far-out. And even if it's true, we'd never run it because it's associated with tabloids and the lunatic fringe."

"It's about how we live our lives. Aside from cleaning up the environment, it [E.T. technology] extricates us from our dependency on oil, gets us out of that Middle East mess and saves the average American household thousands of dollars a year in heating, utilities and gas for their cars. And, globally, it frees up enormous amounts of economic potential. This would be the tide that raises all ships.

*It's a sort of Promethean challenge. **Lawrence Rockefeller** once told me, "It's wonderful that you're going to do this, and it's so important. But I'm afraid you're going to be the court jester that people won't take this seriously." And I said, "Well, we at least have to try."*

Extraterrestrial visitors are waiting for us to grow up and quit destroying Earth and each other. That may be the entry prerequisite for joining the interplanetary club but, we're not there yet. This then is the biggest challenge that the human race must face and come to terms with.

I'm very hopeful. In fact, I have no doubt that the outcome will be peaceful, that these technologies will completely rehabilitate Earth's fortunes and environment and eliminate poverty. And it will happen in our lifetimes. But the question is how much madness has to go on between now and then?"

Releasing Suppressed Technologies to Eliminate Poverty and Pollution

The quest for alien propulsion systems has been likened to the quest for the "***Holy Grail***" or the "***Philosopher's Stone***". Knowing how alien technology works and in particular, what energy sources power and propel alien spacecraft would allow scientists to understand and implement this advanced technology into their own terrestrial technology. Solving how flying saucers are getting here is not only the Holy Grail to the whole UFO mystery, but it is the primary raison d'étre of the military, the military-industrial complex, the intelligence communities of many governments and most space agencies like NASA thereby, giving any nation a greater superiority edge, both technologically and militarily over other nations.

This then becomes the primary motivation among military powers for an aggressive program to track, target and shoot down any Extraterrestrial spacecraft in low orbit or in the atmosphere above their nation. Military brinkmanship has developed beyond nuclear weapons to a whole new level, because of the aggressive acquisition of alien technology which has become one more "toy" for a nation to add to its "toy box" of armaments!

Dr. Greer is able to prove through the testimony of 1000 witnesses, that there have been covert programs that have studied and figured out the energy and propulsion systems behind UFOs. *"We're talking about a whole new type of physics that would enable humans to generate energy from what's called the "quantum vacuum."*

Readers are encouraged to visit Dr. Steven Greer's video lectures on YouTube in which there are hours upon hours of lecture presentations covering a wide range of topics related to everything covered in this book as well as other subjects pertaining to the UFO and ETI phenomenon.

CHAPTER 123

A GLOBAL VISION - HUMANITY MATURES TO BECOME AN EXTRATERRESTRIAL CIVILIZATION

This textbook has proven beyond any doubt that **Unidentified Flying Objects (UFOs)** or **Extraterrestrial Vehicles (ETVs) aka. Extraterrestrial Spacecraft (ETS)** are a reality which are currently visiting this planet. For most Ufologists on the front lines of investigation, the question is no longer "Do you believe in UFOs?" The real investigation is now focused on answers to the questions: "Who are the pilots and navigators of these spacecraft"? "Where do these **Extraterrestrial Intelligences (ETI)** originate from"? Why are these Extraterrestrial intelligences visiting our planet, do they have an agenda"? "How are they getting here"?

"How are they getting here"? is the primary reason behind the coverup, suppression and control of the whole UFO/ETI knowledge by the Military Industrial Complex, by the Intelligence Community, by rogue branches of the government, and by the cabal of oligarchical private industrialists and wealthy corporate elite.

Knowledge is power and those who control knowledge have power to control others and ETVs derived from alien technology represents highly advanced knowledge beyond most terrestrial understanding. and thus, in the hands of a few military groups represents advanced power. This is a power that is jealously guarded and protected away from the reach or knowledge of the common public. It is the reason why targeted flying saucers are shot down and their crash retrievals are accomplished so quickly and secretly before the public even get wind of such an event.

When a solitary scientist or garage inventor tinkers away on a project that could provide a breakthrough in energy generation or a new form of propulsion to various transportation systems, the powers that be, who are ever-watchful, always-listening get wind that someone is about to discover the "Rosetta Stone" to alternative infinite energy (Zero Point Energy) or to an amazing propulsion system, they swoop in and seize all the equipment, papers and documents of that scientist or garage inventor. They may give no reason for their actions other than to say there were gross violations and breaches of the National Security by the inventor. They may say nothing at all or they may threaten and intimate the bewildered inventor who is trying to figure what is happening to him. He may be offered a secret position of employment with some branch arm of the military, government or intelligence agency to carry on in a similar type of work or he could in some rare circumstances, be offered a bribe to keep quiet or be threatened with death just for speaking out about what he was working on.

Such are the ways of the military industrial complex when these types of breakthroughs and discoveries are made in the public sector that they become a part of the **Unacknowledged Special Access Projects (USAPs) and Programs** or they are 'black shelved" or "black vaulted" never to see the light of day again.

No one in the public sector must know or understand the secret of the propulsion system of UFOs or of any alien technology system that has been recovered and reverse engineered. For the general public to have this "insider knowledge" is to upset in the balance of power on this planet and the status quo in the way geo-socio-politics, business and the economy function. The balance of power shifts away from the first world nations to the second and third world nations because the developing and poorer nations are no longer held bankrupted and enslave to the first world wealthy nations.

When you have an advance alien technology integrated and hybridized with terrestrial technology, it has the potential or probability to alleviate mass poverty, resolve the global environmental and pollution problems, generate a sustainable and massive global economic prosperity undreamt or realized before in history, provide clean, cheap infinite energy on demand that could be incorporated into every transportation system, as well as freeing humanity to explore the stars and the cosmos with faster than light speed spacecraft.

Sentience and intelligence appear in many forms as we are learning which can be truly bizarre in appearance to most humans who happen to encounter them. However, their xenomorphology does not equate to a predisposition for hostility or aggressive behaviour unless, of course, it is a matter of self-defence or self-preservation. Not even, if those same qualities typically found in humans become self-justifiable behaviours and responses to hostile situations on another planet. In other words in a truly advanced interstellar civilization that is technically, morally and spiritually developed, self-sacrifice of one's life may be preferable than instigating an all-out confrontation or an interstellar war with another sentient intelligence, no matter how socially and spiritually immature that other species is in their development. Such is the situation upon this Earth, as humans have repeatedly demonstrated their socially immature behaviour to those ET visitors that set foot upon our planet, especially by the military forces of our governments.

This probably means that if ETs don't show aggressive hostile actions to other species then, some high moral code or **prime directive** is in operation that governs their behaviour toward other intelligent species in a set of protocols of behaviours and responses.

What is interesting is that some armchair thinkers say that besides the scientists, the politicians, and the military who would be the first in line to be Earth's Ambassador spokesperson or representative, they also suggests comedians, poets, artists and even a few high ranking clergymen thrown in for good measure. It seems like, almost everyone would be a qualified **"Cosmic Diplomat"**, everyone except you and me, because we don't have enough God-given smarts to know what to do in such human – ETI encounters.

We are just as likely to create an interstellar war between humans and ETs by doing or saying the wrong thing (*remember, the common folk can't keep secrets very well and we tend to lash out irrationally when things don't go right for us*) and yet, I ask you, is that not the very same approach and mentality used by the militaries of the superpower nations to welcome ETs to our planet or in dealing with each other?

We, the people have trusted those officials we have placed into positions of power believing that they have our best interests at heart, never once do we call them to account for their actions or review their performance or the results they have achieved on behalf of the people. It's time for the people to take back their power from those who do not serve the people's interests.

In the business world, when employees are given a set of responsibilities to carry out on behalf of the company and they fail to achieve them in a timely fashion, such employees are usually severely reprimanded or fired from their positions and new people brought in to complete the assigned tasks set by the company! It is time for an accounting and judgment of all people in positions of power and authority to see if they will continue to serve the people!!!

The time has come when we must face up to the fact that the real world is a far more complex place than we realized. No longer can we keep denying the truth that looms before us. We need to grow up and get our heads out of the sands of complacency, apathy, and timidity.

Extraterrestrial life whether on Mars or visiting from another planet originating from outside our Solar System is the proverbial 5 ton elephant in the room. We can ill-afford not to give it our full and immediate attention. We need to do so quickly before it becomes too late and we lose control of the entire situation.

In **Arthur C. Clark's** book "Childhood's End" a poignant principle is conveyed to humanity when it finally encounters and acknowledges the existence of another intelligence, one that is extraterrestrial in origin. People must grow up and leave their adolescent behaviours behind. They must become the mature adults they were destined and meant to be, even if that means that the old generation will not live to see the dawn of a new future where mankind has finally matured!

Once again, parallels can be made to the biblical story to the Ark of salvation that Noah built to save his family from the spiritually bankrupt and moribund generations of his time.

Many resist this necessary evolutionary process that comes to all sentient species. Some people become increasingly uncomfortable with the process which has its impetus originating from a divine wellspring of spiritual guidance that renews itself every thousand years or so. Such a renewal of spiritual values can lead some people to react with great hope while others react with hostility fearing that their old ways of doing things are no longer valid or outdated. In this regard they would be right in that the old ways are being swept aside or rolled up; their usefulness having served its time are now, no longer required. It is not the end of all things, but the beginning of all new things made new, a rebirth of a spiritual springtime as foretold in ancient times and destined to be fulfilled in our time. Any future is always fraught with challenges and uncertainty and to some, it is a fearful confusing time but, if people hold steadfastly with faith and certitude those divine truths that are found in every religion, the future becomes not only hopeful, it can also be the dawning of a golden civilization of mankind. In such a hopeful future our experience as humans grows in quantum leaps as we interact peacefully with other Extraterrestrial civilizations. If we can but seize that bright future, then one day when we leave the confines of our planet and Solar System, we will be perceived by other planetary intelligences and civilizations as the new Extraterrestrials!

The science of consciousness may be one of the cornerstones in the foundation of an advanced Extraterrestrial civilization upon which it has built its whole civilization. In an advanced ET civilization, **Technology Assisted Consciousness (TAC)** or **Conscious Assisted Technology (CAT)**, (terms developed by **Dr. Steven Greer** to describe a planetary society whose civilization is based on **Cosmic Consciousness**, a highly evolved state of mental and spiritual development), would enable that planet's engineers to construct spacecraft, whole cities even, whole artificial planets out of the matrices of ideation and causation which is their true reality, and then, bringing the pure concept of its reality into the physical existence of three-dimensionality, through the mere power of the mind and intention. Literally, producing what would appear to be for all intent and purpose, *something out of nothing*!

When like minds are linked together in harmony and with pure intention, civilizations can produce anything physically and in fact that civilization could easily dispense with all physicality and enter into another altered, higher state of existence or dimensionality from which they can exist and operate, never needing to return to the matrix of the third dimension yet, being quite capable of interacting with our physical reality!

Our true reality and nature is after all pure intellectual, rational mind, soul and spirit energy which is indivisible and eternal and is part of the great One Soul of the universe or as Quantum Physicist Erwin Schrodinger has stated *"The total number of minds in the Universe is One. In fact, Consciousness is a Singularity phasing within All Beings".*

Consciousness or mind is used interchangeably and hence, if there is but one consciousness in the universe there must be one soul. The fact that your mind feels separate from another person is a false perception of the ego, it is an illusion. The people or bodies are real, but the separation of egos is not. Therefore, "There is but one soul in the universe not pieces of soul".

By this is meant that there is an infinite number of bodies in the universe, but only one soul associated with each body. The mind is not the soul, however, every soul does possess a mind. Soul has awareness, while matter does not. In other words, soul possesses mind, hence the mind is one and it must follow that the soul is unified and the soul implies mind, therefore the mind implies unity. Thus, the soul implies unity! The Spiritual Universe: One Physicists Vision of Spirit, Soul, Matter, and Self by Fred Alan Wolf; Hardcover 1996; Simon and Schuster; Needham, MA; ISBN 987-0-9661327-1-7 and ISBN 0-9661327-1-8

This is why it is stated in the Baha'i writings that "we must be as one soul in many bodies", in other words we are all united, we are unified as one soul, though we each have a body and mind, we nevertheless, are in reality one soul. Such a reality is self-evident and also explains the concepts of omnipresence, omnipotence, omniscience and all the psychic abilities latent within each one of us.

The point to this spiritual philosophy is to demonstrate that it isn't just humans who share a common soul, or awareness or sentience, but all extraterrestrial intelligences in the universe share this reality of One Soul and thus, one mind or consciousness. In turn, we are all One and thus, has it been stated by Dr. Greer, *"We are One People, in One Universe"!*

BIBLIOGRAPHY, WEBLIOGRAPHY AND VIDEOGRAPHY

The following list includes all books and major journals, newspapers and web based material, including other reference ebooks and materials such as web links and video links found on the internet in researching this book. Not all chapters use reference material and therefore, these chapter are not listed.

Bibliography listed refers to books marked in RED,
Webliography refers to websites marked in BLUE,
Videography refer to video websites marked in GREEN,
News Service websites are marked in LIGHT BLUE, and
Newspaper websites, magazines and professional papers are marked in PURPLE.

It does not include specific government documents, archival repositories, or various journals or other web based material.

CHAPTER 93
WHAT DO SCIENTISTS REALLY KNOW ABOUT UFOS AND ETI?

Sturrock, Peter A., "An Analysis of the Condon Report on the Colorado UFO Project," Journal of Scientific Exploration, Vol. 1, No. 1, 1987

http://www.ufoevidence.org/topics/science.htm

http://www.ufoevidence.org/documents/doc780.htm

Vallee, J., Confrontations, New York: Ballantine Books, 1990

http://www.ufoevidence.org/topics/SkepticsAnalysis.htm

CE-5: Close Encounters of the Fifth Kind by Richard F. Haines, Ph.D.; 1998; published by Sourcebooks Inc.; Naperville, Illinois, USA; ISBN 1-57071-427-4

http://www.stargate-chronicles.com/site/

http://www.ufoevidence.org/documents/doc1744.htm

Science, Secrecy, and Ufology by Richard M. Dolan; copyright ©2000 by Richard M. Dolan; Published in February/March issue of UFO Magazine

http://www.ufoevidence.org/topics/science.htm

http://www.stargate-chronicles.com/site/

http://www.ufoevidence.org/documents/doc1744.htm

562

http://celticowboy.com/Round%20Aircraft%20Designs.htm

CHAPTER 94
SOME VISIONARY AERONAUTICAL ENGINEERS

http://en.wikipedia.org/wiki/Henri_Coand%C4%83

http://en.wikipedia.org/wiki/Coanda_Effect#Applications

http://en.wikipedia.org/wiki/Ren%C3%A9_Couzinet

http://aerostories.free.fr/constructeurs/couzinet/page3.html

http://www.laesieworks.com/ifo/lib/VTOLdiscs.html

http://www.cufon.org/cufon/couzinet.htm

http://ufologie.patrickgross.org/aircraft/couzinet.htm#sv1

http://www.minijets.org/index.php?id=169

http://en.wikipedia.org/wiki/Thomas_Townsend_Brown

http://www.antigravitytechnology.net/thomas_townsend_brown.html

http://www.piasecki.com/geeps_pa59k.php#

http://en.wikipedia.org/wiki/Flying_car_%28aircraft%29

http://en.wikipedia.org/wiki/Moller_Skycar

http://en.wikipedia.org/wiki/Paul_Moller

http://moller.com/dev/

CHAPTER 95
TRICKS, TRAPS, THREATS AND LEGAL HURDLES FROM THE US PATENT OFFICE IN THE NAME OF "NATIONAL SECURITY"

http://www.apparentlyapparel.com/free-energy.html

http://www.theorionproject.org/en/suppressed.html

http://peswiki.com/index.php/Directory:Suppression#Overview_Documents
https://www.youtube.com/watch?v=48AkxqT16gk

http://www.theorionproject.org/en/documents/Gary_V.pdf

http://peswiki.com/index.php/Directory:Suppression#Statistics

http://www.apparentlyapparel.com/free-energy.html

http://fuel-efficient-vehicles.org/energy-news/?page_id=983

http://blogs.fas.org/secrecy/2010/10/invention_secrecy_2010/

http://www.fas.org/sgp/othergov/invention/35usc17.html

http://www.apparentlyapparel.com/2/post/2012/05/nikola-teslas-wireless-electric-automobile-explained.html

https://fuel-efficient-vehicles.org/energy-news/?page_id=971

http://fuel-efficient-vehicles.org/energy-news/?page_id=952

http://peswiki.com/index.php/Directory:Aether

http://www.apparentlyapparel.com/free-energy.html

http://www.theorionproject.org/en/suppressed.html

https://www.youtube.com/watch?v=48AkxqT16gk

CHAPTER 96
US PATENTS: DISC AIRCRAFT

force field propulsion & space drives)

http://www.rexresearch.com/wingless/wingless.htm

CHAPTER 97
MANMADE FLYING SAUCERS FROM
THE BLACK WORLD OF SCIENCE

http://alienufoparanormal.aliencasebook.com/2008/07/20/flying-saucer-patents--have-you-designed-yours-yet.aspx

Intercept UFO by Renato Vesco (Originally published as a Zebra Book ib1971 under the title "Intercept – But Don't Shoot"); 1974; published by Zebra Publications, Inc., ISBN 0-8468-0010-1

Hitler's Flying Saucers by Henry Stevens; 2003; published by Adventures Unlimited Press; Kempton, Illinois, USA; ISBN 1-931882-13-4

The S.S. Brotherhood of the Bell: The Nazis' Incredible Secret Technology by Joseph P. Farrell; 2006; published by Adventures Unlimited Press; Kempton, Illinois, USA; ISBN 1-931882-61-4

Man-Made UFOs, 1944-1994: Fifty Years of Suppression by Renato Vesco and David Hatcher Childress; 1994; published by Adventures Unlimited Press; Stelle, Illinois, USA; ISBN 0-932813-23-2

http://www.roswellufomuseum.com/research/ufotopics/naziufocrash.html

CHAPTER 98
WELCOME TO THE PLANET EARTH – WHAT'S THE SECRET OF YOUR PROPULSION SYSTEM?

http://www.greatdreams.com/david-adair.htm

http://web.archive.org/web/20040420185409/http://www.flyingsaucers.com/adair3.htm

http://ufologie.patrickgross.org/rw/w/stephenlovekin.htm

http://www.theorionproject.org

http://worldpuja.org/archives.php?list=host&value=steven&rnd=12576

https://www.youtube.com/watch?v=Rqi3jpBSvCc

https://www.youtube.com/watch?v=7zKQe-1BUFQ

http://abcnews.go.com/blogs/headlines/2006/12/can_you_hear_me/

http://www.washingtonpost.com/opinions/snowdens-hypocrisy-on-russia/2014/02/07/23c403c2-8f51-11e3-b227-12a45d109e03_story.html

http://www.huffingtonpost.com/bob-burnett/questioning-authority-edw_b_4744956.html

http://www.theguardian.com/world/2014/jan/29/edward-snowden-nominated-nobel-peace-prize

http://thelostgunmen.blogspot.ca/2008/03/levitation-teleportation-time-travel.html

http://www.stealthskater.com/Documents/Beckwith_02.pdf

http://www.beckwithelectric.com/ber/downloads/Hypotheses.PDF

Hypotheses: Superatoms, Neutrinos, and Extraterrestrials by Bob Beckwith and Drew Craig; 1996 -1998; clearwater, Florida, USA; ISBN 0-9657178-0-2

http://www.beckwithelectric.com/ber/

http://www.bibliotecapleyades.net/disclosure/briefing/disclosure18.htm

The Enterprise Mission

Dr. Steven M. Greer, "Understanding UFO Secrecy"

http://www.ufoevidence.org/documents/doc789.htm

www.cufos.org/UFO_Documents_internet.html

http://www.examiner.com/article/canada-releases-ufo-x-files-to-the-world

http://www.nsa.gov/public_info/_files/ufo/key_to_et_messages.pdf

http://www.disclosureproject.org/countries-releasing-ufo-files.shtml

http://www.educatinghumanity.com/2011/01/list-of-countries-that-have-disclosed.html

CHAPTER 99
OFFICIAL GOVERNMENT DISCLOSURE
VS. THE PEOPLE'S DISCLOSURE

https://www.youtube.com/watch?v=EYzRY2XpLBk

http://www.hillaryclintonufo.net/disclosureefforts.html

http://www.project1947.com/shg/symposium/shgintro.html

http://goldenageofgaia.com/2010/07/lee-spiegel-remembers-his-un-meeting-on-ufos-in-1978/

DISCLOSURE PROJECT BRIEFING DOCUMENT; April 2001 by Dr. Steven M. Greer, Director and Dr. Theodore C. Loder III; The Disclosure Project; PO Box 265, Crozet, VA 22932

http://paradigmresearchgroup.org/dir/chf/un-initiative/

http://www.disclosureproject.org/briefingpoints.shtml

http://www.citizenhearing.org/committee.html

http://www.collective-evolution.com/2014/01/15/obama-appoints-ufo-disclosure-advocate-into-administration/

http://news.yahoo.com/outgoing-obama-adviser-john-podesta-s-biggest-regret-of-2014--keeping-america-in-the-dark-about-ufos-234149498.html

CHAPTER 100
US PRESIDENTIAL QUOTES AND GOVERNMENT COVER-UP

http://www.presidentialufo.com/ufo-quotes

https://www.youtube.com/watch?v=Q-w9-Y2JMZw

http://www.ufoevidence.org/documents/doc1358.htm

CHAPTER 101
QUOTES FROM U. S. MILITARY WITNESSES ABOUT UFOS

http://www.aliensthetruth.com/Aliens_quotes.php?view=1&category=Military#.VS7soZP7OVo

http://www.ufoevidence.org/documents/doc1743.htm

CHAPTER 102
QUOTES FROM INTERNATIONAL MILITARY
WITNESSES ABOUT UFOS

http://www.bibliotecapleyades.net/ciencia/ufo_briefingdocument/quogov.htm

CHAPTER 103
QUOTES FROM U. S. GOVERNMENT WITNESSES ABOUT UFOs

http://www.ufoevidence.org/documents/doc1737.htm

CHAPTER 104
INTERNATIONAL QUOTES FROM WORLD LEADERS
AND OFFICIALS ABOUT UFOS

https://www.britannica.com/biography/Roger-Bacon

CHAPTER 105
QUOTES FROM ASTRONAUTS AND COSMONAUTS ABOUT UFOS

http://www.gravitywarpdrive.com/UFO_Testimonies.htm

CHAPTER 106
QUOTES FROM SCIENTISTS AND SCIENCE ORGANIZATIONS ABOUT UFOS

http://www.bibliotecapleyades.net/ciencia/ufo_briefingdocument/quosci.htm

CHAPTER 107
WHAT IS EXOPOLITICS AND THE INSTITUTE FOR COOPERATION IN SPACE (ICIS)?

http://en.wikipedia.org/wiki/Politics

http://www.bibliotecapleyades.net/exopolitica/esp_exopolitics_v.htm

http://en.wikipedia.org/wiki/Alfred_Webre

http://exopolitics.org/about/welcome/

http://exopolitics.org/about/founder/

http://en.wikipedia.org/wiki/Zoo_hypothesis

CHAPTER 108

CHOOSING THE BEST METHODS TO COMMUNICATE WITH EXTRATERRESTRIAL INTELLIGENCES

No References

CHAPTER 109
SOME CONTROVERSIAL METHODS TO EXTRATERRESTRIAL CONTACT AND COMMUNICATIONS

http://www.thelostfound.com/psychic-abilities.html

http://en.wikipedia.org/wiki/Category:Channelled_entities

http://www.abovetopsecret.com/forum/thread353245/pg1

http://en.wikipedia.org/wiki/List_of_modern_channelled_texts

https://en.wikipedia.org/wiki/Phonon

http://ufodigest.com/article/telepathy-and-how-it-fits-ufo-puzzle

http://en.wikipedia.org/wiki/Astral_projection

http://en.wikipedia.org/wiki/Eckankar

Baha'i World Faith, Selected Writings of Baha'u'llah and 'Abdu'l-Baha; by National Spiritual Assembly of the United States, 1943, 1956; Baha'i Publishing Trust; ISBN 0-87743-043-8; pg. 326

The Seven Valleys and the four Valleys, by Baha'u'llah; NSA of the USA; 1945 and 1952; Baha'i Publishing Trust; pg. 32-33

http://en.wikipedia.org/wiki/Telepathy

http://www.wired.com/dangerroom/2009/05/pentagon-preps-soldier-telepathy-push/

http://www.bibliotecapleyades.net/vision_remota/esp_visionremota_9c.htm

http://en.wikipedia.org/wiki/Psychotronics

http://www.irva.org/remote-viewing/definition.html

http://www.princeton.edu/~pear/pdfs/1979-precognitive-remote-viewing-stanford.pdf

http://www.lfr.org/lfr/csl/library/AirReport.pdf

http://en.wikipedia.org/wiki/Remote_viewing

http://www.scientificexploration.org/journal/jse_10_1_puthoff.pd

* Ingo Swann interview on 'Dreamland' transcribed organization, University of Wisconsin, 12 December 1996. Quoted from "Remote Viewing and the US Intelligence Community" Armen Victorian (Lobster magazine June 1996 No. 31)

http://ia600605.us.archive.org/30/items/PenetrationTheQuestionOfExtraterrestrialAndHumanTelepathy/Penetration_Ingo_Swann.pdf

http://www.princeton.edu/~pear/pdfs/1979-precognitive-remote-viewing-stanford.pdf

http://www.themindunleashed.org/2013/10/shocking-discoveries-made-studies.html

http://www.bibliotecapleyades.net/sociopolitica/hambone_info/People1.html#Joseph_McMoneagle

http://www.bibliotecapleyades.net/vision_remota/esp_visionremota_35.htm

http://www.synchrosecrets.com/synchrosecrets/?p=17114

http://en.wikipedia.org/wiki/Joseph_McMoneagle

https://www.youtube.com/watch?v=IBcQ8RDIe9w

http://voices.yahoo.com/what-astral-projection-bilocation-796569.html

http://ufodigest.com/news/0708/sixto.html

http://www.contactunderground.com/

https://www.facebook.com/ContactUnderground

http://ufodigest.com/news/0708/sixto.html

http://ufodigest.com/news/0708/sixto2.html

http://subversivethinking.blogspot.ca/2010/11/sixto-paz-wells-contemporary-contactee.html

http://www.youtube.com/watch?v=TkC7ShSdQ0I

CHAPTER 110
THE SCIENCE OF CONSCIOUSNESS OR HIGHER CONSCIOUSNESS

http://en.wikipedia.org/wiki/Consciousness

https://www.deepakchopra.com/blog/view/1288/cosmic_consciousness

http://en.wikipedia.org/wiki/Siddhi

http://en.wikipedia.org/wiki/Higher_consciousness

http://en.wikipedia.org/wiki/Cosmic_consciousness

http://en.wikipedia.org/wiki/Universal_mind

http://www.mind-your-reality.com/universal_mind.html#Part_2

https://www.youtube.com/watch?v=tEH6sQJd7CE

CHAPTER 111
IN WHAT LANGUAGE DO EXTRATERRESTRIAL INTELLIGENCES COMMUNICATE?

http://en.wikipedia.org/wiki/Alien_language

http://www.uk-ufo.org/condign/hist1916.htm

https://www.youtube.com/watch?v=KTc3PsW5ghQ

http://en.wikipedia.org/wiki/Crop_circle

CHAPTER 112
WHY ARE ETS VISITING THE EARTH – IS THERE AN ALIEN AGENDA?

Alien Agenda: Investigating the Extraterrestrial Presence among Us by Jim Marrs; 1997; published by HarperCollins Publishers; New York, USA; ISBN 0-06-018642-9

The Interrupted Journey; 1966 by John G. Fuller; published by Berkley Medallion Books, New York, USA; SBN 425-02572-1

http://prestondennett.weebly.com/ufo-healings.html

http://www.cseti.org/crashes/089.htm

CHAPTER 113
HUMAN AGGRESSION AND HOSTILITY - A FEW CLASSIC CE-3 CASES REVISITED

http://inexplicata.blogspot.ca/2013/12/venezuela-classic-ce3k-revisited-1954.html

Flying Saucers: The Startling Evidence of the invasion from Outer Space by Corel E. Lorenzen; 1966; published by Signet Books; New York, USA

http://www.ufocasebook.com/Kelly-Hopkinsville.html
http://williamson-labs.com/ufo.htm

http://en.wikipedia.org/wiki/Opium_of_the_people

God Passes By; by Shoghi Effendi; 1944; published by the Baha'i Publishing Trust; Wilmette, Illinois, USA; ISBN 0-87743-020-9

Gleanings from the Writings of Baha'u'llah; translated by Shoghi Effendi NSA of the Baha'is of the US); 1939 and 1952 published by Baha'i Publishing Trust; Wilmette, Illinois, USA; ISBN 52-24896

CHAPTER 114
PREPARATION FOR CONTACT AND
SOME "INALIENABLE TRUTHS"!

CE-5: Close Encounters of the Fifth Kind by Richard F. Haines, Ph.D.; 1998; published by Sourcebooks Inc.; Naperville, Illinois, USA; ISBN 1-57071-427-4

https://www.youtube.com/watch?v=8DycG5ZMjTo

http://www.youtube.com/watch?v=WfDyr_FDZiI

http://www.openminds.tv/radio

http://en.wikipedia.org/wiki/Messianic_Age

http://en.wikipedia.org/wiki/Realized_eschatology

http://en.wikipedia.org/wiki/Parusia

http://bci.org/prophecy-fulfilled/hindutim.htm

http://en.wikipedia.org/wiki/Maitreya

http://bci.org/prophecy-fulfilled/buddhasa.htm

Baha'u'llah, Baha'i World Faith, Selected Writing of Baha'u'llah and Abdu'l-Baha, p.9-11; by the National Spiritual Assembly of the Unites States; 1945, 1956; Baha'i Publishing Trust; ISBN 0-87743—043-8

The Bahá'í Revelation, Selections from the Bahá'í Holy Writings and Talks by 'Abdu'l-Bahá; first published in 1955; Bahá'í Publishing Trust; London, U.K.; Library Reference 299.15 Bahá'í

Prisoner and the Kings by William Sears, 1971; General Publishing Co. Ltd., Toronto, Canada; Revised Edition (Jan. 1 2007); U.S. Baha'i Publishing Trust; Wilmette, Illinois, USA

Promised Day is Come by Shoghi Effendi; 1941; published by Baha'i Publishing Trust; Wilmette, Illinois, USA; ISBN 0-87743-132-9

Proclamation of Baha'u'llah by Baha'u'llah; 1978 reprint; US Bahá'í Publishing Trust, Wilmette, Illinois, USA

The Bahá'í Revelation, Selections from the Bahá'í Holy Writings and Talks by 'Abdu'l-Bahá; first published in 1955; Bahá'í Publishing Trust; London, U.K.; Library Reference 299.15 Bahá'í

The Promulgation of Universal Peace; p. 18. by 'Abdu'l-Bahá; published in 1982 (originally published 1922-1925) by the National Spiritual Assembly of the United States; BP360.A375 1982 and ISBN 0-87743-172-8

CHAPTER 115
GLOBAL UNITY BEFORE GALACTIC UNITY

Abdu'l-Baha: Paris Talks, (Page: 22)

'Abdu'l-Baha: Promulgation of Universal Peace, (Pages: 204-205)

(1981) [1904-06]. Some Answered Questions. Wilmette, Illinois, USA: Bahá'í Publishing Trust. ISBN 0-87743-190-6.

'Abdu'l-Bahá (1982) [1912]. The Promulgation of Universal Peace (Hardcover ed.). Wilmette, Illinois, USA: Bahá'í Publishing Trust. ISBN 0-87743-172-8.

Bahá'u'lláh (1976). Gleanings from the Writings of Bahá'u'lláh. Wilmette, Illinois, USA: Bahá'í Publishing Trust. ISBN 0-87743-187-6

Effendi, Shoghi (1944). God Passes By. Wilmette, Illinois, USA: Bahá'í Publishing Trust. ISBN 0-87743-020-9.

http://en.wikipedia.org/wiki/United_Nations

http://en.wikipedia.org/wiki/United_Nations_Security_Council_veto_power

CHAPTER 116
GLOBAL UNITY - A PREREQUISITE FOR ADMISSION
INTO GALACTIC CIVILIZATION

http://reference.bahai.org/en/t/b/GWB/gwb-117.html

http://reference.bahai.org/en/t/ab/SDC/sdc-4.html

http://reference.bahai.org/en/t/ab/SAB/sab-228.html

http://reference.bahai.org/en/t/se/WOB/wob-56.html

http://reference.bahai.org/en/t/se/WOB/wob-19.html

http://reference.bahai.org/en/t/se/WOB/wob-22.html

http://new.cseti.org/position-papers/14-position-papers/42-art-one-universe-one-people.html

Gleanings from the Writings of Baha'u'llah; translated by Shoghi Effendi NSA of the Baha'is of the US); 1939 and 1952 published by Baha'i Publishing Trust; Wilmette, Illinois, USA; ISBN 52-24896

Gleanings from the Writings of Baha'u'llah; translated by Shoghi Effendi NSA of the Baha'is of the US); 1939 and 1952 published by Baha'i Publishing Trust; Wilmette, Illinois, USA; ISBN 52-24896

Kitab-i-Iqan The Book of Certitude; Baha'u'llah translated by Shoghi Effendi ; 1931, rev. edn. 1954; Baha'i Publishing Trust; Wilmette, Illinois, USA; ISBN 0-87743-022-5

(Baha'u'llah, Arabic Hidden Words)

(Abdu'l-Baha, Divine Philosophy)

(Abdu'l-Baha, Divine Philosophy 178)

http://www.inspiremore.com/nasa-just-released-the-largest-photo-ever-taken-what-it-shows-will-shake-you-up/

http://www.bibliotecapleyades.net/ciencia/ciencia_consciousuniverse43.htm

http://www.gestaltreality.com/articles/favorite-quotations/

Gleanings from the Writings of Baha'u'llah; translated by Shoghi Effendi NSA of the Baha'is of the US); 1939 and 1952 published by Baha'i Publishing Trust; Wilmette, Illinois, USA; ISBN 52-24896

CHAPTER 117
CAN YOU PUT THAT IN WRITING? XENOLINGUISTICS - EXTRATERRESTRIAL SCRIPTS AND SYMBOLS

https://www.youtube.com/watch?v=QnSGE_plJVM

CHAPTER 118
WHO BEST SPEAKS FOR THE PEOPLE OF EARTH WHEN CONTACT WITH EXTRATERRESTRIAL INTELLIGENCES IS IMMINENT?

http://en.wikipedia.org/wiki/Potential_cultural_impact_of_extraterrestrial_contact

"What Role will Extraterrestrials Play in Humanity's Future?"; by Allen Tough; (1986); Journal of the British Interplanetary Society 39: 491–498. Bibcode:1986JBIS...39..491T.

http://en.wikipedia.org/wiki/Islamic_religious_leaders

Making Contact, A Serious Handbook for Locating and Communicating with Extraterrestrials" edited by Bill Fawcett; 1997; published by William Morrow and Co. Inc. New York, NY. USA; ISBN 0-688-14486-1

"Hidden Truth, Forbidden Knowledge: It's time for You to Know" by Steven M. Greer M.D.; 2006; published by Crossing Point, Inc. Publication; Crozet, Va., USA; ISBN 0-9673238-2-7

http://www.telegraph.co.uk/science/space/8025832/UN-to-appoint-space-ambassador-to-greet-alien-visitors.html

http://en.wikipedia.org/wiki/Mazlan_Othman

http://www.aolnews.com/2010/09/27/mazland-othman-is-not-earths-alien-ambassador-after-all/

http://www.foxnews.com/scitech/2010/09/27/appoints-contact-visiting-space-aliens/

Contact with Alien Civilizations: *Our Hopes and Fears about Encountering Extraterrestrials* by Michael A. G. Michaud; 2007; published by Copernicus Books; New York, NY, United States; ISBN 978-0-387-28598-6. Archived from the original on 24 December 2012.

"Contact" Countdown to Transformation – The CSETI Experience from 1992 - 2009 by Steven M. Greer, M.D. 2009; publish by 123PrintFinder, Inc.; Ladera Ranch, VA, USA; ISBN 9780967323831

http://en.wikipedia.org/wiki/Potential_cultural_impact_of_extraterrestrial_contact

https://www.youtube.com/watch?v=hx9i-KRMCCc

https://www.youtube.com/watch?v=IVV4zRuE1mw

http://www.dailygalaxy.com/my_weblog/2010/03/if-et-calls-who-speaks-for-humanity-.html

http://www.theguardian.com/science/2011/jan/10/earth-close-encounter-aliens-extraterrestrials

http://en.wikipedia.org/wiki/British_Empire

CHAPTER 119
THE ROSETTA STONE OF EXTRATERRESTRIAL COMMUNICATIONS

Making Contact, A Serious Handbook for Locating and Communicating with Extraterrestrials" edited by Bill Fawcett; 1997; published by William morrow and Co. Inc. New York, NY. USA; ISBN 0-688-14486-1

"CE-5 Close Encounters of the Fifth Kind" by Richard F. Haines; 1999; published by Sourcebooks, Inc.; Naperville, Illinois, USA; ISBN 1-57071-427-4

"Hidden Truth, Forbidden Knowledge: It's time for You to Know" by Steven M. Greer M.D.; 2006; published by Crossing Point, Inc. Publication; Crozet, Va., USA; ISBN 0-9673238-2-7

http://www.siriusdisclosure.com/ce-5-initiative/past-events/cseti-history/

Dr. Greer, from The Disclosure Project's 4 hr. Witness Testimony

http://www.newdawnmagazine.com/articles/sirius-the-film-disclosures-next-step

http://www.siriusdisclosure.com/evidence/atacama-humanoid/#sthash.bWe240nZ.dpuf

http://www.siriusdisclosure.com/evidence/atacama-humanoid/

www.disclosureproject.org.

www.siriusdisclosure.com.

CHAPTER 120
BECOMING AN AMBASSADOR - COSMIC DIPLOMAT
TO THE UNIVERSE/

http://cseti.org/

http://siriusdisclosure.org

CHAPTER 121
UFO AND ETI PHENOMENON OBSERVED BY CSETI FIELD TEAMS

http://cseti.org/

http://siriusdisclosure.org)

The above examples of UFO/ET phenomena were compiled by T. Loder (Dec. 4, 1998) and updated May 11, 1999 with help from L. Willitts, T. Guyker, T. Craddock, D. Foch and others Copyright 1998 CSETI

"Contact" Countdown to Transformation – The CSETI Experience from 1992 - 2009 by Steven M. Greer, M.D. 2009; publish by 123PrintFinder, Inc.; Ladera Ranch, VA, USA; ISBN 9780967323831

http://disclosureproject.com/

"The Holographic Universe" by Michael Talbot; 1991; published by HarperCollins Publishers, Inc.; New York, N.Y., USA; ISBN 0-06-092258-3

CHAPTER 122
SOME ANSWERED QUESTIONS ABOUT MILITARY COVER UPS, UFOS, ETI, ZERO POINT ENERGY

Extraterrestrial Contact: The Evidence and Implications by Steven M. Greer, MD p. 178 – 180

http://www.siriusdisclosure.com/cseti-papers/the-imperative-of-consciousness/
PureEnergySystems.com

Witness Testimony Overview - "The Greatest Secret in Modern History" - excerpts from Disclosure: Military and Government Witnesses Reveal the Greatest Secrets in Modern History by Steven M. Greer, MD.

PureEnergySystems.com

CHAPTER 123
A GLOBAL VISION - HUMANITY MATURES TO BECOME AN EXTRATERRESTRIAL CIVILIZATION

The Spiritual Universe: One Physicists Vision of Spirit, Soul, Matter, and Self by Fred Alan Wolf; Hardcover 1996; Simon and Schuster; Needham, MA; ISBN 987-0-9661327-1-7 and ISBN 0-9661327-1-8

INDEX (VOLUME SIX)

H

S

594

About the Author

Terry Tibando's background experience and understanding of this phenomenon spans 65 years of personal UFO sightings and ET contact that began at the age of five years. This childhood experience initiated a lifetime of many other-worldly sightings and encounters into a mysterious universe of Unidentified Flying Objects, Extraterrestrial Intelligence and the paranormal. As an experiencer, researcher, and investigator in Ufology he brings a unique and refreshing perspective on this subject based on a world view.

While attending Victoria High School in the mid sixties, Terry began attending UFO lectures meeting such people as Dr. Edward Edwards, a linguist from the University of Victoria and a fellow member of APRO (Aerial Phenomenon Research Organization and also Daniel Fry from New Mexico, USA, well known contactee and UFO author.

Terry attended the University of Victoria majoring in astronomy, physics, math and other sciences. During those university years other alien craft were sighted near his family's home in Victoria leading Terry to theorized that a possible undersea ET base existed off the coast of Vancouver Island which may account for the numerous UFO sightings seen over the Island.

He was a former member of APRO and its Canadian sister organization CAPRO during the sixties. His investigative research culminated back in the summer of 1996 when he met with Dr. Steven M. Greer during a one week "Ambassadors to the Universe" training seminar. They soon discovered that they shared similar UFO/ETI experiences during their early life.

Terry was a speaker at the Bellingham UFO Group (BUFOG) UFO seminar in 1996, and as a panel speaker along with Peter Davenport from NUFORC and Sharon Filip, alien abduction researcher.

He has talked on the Grimerica blog talk radio and been interviewed on the Discovery Channel during their "Alien Week" series in 1997 which had two ET spacecraft show up during the TV interview; he has been interviewed on

BCTV News, and appeared briefly in Dr. Greer's successful documentary movie "Sirius" and was a major financial contributor to the current documentary "Unacknowledged"!

He was instrumental in coordinating, hosting and emceeing the first Disclosure Project event on UFOs and ETS in Canada as a part of Dr. Greer's Disclosure Witness Tour held at Simon Fraser University in Vancouver on September 9, 2001, which included guest speakers Dr. Steven Greer, Dr. Carol Rosin and Dr, Alfred Webre.

 For the last 25 years, Terry has been the field coordinator of CSETI Vancouver leading teams of people on field expeditions to successfully establish contact and communications with extraterrestrial intelligences visiting the Earth. Currently, he is finishing the remaining four volumes in this series in preparation for publishing and printing.